Pietro Giuseppe Fré

Discrete, Finite and Lie Groups

Also of Interest

Groups and Manifolds
Lectures for Physicists with Examples in Mathematica
Pietro Giuseppe Fré, Alexander Fedotov, 2017
ISBN 978-3-11-055119-8, e-ISBN (PDF) 978-3-11-055120-4

Algebraic Quantum Physics
Volume 1 Quantum Mechanics via Lie Algebras
Arnold Neumaier, Dennis Westra, 2022
ISBN 978-3-11-040610-8, e-ISBN (PDF) 978-3-11-040620-7

Quantum Mechanics
An Introduction to the Physical Background and Mathematical Structure
Gregory L. Naber, 2021
ISBN 978-3-11-075161-1, e-ISBN (PDF) 978-3-11-075194-9

The Structure of Compact Groups
A Primer for the Student – A Handbook for the Expert
Karl H. Hofmann, Sidney A. Morris, 2020
ISBN 978-3-11-069595-3, e-ISBN (PDF) 978-3-11-069599-1

Topics in Infinite Group Theory
Nielsen Methods, Covering Spaces, and Hyperbolic Groups
Benjamin Fine, Anja Moldenhauer, Gerhard Rosenberger, Leonard Wienke, 2021
ISBN 978-3-11-067334-0, e-ISBN (PDF) 978-3-11-067337-1

Pietro Giuseppe Fré

Discrete, Finite and Lie Groups

Comprehensive Group Theory in Geometry and Analysis

DE GRUYTER

Mathematics Subject Classification 2020
11E57, 11F22, 11F23, 15A03, 17B01, 17B05, 17B25, 17B10, 17B56, 20-01, 20B05, 20B30, 20B35, 20C15, 20E32, 20E45, 20F28, 20F05, 20F55, 20J06, 20K35, 22-01, 22E25, 22E60, 22E10, 22F30, 22E46, 22E70, 22F50, 30-01, 34-01, 34A26, 83E50, 68T01, 58A05, 58A10, 55A12, 55P10, 55R10, 83C20, 33C45

Author
Prof. Pietro Giuseppe Fré
Università degli Studi di Torino
Via P. Giuria 1
10125 Torino
Italy
pietro.fre@esteri.it

ISBN 978-3-11-120075-0
e-ISBN (PDF) 978-3-11-120153-5
e-ISBN (EPUB) 978-3-11-120277-8

Library of Congress Control Number: 2023935709

Bibliographic information published by the Deutsche Nationalbibliothek
The Deutsche Nationalbibliothek lists this publication in the Deutsche Nationalbibliografie; detailed bibliographic data are available on the Internet at http://dnb.dnb.de.

Cover image: bin kontan / iStock / Getty Images Plus
Typesetting: VTeX UAB, Lithuania
Printing and binding: CPI books GmbH, Leck

www.degruyter.com

This book is dedicated by the author to his beloved daughter Laura

Preface

In Germany, at the end of the nineteenth century, relying on his own prestige as an outstanding scientist, Felix Klein promoted a campaign in order to stimulate the diffusion, at all levels of teaching and in all types of schools, of an education solidly based on *functional thinking*. The concept of function was already almost three centuries old, with differential and integral calculus being not too much younger, and the theory of analytic functions of one complex variable had been fully developed in the course of the current century, yet *functional thinking* had not yet become the automatic and customary way of thinking for any educated person in whatever professional field.

About 120 years later, *functional thinking* was relatively well established. In the whole world the basic theory of functions and integral-differential calculus are taught at the level of the final years of high school and are an essential part of most university curricula, at least of those whose character is technical-scientific like physics, chemistry, biology, engineering in all of its declensions, statistics, and economics. The most important aspect of this evolution is that, while teaching and learning such mathematical topics, only the abstract concepts are dealt with: no one doubts what a function $f(x)$ and its derivative $\partial_x f(x)$ might represent, what they might be good for, and so on. It is just clear, both to the students and to the teachers, that $f(x)$ might describe anything in whatever field; so all efforts are focused on *abstract constructions and manipulations*. In this way calculus has finally reached the same status as *grammar and syntaxis* already did since antiquity. The educational process, elaborated through a long historical evolution that led to the medieval crystallization in the *trivium* (*grammar, logic, and rhetoric*; see Figure 1), focused on the formal structure of the language and not on its actual content. Everyone perfectly knew what language was good for and the discussion of the cases in declension was not preceded by any illustration of the practical advantages or of the concrete applications of such abstract notions. Assimilated and examined by each individual, grammar and logic became an automatic habit forging one's thinking.

The contemporary assimilation and analysis of *functional thinking* is a great and valuable progress although even now it is not yet a full achievement: indeed there are still countries like Italy where, in the twenty-first century, you can become a medical doctor while ignoring functions and derivatives.

The reader has certainly noted that, while praising the educational virtues of *grammar and syntaxis*, I utilized the past tense. It was not without a reason. Indeed, while *functional thinking* was progressing, the centrality of *trivium* education started losing ground everywhere in the world with a wide spectrum of different velocities, but a uniform negative sign. The earliest disruption of classical education occurred in the Soviet Union with the October Revolution, and the outcome of the complete obliteration of Latin and Greek teaching in a country whose language is inflectional, like Latin is, is visible nowadays after one century of absence. By personal experience I know how difficult it is to argue in grammatical terms about the Russian language with Russian educated people, even academicians. Furthermore, present-day Russia is a country where

https://doi.org/10.1515/9783111201535-201

Figure 1: Priscian or "The Grammar" relief by Luca della Robbia from Giotto's Bell Tower in Florence.

being *gramatny*, namely, speaking and writing in the native language without introducing macroscopic mistakes, is a matter of distinction carefully boasted by those few who possess such an ability. The same decline, however, is rapidly occurring also in Europe and the United States through many ill-conceived reforms of the educational systems that have affected all the countries and are generally marked by a tendency to privilege an alluvion of *disorganized information* with respect to *formation*. So the balance of *abstract thinking*, which is proper to *mathematics*, yet not exclusive to it, and is highly promoted by an early and serious *grammatical and logical education*, is still shaking.

A non-marginal responsibility of this state of affairs is located in the economical structure of the contemporary world, dominated by the short time horizons of private investors, who naturally and legitimately privilege *applied science* with respect to *fundamental science*, this dominance not being sufficiently counterbalanced by public (national and international) long time horizon investment policies.

Nowadays, in the correct balance of the *abstract thinking* diffusion, one notices a new impellent urgency. Just as more than one century ago Felix Klein envisaged the urge to promote *functional thinking*, I deem that priority number one is at the present time that of promoting the diffusion of *group thinking*, which should become an essential part of one's education forging one's mental framework, at least among scientists and engineers of all specializations.

Group theory codifies the concept of symmetry and, as such, it is the backbone of mathematics, physics, and chemistry, which, up to a certain degree, is just that branch of physics that focuses on atoms and molecules. The notion of what is a group developed slowly, from the end of the eighteenth century with the contributions of Lagrange and others, which were mainly focused on the group of permutations, named at the time *substitutions*, through the whole span of the nineteenth century. The first very significant results on group theory were those obtained by Galois before his death in 1832 and published by Liouville in 1846 [1]. Galois groups are the symmetry groups of algebraic equations and, as such, they are finite groups, subgroups of the permutation group S_n, where n is the degree of the considered algebraic equation. In the middle of the nineteenth century the first attempts at a rigorous abstract definition of the group structure were developed by Arthur Cayley [2]. Finite group theory came to ripeness by the end of the nineteenth century and the beginning of the twentieth century through the work of Camille Jordan [3], Enrico Betti [4], Adolf Hurwitz [5], Felix Klein [6], Issai Schür [7], and others. An essential allied topic and ingredient of group theory was *linear algebra and matrix theory*, which were slowly developed in the central part of the nineteenth century most intensively by Arthur Cayley [8] and James Sylvester [9, 8, 10]. It is mandatory to stress that the key notion in linear algebra, namely, that of *vector space*, although correctly conceived and explained by Grassmann since 1844 [11], had a hard time to be accepted and metabolized by the mathematical scientific community and became part of shared knowledge only after the admirable lectures by Giuseppe Peano published in 1888 [12].

Along a parallel route, starting from the 1870s, the theory of continuous analytic groups, namely, Lie groups with their associated Lie algebras, was firmly established, before the end of the nineteenth century, through the work of Sophus Lie [13, 14, 15, 16, 17], of Wilhelm Killing [18, 19], and finally of Élie Cartan [20, 21]. In this way modern group theory was essentially accomplished already more than 120 years ago and underwent some important additional refinements through the work of Hermann Weyl [22], Coxeter [23], and Dynkin [24] just by the end of World War II.

The complete merge of geometry with group theory, started by the monumental work of Élie Cartan on symmetric homogeneous spaces in the 1930s and continued in between the two wars by the development of the notion of fiber bundles and their classification by means of characteristic classes (Chern and Weil), was finally achieved in the 1950s by the parallel mathematical work on connections conducted by Ehresmann [25, 26] and by the introduction in physics of Yang–Mills gauge theories [27]. Although physicists already knew groups and used them in elaborating their theories, it took a long while before the complete identity of *gauge fields* with *connections on principal bundles* was recognized and a decisive upgrade in the mathematical education of younger generation physicists came into being. I think that the complete inclusion into fundamental physics of the fields of group theory, algebraic topology, and differential and algebraic geometry was accomplished only in the last two decades of the twentieth century. In those years this inclusion already produced magnificent fruits in both directions: from mathematics to physics and vice versa.

Nowadays theoretical physicists, like mathematicians, have assimilated and examined *group thinking*. It could not be different since the present status of the *episteme* concerning fundamental interactions and fundamental constituents of matter can be summarized as I did in the historical-philosophical book [28] with a rather simple and universal scheme of interpretation based on the following few principles:

(A) The categorical reference frame is provided by field theory defined by some action $\mathscr{A} = \int_{\mathcal{M}} \mathscr{L}(\Phi, \partial\Phi)$, where $\mathscr{L}(\Phi, \partial\Phi)$ denotes some Lagrangian depending on a set of fields $\Phi(x)$.

(B) All fundamental interactions are described by *connections* **A** on principal fiber bundles $P(\mathrm{G}, \mathcal{M})$, where G is a Lie group and the base manifold $\mathcal{M}$ is some *space-time* in $d = 4$ or in higher dimensions.

(C) All the fields Φ describing fundamental constituents are *sections* of *vector bundles* $B(\mathrm{G}, V, \mathcal{M})$ associated with the principal one $P(\mathrm{G}, \mathcal{M})$ and determined by the choice of suitable *linear representations* $D(\mathrm{G}) : V \to V$ of the structural group G.

(D) The spin zero particles, described by scalar fields ϕ^I, have the additional feature of admitting non-linear interactions encoded in a scalar potential $\mathscr{V}(\phi)$ for whose choice general principles supported by experimental confirmation have not yet been determined.

(E) Gravitational interactions are special among the others and universal since they deal with the tangent bundle $T\mathcal{M} \to \mathcal{M}$ to space-time. The relevant connection is in this case the Levi-Civita connection (or some of its generalizations with torsion) which is determined by a metric g on $\mathcal{M}$.

The above six principles make sense only within the theory of principal and associated vector bundles, so that *group thinking* is indeed mandatory for physicists as I already stated.

Yet *group theory and group thinking* are transversal to most sciences including the various subsectors of mathematics itself. This fact emerges constantly. Chronologically the most recent recognition of the group theoretical foundations of one's work happened in the field of artificial intelligence, steering a lot of interest and talking [29, 30, 31, 32, 33, 34]. In what now goes under the name of *geometric deep learning* it was recognized and stated that principal fiber bundles and associated vector bundles are the appropriate mathematical playground for *convolutionary neural networks* that can be conceptualized as maps from the space of sections $\Gamma[G, \mathcal{M}, V_1]$ of a vector bundle into another one $\Gamma[G, \mathcal{M}, V_2]$ sharing the same structural group and the same base, yet having different standard fibers (see Section 6.7): in short exactly the same mathematical setup of fundamental quantum field theories.

Another macro-area where symmetries and group theory play a fundamental role is the compound of chemistry/physics, crystallography, molecular biology, and genetics. Here the point and space groups associated with discrete rotations and reflections, lattices, and space-tessellations that were classified about 130 years ago by Fyodorov [35] play the role of primary actors, yet the nomenclature (typically the 100-year-old Schönflies one) and the mathematical setups and techniques generally employed in the reference scientific community are quite obsolete, regardless of the one-century advances in representation and group extension theories, finally deprived of the necessary degree of abstraction to recognize general underlying patterns.

Discrete and continuous groups appear also in probability theory, in graph theory, in the theory of games, and hence in economical and statistical sciences.

These scattered remarks are an illustration of the above advocated transversality and hence of the need for promoting *group thinking* at large in the education and self-education of scientists, technicians, and engineers. The hallmark of this promotion is the separation of the abstract concepts from their possible applications in order to develop a higher degree of abstraction. Like it happens for functions $f(x)$, one should no longer refer to any predetermined application when talking about fiber bundles, connections, homogeneous spaces, or crystallographic groups.

I was repeatedly asked by the editors if this book was going to be a textbook. At the moment it is not because in most universities a transversal course like that which could be taught on the basis of the present monograph does not exist. Yet it should be introduced and it should be delivered to students of a wide spectrum of majors like it happens for calculus. Indeed, *group thinking* is very much needed in the current world.

The present book aims at a self-contained, yet architecturally exhaustive presentation of group theory in its multifaceted aspects. Groups are intrinsically related with differential geometry whose conceptual foundations are also covered in proper logical order.

In existing textbooks the various strongly correlated components of this vast, yet unitary subject are typically treated separately and moreover in different approaches, targeted to specific fields of application.

Here I aim instead at unitarity in a logical order. First I present the intuitive idea of groups as closed sets of transformations where the important things are not the objects that are transformed, but rather the relations satisfied by the operations to be performed (Chapter 1). Then I make a quick review of the hierarchy of algebraic structures (Chapter 2), providing next noticeable examples of both continuous and discrete groups (Chapter 3). In Chapters 4 and 5 I complete the basic elements in the theory of finite groups and of their linear representations, addressing the classification of the finite subgroups of the rotation group in three dimensions. This is the fascinating ADE classification, which provides on one side the basis for crystallography, while on the other side it turns out to be isomorphic to the classification of simple, simply laced Lie algebras, as it will be explained in later chapters.

Chapter 6 develops the essential items of basic differential geometry with the definition of manifolds, tangent bundles, and cotangent bundles, and then introduces first Lie groups and then principal and associated fiber bundles. The discussion of manifolds and differential forms provides the opportunity to introduce the basic notions of homology, cohomology, and homotopy. So groups are now seen in a different capacity as homology (or cohomology) and homotopy groups, which encode, in an intrinsic way, topological properties of the manifolds they are associated with. These general concepts are next illustrated with a revisitation of the theory of holomorphic functions of one variable and of their integrals. This territory, supposedly already known to the reader, is a perfect playground to improve and test one's comprehension of the general scheme. The group of automorphisms of the complex plane extended with the point at infinity is also very handy for illustration of the general concepts. The short Chapter 7 follows, where I derive in detail the relation between a Lie group manifold G and its Lie algebra $\mathbb{G}$.

Since the structural theory of Lie algebras with a complete exposition of the roots and Dynkin diagrams lore leads to representation theory based on weight lattice theory, before addressing such topics in Chapters 10, 11, and 12, I make an intermission with Chapter 8 entirely devoted to lattices and crystallographic groups in general. In this chapter particular attention is dedicated to the theory of group extensions and its application to the construction of space groups, which are discrete and include the lattice as a normal infinite subgroup. Chapter 9, dedicated to differential equations and monodromy groups, shows an example of such infinite discrete groups in the context of monodromy, which is another capacity in which groups make their appearance in analysis.

Chapter 13 provides a glimpse at the peculiarities of exceptional Lie algebras presenting the structure and fundamental representations of both $\mathfrak{g}_2$ and $\mathfrak{f}_4$.

Chapter 14 provides an in depth study of a finite discrete group PSL(2, 7) which is simple and crystallographic in seven dimensions. The rich structure provided by this example constitutes an illustration of the need for a case-by-case study of simple groups, whose irreducible representations have to be derived with some art in want of a general scheme. Furthermore, the crystallographic nature of PSL(2, 7) in seven dimensions allows to study the orbits of the group on the lattice, in this case the root lattice of $\mathfrak{a}_7$.

Chapter 15 develops in full the theory of connections of principal fiber bundles $P(\mathrm{G}, \mathcal{M})$ and Riemannian geometry. The full theory of isometries and of homogeneous spaces G/H geometry is discussed in the subsequent Chapter 16.

The last developed Chapter 17 is an introduction to functional spaces and illustrates with a worked out example the general property that unitary representations of a non-compact Lie group or Lie algebra are necessarily infinite-dimensional.

Finally, Chapter 18 gives some glimpses of harmonic analysis and contains the final remarks of the author.

Appendix A contains the list of Wolfram MATHEMATICA NoteBooks that the reader can download to practice in group theory calculations and explore in a factive way all the material presented in this book or perform new original calculations inspired by his/her fantasy.

Pietramarazzi (Alessandria), Italy
June 2023

Pietro Giuseppe Fré
Emeritus Full Professor
of Theoretical Physics
University of Torino

Some remarks about notations

Concerning cyclic groups we use alternatively the notation $\mathbb{Z}_k$ or C_k. In particular we give preference to the second notation when we look at the cyclic group as an abstract group, while the first notation is preferably used when we think of it as given by integer numbers *mod k*.

In general when discussing Lie groups we utilize the notation G for the abstract Lie group and $\mathbb{G}$ for the corresponding Lie algebra. In several cases for specific Lie algebras like the exceptional ones we also use the notation $\mathfrak{e}_{6,7,8}$, $\mathfrak{g}_2$, $\mathfrak{f}_4$ or similarly $\mathfrak{a}_\ell$, $\mathfrak{b}_\ell$, $\mathfrak{c}_\ell$, $\mathfrak{d}_\ell$ for the classical Lie algebras.

According to a quite standard mathematical notation, by $\mathrm{Hom}(V, V)$ we denote the group of linear homomorphisms of a vector space V into itself.

We use throughout the book the standard convention that repeated indices are summed over. Depending on the case and on graphical convenience we utilize either the 8 to 2 or the 10 to 4 convention, and sometimes also the repetition at the same level of the same index.

The *Euclidean group* in n dimensions includes rotations, translations, and reflections and it is denoted Eucl_n. It includes the roto-translation group ISO(n) as a proper subgroup.

In relation with the number of elements contained in a set $\mathscr{S}$ we use alternatively either the notation $|\mathscr{S}|$, typically reserved to the case when $\mathscr{S}$ is a finite group G, or the notation $\mathrm{card}\,\mathscr{S}$, when $\mathscr{S}$ is some type of set, for instance the set of all roots of a simple Lie algebra. When we consider specific types of objects we also utilize the often employed notation # of *objects such and such*.

Let us clarify notations for elements of a Lie algebra. In Chapter 10 and earlier in Chapter 2, while discussing general theorems on Lie algebras we denote elements $\mathbf{X} \in \mathbb{G}$ of a generic Lie algebra $\mathbb{G}$ by boldfaced letters to emphasize their nature of vectors in a vector space and to facilitate the distinction with respect to groups, algebras, and subalgebras. In Chapters 11, 12, and 13, where we develop and extensively use the formalism of roots and weights, we have suppressed the boldfaced notation for Lie algebra elements in order to come in touch with the standard notations utilized for generators in the Cartan–Weyl basis and also to avoid the excess of boldfaced symbols that would have resulted.

In the chapters devoted to differential geometry and group manifolds, namely, Chapters 6 and 15, vector fields are typically denoted by boldfaced capital letters; in particular we use such a notation for the left- or right-invariant vector fields that generate the Lie algebra of a Lie group.

Concerning the tensor product of vector spaces, when it is convenient for us, we utilize the abbreviation $V_{m|n}$, which denotes the tensor product of m copies of a vector space V_n of dimension n.

https://doi.org/10.1515/9783111201535-202

Contents

1 Groups: the intuitive notion

The first is that one looks at altogether very much scattered things and reduces them to an idea, so that every time, defining, he makes manifest the thing which he wants to teach.

Plato – from Phaedrus

The essentials of group theory can be summarized in few mathematical definitions that admit a description in relatively simple words.

A group G is first of all a set of elements. There are three cases:

1. The set G contains a finite number r of elements $\{\gamma_1, \gamma_2, \dots, \gamma_r\}$. In this case G is a finite group and the number r, usually denoted $|G|$, is named the order of the group G (see Figure 1.1).
2. The set G contains an infinite number of elements, but it is denumerable, namely, we can count the elements as $\{\gamma_1, \gamma_2, \gamma_3, \dots, \gamma_\infty\}$. In this case the group is infinite but discrete.
3. The set G is a continuous space, for instance a plane, a sphere, or some higher-dimensional variety. In this case the group G is named a *continuous group* (see Figure 1.2), and when additional properties of analyticity are satisfied it is a *Lie group*.

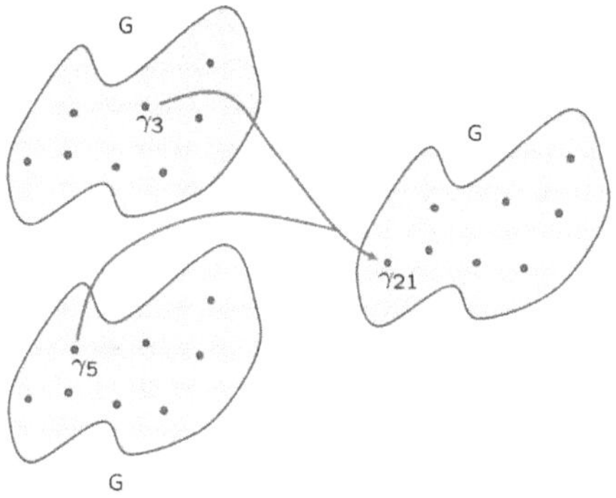

Figure 1.1: A finite group G is a set with a finite number of elements and an internal binary operation named the product. In the above picture we imagine a finite group where the product of the element γ_3 with the element γ_5 produces the element γ_{21}.

The feature that promotes a set G (falling in one of the above specified cases) to the status of a group is the existence of a binary operation:

$$p : G \times G \longrightarrow G. \tag{1.1}$$

Modern mathematics has at its center the notion of *map*. In simple words a map φ is a correspondence between two sets A and B,

$$\varphi : A \longrightarrow B, \tag{1.2}$$

https://doi.org/10.1515/9783111201535-001

that, with each element $a \in A$ of the first set, associates an element $\varphi(a) \in B$ of the second set. The element $\varphi(a)$ is named the image of a in B. On the other hand, any element $a \in A$ whose image is a given element $b \in B$ is said to be in the preimage $\varphi^{-1}(b)$. In general the preimage $\varphi^{-1}(b)$ can contain more than one element.

The binary product p of a group G is a map from the set of ordered pairs $\{a, b\}$, where $a, b \in G$ are elements of the group, to the group G. The image of the pair

$$p(a,b) \equiv a \cdot b \in G \tag{1.3}$$

is an element of the same set G and it is named the product of a with b.

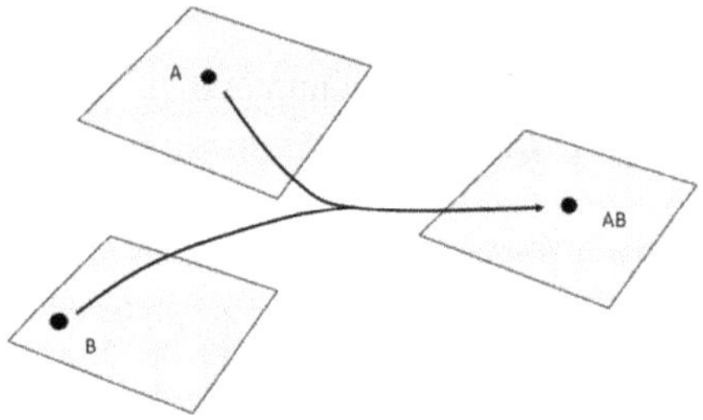

Figure 1.2: A continuous group G is a continuous (topological) space, like a plane or some other higher-dimensional manifold, whose points can be labeled by coordinates and which is endowed with an internal binary operation named the product.

In order for G to deserve the name of *group*, the *product* must have the following necessary properties:

(a) The set G must include a specific element $\mathbf{e} \in G$, named the *identity*, which multiplied either on the left or on the right with any element $x \in G$ reproduces the latter, i. e.,

$$x \cdot \mathbf{e} = \mathbf{e} \cdot x = x. \tag{1.4}$$

(b) Chosen any element x belonging to the group G, the latter must contain also a unique element x^{-1}, named the inverse of x, such that

$$x \cdot x^{-1} = x^{-1} \cdot x = \mathbf{e}, \tag{1.5}$$

where $\mathbf{e}$ is the previously introduced identity element.

1.1 Examples

A familiar example of an infinite but discrete group is provided by the integer relative numbers $\mathbb{Z}$. In this case the product is simply the sum $a + b$, the identity element is the number zero 0, and the inverse of any element a is just $-a$. A very simple example of a continuous group is provided by the complex numbers deprived of the number 0, namely, $\mathbb{C}^\star = \mathbb{C} - \{0\}$. In this case the product is the ordinary product, the identity element is the number 1, and the inverse of $z \in \mathbb{C}^\star$ is the reciprocal $\frac{1}{z}$, which always exists, since we have excluded $z = 0$. The simplest example of a finite group is $\mathbb{Z}_2$ formed by the two-element set $\{1, -1\}$. The product is the ordinary one, the identity element is 1, and the inverse of -1 is just the same element, since $(-1) \times (-1) = 1$.

1.2 Groups as transformation groups

In all the examples quoted above the product operation is commutative, namely, the product $a \cdot b$ of the element a with the element b yields the same result as the product $b \cdot a$ taken in the reverse order. This is not the general case, and the groups that possess such a property form the subclass of *Abelian groups*. The generic case is that of *non-Abelian groups*.

To understand how the apparently unfamiliar situation $a \cdot b \neq b \cdot a$ enters the stage we have to think of the groups not as sets of numbers, but rather as sets of *transformations* that act on another set S, which can be either finite, or infinite discrete, or continuous. In other words, every element $\gamma \in G$ of a given group G is viewed as a map,

$$\gamma : S \longrightarrow S, \tag{1.6}$$

that associates an image $\gamma(a) \in S$ in the same set with every element a of the set S. The product $\gamma_2 \cdot \gamma_1$ of two group elements is just the transformation of the set S into itself that is obtained by applying first the transformation γ_1 and then the transformation γ_2 in the specified sequence. This fundamental idea is illustrated in Figure 1.3 with the example of the rotations in three-dimensional space. The set of such rotations is the rotation group that has the mathematical name SO(3). It is evident from familiar experience that, once thought of in this way, the group product can be non-commutative. The result of performing first a rotation around the x-axis and then a rotation around the y-axis on any three-dimensional object is typically different from the result obtained by performing the same rotations in reversed order.

It is precisely in the capacity of sets of transformations that groups became the pivot in the modern conception of symmetry originating from the fundamental work of Galois. The above summarized concept of group came into being through a rather long historical process.

The first group to be considered, which is also at the basis of Galois' work and which actually encompasses all the other finite groups as subgroups, was the *permutation group* of n objects. The elements of the latter, denoted S_n and named the *n-th symmetric group*, are the permutations of an array of n objects into a different order, like, for instance,

$$\pi : \{\spadesuit, \heartsuit, \blacksquare, \clubsuit\} \longrightarrow \{\heartsuit, \spadesuit, \clubsuit, \blacksquare\}, \tag{1.7}$$

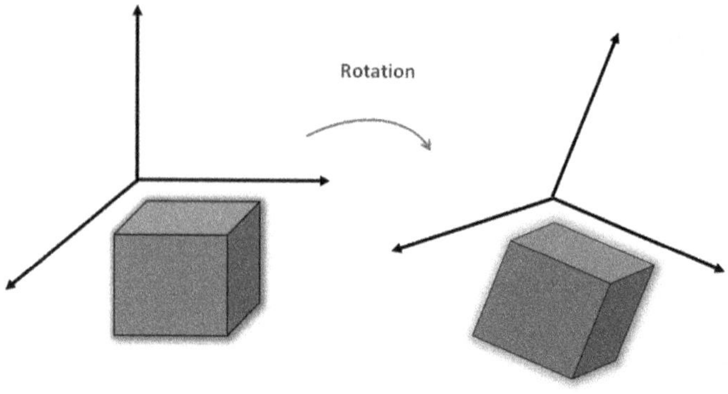

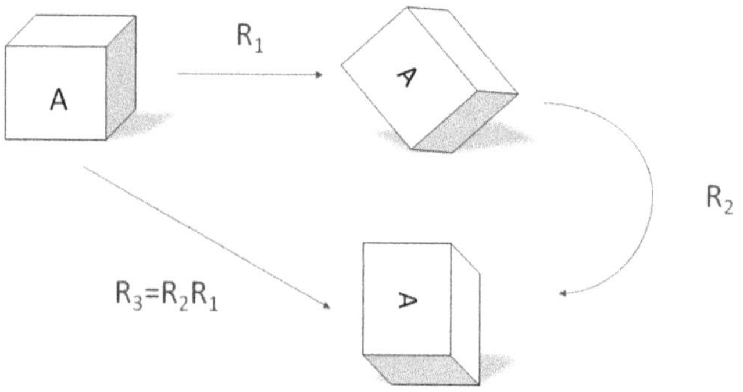

Figure 1.3: In this picture we illustrate the notion of group product with the example of familiar three-dimensional rotations. A finite rotation of a three-dimensional solid object is effected around some axis through the extension of some angle θ. After performing a rotation R_1, we can perform a second rotation R_2. The net result of the sequence of the two transformations is a new rotation R_3 around some new axis and through the extension of some new angle.

which is an element of S_4. As probably already known to most readers, the total number of permutations of n objects is $n! = n \times (n-1) \times (n-2) \times \cdots \times 2 \times 1$, which is the order $|S_n|$ of the symmetric group S_n. In the case of four objects, like the playing card suits, the number of permutations, and as such the order of the corresponding symmetric group, is just 24.

An example of a product of permutations is provided in Figure 1.4.

1.3 Representations of a group

Once the idea of transformation is absorbed, it becomes evident that every group G acts as a transformation group on itself, since each of its elements $g \in G$ acts, via the product, on all the group elements $\gamma \in G$ (the same g included) and maps them in other

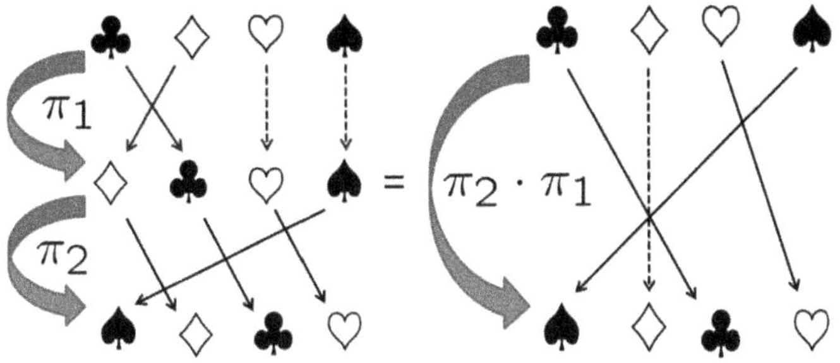

Figure 1.4: Focusing on the case of four objects in the above figure we exemplify the product law within the permutation group.

elements of G. Furthermore, it is evident that the same group can operate as a transformation group on different spaces. Each of these incarnations of G is named one of its *representations* and a central issue of group theory is the classification of all possible representations of each G.

To be more precise, we ought to rely on the notion of *homomorphism*. A homomorphism is a map from one group G to another group Γ that respects the product law of the two groups. Let us denote by $\cdot$ the product in the first group G and by $\diamond$ the product in the second group Γ. A map

$$h : G \longrightarrow \Gamma \tag{1.8}$$

is a group homomorphism if and only if the product of two images is equal to the image of the product, namely, for any $a, b \in G$ we must have

$$h(a) \diamond h(b) = h(a \cdot b). \tag{1.9}$$

Given a group G and a group of transformations $\mathscr{T}$ acting on some space S, we say that $\mathscr{T}$ is a representation of G if there exists a homomorphism $h : G \longrightarrow \mathscr{T}$.

Among the possible representations a distinctive role is played by the privileged linear ones. What do we mean by this? To answer such a question we need the notion of *vector space*. This is the generalization to arbitrary dimension of the familiar notion of three-dimensional vectors.

1.3.1 Vector spaces

Let us consider Figure 1.5, which displays two vectors $\mathbf{v}$ and $\mathbf{w}$ in the ordinary three-dimensional space $\mathbb{R}^3$. The basic property of the space of vectors is that vectors can be summed and the sum is a vector in the same space. For instance, the new vector $\mathbf{v} + \mathbf{w}$ is displayed in the figure. A vector $\mathbf{v}$ can be multiplied also by real numbers $\lambda \in \mathbb{R}$, yielding

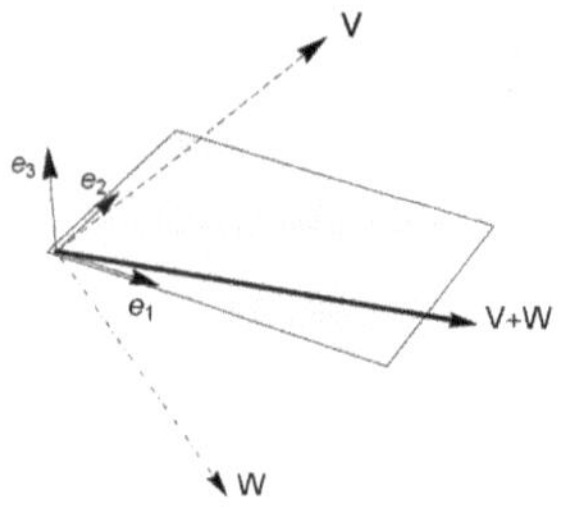

Figure 1.5: In this picture we display two vectors **v** and **w** in the ordinary three-dimensional vector space $V_3 \simeq \mathbb{R}^3$ and we show their sum. For reference we display also the three unit vectors $\mathbf{e}_{1,2,3}$ respectively aligned with the *x*-, *y*-, and *z*-axes. Every vector in V_3 is a linear combination of the basis vectors $\mathbf{e}_{1,2,3}$.

a vector $\lambda\mathbf{v}$ that has the same direction if $\lambda > 0$ but length $\lambda \times |\mathbf{v}|$, where the latter symbol denotes the length of $\mathbf{v}$. In the case $\lambda < 0$ the vector $\lambda\mathbf{v}$ has direction opposite to the direction of $\mathbf{v}$ and length $-\lambda|\mathbf{v}|$.

Actually the entire vector space V_3 of three-dimensional vectors can be viewed as the set of all possible linear combinations of three linear independent vectors $\mathbf{e}_{1,2,3}$, such as the orthonormal vectors displayed in Figure 1.5:

$$V_3 = \left\{ \bigoplus_{i=1}^{3} \lambda^i \mathbf{e}_i \;\middle|\; \lambda_i \in \mathbb{R} \right\}. \tag{1.10}$$

The essential point is that the basis of a given vector space is not uniquely defined. In the case of V_3 any other triplet $\boldsymbol{\epsilon}_{1,2,3}$ of three vectors that do not lie in the same plane (this is the concept of linear independence) provides an equally good basis as the orthonormal set $\mathbf{e}_{1,2,3}$. This is illustrated in Figure 1.6, which displays the same vectors $\mathbf{v}$ and $\mathbf{w}$ already displayed in Figure 1.5 but emphasizes that they, as any other vector in the same vector space, can be expressed also as linear combinations of the non-orthonormal triplet $\boldsymbol{\epsilon}_{1,2,3}$.

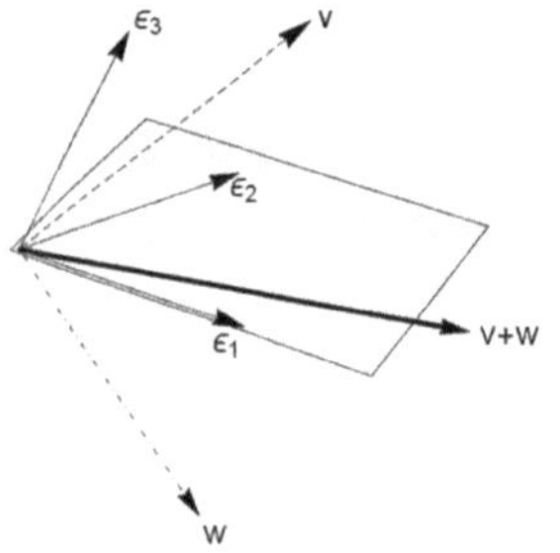

Figure 1.6: The vectors **v** and **w** (and their sum) displayed in Figure 1.5 can be expressed as linear combinations with real coefficients also of the non-orthonormal basis vectors $\boldsymbol{\epsilon}_{1,2,3}$ displayed in this figure.

The general concept of vector space emerges therefore from such a discussion. A vector space V is first of all a commutative group with respect to an operation that we can name vector addition. The identity element is the $\mathbf{0}$-vector and the inverse of any element (vector) $\mathbf{v}$ is just $-\mathbf{v}$. In addition, the vector space has a second operation that is a map:

$$\mathfrak{s} : \mathbb{K} \times V \longrightarrow V, \tag{1.11}$$

where $\mathbb{K}$ is a field, typically $\mathbb{R}$ – in this case V is a real vector space – or $\mathbb{C}$ – in this case V is a complex vector space. The vector space is of finite dimension $n < \infty$ if the maximal number of vectors $\mathbf{v}_{i=1,\dots,n}$ that can be linearly independent is n. Linear independence of r vectors $\mathbf{v}_{i=1,\dots,r}$ means that the equation

$$\sum_{i=1}^{r} \lambda^i \mathbf{v}_i = 0 \tag{1.12}$$

has the unique solution $\lambda_1 = \lambda_2 = \cdots = \lambda_r$. This is the generalization to higher dimension of the condition applying to the $n = 3$ case that three linear independent vectors cannot lie in the same plane. Surprisingly, it took about a century to arrive at the above four definitions, which appear at first sight extremely simple and natural.

In any n-dimensional vector space we can always choose a basis formed by n linearly independent vectors $\boldsymbol{\epsilon}_{i=1,\dots,n}$ and express all the others as linear combinations thereof, as in equation (1.10):

$$V_{\mathbf{n}} = \left\{ \bigoplus_{i=1}^{n} \lambda^i \boldsymbol{\epsilon}_i \,\middle|\, \lambda_i \in \mathbb{K} \right\}. \tag{1.13}$$

It emerges from the present short discussion that the notion of group is fundamental because it corresponds to the notion of calculus of symmetry operations and it involves the fundamental new idea that the nature of *things* of any kind of *thing* in any science is best understood in terms of its *symmetry*, namely, in terms of the set of transformations that leave that thing *invariant*. To be precise, it was the yield of Galois' intellectual revolution that what matters is the *algebraic structure* of the group of symmetry G of that *thing*. It does not matter whether the thing under investigation is a molecule, a protein, a crystal, an algorithm, a physical interaction, or an economic process. Two things that have the same group of symmetry G and are invariant or transform in the same way with respect to the operations of G are very close relatives, and knowledge of properties of the first maps into a similar knowledge of properties of the second. Different things are classified into equivalence classes that are the possible representations of their group of symmetries. This is the reason why group theory is the fundamental backbone of many different sciences and sits at the center of mathematical sciences.

It should be evident from the above intuitive remarks that in order to approach group theory we first need to review the fundamental concepts of algebra in a hierarchical presentation. This is the task addressed in the following chapter.

2 Fundamental notions of algebra

Quando chel cubo con le cose appresso
Se agguaglia a qualche numero discreto
Trouan dui altri differenti in esso.

Niccoló Tartaglia

2.1 Historical remarks

Any introductory mathematical course for mathematics, physics, chemistry, engineering, economics, and information theory majors makes extensive use of matrices; from the point of view of contemporary students we rightly consider such material as simple and elementary, yet this fact, which is true, should not induce us into the erroneous assumption that what nowadays we call *linear algebra* is something naturally obvious for the human mind. Historically it took a rather long time before the fundamental concepts of linear algebra were consolidated and settled down to the apparently simple shape used in current textbooks and lecture courses. The same is true for the abstract notion of a group.

Reconsidering the conceptual history of these ideas is very useful in order to fully appreciate the degree of abstraction which is tacitly involved in our current way of thinking in all fundamental and applied sciences, a degree of abstraction which has percolated down the generation tree and currently makes part of the educational process. As a result, many twentieth- and twenty-first-century students already incorporate such categorical structures as part of their logical thinking, which is a non-trivial advance.

2.1.1 Cayley and Sylvester: a short account of their lives

The main figures in the early history of *linear algebra* are Arthur Cayley and Joseph Sylvester, who became lifelong friends and whose lives often intersected. Hence we start our historic outline with a summary of their biographies.

Cayley was first educated at King's College School in London and then entered Cambridge University, where he studied mathematics. At the beginning he could not continue an academic career in Cambridge since he refused to take the minor orders of the Church of England. Having turned to law, and having worked for 14 years as an attorney in the City of London, during which he never stopped doing research in mathematics, at the age of 42 Cayley was elected Sadleirian Professor of Mathematics in Cambridge, a position that he occupied until his death. He has been one of the most prolific mathematicians of history, making extensive contributions to different fields of algebra, geometry, and analysis.

https://doi.org/10.1515/9783111201535-002

James Joseph Sylvester studied mathematics at St John's College, Cambridge. He was not awarded a Cambridge degree since, to that purpose, he had to renounce his Jewish religion and accept the Thirty-Nine Articles of the Church of England, which he refused to do. After holding for some time a teaching position in London and then obtaining a degree from Trinity College in Dublin, he became professor in the United States in Virginia. He stayed there briefly and came back to London, where he studied law; the following 10 years he worked in an insurance company. In London he met Cayley, and a lifelong interaction between the two mathematicians started which was very fruitful for both. Then he crossed once again the Atlantic and for several years he was professor of mathematics at John Hopkins University in Maryland. In 1883 he returned to England, where he was appointed Savilian Professor of Geometry at Oxford University.

Cayley and Sylvester have been nicknamed the *Invariant Twins* for their extensive and outstanding contributions to the theory of invariants (see Figure 2.1).

Figure 2.1: Arthur Cayley (Richmond 1821–Cambridge 1895) and James Sylvester (London 1814–London 1897).

2.1.1.1 The contributions of Cayley and Sylvester to linear algebra

The main focus of both Sylvester's and Cayley's mathematical studies were matrices; Most of the basic concepts in matrix theory are due to one or the other. In particular, Cayley is responsible for one of the main theorems [8], namely, the one which states that the eigenvalues of any matrix are the roots of the secular equation. Notwithstanding their vast contributions to the theory of matrices, the latter were just conceived as arrays of numbers by both Cayley and Sylvester. This might seem surprising, but it is natural since the fundamental concept of vector spaces was not assimilated by the mathematical community before the end of the nineteenth century and the famous lectures by Peano [12] published in 1888 in Torino, where the basic work by Hermann Grassmann, who, in his astonishing book of 1844 [11], had anticipated the concept of vector spaces and basis vectors, was finally understood and given the due recognition (see Figure 2.2).

Figure 2.2: On the left, Hermann Günther Grassman (Stettin 1809–Stettin 1877). On the right, Giuseppe Peano (Cuneo 1858–Torino 1932).

2.1.2 Galois and the advent of group theory

Everything is exceptional about *Évariste Galois* (see Figure 2.3), including his mathematical achievements and his short unlucky personal life. A more romantic and tragic cradle for the theory of groups could not have been invented by capricious destiny. His career and his relationship with other humans were full of contradictory aspects, so his first-class mathematical results present two quite contrasting faces. The theory named after him and the theorem within this theory which constitutes Galois' major result are rather difficult, at the level of both the definitions and the proofs; in addition, one can honestly say that Galois theory of the solubility of algebraic equations is a rather specialized topic which, nowadays, finds relevant applications eminently in number theory and associated topics, but not too many in geometry at large and in physics. On the contrary, the weapon that Galois developed to obtain his own results, namely, the theory of groups, has proved of extraordinary conceptual relevance and fertility, being the starting point for an entirely new vision of mathematics and in particular symmetry.

Évariste Galois was born on 25 October 1811 in the small town of Bourg-la-Reine. He died at the dawn of 31 May 1832 from the wounds received the day before in a duel. During the 21 years of his life he suffered all types of misfortunes and blows, mainly caused by the incomprehension and stiff stupidity of his teachers, by the political turmoil of the time, and by his naiveness. Both his mother and his father were highly educated persons, committed to revolutionary ideals and fierily opposed to the Restoration. Galois' father served as Mayor of Bourg-la-Reine, and in 1827 he fell victim to a clerical conspiracy organized by a priest who circulated a false poem, full of obscenities, that he pretended written by the Mayor; Galois' father, full of rage and shame, escaped to Paris and committed suicide in a hotel room. During the funeral Évariste suffered the aggra-

Figure 2.3: Évariste Galois (1811–1832).

vating sorrow to see his father's coffin at the center of a violent brawl between clericals and liberals.

In the *Lycée Louis-le-Grand* where he was studying, Galois met quite stiff and stupid teachers, who did not understand his exceptional talents for mathematics and treated him as an idiot. Twice he was rejected at the entrance exams to the *École Polytechnique* where he ardently desired to enroll, not only for the excellent tuition there available, but also for the democratic ideals that inflamed all of the *Polytechnique* students. Notwithstanding these adversities, Galois studied mathematics by himself, directly reading books and articles by Legendre, Fourier, Abel, and Gauss, and at the age of 17, he was already well advanced in the development of his own theory of algebraic equations. He wrote his results in a paper that he wanted to submit to the Academy and, for that purpose, he managed to give it to Cauchy, who promised to support its publication. Unfortunately Cauchy lost Galois' manuscript.

In 1830 Galois tried once again to publish his own results by giving a new paper to the scientific secretary of the Academy, who brought Galois' manuscript home to read it, but the very same night he unexpectedly died; Galois' work was once again lost. Disappointed and disgusted by life, Galois entered the political agon, just at the eve of the July Revolution, supporting the Republicans.

In the last two years of his life Galois was twice arrested as a subversive, spent some months in prison, was released, participated in other political quarrels, had a love affair with a girl of vulgar personality, who disgusted him also on that front, and finally was involved in a stupid debate with a political exponent of opposite views, which ended in the duel which caused his death. Perfectly aware of his almost sure death, the night before the duel, Évariste wrote a 60-page exposition of all his mathematical results, which he gave to his loyal friend Auguste Chevalier, who, fortunately, did not lose the work, and in 1846 Galois' main theorem was finally published [1] in the *Journal de Mathématiques Pures et Appliquées*, with praising comments of its main editor, *Joseph Liouville*.

2.2 Summary of the content of this chapter

The present chapter summarizes the formal definitions of all algebraic structures we are going to use in the sequel. These algebraic structures are presented in their natural hierarchical order of increasing complexity:
1 groups
2 rings
3 fields
4 vector spaces
5 algebras
6 Lie algebras, and
7 modules and representations

The purpose of this chapter is to establish the mathematical language which will be our reference point in all subsequent discussions.

In the above list of algebraic structures, groups are the simplest, since they involve just one internal binary operation, while all the other structures involve more than one, typically a *sum* and a *product*, or also a multiplication by *scalar coefficients*, as is the case for vector spaces and modules. Yet groups constitute the most important algebraic structure because of their basic interpretation as a *set of transformations* acting on some *space* composed of *objects* which can have the most different nature. However, in developing group theory, in particular the theory of group representations, all the other algebraic structures of the above list come into play, and for this reason we need to review them here. Furthermore, in order to deal with *continuous groups* and their relation with the differential geometry of *homogeneous spaces* we have to introduce the notions of *manifolds, fiber bundles*, and *differential forms*. This will be done in later chapters.

2.3 Groups

In this section we introduce the algebraic definition of a group and we illustrate it with several examples. Before moving to the next algebraic structure, namely, *rings*, we also fix the basic concepts of order of a group, homomorphisms, isomorphisms, automorphisms, generators, and relations. The analysis of the general algebraic properties of the group structure and the related fundamental theorems are postponed to Chapter 4, where they are considered in full generality and in depth.

Definition 2.3.1. A group G is a set equipped with a binary operation $\cdot$ such that:
1. $\forall a, b, c \in G$, $a \cdot (b \cdot c) = (a \cdot b) \cdot c$, i. e., *associativity* holds.
2. $\exists e \in G$ such that $a \cdot e = e \cdot a = a \; \forall a \in G$, i. e., there is an *identity* element e.
3. $\forall a \in G \; \exists b \in G$ such that $a \cdot b = b \cdot a = e$, i. e., each element a admits an *inverse* b, which is usually denoted as a^{-1}.

In the following, we will often indicate the group product simply as ab, or we may indicate it with different symbols, e. g., as $a+b$, when the product law is in fact the usual addition. To stress a specific choice of the product law, we may also indicate the group as, for instance, $(G,+)$.

2.3.1 Some examples

- Consider the set $\{0,1\}$ with the group product being the usual addition defined mod 2; this is a group, usually denoted as $\mathbb{Z}_2$.
- The set $\{1,-1\}$ equipped with the usual multiplication is a group. This group is isomorphic with, i. e., has the same abstract structure as, the group $\mathbb{Z}_2$ considered in the previous example.
- The set of real numbers $\mathbb{R}$, with the group law being addition, is a group.
- The set $\mathrm{U}(1) \equiv \{e^{i\theta}; \theta \in [0,2\pi]\}$ with the usual multiplication is a group.
- The groups $\{e, a, a^2 \equiv a \cdot a, \ldots, a^{k-1}\}$ containing all the powers of a single generator a with respect to the group product, with the single extra relation $a^k = e$, are named *cyclic groups* and are denoted as $\mathbb{Z}_k$.
- The set of permutations of three numbered objects 1, 2, 3 forms a group called S_3, the product law being the composition of the permutations. This group has order 6: the identical permutation, three exchanges ($p_{12} = (1 \leftrightarrow 2)$, $p_{13} = (1 \leftrightarrow 3)$, and $p_{23} = (2 \leftrightarrow 3)$), and two cyclic permutations ($p_{123} = (1 \to 2 \to 3 \to 1)$ and $p_{132} = (1 \to 3 \to 2 \to 1)$).

2.3.2 Abelian groups

The group product is *not* required to be commutative. When the product *is* commutative, the group is called an *Abelian group*:

$$G \text{ Abelian:} \quad ab = ba \quad \forall a, b \in G. \tag{2.1}$$

Abelian groups are of course the simplest types of groups. All the groups of the previous examples are in fact Abelian, except the permutation group S_3.

2.3.3 The group commutator

Two elements g, h of a group commute if for the group product $gh = hg$, i. e., $ghg^{-1}h^{-1} = e$ (e being the identity). Then the *group commutator* of g and h, defined as

$$ghg^{-1}h^{-1}, \tag{2.2}$$

indicates if (and how) the two elements fail to commute.

2.3.4 Conjugate elements

Two group elements h and h' are said to be *conjugate* to each other if

$$\exists g \in G \quad \text{such that } h' = g^{-1}hg. \tag{2.3}$$

If the group is Abelian, each element is conjugate only to itself.

2.3.5 Order and dimension of a group

The number of elements of a group G can be finite, infinite but denumerable, or continuously infinite.

In the first two cases, the number of elements of G is named the *order* of the group, denoted as $|G|$ (in the second case, $|G| = \infty$). A group of finite order is called a *finite group*.

2.3.5.1 Examples

- The cyclic group $\mathbb{Z}_k$ is a finite group of order k.
- The integers $\mathbb{Z}$ with the group product being the addition form a group of infinite order.
- The set of real numbers (with the zero excluded) $\mathbb{R}\setminus\{0\}$, equipped with the ordinary product, is a continuous group.

2.3.6 Order of an element

If $a \in G$, then $a^2, a^3, \ldots \in G$. If all the powers of a are distinct, then a is an element of *infinite order* (and then, of course, G cannot be a group of finite order). If some of the powers of a coincide, this means that there exists some integer m such that $a^m = e$. Let n be the smallest of such positive integers m. Then a is said to be an element *of order n*.

2.3.6.1 Observations

- In a finite group G all elements have order $\leq |G|$.
- If G is a cyclic group of order n, then the order of any element is a divisor of n.
- In a finite group, the inverse of any element a is a power of a; indeed, $a^{-1} = a^{n-1}$, where n is the order of a.

2.3.7 The multiplication table of a finite group

A group is abstractly defined by describing completely (i. e., $\forall g_1, g_2 \in G$) the product law $(g_1, g_2) \mapsto g_1 g_2$, namely, by identifying the result of each possible product. For finite groups this can be encoded explicitly in a *multiplication table* whose entry in the i-th row and j-th column describes which element is the result of the product $g_i g_j$ (so we use the convention that $g_1 = e$):

e	g_2	g_3	$\dots$
g_2	$(g_2)^2$	$(g_2 g_3)$	$\dots$
g_3	$(g_3 g_2)$	$(g_3)^2$	$\dots$
$\vdots$	$\vdots$	$\vdots$	$\ddots$

Notice that all the elements appearing in a line of the multiplication table are different (and therefore all elements appear in each line). Indeed, if we had $g_i g_j = g_i g_k$ we could conclude that $g_j = g_k$. The same applies to each column. These properties constrain the possible multiplication tables, especially at low orders.

A finite group is abstractly defined by its multiplication table, up to relabeling of the elements (i. e., up to rearrangement of the rows and columns). A given table, i. e., a given group, may have different concrete realizations. We will shortly make more precise what we mean by this observation.

2.3.8 Homomorphisms, isomorphisms, and automorphisms

We have seen that we can have different realizations of a given abstract group. We are interested in concrete realizations of the group, where the elements of the group acquire an explicit meaning as numbers, matrices, symmetry operations, or other quantities with which we can perform explicit computations.

In precise terms, finding a different realization G' of a given group G means to find an *isomorphic mapping* (or *isomorphism*) between G and G'. Let us explain this terminology.

2.3.8.1 Homomorphisms

A map ϕ from a group G to a group G' is called a *homomorphism iff* it preserves the group structure, namely, *iff*

$$\forall g_1, g_2 \in G, \quad \phi(g_1 g_2) = \phi(g_1)\phi(g_2), \tag{2.4}$$

where the products $g_1 g_2$ and $\phi(g_1)\phi(g_2)$ are taken with the group product law of G and G', respectively. The map is not required to be invertible, i. e., one-to-one.

2.3.8.2 Isomorphisms

A homomorphism $\phi : G \to G'$ which is also *invertible* is called an *isomorphism.* Two groups G and G' such that there exists an isomorphism $\phi : G \to G'$ are said to be *isomorphic.* They correspond to different realizations of the same abstract group structure.

2.3.8.3 Automorphisms

An isomorphic mapping σ from a group G to itself is called an *automorphism* of G. The set of all automorphisms of a given group G is a group called Aut(G), the product law being the composition of mappings. Indeed, the composition of two automorphisms σ_1, σ_2 is still an automorphism:

$$\sigma_1(\sigma_2(h_1h_2)) = \sigma_1(\sigma_2(h_1)\sigma_2(h_2)) = \sigma_1(\sigma_2(h_1))\sigma_1(\sigma_2(h_2)). \tag{2.5}$$

Of course, the composition of mappings is associative, there is an identical automorphism, and any automorphism is assumed to be invertible by definition so that the group axioms are satisfied.

For finite groups, the automorphisms of G are particular permutations of the elements of G, namely, Aut(G) is a subgroup of $S_{|G|}$. They correspond to symmetries of the multiplication table of G, in the following sense. As we remarked, the ordering of the rows and columns of the multiplication table is irrelevant. If we apply a given map $\sigma : g \mapsto \sigma(g)$ to labels and entries of the multiplication table it may happen that the resulting table corresponds just to a rearrangement of rows and columns of the original table. In this case the permutation σ is an automorphism.

An important class of automorphic mappings is that corresponding to conjugation by a fixed element of a group:

$$\sigma_g : h \in G \mapsto \sigma_g(h) = g^{-1}hg \in G. \tag{2.6}$$

Automorphisms that correspond to conjugations are called *inner automorphisms*, and automorphisms that do not are called *outer automorphisms*. Notice that for Abelian groups the only non-trivial automorphisms are the outer ones.

2.3.9 Rank, generators, and relations (a first bird's-eye view)

Let the elements of a finite group G be $g_1 = e, g_2, g_3, \ldots$. All elements can be written in the form

$$g_k = g_{i_1}g_{i_2}\cdots g_{i_s}, \tag{2.7}$$

for suitable elements g_{i_j} (at worst the only possibility is $g_k = g_k$). A set of group elements (different from the identity e) which multiplied in all possible ways give *all* the elements of G is said to *generate* the group G. The minimal such set is a set of *generators* of G. The minimal number of generators is the *rank* of G.

If a group G is denumerable rather than finite, then we say that it is generated by a finite subset B of elements iff every element $g \in G$ can be written as

$$g = g_{i_1}^{\pm 1} g_{i_2}^{\pm 1} \dots g_{i_s}^{\pm 1}, \tag{2.8}$$

with all the g_{i_l} belonging to B. So, with respect to the case of finite groups, now not only positive but also negative powers of generators may appear. In the case of finite groups, all generators have finite order, so their negative powers can be reexpressed in terms of the positive ones; this is not true in general.

2.3.9.1 Presentation of a group: relations

A group can be described by giving its *generators* and (if present) the *relations* that they satisfy. Such a description of a group is called a *presentation* of the group.

Indeed, starting with the generators g_a $(a = 1, \dots, \text{rank}\, G)$ one can construct *words* of increasing length: g_a, then $g_a g_b$, etc. In this way one obtains the further elements of G. In the process, however, me must take into account the *relations* to which the generators may be subject, which may be cast in the form $R_i(\{g_a\}) = e$, with R_i being a set of specific words. If G is a finite group, then each generator must be of finite order and therefore we have at least the relations $g_a^{n_a} = e$, where n_a is the order of g_a; there can be others.

We come back to these issues in Chapter 4.

2.3.9.2 Examples

- The group S_3 is defined by its presentation consisting of two generators $s = p_{12}$ and $t = p_{13}$ subject to the relations $s^2 = e$, $t^2 = e$, and $(st)^3 = e$.

For continuous groups, the analog of generators is given, as we will see, by the existence of a set of *infinitesimal generators* closing a *Lie algebra*.

2.3.10 Subgroups

A subset $H \subset G$ is a *subgroup* of G if it is a group, with the same product law defined in G. To this effect it is sufficient that:

(i) $\forall h_1, h_2 \in H, h_1 h_2 \in H$;

(ii) $\forall h_1 \in H, h_1^{-1} \in H$.

While for an infinite (or continuous) group both requirements have to be separately checked, if G is a finite group, then item (i) is enough since all elements are of finite order.

The relation "being a subgroup of" is transitive:

$$\begin{cases} H \subset G, \\ K \subset H \end{cases} \Rightarrow \quad K \subset G, \tag{2.9}$$

where $H \subset G$ means H *is a subgroup of* G. In general, a given group G admits chains of subgroups

$$G \supset H_1 \supset H_2 \cdots \supset \{e\}. \tag{2.10}$$

G itself and the group containing only the identity e are *trivial* subgroups of G; other subgroups are called *proper* subgroups. One of the most important problems in group theory is the determination (up to conjugation) of all proper subgroups of a given group.

Infinite groups may admit infinite sequences of subgroups.

2.4 Rings

Let us now turn to the definition of rings.

Definition 2.4.1. A ring $\mathscr{R}$ is a set equipped with two binary composition laws respectively named the *sum* and the *product*:

$$\begin{aligned} &\forall x, y \in \mathscr{R}: \quad x + y \in \mathscr{R}, \\ &\forall x, y \in \mathscr{R}: \quad x \cdot y \in \mathscr{R}, \end{aligned} \tag{2.11}$$

which satisfy the following properties.

R1 With respect to the *sum*, $\mathscr{R}$ is an Abelian group, namely,

$$x + y = y + x,$$

and there exist a neutral element $0 \in \mathscr{R}$ such that $0 + x = x$, $\forall x \in \mathscr{R}$, and an inverse $-x$ of each element $x \in \mathscr{R}$ such that $x + \{-x\} = 0$.

R2 The product has the *distributive* property with respect to the sum:

$$\forall x, y, z \in \mathscr{R}: \quad x \cdot (y + z) = x \cdot y + x \cdot z.$$

R3 The product is *associative*:

$$\forall x, y, z \in \mathscr{R}: \quad (x \cdot y) \cdot z = x \cdot (y \cdot z).$$

R4 There exists an identity element $e \in \mathscr{R}$ for the product such that

$$\forall x \in \mathscr{R}: \quad x \cdot e = e \cdot x = x.$$

A typical example of a ring is provided by the relative integers $\mathbb{Z}$. The identity element for the product is obviously the number 1. Note that with respect to the product the existence of an inverse element is not required. This is precisely the case of integer numbers. The inverse of an integer is not integer in general; indeed, the only exception is precisely the number 1.

Another example of a ring is provided by the continuous and finite functions over an interval, for instance $[0, 1]$.

2.5 Fields

Next we introduce the definition of a field.

Definition 2.5.1. A field $\mathbb{K}$ is a ring with the additional property that:

K $\forall x \in \mathbb{K}$ different from the neutral element with respect to the sum $x \neq 0$ there is an inverse element, namely, $\exists x^{-1} \in \mathbb{K}$, such that

$$x \cdot x^{-1} = x^{-1} \cdot x = e.$$

If the product is commutative, then we say that $\mathbb{K}$ is a commutative field.

Examples of commutative fields are provided by the set of rational numbers $\mathbb{Q}$, by the set of real numbers $\mathbb{R}$, and by the set of complex numbers $\mathbb{C}$.

There are also fields with a finite number of elements. For instance, $\mathbb{Z}_q$, the set of integer numbers modulo a prime q, is a commutative field with exactly q elements. An element of $[a] \in \mathbb{Z}_q$ is the entire equivalence class of integers of the form

$$[a] = a + nq; \quad a, n, q \in \mathbb{Z}. \tag{2.12}$$

With respect to the sum, $\mathbb{Z}_q$ is an Abelian cyclic group as we have already seen. It suffices to note that the product operation goes to the quotient with respect to the equivalence classes. Indeed,

$$(a + qn)(b + mq) = ab + q(ma + nb + qnm) \sim [ab]. \tag{2.13}$$

Furthermore, for each equivalence class $[a]$ there is always another equivalence class $[b] = [a]^{-1}$ such that $ab = 1 \bmod q$, namely, the inverse element. As an example consider $\mathbb{Z}_5$. With respect to the product we have the following multiplication table of equivalence classes:

	1	2	3	4
1	1	2	3	4
2	2	4	1	3
3	3	1	4	2
4	4	3	2	1

(2.14)

2.6 Vector spaces

Definition 2.6.1. A vector space is defined giving a set V, whose elements are named vectors $\vec{v} \in V, \vec{w} \in V, \ldots$, and a commutative field $\mathbb{K}$ (either $\mathbb{R}$ or $\mathbb{C}$), whose elements are named scalars. Two binary operations are then introduced. The first, named sum +, is a map:

$$V \times V \to V. \tag{2.15}$$

With respect to the sum, V is a commutative group:

V1 $\forall \vec{v}, \vec{w} \in V$, we have $\vec{v} + \vec{w} = \vec{w} + \vec{v} \in V$;

V2 $\exists \vec{0} \in V$ such that $\forall \vec{v} \in V$ the following is a true statement: $\vec{0} + \vec{v} = \vec{v} + \vec{0}$;

V3 $\forall \vec{v} \in V$ there exists the opposite element $-\vec{v} \in V$ such that $\vec{v} + (-\vec{v}) = \vec{0}$.

The second operation, named multiplication times scalars, associates with each pair $(\lambda, \vec{v})$ of a scalar $\lambda \in \mathbb{K}$ and a vector $\vec{v} \in V$ a new vector simply denoted $\lambda\vec{v} \in V$, namely, it is the map

$$\mathbb{K} \times V \to V. \tag{2.16}$$

Naming 1 and 0 the neutral elements of the field respectively for its internal multiplication and addition, the operation (2.16) has the following defining properties:

V4 $\forall \vec{v} \in V : 1\vec{v} = \vec{v}$;

V5 $\forall \lambda, \mu \in \mathbb{K}$ and $\forall \vec{v} \in V : \lambda(\mu\vec{v})\vec{v} = (\lambda\mu)\vec{v}$;

V6 $\forall \lambda, \mu \in \mathbb{K}$ and $\forall \vec{v} \in V : (\lambda + \mu)\vec{v} = \lambda\vec{v} + \mu\vec{v}$;

V7 $\forall \lambda \in \mathscr{K}$ and $\forall \vec{v}, \vec{w} \in V : \lambda(\vec{v} + \vec{w}) = \lambda\vec{v} + \lambda\vec{w}$.

From the above axioms it follows that $0\vec{v} = \vec{0}$.

Familiar examples of vector spaces are $\mathbb{R}^n$ and $\mathbb{C}^n$.

We also recall that vector spaces can be of finite or infinite dimension. Finite-dimensional vector spaces are those V for which a set of m linearly independent vectors $\mathbf{e}_i$ $(i = 0, 1, \ldots, m)$ can be found such that:

B1) $\sum_{i=1}^m \alpha^i \mathbf{e}_i = \mathbf{0}$ with $\alpha^i \in \mathbb{F}$ necessarily implies $\alpha_i = 0$;

B2) $\forall \mathbf{v} \in V\ \exists v^i \in \mathbb{F}$ such that $v = \sum_{i=1}^m v^i \mathbf{e}_i$.

The number m is named the dimension of the vector space.

2.6.1 Dual vector spaces

Definition 2.6.2. A linear functional f on a vector space V is a linear map $f : V \to \mathbb{F}$ such that

$$f(\alpha\mathbf{v}_1 + \beta\mathbf{v}_2) = \alpha f(\mathbf{v}_1) + \beta f(\mathbf{v}_2).$$

Definition 2.6.3. Linear maps on a vector space V can be linearly composed $(af_1 + \beta f_2)(\mathbf{v}) = af_1(\mathbf{v}) + \beta f_2(\mathbf{v})$ and therefore they form a vector space which is denoted $V^\star$ and it is named the dual vector space.

For finite-dimensional vector spaces V, the dual space $V^\star$ has the same dimension:

$$\dim V = \dim V^\star. \tag{2.17}$$

Indeed, it suffices to note that given a basis $\mathbf{e}_i$ $(i = 1, \ldots, n = \dim V)$ of V, we can introduce a dual basis for $V^\star$ composed of the n linear functionals defined by

$$f^j(\mathbf{e}_i) = \delta^j_i. \tag{2.18}$$

2.6.2 Inner products

Definition 2.6.4. Let V be a finite-dimensional vector space over a field $\mathbb{F} = \mathbb{R}$ or $\mathbb{C}$. An inner product on V is a map

$$\langle , \rangle : V \times V \to \mathbb{F} \tag{2.19}$$

fulfilling the following properties:

(a) conjugate symmetry:

$$\forall x, y \in V : \quad \langle x, y\rangle = \overline{\langle y, x\rangle};$$

(b) linearity in the second argument:

$$\forall x, y \in V,\ \forall a, b \in \mathbb{F} : \quad \langle x, ay + bz\rangle = a\langle x, y\rangle + a\langle x, z\rangle;$$

(c) non-degeneracy:

$$\forall y \in V : \quad \langle x, y\rangle = 0 \quad \Rightarrow \quad x = 0 \in V.$$

If in addition to property (c) we also have the following one,

(d) positive definiteness:

$$\forall x \in V : \quad \langle x, x\rangle \geq 0; \quad \langle x, x\rangle = 0 \quad \Rightarrow \quad x = 0 \in V,$$

we say that the inner product is positive definite.

Given a basis $\mathbf{e}_i$ of the vector space, the inner product is completely characterized by the following matrix:

$$M_{ij} \equiv \langle \mathbf{e}_i, \mathbf{e}_j\rangle, \tag{2.20}$$

which by force of the above axioms obeys the following properties:

$$M_{ji} = \overline{M_{ij}} \quad \Leftrightarrow \quad M = M^{\dagger}, \quad \det M \neq 0. \tag{2.21}$$

In other words, the inner product is characterized by a *non-degenerate Hermitian matrix*. The inner product is positive definite if all the eigenvalues of the matrix M are positive.

2.7 Algebras

Definition 2.7.1. An algebra $\mathscr{A}$ is a vector space over a field $\mathbb{K}$ on which a further binary operation of internal composition named product is defined:

$$\cdot : \mathscr{A} \otimes \mathscr{A} \to \mathscr{A}. \tag{2.22}$$

The product is distributive with respect to the operations characterizing a vector space:
A1) $\forall a, b, c \in \mathscr{A}$ and $\forall \alpha, \beta \in \mathbb{K}$,

$$a \cdot (\alpha b + \beta c) = \alpha a \cdot b + \beta a \cdot c. \tag{2.23}$$

If one imposes the extra property of associativity,

$$(a \cdot b) \cdot c = a \cdot (b \cdot c), \tag{2.24}$$

one obtains an associative algebra.

For instance, the $n \times n$ matrices form an associative algebra. Indeed, the linear combination with real or complex numbers of two matrices is still a matrix, which shows that matrices form a vector space. Furthermore, matrix multiplication provides the additional binary composition law, which completes the axioms of an algebra.

2.8 Lie algebras

A Lie algebra $\mathbb{G}$ is an algebra where the product is endowed with special properties.

Definition 2.8.1. A Lie algebra $\mathbb{G}$ is an algebra over a commutative field $\mathbb{K}$ where the product operation $\mathbb{G} \times \mathbb{G} \to \mathbb{G}$, named *Lie product* or *Lie bracket* and denoted as

$$\forall x, y \in \mathbb{G}, \quad x, y \mapsto [x, y], \tag{2.25}$$

has the following properties:

(i) linearity:

$$[x, \alpha x + \beta z] = \alpha [x, y] + \beta [x, z], \tag{2.26}$$

where $\alpha, \beta \in \mathbb{K}$;

(ii) antisymmetry:

$$[x, y] = -[y, x]; \tag{2.27}$$

(iii) Jacobi identity:

$$[x, [y, z]] + [y, [z, x]] + [z, [x, y]] = 0. \tag{2.28}$$

Notice that items (i) and (ii) imply the linearity also in the first argument and that the Lie product is *not associative*. Indeed, we have $[x, [y, z]] \neq [[x, y], z]$, as from the Jacobi identity we get $[x, [y, z]] - [[x, y], z] = -[y, [z, x]]$, which is generically non-zero.

The dimension of $\mathbb{G}$ as a vector space is also named the *dimension* or order of the Lie algebra, and it is denoted $\dim \mathbb{G}$.

Fixing a basis $\{t_i\}$ of vectors on which every element of $\mathbb{G}$ can be expanded, $x = x^i t_i$, the Lie product structure of $\mathbb{G}$ is encoded in a set of $(\dim \mathbb{G})^3$ constants $c_{ij}{}^k$, called the *structure constants* of the algebra. The basis vectors t_i are called the *generators* of the Lie algebra.

As a consequence of the properties of the Lie product, the structure constants obey the following relations:

(i) antisymmetry:

$$c_{ij}{}^k = -c_{ji}{}^k; \tag{2.29}$$

(ii) Jacobi identity:

$$c_{im}{}^s c_{jk}{}^m + c_{jm}{}^s c_{ki}{}^m + c_{km}{}^s c_{ij}{}^m = 0. \tag{2.30}$$

The explicit set of structure constants for a given Lie algebra is unique only up to the effect of changes of basis.

2.8.1 Homomorphism

Two Lie algebras $\mathbb{G}$ and $\mathbb{K}$ are *homomorphic* if there exists an *homomorphism* between the two, namely, a map $\phi : \mathbb{G} \to \mathbb{K}$ that preserves the Lie product:

$$[\phi(x), \phi(y)] = \phi([x, y]). \tag{2.31}$$

2.8.2 Subalgebras and ideals

Definition 2.8.2. A vector subspace $\mathbb{H} \subset \mathbb{G}$ of a Lie algebra $\mathbb{G}$ is named a *subalgebra* iff $\forall \mathbf{X}, \mathbf{Y} \in \mathbb{H}$ we have $[\mathbf{X}, \mathbf{Y}] \in \mathbb{H}$, which is usually abbreviated as $[\mathbb{H}, \mathbb{H}] \subset \mathbb{H}$.

Definition 2.8.3. A subalgebra $\mathbb{I} \subset \mathbb{G}$ of a Lie algebra $\mathbb{G}$ is named an *ideal* iff $\forall \mathbf{X} \in \mathbb{I}$ and $\forall \mathbf{Y} \in \mathbb{G}$ we have $[\mathbf{X}, \mathbf{Y}] \in \mathbb{I}$, which is usually abbreviated as $[\mathbb{I}, \mathbb{G}] \subset \mathbb{I}$.

2.8.3 Representations

A Lie algebra can be realized by a set of square matrices (forming a vector space), the Lie product being defined as the commutator in the matrix sense. If a Lie algebra $\mathbb{G}$ is *homomorphic* to some *matrix Lie algebra* $\mathscr{D}(\mathbb{G})$, then $\mathscr{D}(\mathbb{G})$ is said to give a matrix *representation* of $\mathbb{G}$.

2.8.4 Isomorphism

Two Lie algebras between which there exists a *homomorphism* that is *invertible*, i. e., an *isomorphism*, are said to be *isomorphic*. They correspond to two different realizations of the same abstract Lie algebra. An isomorphism of $\mathbb{G}$ into some matrix Lie algebra $\mathscr{D}(\mathbb{G})$ is called a *faithful* representation.

2.8.5 Adjoint representation

We have seen that, given a Lie algebra $\mathbb{G}$, we can associate with it a set of structure constants (unique only up to changes of basis). The converse is also true. Indeed, given a set of n^3 constants $c_{ij}{}^k$, the conditions in (2.29) and (2.30) are necessary and sufficient for the set of $c_{ij}{}^k$ to be the structure constant of a Lie algebra. The necessity was argued before. That it is sufficient is exhibited by constructing explicitly a *matrix* Lie algebra with structure constants $c_{ij}{}^k$. Indeed, we define the $n \times n$ matrices $\mathbf{T}_i$, with matrix elements

$$(\mathbf{T}_i)_j{}^k \equiv c_{ji}{}^k. \tag{2.32}$$

The matrix commutator of two such matrices can be computed by making use of the antisymmetry and of the Jacobi identity:

$$[\mathbf{T}_i, \mathbf{T}_j]_p{}^q = (\mathbf{T}_i)_p{}^s(\mathbf{T}_j)_s{}^q - (i \leftrightarrow j) = \cdots = c_{ij}{}^k(\mathbf{T}_k)_p{}^q. \tag{2.33}$$

Thus, the matrices obtained as real linear combinations of the $\mathbf{T}_i$'s form a Lie algebra of dimension n, with structure constants $c_{ij}{}^k$.

The above construction tells us also that, given a Lie algebra $\mathbb{G}$ with structure constants $c_{ij}{}^k$, it always possesses a faithful representation in terms of $\dim \mathbb{G} \times \dim \mathbb{G}$ matrices, called the *adjoint representation*, generated by the matrices $\mathbf{T}_i$ of (2.32).

2.9 Moduli and representations

Next we come to the proper mathematical setup underlying the notion of Lie algebra representations, which is that of *module.*

Definition 2.9.1. Let $\mathbb{L}$ be a Lie algebra and let V be a vector space over the field $\mathbb{F} = \mathbb{R}$ or $\mathbb{C}$, endowed with an operation

$$\mathbb{L} \otimes V \mapsto V,$$

denoted $(x, v) \mapsto x.v$. The vector space V is named an $\mathbb{L}$-module if the conditions provided by the following axioms are satisfied:

(M1) $\forall x, y \in \mathbb{L}$, $\forall v \in V$, and $\forall a, b \in \mathbb{F}$,

$$(ax + by).v = a(x.v) + b(y.v);$$

(M2) $\forall x \in \mathbb{L}$, $\forall v, w \in V$, and $\forall a, b \in \mathbb{F}$

$$x.(av + bw) = a(x.v) + b(x.w);$$

(M3) $\forall x, y \in \mathbb{L}$, $\forall v \in V$,

$$[x, y].v = x.(y.v) - y.(x.v).$$

As we see from axiom (M2), with each element $x \in \mathbb{L}$ one associates a linear map $\mathscr{D}(x) \in \mathrm{Hom}(V, V)$:

$$\forall x \in \mathbb{L} \quad \mathscr{D}(x) : V \to V. \tag{2.34}$$

Choosing a basis $\mathbf{e}_1, \ldots \mathbf{e}_m$ of the vector space V, we obtain

$$x.\mathbf{e}_i = \mathbf{e}_j \mathscr{D}_{ji}(x), \tag{2.35}$$

where $\mathscr{D}_{ij}(x)$ is an $m \times m$ matrix. Hence, with each element $x \in \mathbb{L}$ we associate a matrix $\mathscr{D}(x)$ and by force of axiom (M3) we have

$$\mathscr{D}([x, y]) = \mathscr{D}(x)\mathscr{D}(y) - \mathscr{D}(y)\mathscr{D}(x) = [\mathscr{D}(x), \mathscr{D}(y)], \tag{2.36}$$

which shows the equivalence of the previously introduced notion of linear representation of a Lie algebra to that of module. The notion of module is mathematically more precise since it does not make reference to any particular basis in the vector space V. Two linear representations $\mathscr{D}_1$ and $\mathscr{D}_2$ of the same Lie algebra are equivalent if they can be related to one another by an overall similarity transformation $\mathscr{S}$:

$$\forall x \in \mathbb{L} \quad \mathscr{D}_1(x) = \mathscr{S}\mathscr{D}_2(x)\mathscr{S}^{-1}. \tag{2.37}$$

The above happens when $\mathcal{D}_1$ and $\mathcal{D}_2$ are obtained from the same module in two different systems of basis vectors.

Summarizing, the notion of module encodes the notion of equivalence classes of equivalent linear representations of a Lie algebra.

2.9.1 Fields, algebraic closure, and division algebras

The notion of *field* was implicitly used by *Abel* and *Galois*; then the concept developed steadily through the work of *Karl von Staudt*, *Richard Dedekind*, who introduced the German denomination *Körper*, *David Hilbert*, and finally *Heinrich Weber*, who provided the first axiomatic definition of a *field*.

In the meantime, the idea of field extension and *algebraic closure* was to be correlated with the idea of *division algebra*.

Let us start with algebraic closure.

A field $\mathbb{K}$ is algebraically closed if and only if it contains a root for every non-constant polynomial $\mathfrak{P}(x) \in \mathbb{K}[x]$, the ring of polynomials in the variable x with coefficients in $\mathbb{K}$.

In other words, a field is closed if all algebraic equations $\mathfrak{P}(x) = 0$ can be solved by means of special elements x_ℓ of the same field $\mathbb{K}$ wherefrom the coefficients a_i of the polynomial

$$\mathfrak{P}(x) \equiv \sum_{i=0}^{n} a_i x^i$$

are taken, in such a way that $\mathfrak{P}(x_\ell) = 0$ for all $\ell = 1, 2, \ldots, n$. The numbers $x_\ell \in \mathbb{K}$ are named the *roots* of the considered polynomial.

The field of real numbers $\mathbb{R}$ is not closed and just for this reason we have to extend it to the complex number field $\mathbb{C}$, which is closed.

Recalling the definition of algebras, we have the following.

Definition 2.9.2. Let $\mathcal{D}$ be an algebra over a field $\mathbb{K}$ and let us assume that $\mathcal{D}$ does not just consist of its zero element. We call $\mathcal{D}$ a division algebra if $\forall a \in \mathcal{D}$ and for any non-zero element $b \in \mathcal{D}$, $\exists x \in \mathcal{D}$ such that $a = bx$ and $\exists y \in \mathcal{D}$ such that $a = yb$.

The name given to this type of algebras is clearly justified by their definition. We can always define the ratio a/b of any two elements a, b of the algebra except for the case where $b = 0$.

2.9.1.1 The complex numbers as a division algebra

Another way of looking at complex numbers is the following: $\mathbb{C}$ is nothing but a division algebra of dimension two over $\mathbb{R}$, the field of real numbers. This way of thinking is the result of an almost 2000-year-long constructive process. Descartes had difficulties to understand imaginary numbers, but he had no problem with the pairs of real numbers (x, y), namely, with a two-dimensional vector space, especially taking into account the fact that he had himself invented such a notion. The key aspect to which Descartes had not given due attention is that pairs of real numbers can not only be summed and subtracted (they form a vector space), but they can also be multiplied among themselves:

$$(x, y) \cdot (u, v) \equiv (xu - yv, xv + yu). \tag{2.38}$$

This promotes the complex numbers to an algebra. The marvelous point is that such an algebra is a *division algebra*! Indeed, for any pair of real numbers different from $(0, 0)$ we can construct its inverse:

$$(x, y)^{-1} = \left(\frac{x}{x^2 + y^2}, -\frac{y}{x^2 + y^2} \right). \tag{2.39}$$

Equation (2.39) constitutes a miracle: given a vector $\vec{v}$ it makes sense to talk about its inverse $\vec{v}^{-1}$, which is clearly nonsense in most vector spaces.

The algebraic closed field of the complex numbers $\mathbb{C}$ is a division algebra of dimension two constructed over the field of real numbers $\mathbb{R}$. It is obtained by introducing one imaginary unit i.

2.10 There are only two other division algebras: the quaternions and the octonions

We cannot dwell on this issue, yet let us mention that besides the complex numbers $\mathbb{C}$, there are other two division algebras: the *quaternions* $\mathbb{H}$, discovered by William Hamilton (the same who invented Hamiltonian mechanics), and the *octonions* $\mathbb{O}$, independently found by Arthur Cayley and by a learned friend of Hamilton, the amateur mathematician John. T. Graves.

The quaternion division algebra is associative, but not commutative, while the octonion division algebra is neither associative nor commutative.

2.10.1 Frobenius and his theorem

Ferdinand Georg Frobenius is known for his important contributions to the development of finite group theory and to some aspects of differential geometry. It was reserved to Ferdinand Georg Frobenius to prove by means of a theorem published in 1877 that, up to isomorphism, the only *associative, normed division algebras* over the reals are $\mathbb{R}$, $\mathbb{C}$, and $\mathbb{H}$, of dimensions one, two, and four, respectively [36].

By norm here we mean a quadratic non-degenerate form over the algebra $\mathscr{D}$:

$$\begin{aligned} &\mathrm{N}:\mathscr{D}\to\mathbb{R},\\ &\mathrm{N}(\lambda a)=\lambda\mathrm{N}(a)\quad\forall\lambda\in\mathbb{R},\forall a\in\mathscr{D},\\ &\mathrm{N}(ab)=\mathrm{N}(a)\mathrm{N}(b)\quad\forall a,b\in\mathscr{D}.\end{aligned}\tag{2.40}$$

The norm corresponds to the modulus square of the real, complex, or quaternionic number.

If we relax the hypothesis of associativity, the landscape is not too much enlarged. Using for instance the very powerful *Bott periodicity theorem*, it can be shown that any *real normed division algebra* must be isomorphic to either the real numbers $\mathbb{R}$, the complex numbers $\mathbb{C}$, the quaternions $\mathbb{H}$, or the octonions $\mathbb{O}$.

2.10.2 Galois and field extensions

What had Galois actually done? He had established a criterion for the solubility of algebraic equations in terms of radicals. Permutations had already been used by Ruffini and Lagrange, but he grasped their essence, which is the property of being elements of a group, and he started building on that fact. Let us outline his arguments.

Consider an algebraic equation of degree n with rational coefficients $u_i \in \mathbb{Q}$:

$$\mathscr{P}_n(x)=x^n+u_1x^{n-1}+\cdots+u_n=0.\tag{2.41}$$

The set of rational numbers $\mathbb{Q}$ fulfills the axioms of a field. Using a more abstract viewpoint we can say that the field $\mathbb{F}=\mathbb{Q}$ to which the coefficients belong is provided by the set of rational functions with rational coefficients of the coefficients u_i. Since a rational function of a rational number is a rational number, this definition seems rather tautological, yet it is convenient to envisage the next step, which is the field extension.

Let α_i $(i = 1,\dots n)$ be the n roots of the considered polynomial (2.41). Define

$$\mathbb{K}\equiv\mathbb{F}(\alpha_1,\dots,\alpha_n),\tag{2.42}$$

the field obtained by adjoining to $\mathbb{F}$ the roots. This means that every element $a \in \mathbb{K}$ can be written as a rational function, with coefficients in $\mathbb{F}$, of the roots: $a = \mathfrak{r}(\alpha_1,\dots\alpha_n)$. Consider a permutation of the roots:

$$P:(\alpha_1,\dots,\alpha_n)\mapsto(P\alpha_1,\dots,P\alpha_n).\tag{2.43}$$

The action of the permutation P can be defined on the field $\mathbb{K}$, by setting

$$P:a=\mathfrak{r}(\alpha_1,\dots\alpha_n)\mapsto Pa\equiv\mathfrak{r}(P\alpha_1,\dots,P\alpha_n).\tag{2.44}$$

A given element of $\mathbb{K}$ has different representations in terms of rational functions of the roots. If $a = \mathfrak{r}_1(\alpha_1, \ldots) = \mathfrak{r}_2(\alpha_1, \ldots)$ are two such representations, then equation (2.44) defines a consistent action of the permutation on the field $\mathbb{K}$ only if $\mathfrak{r}_1(P\alpha_1, \ldots) = \mathfrak{r}_2(P\alpha_1, \ldots)$. This amounts to requiring that the two internal operations of the field $\mathbb{K}$ are preserved by the map P:

$$\begin{aligned} \forall a, b \in \mathbb{K}: \quad & P(a+b) = P(a) + P(b), \\ \forall a, b \in \mathbb{K}: \quad & P(ab) = P(a)P(b). \end{aligned} \tag{2.45}$$

Maps P satisfying (2.45) are named *automorphisms* of the extended field $\mathbb{K}$ *relative to the base field* $\mathbb{F}$. They form a group $\mathfrak{G}(\mathbb{K})$ (the *Galois group* of the original polynomial) since the product of any two elements $P_1, P_2 \in \mathfrak{G}(\mathbb{K})$ is still an element in the same set which contains also an identity element and the inverse of any element in the set. By product of two permutations we mean here the result of performing P_1 first and then P_2.

A solvable group by definition is one that possesses a chain of normal subgroups ending in the trivial group composed only of the identity. In order to understand such a definition the reader should study the next chapter on finite group theory; here we should simply take note of the name and focus on the fact that referring to such a specific structure of the possible Galois group we can state the Galois theorem. He showed that an algebraic equation such as that in (2.41) is *solvable by radicals* if and only if *its Galois group is solvable*. This theorem is remarkable for the remarkable change of perspective from that in Chapter 1, which we have named the Galois algebraic revolution, namely, the recognition that the symmetry of things is more important than the things themselves and unveils their true nature. In the next subsection we briefly outline the implications of Galois' approach.

2.10.2.1 Conceptual analysis of Galois' heritage for modern mathematical thinking

Let us analyze the revolutionary content of what Galois did in the field of mathematics.

(1) First of all, he changed perspective, and rather than analyzing the properties of static mathematical objects, he emphasized the relevance of the transformations one can operate on them.

(2) Secondly, he put to the forefront the notion of group: the transformations one can consider are combined by an operation, the product, and with respect to that operation they form an algebraic structure, the group G.

(3) Thirdly, he put in evidence that once a group G is introduced, its action, originally defined on something (the roots in this case), can be extended to something else (the field extension $\mathbb{K}$). This was the beginning of representation theory.

(4) Next, by means of the concept of automorphism he showed that given a group G which acts on some *space* **V** (the extended field $\mathbb{K}$ in this case), the most important things to study are the properties of objects contained in **V** that are left unchanged by G-transformations. This is the beginning of the theory of invariants, which was

central to mathematics for the whole nineteenth century and still is essential in contemporary scientific thought.

(5) Last but not least, by means of his very theorem he showed that the key information about the structure of a mathematical object (in this case the algebraic equation) operated on by a group of transformations G (in this case the Galois group acting on the roots) resides in the algebraic structure of G. In this case the equation is solvable if G is solvable.

2.11 Bibliographical note

The basic notions of algebra summarized in the present chapter are covered in various combinations in many elementary and less elementary textbooks. A short list of suggestions for further reading is the following one:

1. Gilmore's textbook [37].
2. Helgason's monograph [38].
3. Jacobson's monograph [39].
4. Cornwell's treatise [40].
5. Guido Fano's textbook [41] for a comprehensive introduction to vector spaces and the basic notions of topology.
6. Nakahara's advanced textbook [42] for a review of many different items.

3 Groups: noticeable examples and some developments

Esse quam videri

Arthur Cayley

3.1 Survey of the contents of this chapter

In Section 2.3 we introduced the notion of a group, we illustrated it with a few simple examples, and we listed the main concepts related to the discussion of such an algebraic structure. In the present chapter we introduce several classes of groups which are very frequently encountered in several branches of science. This *substantiates* the definitions given in the previous two chapters and provides a set of important concrete examples with which we will work extensively in the sequel. Here, "concrete" means written in terms of matrices.

3.2 Groups of matrices

In this section we display a finite list of groups G whose elements are just $n \times n$ matrices M characterized by some defining property that is preserved by matrix multiplication. All these matrix groups are instances of what we shall name *Lie groups*, namely, groups that are also differentiable analytic manifolds. Actually, when we come to the classification of Lie groups via the classification of *Lie algebras* we will see that the classical matrix groups exhaust the list of non-exceptional Lie algebras. Matrix groups appear ubiquitously in all branches of mathematics, physics, engineering, chemistry, and applied sciences, including IT.

Actually, from another viewpoint, groups described as a collection of matrices encompass almost the whole landscape of possible groups, since all finite and most discrete infinite and continuous Lie groups admit linear representations that, once a basis is given, associate a well-defined matrix with each element of the abstract group.

Here we consider what we announced, namely, groups given by the collection of all matrices satisfying a specified property. In Weyl's terminology these are also named the classical groups.

3.2.1 General linear groups

The group of *all* $n \times n$ invertible matrices with complex entries is called the (complex) *general linear group* in n dimensions and is denoted as $\mathrm{GL}(n, \mathbb{C})$. If the entries are real, we have the real general linear group $\mathrm{GL}(n, \mathbb{R})$, which is a subgroup of the former.

https://doi.org/10.1515/9783111201535-003

An element of $\mathrm{GL}(n, \mathbb{C})$ is parameterized by n^2 complex numbers, the entries of the matrix (of course the n^2 parameters are real for $\mathrm{GL}(n, \mathbb{R})$).

One can define further matrix groups by placing restrictions, typically in the form of matrix equations or of conditions on the determinant, that are preserved by the matrix product.

3.2.2 Special linear groups

The group of all $n \times n$ matrices with complex entries and *determinant equal to 1* is named the *special linear group* and is indicated as $\mathrm{SL}(n, \mathbb{C})$. It is obviously a subgroup of $\mathrm{GL}(n, \mathbb{C})$: the condition of having unit determinant is preserved by the product. Similarly, one defines $\mathrm{SL}(n, \mathbb{R})$. One can also define the group $\mathrm{SL}(n, \mathbb{Z})$; indeed, the inverse matrices also have integer entries since the determinant that would appear in the denominator of these entries is just 1.

An element of $\mathrm{SL}(n, \mathbb{C})$ depends on n^2-1 complex parameters, as the relation $\det M = 1$ has to be imposed on the n^2 entries of any matrix M. Such parameters are real (integers) for $\mathrm{SL}(n, \mathbb{R})$ or $\mathrm{SL}(n, \mathbb{Z})$.

3.2.3 Unitary groups

The group of *unitary matrices* $\mathrm{U}(n, \mathbb{C}) \subset \mathrm{GL}(n, \mathbb{C})$ contains all the complex matrices U such that

$$U^\dagger U = \mathbf{1}. \tag{3.1}$$

This property is preserved by matrix multiplication and this guarantees that $\mathrm{U}(n, \mathbb{C})$ is a group. Similarly, one could define $\mathrm{U}(n, \mathbb{R}) \subset \mathrm{GL}(n, \mathbb{R})$, containing real unitary matrices, $U^\dagger U = U^T U = \mathbf{1}$, but these are nothing else than real orthogonal matrices, to be introduced shortly. So the group of complex unitary matrices is usually simply denoted as $\mathrm{U}(n)$. Complex unitary matrices are parameterized by $2n^2 - n^2 = n^2$ *real* parameters (we have to subtract the n^2 real conditions corresponding to the entries of the equation $U^\dagger U - \mathbf{1} = 0$ from the n^2 complex parameters of the matrix U). So the condition of unitarity halves the number of parameters, with respect to a generic complex matrix.

3.2.4 Special unitary groups

The subgroup $\mathrm{SU}(n, \mathbb{C}) \subset \mathrm{U}(n, \mathbb{C})$ contains the unitary matrices with unit determinant. Usually it is simply denoted by $\mathrm{SU}(n)$. It is determined by $n^2 - 1$ real parameters. We have to subtract the real condition of having determinant 1 from the parameters of a unitary matrix; recall that by itself the determinant of a unitary matrix can assume a continuous range of values $\exp(2\pi i\theta)$, $\theta \in [0, 1]$.

3.2.5 Orthogonal groups

The group of *orthogonal matrices* $\mathrm{O}(n,\mathbb{C}) \subset \mathrm{GL}(n,\mathbb{C})$ contains all the complex matrices O such that

$$O^T O = \mathbf{1}. \tag{3.2}$$

More frequently encountered are the *real* orthogonal matrices $\mathrm{O}(n,\mathbb{R}) \subset \mathrm{GL}(n,\mathbb{R})$. Usually these groups are simply denoted as $\mathrm{O}(n)$. Real orthogonal matrices are parameterized by $n(n-1)/2$ real numbers. Indeed, from the n^2 parameters of a general real matrix, we have to subtract the $n(n+1)/2$ conditions given by the entries of the matrix condition $O^T O = \mathbf{1}$, which is *symmetric*.

3.2.6 Special orthogonal groups

The group $\mathrm{SO}(n)$ contains the real orthogonal matrices with unit determinant. Analogously, one could define its complex extension $\mathrm{SO}(n,\mathbb{C})$. They have the same number of parameters, $n(n-1)/2$, as the orthogonal matrices. Indeed, the determinant of an orthogonal matrix O can have only a finite set of values: $\det O = \pm 1$; imposing $\det O = 1$ does not alter the dimensionality of the parameter space.

3.2.7 Symplectic groups

The group of *symplectic matrices* $\mathrm{Sp}(n,\mathbb{C})$ contains the $2n \times 2n$ matrices A that preserve the "symplectic[1] form" Ω, namely, the matrices such that

$$A^T \Omega A = \Omega, \quad \Omega = \begin{pmatrix} \mathbf{0} & \mathbf{1} \\ -\mathbf{1} & \mathbf{0} \end{pmatrix}. \tag{3.3}$$

One defines $\mathrm{Sp}(n,\mathbb{R})$ similarly. Since the restriction (3.3) is an antisymmetric matrix expression, it imposes $(2n)(2n-1)/2$ conditions. Thus, the symplectic matrices depend on $(2n)^2 - (2n)(2n-1)/2 = n(2n+1)$ parameters (complex or real for $\mathrm{Sp}(n,\mathbb{C})$, respectively $\mathrm{Sp}(n,\mathbb{R})$).

The groups $\mathrm{U}(n)$, $\mathrm{O}(n)$, and $\mathrm{Sp}(n)$ form the three families of *classical matrix groups*.

---

1 The symplectic form is the non-positive quadratic form often appearing in analytical mechanics, e. g., in the definition of the Poisson brackets, $\{F,G\} = \frac{\partial F}{\partial y^I}\Omega^{IJ}\frac{\partial F}{\partial y^J}$, where $y^I = (q^i, p^i)$ are the phase-space coordinates.

3.2.8 Groups of transformations

Square $n \times n$ matrices represent homomorphisms of an n-dimensional vector space V into itself. This is what, according to the tale told in [28], took such a long time to be recognized.

Thus, the matrix groups are in fact groups of linear transformations of vector spaces. In most applications in all branches of science, the elements of G are interpreted as transformations $\tau \in \mathrm{Hom}(V, V)$ acting on some space V,

$$\tau \in G : \; V \ni v \mapsto \tau(v) \in V, \tag{3.4}$$

and the group composition is the composition of these transformations,

$$\tau_1\tau_2 \in G : \; V \ni v \mapsto \tau_1(\tau_2(v)) \in V. \tag{3.5}$$

In this case, the associativity is automatically satisfied. Notice the convention that in the product $\tau_1\tau_2$ one acts first with τ_2 and then with τ_1: $v \overset{\tau_2}{\mapsto} \tau_2(v) \overset{\tau_1}{\mapsto} \tau_1(\tau_2(v)) \equiv \tau_1\tau_2(v)$. Of course this is the same convention that arises in taking the matrix product as the group product law for groups of linear transformations on vector spaces.

A major conceptual step forward is the identification of the group or of another algebraic structure with its isomorphism class. Two isomorphic structures are just the same structure, in the sense that there is a sort of abstract *platonic idea* of the structure and the actual instances of its realizations by means of matrices or transformations are just *representations*.[2]

Transformations groups are Abelian when the order in which two transformations are subsequently performed does not affect the final result.

3.3 Some examples of finite groups

As already emphasized, all the groups considered in the previous section are instances of continuous (actually Lie) groups where the group elements fill a continuous space and cannot be enumerated. In this section we consider instead instances of *discrete groups*, whose elements can be enumerated. Discrete groups are subdivided into two main classes: *discrete infinite groups*, which contain a denumerable infinity of elements, and *finite groups*, whose elements constitute a finite set. The theory of finite groups is a very rich and fascinating branch of mathematics with extremely relevant applications in physics, chemistry, and other branches of science. We will come back to it in Chapter 4; here we begin by presenting some primary examples.

2 For further details on the conceptual history of the notion of group, see [28].

3.3.1 The permutation groups S_n

Consider a finite set A. The automorphisms (i. e., the *bijective mappings* $P : A \leftrightarrow A$) form a group $S(A)$, called the *symmetric group* of A. The nature of the objects in the set A does not matter, only their number $|A|$ does. So, if $|A| = n$, we can label the objects with the integer numbers $1, 2, \ldots n$ and indicate the symmetric group, also called the *permutation group* on n objects, as S_n. An element of this group, a *permutation* P, is explicitly defined by its action on the elements $1, \ldots, n$ of the set:

$$P = \begin{pmatrix} 1 & 2 & \cdots & n \\ P(1) & P(2) & \cdots & P(n) \end{pmatrix}. \tag{3.6}$$

The product law of the symmetric group is the composition of permutations with the convention described above: PQ means effecting first the permutation Q and then the permutation P. While S_2 is Abelian, all S_n with $n > 2$ are not Abelian. The symmetric group S_n has $n!$ elements.

We can give an explicit expression to a permutation $P \in S_n$ as an $n \times n$ matrix defined by

$$(P)_{ij} = \delta_{i,P(j)}, \quad i, j = 1, \ldots n. \tag{3.7}$$

In this way the composition of permutations corresponds to the product of the defining matrix representatives (3.7):

$$(PQ)_{ij} = \sum_k \delta_{i,P(k)} \delta_{k,P(j)} = \sum_k \delta_{i,P(k)} \delta_{P(k),P(Q(j))} = \delta_{i,P(Q(j))}. \tag{3.8}$$

Notice that the matrix representatives of permutations are unitary and real, that is, they are orthogonal matrices. So S_n can be seen as a subgroup of $\mathrm{O}(n)$.

3.3.1.1 Cycle decomposition

Let us illustrate the notion of *cycle* of a permutation by means of an example. Consider the following permutation $P \in S_8$:

$$P = \begin{pmatrix} 1 & 2 & 3 & 4 & 5 & 6 & 7 & 8 \\ 2 & 3 & 1 & 5 & 4 & 7 & 6 & 8 \end{pmatrix}. \tag{3.9}$$

Let us check what would happen to the various elements $1, \ldots, 8$ if we were to apply the permutation repeatedly. We would have $1 \to 2 \to 3 \to 1$, after which everything repeats again. We say that $1, 2, 3$ form a cycle of order 3 in P, and we denote this cycle compactly as (123). We also have $4 \to 5 \to 4$, so we have a cycle (45). Also, $6 \to 7 \to 6$, and thus there is the cycle (67). Finally, 8 is invariant, i. e., it is a trivial cycle (8). The permutation P has the *cycle decomposition*

$$P = (123)(45)(67)(8). \tag{3.10}$$

Often the cycles of length 1, such as (8) above, are omitted when writing the cycle decomposition. It is quite evident that the example we took is not limited in any way, and every permutation of any S_n group will admit a cycle decomposition. As we will see, the type of cycle decompositions of permutations is of fundamental importance in the analysis of permutation groups. Let us make some simple observations.

(i) The sum of the lengths of the cycles for a $P \in S_n$ equals n: every element is in some cycle, and cycles do not have elements in common.
(ii) Having no common elements, two cycles in the decomposition of a given permutation commute. For instance, in (3.10), $(123)(45) = (45)(123)$.
(iii) Just by their definition, the cycles can be shifted freely without affecting them: $(123) = (231) = (312)$ (but $(123) \neq (213)$!).
(iv) Every cycle can in turn be decomposed into a product of cycles of order 2, called also *transpositions* or *exchanges*. However, the latter have now elements in common. For instance, $(123) = (13)(12)$. In general, $(12\dots n) = (1n)(1, n-1)\dots(13)(12)$.

3.3.1.2 Odd and even permutations

As discussed above, every permutation can be decomposed into a product of transpositions. A permutation is called *odd* or *even* depending on whether in such a decomposition an odd or an even number of transpositions appear.

3.3.1.3 The alternating groups

The *even* permutations form a subgroup of S_n (the odd ones clearly do not form a subgroup) called the *alternating group* on n elements, denoted as A_n. Its order is $|A_n| = |S_n|/2 = n!/2$.

3.4 Groups as the invariances of a given geometry

After 2000 years of Euclidean geometry, elevated by Immanuel Kant to the rank of an *a priori fundament of all perceptions*, in the middle of the nineteenth century non-Euclidean geometries finally worked their way into mathematics and philosophy. As discussed at length in [28], in a famous program named the *Erlangen Program*, after the German town where it was announced, Felix Klein proposed a new vision where all possible geometries are classified according to the group of transformations with respect to which the relations and the propositions existing in that geometry are kept invariant. Euclidean geometries are simply the geometries invariant with respect to the Euclidean groups.

Something that is important to notice is that the concept of geometry does not apply only to physical space (i. e., the arena of physical phenomena), but also to all sorts of continuous varieties and manifolds $\mathcal{M}$[3] whose points have quite different interpretation, depending on the science in which we work. The points $p \in \mathcal{M}_n$ in a higher-dimensional manifold can describe the dynamical status of a complex biological or chemical system. Alternatively, they can represent the status of an economic system or of a thermodynamical one. As long as the set of states is a continuous space, namely, a differentiable manifold, the space of all states is liable to be endowed with some kind of geometric structure and such structure, in turn, typically admits a group of symmetries or isometries.[4]

3.4.1 The Euclidean groups

The so-called Euclidean group in dimension d is the group of *isometry* transformations in a Euclidean space $\mathbb{R}^d$. It consists of translations, rotations around some axis (proper rotations), and reflections with respect to hyperplanes. All such transformations leave the Euclidean distance between any two points of $\mathbb{R}^d$ unaltered.

3.4.1.1 The Euclidean group in two dimensions

Consider the transformations of a plane $\mathbb{R}^2$ into itself given by rigid rotations around a perpendicular axis through the origin. They clearly form a group. The elements R_θ of the group are identified by an angle θ (the angle of, e. g., clockwise, rotation) defined mod 2π; that is, the elements of the group correspond to the points of a circle S^1. The composition of two rotations results in $\mathcal{R}(\theta_1)\mathcal{R}(\theta_2) = R(\theta_1+\theta_2)$; the group is Abelian. We can describe a transformation $R(\theta)$ via its effects on the Cartesian coordinates $\mathbf{x} \equiv (x,y)$ of a point, $\mathbf{x} \overset{\mathcal{R}(\theta)}{\mapsto} \mathbf{x}'$, with

$$\begin{cases} x' = \cos\theta x + \sin\theta y, \\ y' = -\sin\theta x + \cos\theta y, \end{cases} \tag{3.11}$$

that is, $\mathbf{x}' = \mathcal{R}(\theta)\mathbf{x}$, with $\mathcal{R}(\theta)$ being an orthogonal 2×2 matrix with unit determinant, $\mathcal{R}(\theta) \in \mathrm{SO}(2)$. In fact, this correspondence between rotations and matrices of SO(2) is an isomorphism. Thus, with a slight abuse of language, we can say that SO(2) is the (proper) rotation group in two dimensions.

Also the translations $T(\mathbf{v})$ by a 2-vector $\mathbf{v}$ acting on the Euclidean space $\mathbb{R}^2$, $T(\mathbf{v}) : \mathbf{x} \mapsto \mathbf{x} + \mathbf{v}$, $\forall \mathbf{x} \in \mathbb{R}^2$, form a group, with the composition of two translations resulting in $T(\mathbf{v}_1)T(\mathbf{v}_2) = T(\mathbf{v}_1 + \mathbf{v}_2)$; this group is Abelian.

3 See Chapters 6 and 15 for the introduction to differential geometry and differentiable manifolds.

4 The reader will thoroughly understand the notion of isometry after reading and studying the sections of this book related to connections and metrics (see Chapter 15).

Consider now the group of all transformations of $\mathbb{R}^2$ onto itself given by simultaneous rotations and/or translations with arbitrary parameters. Let us denote such transformations as $(\mathscr{R}(\theta), T(\mathbf{v}))$. They act on the coordinate vectors by

$$(\mathscr{R}(\theta), T(\mathbf{v})) : \mathbf{x} \mapsto \mathscr{R}(\theta)\mathbf{x} + \mathbf{v}. \tag{3.12}$$

Notice that the translation parameters $\mathbf{v}$, being vectors, are acted on by the rotations. You can easily verify that the product law resulting from the composition of two such transformations is

$$(\mathscr{R}(\theta_1), T(\mathbf{v}_1)) \cdot (\mathscr{R}(\theta_2), T(\mathbf{v}_2)) = (\mathscr{R}(\theta_1 + \theta_2), T(\mathbf{v}_1 + \mathscr{R}(\theta_1)\mathbf{v}_2)). \tag{3.13}$$

With respect to this composition law the set of elements $(\mathscr{R}(\theta), T(\mathbf{v}))$ close a group which is named the *inhomogeneous rotation group* in two dimensions, denoted ISO(2).

A proper rotation maps an oriented orthogonal frame into a new orthogonal frame with the same orientation. Inversions (or reflections) of the plane with respect to a line through the origin also map it to an orthogonal frame, but with the opposite orientation. For instance, reflection with respect to the x-axis maps (x, y) to $(x, -y)$. The set of all transformations obtained as compositions of proper rotations and inversions is a larger group. The inversion with respect to a direction forming an angle θ with the x-axis is effected by the matrix

$$\mathscr{I}(\theta) = \begin{pmatrix} \cos 2\theta & \sin 2\theta \\ \sin 2\theta & -\cos 2\theta \end{pmatrix} = \mathscr{I}(0)\mathscr{R}(2\theta), \tag{3.14}$$

where $\mathscr{I}(0) = \mathrm{diag}(1, -1)$ is orthogonal but with determinant -1. Thus, rotations plus inversions of the plane are represented by all orthogonal 2×2 matrices, i. e., by elements of O(2).

Translations, rotations, and inversions form a group, $\mathrm{Eucl}_2 \supset \mathrm{ISO}(2)$, which is named the *Euclidean group* in two dimensions.

3.4.1.2 The Euclidean group in three dimensions

The translations acting on $\mathbb{R}^3$ form an Abelian group (there is no difference with respect to the $\mathbb{R}^2$ case).

Now consider the rotations around any axis through the origin. Such transformations form the group of *proper orthogonal rotations* in three dimensions; the effect of performing two subsequent rotations around two different axes is a new rotation around a third axis.

The proper rotations map an orthonormal frame into a new orthonormal frame with the same orientation and are represented on the vectors by means of orthogonal 3×3 matrices with unit determinant. In fact, the group of proper rotations is isomorphic to SO(3). The group of three-dimensional rotations SO(3) is *non-Abelian.*

The group ISO(3) of roto-translations in three dimensions can be defined with no formal modifications with respect to the two-dimensional case.

Reflections with respect to a plane and total spatial reflection $\mathbf{x} \mapsto -\mathbf{x}$ map orthogonal frames to orthogonal frames with the opposite orientation. They are represented by orthogonal 3×3 matrices with determinant -1. Thus, rotations and reflections form a group, which is isomorphic to the group $\mathrm{O}(3)$ of orthogonal matrices acting on the vectors.

Translations, rotations, and reflections form the Euclidean group in three dimensions Eucl_3.

3.4.2 Projective geometry

The primary example of a non-Euclidean geometry is provided by Lobachevsky hyperbolic geometry analytically realized on the upper complex plane named in this context after *Henri Poincaré*. This will be the subject of Section 15.8.2. Here we consider the wider case of projective geometry. Figures in complex planes can change their size and shape by compression and stretching, yet the angles between the tangents to any two lines composing the figure have to be preserved. This is done by the group of Möbius transformations.

3.4.2.1 The Möbius group

Consider the *conformal transformations* of the compactified complex plane (or Riemann sphere) $\tilde{\mathbb{C}} \equiv \mathbb{C} \cup \{\infty\}$. Conformal mappings $z \mapsto w$ are represented by *analytic* functions $w(z)$. We require the transformations to be invertible, i. e., one-to-one. Hence the function $w(z)$ can have at most one simple pole; otherwise, the point at infinity would have several preimages and for analogous reasons at most one zero; furthermore, the Jacobian $\partial w/\partial z$ must not vanish. Thus, the most general transformation M fulfilling all requirements is of the following form:

$$z \overset{M}{\mapsto} w(z) = \frac{az+b}{cz+d}, \quad a,b,c,d \in \mathbb{C},\ ad - bc \neq 0, \tag{3.15}$$

where the last condition follows from the invertibility requirement $\partial w/\partial z \neq 0$. Notice that all transformations of parameters (ka, kb, kc, kd) with $k \in \mathbb{C} \setminus \{0\}$ are equivalent. This scale invariance can be used to fix $ad - bc = 1$. The transformations in (3.15) are known as *fractional linear transformations*, *projective transformations*, or *Möbius transformations*.

The composition of two Möbius transformations M, M' is again a Möbius transformation M'':

$$z \overset{M}{\mapsto} w = \frac{az+b}{cz+d} \overset{M'}{\mapsto} x = \frac{a'w+b'}{c'w+d'} = \frac{(a'a+b'c)z + a'b + b'd}{(c'a+d'c)z + c'b + d'd} = \frac{a''z+b''}{c''z+d''}. \tag{3.16}$$

The product transformation $z \overset{M''}{\mapsto} x$ satisfies $a''d'' - b''c'' = 1$.

Thus, it is natural to associate a Möbius transformation M with a 2×2 matrix $\mathscr{M}$ with unit determinant

$$\mathscr{M} = \begin{pmatrix} a & b \\ c & d \end{pmatrix}, \quad \det \mathscr{M} = ad - bc = 1, \tag{3.17}$$

namely, with an element of $\mathrm{SL}(2, \mathbb{C})$. To the product of two transformations $M'' = M'M$ corresponds a matrix which is the matrix product of the two factors: $\mathscr{M}'' = \mathscr{M}'\mathscr{M}$. However, the mapping from the group of Möbius transformations to $\mathrm{SL}(2, \mathbb{C})$ is one-to-two, since the matrices $\pm\mathscr{M}$ correspond to the same Möbius transformation M. We will come back to the Möbius group in later sections and derive its structure in a rigorous way as a group of automorphisms of the extended complex plane.

3.5 Matrix groups and vector spaces

We turn to analyzing classical groups as groups made of homomorphisms $\mu : V \to V$ of a vector space into itself.

3.5.1 General linear groups and basis changes in vector spaces

As already recalled, the non-singular $n \times n$ matrices A with real or complex entries, i. e., the elements of $\mathrm{GL}(n, \mathbb{R})$ or $\mathrm{GL}(n, \mathbb{C})$, are *automorphisms* of real or complex vector spaces. Namely, they describe the possible changes of basis in an n-dimensional vector space $\mathbb{R}^n$ or $\mathbb{C}^n$: $\mathbf{e}_i{}' = \mathbf{e}_j A^j{}_i$ (in matrix notation, $\mathbf{e}' = \mathbf{e}A$). Two subsequent changes of basis, first with A and then with B, result in a change described by the matrix product AB. The basis elements $\mathbf{e}_i$ are said to transform *contravariantly*. The components v^i of a vector $\mathbf{v} = v^i \mathbf{e}_i$ transform instead *covariantly*: from $\mathbf{v} = v^{i'} \mathbf{e}_i{}' = v^i \mathbf{e}_i$ it follows that $v^i = A^i{}_j v^{j'}$ (in matrix notation, $v = Av'$, or $v' = A^T v$).

3.5.2 Tensor product spaces

Starting from two vector spaces, say V_1 (n-dimensional, with basis $\mathbf{e}_i$) and V_2 (m-dimensional, with basis $\mathbf{f}_j$), one can construct bigger vector spaces, for instance by taking the *direct sum* $V_1 \oplus V_2$ of the two spaces or their *direct product* $V_1 \otimes V_2$. The direct product has dimension mn and basis $\mathbf{e}_i \otimes \mathbf{f}_j$. A change of basis A in V_1 and B in V_2 induces a change of basis in the direct product space described[5] by $(\mathbf{e}_i \otimes \mathbf{f}_j)' = A_i{}^k B_j{}^l \mathbf{e}_k \otimes \mathbf{f}_l$. An element of a direct product vector space is called a *tensor*.

5 The subgroup of all possible changes of basis obtained in this way is the *direct product* $G_1 \otimes G_2$ (a concept we will deal with in the next section) of the general linear groups (groups of basis changes) G_1 and G_2 of V_1 and V_2, respectively.

In particular, one can take the direct product of a vector space V of dimension n with itself (in general, m times), $V \otimes \cdots \otimes V$, which we may indicate for shortness as $\otimes^m V$. Its n^m basis vectors are $\mathbf{e}_{i_1} \otimes \cdots \otimes \mathbf{e}_{i_m}$. An element of $\otimes^m V$ is called a *tensor of order m*. An element A of the general linear group G of V (a change of basis in V) induces a change of basis on $\otimes^m V$:

$$\mathbf{e}'_{i_1} \otimes \cdots \otimes \mathbf{e}'_{i_m} = A_{i_1}^{\ j_1} \cdots A_{i_m}^{\ j_m} \mathbf{e}_{j_1} \otimes \cdots \otimes \mathbf{e}_{j_m}. \tag{3.18}$$

Such changes of bases on $\otimes^m V$ form a group which is the direct product of G with itself, m times, $\otimes^m G$. Its elements are $n^m \otimes n^m$ matrices, and the association of an element A of G with an element of $\otimes^m G$ given in (3.18) preserves the product; thus, $\otimes^m G$ is a representation of G, called an m-th order tensor product representation. As we are going to see in the sequel, this way of constructing representations by means of tensor products is of utmost relevance.

3.5.3 (Anti)symmetrized product spaces

Within the direct product space $V \otimes V$, we can single out two subspaces respectively spanned by the *symmetric* and by the *antisymmetric* combinations[6] of the basis vectors $\mathbf{e}_i \otimes \mathbf{e}_j$:

$$\begin{aligned} \mathbf{e}_i \vee \mathbf{e}_j &\equiv \mathbf{e}_i \otimes \mathbf{e}_j + \mathbf{e}_j \otimes \mathbf{e}_i = (\mathbf{1} + (12))(\mathbf{e}_i \otimes \mathbf{e}_j), \\ \mathbf{e}_i \wedge \mathbf{e}_j &\equiv \mathbf{e}_i \otimes \mathbf{e}_j - \mathbf{e}_j \otimes \mathbf{e}_i = (\mathbf{1} - (12))(\mathbf{e}_i \otimes \mathbf{e}_j), \end{aligned} \tag{3.19}$$

where in the second equality by $\mathbf{1}$ and (12) we have respectively denoted the identity and the exchange element of the symmetric group S_2 acting on the positions of the two elements in the direct product basis. The symmetric subspace has dimension $n(n+1)/2$, and the antisymmetric one has dimension $n(n-1)/2$.

Similarly, within $\otimes^m V$ the *fully symmetric* and *fully antisymmetric* subspaces can easily be defined. Their basis elements can be written as

$$\begin{aligned} \mathbf{e}_{i_1} \vee \cdots \vee \mathbf{e}_{i_m} &\equiv \sum_{P \in S_m} P(\mathbf{e}_{i_1} \otimes \cdots \otimes \mathbf{e}_{i_m}), \\ \mathbf{e}_{i_1} \wedge \cdots \wedge \mathbf{e}_{i_m} &\equiv \sum_{P \in S_m} (-)^{\delta_P} P(\mathbf{e}_{i_1} \otimes \cdots \otimes \mathbf{e}_{i_m}), \end{aligned} \tag{3.20}$$

where δ_P is 0 or 1 if the permutation P is even or odd, respectively. The fully symmetric subspace has dimension $n(n+1)\dots(n+m-1)/m!$, the fully antisymmetric one has dimension $n(n-1)\dots(n-m+1)/m! = \binom{n}{m}$.

6 The notation $\mathbf{e}_i \vee \mathbf{e}_j$, though logical, is not used too much in the literature. Often, with a bit of abuse, a symmetric combination is simply indicated as $\mathbf{e}_i \otimes \mathbf{e}_j$; we will sometimes do so, when no confusion can arise.

3.5.4 Exterior forms

Let W be a vector space of finite dimension over a field $\mathbb{F}$ ($\mathbb{F}$ can be either $\mathbb{R}$ or $\mathbb{C}$, depending on the case). In this section we show how we can construct a sequence of vector spaces $\Lambda_k(W)$ with $k = 0, 1, 2, \ldots, n = \dim W$ defined in the following way:

$$\begin{aligned} \Lambda_0(W) &= \mathbb{F}, \\ \Lambda_1(W) &= W^\star, \\ &\vdots \\ \Lambda_k(W) &= \text{vector space of } k\text{-linear antisymmetric functionals over } W. \end{aligned} \tag{3.21}$$

The spaces $\Lambda_k(W)$ contain the linear functionals on the k-th exterior powers of the vector space W. Such functionals are called *exterior forms* of degree k on W.

Let $\phi^{(k)} \in \Lambda_k(W)$ be a k-form. It describes a map,

$$\phi^{(k)} : \underbrace{W \otimes W \otimes \cdots \otimes W}_{k \text{ times}} \to \mathbb{F}, \tag{3.22}$$

with the following properties:

$$\begin{aligned} &(i) \quad \phi^{(k)}(w_1, \ldots, w_i, \ldots, w_j, \ldots, w_k) = -\phi^{(k)}(w_1, \ldots, w_j, \ldots, w_i, \ldots, w_k), \\ &(ii) \quad \phi^{(k)}(w_1, \ldots, \alpha x + \beta y, \ldots, w_k) = \alpha\phi^{(k)}(w_1, \ldots, x, \ldots, w_k) + \beta\phi^{(k)}(w_1, \ldots, y, \ldots, w_k), \end{aligned} \tag{3.23}$$

where $\alpha, \beta \in \mathbb{F}$ and $w_i, x, y \in W$.

The first of the properties in (3.23) guarantees that the map $\phi^{(k)}$ is antisymmetric in any two arguments. The second property states that $\phi^{(k)}$ is linear in each argument.

The sequence of vector spaces $\Lambda_k(W)$,

$$\Lambda(W) \equiv \bigcup_{k=0}^{n} \Lambda_k(W), \tag{3.24}$$

can be equipped with an additional operation, named exterior product, which with each pair of a k_1- and a k_2-form $(\phi^{(k_1)}, \phi^{(k_2)})$ associates a new $(k_1 + k_2)$-form. Namely, we have

$$\wedge : \quad \Lambda_{k_1} \otimes \Lambda_{k_2} \to \Lambda_{k_1+k_2}. \tag{3.25}$$

More precisely, we set

$$\phi^{(k_1)} \wedge \phi^{(k_2)} \in \Lambda_{k_1+k_2}(W), \tag{3.26}$$

and we write

$$\phi^{(k_1)} \wedge \phi^{(k_2)}(w_1, w_2, \ldots, w_{k_1+k_2})$$
$$= \sum_P (-)^{\delta_P} \frac{1}{(k_1+k_2)!}(\phi^{(k_1)}(w_{P(1)}, \ldots, w_{P(k_1)})\phi^{(k_2)}(w_{P(k_1+1)}, \ldots, w_{P(k_1+k_2)})), \tag{3.27}$$

where P are the permutations of $k_1 + k_2$ objects, namely, the elements of the symmetric group $\mathscr{S}_{k_1+k_2}$, and δ_P is the parity of the permutation P ($\delta_P = 0$ if P contains an even number of exchanges with respect to the identity permutation, while $\delta_P = 1$ if this number is odd).

In order to make this definition clear, let us consider the explicit example where $k_1 = 2$ and $k_2 = 1$. We have

$$\phi^{(2)} \wedge \phi^{(1)} = \phi^{(3)}, \tag{3.28}$$

and we find

$$\begin{aligned}\phi^{(3)}(w_1, w_2, w_3) &= \frac{1}{3!}(\phi^{(2)}(w_1, w_2)\phi^{(1)}(w_3) - \phi^{(2)}(w_2, w_1)\phi^{(1)}(w_3)\\ &\quad - \phi^{(2)}(w_1, w_3)\phi^{(1)}(w_2) - \phi^{(2)}(w_3, w_2)\phi^{(1)}(w_1)\\ &\quad + \phi^{(2)}(w_2, w_3)\phi^{(1)}(w_1) + \phi^{(2)}(w_3, w_1)\phi^{(1)}(w_2))\\ &= \frac{1}{3}(\phi^{(2)}(w_1, w_2)\phi^{(1)}(w_3) + \phi^{(2)}(w_2, w_3)\phi^{(1)}(w_1) + \phi^{(2)}(w_3, w_1)\phi^{(1)}(w_2)).\end{aligned}$$

The exterior product we have just defined has the following formal property:

$$\phi^{(k)} \wedge \phi^{(k')} = (-)^{kk'}\phi^{(k')} \wedge \phi^{(k)}, \tag{3.29}$$

which can be immediately verified starting from the definition (3.27). Indeed, assuming for instance that $k_2 > k_1$, it is sufficient to consider the parity of the permutation,

$$\Pi = \begin{pmatrix} 1, & 2, & \ldots, & k_1, & k_1+1, & \ldots, & k_2, & k_2+1, & \ldots, & k_1+k_2 \\ k_1, & k_1+1, & \ldots, & k_1+k_1, & 2k_1+1, & \ldots, & k_1+k_2, & 1, & \ldots, & k_1 \end{pmatrix}, \tag{3.30}$$

which is immediately seen to be

$$\delta_\Pi = k_1 k_2 \bmod 2. \tag{3.31}$$

Setting $P = P'\Pi$ (which implies $\delta_P = \delta_{P'} + \delta_\Pi$), we obtain

$$\begin{aligned}&\phi^{(k_2)} \wedge \phi^{(k_1)}(w_1, \ldots, w_{k_1+k_2})\\ &= \frac{1}{(k_1+k_2)!}\sum_P (-)^{\delta_P}\phi^{(k_2)}(w_{P(1)}, \ldots, w_{P(k_2)})\phi^{(k_1)}(w_{P(k_2+1)}, \ldots, w_{P(k_1+k_2)})\\ &= \frac{1}{(k_1+k_2)!}\sum_{P'} (-)^{\delta_{P'}+\delta_\Pi}\phi^{(k_2)}(w_{P'\Pi(1)}, \ldots, w_{P'\Pi(k_2)})\phi^{(k_1)}(w_{P'\Pi(k_2+1)}, \ldots, w_{P'\Pi(k_2+k_1)})\end{aligned}$$

$$
\begin{aligned}
&= \frac{1}{(k_1+k_2)!}(-)^{\delta_\Pi}\sum_{P'}(-)^{\delta_{P'}}\phi^{(k_2)}(w_{P'(k_1+1)},\dots,w_{P'(k_1+k_2)})\phi^{(k_1)}(w_{P'(1)},\dots,w_{P'(k_1)})\\
&= (-)^{\delta_\Pi}\phi^{(k_1)}\wedge\phi^{(k_2)}(w_1,\dots,w_{k_1+k_2}).
\end{aligned} \tag{3.32}
$$

3.5.5 Special linear groups as volume-preserving transformations

In the particular case of $\otimes^n V$ (with V of dimension n), the fully antisymmetric subspace has dimension $\binom{n}{n} = 1$, and its basis element is $\mathbf{e}_1 \wedge \mathbf{e}_2 \cdots \wedge \mathbf{e}_n$. It is named[7] the *volume element*. A change of basis A on V induces a transformation of the volume element that can occur only by means of a multiplicative factor:

$$
(\mathbf{e}_1 \wedge \mathbf{e}_2 \cdots \wedge \mathbf{e}_n)' = (\det A)\mathbf{e}_1 \wedge \mathbf{e}_2 \cdots \wedge \mathbf{e}_n, \tag{3.33}
$$

where the determinant arises from the application of (3.20):

$$
\det A = \sum_{P\in S_n} (-1)^{\delta_P} A_1^{j_1} A_2^{j_2} \cdots A_n^{j_n}. \tag{3.34}
$$

The permutations P act by exchanging the indices j_i. We have defined the *special linear group* SL(n) (real or complex) to be the subgroup of the general linear group GL(n) (real or complex) containing the matrices A such that $\det A = 1$. We see now by (3.33) that the special linear group is the subset of basis changes on V that *preserve the volume* element of $\otimes^n V$.

3.5.6 Metric-preserving changes of basis

A *metric* on a vector space V is a functional from $V \otimes V$ into the field $\mathbb{F}$ ($\mathbb{R}$ or $\mathbb{C}$ for us) associated with the vector space V. That is, a metric is the assignment of a value $(\mathbf{v}_1, \mathbf{v}_2) \in \mathbb{F}$ to every pair of vectors, namely, $\forall \mathbf{v}_{1,2} \in V$. The metric can be required to be *bilinear*, in which case

$$
\begin{aligned}
(\mathbf{v}_1, \alpha\mathbf{v}_2 + \beta\mathbf{v}_3) &= \alpha(\mathbf{v}_1, \mathbf{v}_2) + \beta(\mathbf{v}_1, \mathbf{v}_3),\\
(\alpha\mathbf{v}_1 + \beta\mathbf{v}_2, \mathbf{v}_3) &= \alpha(\mathbf{v}_1, \mathbf{v}_3) + \beta(\mathbf{v}_2, \mathbf{v}_3),
\end{aligned} \tag{3.35}
$$

or *sesquilinear*, in which case

$$
\begin{aligned}
(\mathbf{v}_1, \alpha\mathbf{v}_2 + \beta\mathbf{v}_3) &= \alpha(\mathbf{v}_1, \mathbf{v}_2) + \beta(\mathbf{v}_1, \mathbf{v}_3),\\
(\alpha\mathbf{v}_1 + \beta\mathbf{v}_2, \mathbf{v}_3) &= \alpha^*(\mathbf{v}_1, \mathbf{v}_3) + \beta^*(\mathbf{v}_2, \mathbf{v}_3).
\end{aligned} \tag{3.36}
$$

7 More precisely such a name is appropriate for the *dual* basis element of $\Lambda_n(V)$, the space of *n-forms* or n-linear antisymmetric functionals on V that we have discussed in the previous section.

Bilinearity and sesquilinearity are different only if the field $\mathbb{F}$ is $\mathbb{C}$ rather than $\mathbb{R}$.

The above notion of metric is identical to the notion of inner product introduced in Section 2.6.2. As noted there, a metric, or inner product, is specified by its action on a pair of basis vectors. Let us denote

$$(\mathbf{e}_i, \mathbf{e}_j) = g_{ij}. \tag{3.37}$$

For a sesquilinear metric we have $(\mathbf{v}, \mathbf{u}) = v^{i*} g_{ij} u^j$. We assume that the metric is *non-degenerate*, namely, that $\det g \neq 0$ (where g is the matrix of the elements g_{ij}). Under a change of basis A, a sesquilinear metric transforms as follows:

$$g'_{ij} = (A_i{}^k)^* g_{kl} A_j{}^l. \tag{3.38}$$

Notice that g_{ij} transforms covariantly.

If a (sesquilinear) metric is *Hermitian*, $g_{ij} = g^*_{ji}$, then it is always possible to find a basis change that puts it into a canonical form,

$$g_{ij} \to \mathrm{diag}(\underbrace{1, \ldots, 1}_{p}, \underbrace{-1, \ldots, -1}_{q}). \tag{3.39}$$

Indeed, we can first diagonalize g_{ij} to $\lambda_i \delta_{ij}$, changing basis with its eigenvector matrix $\mathbf{e}_i{}' = S_i{}^j \mathbf{e}_j$. Since g_{ij} is Hermitian by hypothesis, the eigenvalues λ_i are all real. Then we can further change basis by rescaling the basis vectors to $\mathbf{f}_i = |\lambda_i|^{-1/2} \mathbf{e}_i{}'$ so as to obtain (3.39). A metric of canonical form (3.39) is said to have *signature* (p, q).

The procedure outlined above is the content of a theorem proved in 1852 by James Joseph Sylvester and named by him the *law of inertia of quadratic forms*[10]. Indeed, according to Sylvester's theorem, a symmetric non-degenerate $m \times m$ matrix A can always be transformed into a diagonal one with ± 1 entries by means of a substitution $A \mapsto B^T \cdot A \cdot B$. On the other hand, no such transformation can alter the signature $(p, m - p)$, which is intrinsic to the matrix A. This result in matrix algebra turned out to be very important for differential geometry, as we are going to discuss in Section 15.5.1.

Also a bilinear *antisymmetric* metric $g_{ij} = -g_{ij}$ can be put into a canonical form. Such a metric is non-degenerate only if the dimension n of the space is even, $n = 2m$. Indeed, $\det g = \det g^T = \det(-g) = (-)^n \det g$. If $n = 2m$, g_{ij} can be first *skew-diagonalized* by a change of basis,

$$g_{ij} \to \begin{pmatrix} \begin{matrix} 0 & \lambda_1 \\ -\lambda_1 & 0 \end{matrix} & \mathbf{0} & \cdots \\ \mathbf{0} & \begin{matrix} 0 & \lambda_2 \\ -\lambda_2 & 0 \end{matrix} & \cdots \\ \vdots & \vdots & \ddots \end{pmatrix}, \tag{3.40}$$

and then brought to a canonical form by suitably rescaling the basis vectors. The canonical form can be that of (3.40), with all λ_i reduced to 1, or the so-called symplectic form Ω already introduced in (3.3), obtained by a further reordering of the basis vectors.

Having endowed a vector space with a metric, we can consider those automorphisms that *preserve that metric*. It is not difficult to see that such changes of basis form a subgroup of the general linear group. Indeed, if A and B are two changes of bases that preserve the metric, then the product change of basis AB also preserves it, so closure is verified. Also the inverse of a metric-preserving automorphism preserves it, and the identity certainly does.

We can now identify the classical matrix groups as those subgroups of the general linear group that preserve certain types of metrics.

- The *pseudo-unitary group*[8] $\mathrm{U}(p,q;\mathbb{C})$ is the subgroup of $\mathrm{GL}(p+q,\mathbb{C})$ that preserves a *Hermitian* sesquilinear metric of signature (p,q). The prefix pseudo- is dropped when the metric is positive definite, i. e., when $q = 0$. In this case the metric in canonical form $g = \mathbf{1}$ is preserved by a basis change U *iff* $\mathrm{U}^\dagger \mathrm{U} = \mathbf{1}$.
- The *pseudo-orthogonal group*[9] $\mathrm{O}(p,q;\mathbb{R})$ is the subgroup of $\mathrm{GL}(p+q,\mathbb{R})$ that preserves a *symmetric* bilinear metric of signature (p,q). For a positive definite metric, the condition to be preserved by a basis change O is just $O^T O = \mathbf{1}$.
- The *symplectic group* $\mathrm{Sp}(m,\mathbb{R})$ is the subgroup of $\mathrm{GL}(2m,\mathbb{R})$ that preserves an *antisymmetric* bilinear metric (also called symplectic form). One can similarly define $\mathrm{Sp}(m,\mathbb{C})$.

We may further restrict the automorphisms to preserve the volume element, i. e., to have unit determinant. In this case the various groups acquire the denomination “special” and an S is prepended to their notation. For instance, $\mathrm{SL}(n,\mathbb{C}) \cap \mathrm{U}(n) = \mathrm{SU}(n)$, the special unitary group.

3.5.6.1 Example

Let us compare the groups SO(2) and SO(1, 1). A generic matrix $R \in \mathrm{SO}(2)$, namely, a matrix satisfying $A^T A = \mathbf{1}$ and $\det A = 1$, can be parameterized by an angle θ as

$$R = \begin{pmatrix} \cos\theta & \sin\theta \\ -\sin\theta & \cos\theta \end{pmatrix}. \tag{3.41}$$

A matrix $\Lambda \in \mathrm{SO}(1,1)$ must satisfy the following equations: $\Lambda^T \eta \Lambda = \eta$, where $\eta = \mathrm{diag}(-1,1)$, and $\det\Lambda = 1$. Writing Λ as a generic real 2×2 matrix $\Lambda = \left(\begin{smallmatrix} a & b \\ c & d \end{smallmatrix}\right)$, these equations read

8 One usually writes simply U(p, q), as the real unitary groups coincide with orthogonal real groups (on real vector spaces there is no difference between sesquilinear and bilinear).

9 One usually writes simply O(p, q), as the complex orthogonal groups are not so frequently used.

$$\begin{pmatrix} a & c \\ b & d \end{pmatrix} \begin{pmatrix} -1 & 0 \\ 0 & 1 \end{pmatrix} \begin{pmatrix} a & b \\ c & d \end{pmatrix} = \begin{pmatrix} -a^2 + c^2 & -ab + cd \\ -ab + cd & -b^2 + d^2 \end{pmatrix} = \begin{pmatrix} -1 & 0 \\ 0 & 1 \end{pmatrix} \tag{3.42}$$

and $ad - bc = 1$. The solution to these constraints turns out to be the following:

$$\Lambda = \begin{pmatrix} \cosh \nu & \sinh \nu \\ -\sinh \nu & \cosh \nu \end{pmatrix}, \tag{3.43}$$

with ν being a real parameter, called the "rapidity." A possible alternative is to introduce a parameter β related to the rapidity by $\cosh \nu = 1/\sqrt{1-\beta^2}$, $\sinh \nu = \nu/\sqrt{1-\beta^2}$.

3.5.7 Isometries

The Euclidean group Eucl_d in d dimensions is the group of invertible transformations of the Euclidean space $\mathbb{R}^d$ into itself that preserve the Euclidean distance: for any transformation $\mathscr{E} \in \mathrm{Eucl}_d$, if $\mathbf{x}_1'$, $\mathbf{x}_2'$ are the images under $\mathscr{E}$ of $\mathbf{x}_1$, $\mathbf{x}_2$, we have $|\mathbf{x}_2' - \mathbf{x}_1'| = |\mathbf{x}_2 - \mathbf{x}_1|$, for any couple $\mathbf{x}_1$, $\mathbf{x}_1$. Notice that the Euclidean distance is that arising from having endowed the vector space $\mathbb{R}^d$ with a symmetric bilinear positive definite metric $g_{ij} = \delta_{ij}$. This defines the scalar product $(\mathbf{x}, \mathbf{y}) = x^i g_{ij} y^j = \mathbf{x} \cdot \mathbf{y}$, and hence the distance $|\mathbf{x} - \mathbf{y}| = \sqrt{(\mathbf{x} - \mathbf{y}, \mathbf{x} - \mathbf{y})}$.

More generally, a notion of distance can be introduced not only on vector spaces, but also on manifolds. Then one talks of Riemannian manifolds, i. e., differentiable manifolds of dimension d equipped with a positive definite quadratic form, that is, a metric locally expressible as

$$ds^2 = g_{\alpha\beta} dx^\alpha \otimes dx^\beta, \quad g_{\alpha\beta} = g_{\beta\alpha} \quad (\alpha, \beta = 1, \ldots d), \tag{3.44}$$

with $g_{\alpha\beta}(x)$ being a differentiable function of the coordinates (that transform as a 2-tensor under coordinate changes). The line element ds^2 defines the square length of the minimal arc connecting two points whose coordinates differ by dx^α.

We will discuss the notion of metric in a differentiable manifold extensively in Chapter 15.

For Euclidean spaces $\mathbb{R}^d$, the metric can always be chosen to be constant. Thus, the possible transformations are of the form $x'^\alpha = \mathscr{R}^\alpha{}_\beta x^\beta + v^\alpha$, that is, in matrix notation, $\mathbf{x}' = \mathscr{R}\mathbf{x} + \mathbf{v}$. It is easily seen that the metric, which in matrix notation is written as $ds^2 = d\mathbf{x}^T g d\mathbf{x}$, is invariant iff

$$\mathscr{R}^T g \mathscr{R} = g. \tag{3.45}$$

In a coordinate choice where $g_{\alpha\beta} = \delta_{\alpha\beta}$, the isometry condition becomes simply $\mathscr{R}^T \mathscr{R} = \mathbf{1}$, namely, $\mathscr{R} \in \mathrm{O(d)}$. We retrieve thus the description of Euclidean isometries as products of translations and orthogonal transformations (rotations plus inversions).

3.6 Bibliographical note

The introductory topics related to the notion of a group and the diverse examples collected and illustrated in this chapter are covered in a vast body of literature on both mathematics and physics.

As explained at length in the historical book [28], the very notions of a group and of a vector space were slowly developed throughout a historical process that embraced almost an entire century, from Lagrange and Galois until the book *Classical Groups* of Hermann Weyl [22]. Fundamental contributions in this process were made by Jordan [3], Cayley [2], Sylvester [10], Klein [43, 6], Hurwitz [5], Lie [13, 14, 15, 16, 17], Frobenius [36], Killing [18], and Cartan [20, 21], as far as finite and continuous groups are concerned. As far as vector spaces and matrices are concerned, the main contributors to the development of these by now so familiar mathematical structures are Cayley [8], Sylvester [10], Grassmann [11], and finally Peano [12].

Modern and less modern textbooks for further reading on the topics touched upon in this chapter are:

1. On the use of group theory in quantum mechanics, [44, 45, 46, 47].
2. On the multifaceted applications of group theory in solid state physics, [48, 49, 50, 51].
3. On the use of group theory in particle physics, [52, 53, 54, 55].
4. For general illustrations of the use of groups while treating the symmetries of physical systems, [56, 57, 58, 59, 60, 61, 62, 63, 64, 65, 66, 67, 68].

4 Basic elements of finite group theory

Ancora indietro un poco ti rivolvi,
diss'io, là dove di' ch'usura offende
la divina bontade, e 'l groppo solvi.

Dante, Inferno XI, 94

4.1 Introduction

In Chapter 3 we already presented many different examples of groups, both continuous and discrete.

Once the conception of groups as abstract mathematical objects is completely integrated into the fabrics of mathematics and of its application to other sciences, the next two natural questions are:

(a) The classification issue. Which are the possible groups G?
(b) The representation issue. Which are the possible explicit realizations of each given abstract group G?

Let us comment on the first issue. If the *classification* of groups were accomplished, given a specific realization of a group arising, e. g., in some system (physical, chemical, dynamical, or anything else), one might just identify its isomorphic class in the general classification and know a priori, via the isomorphism, all the relevant symmetry properties of the considered systems. For finite groups, this means the classification of all possible distinct (non-isomorphic) groups of a given finite order n. Such a goal is too ambitious and cannot be realized, yet there is a logical way to proceed.

One is able to single out certain types of groups (the so-called *simple groups*) which are the *hard core* of the possible different group structures. The complete classification of simple finite groups is a notable achievement of modern mathematics in the 1970s and 1980s, which became possible thanks to massive computer calculations. We do not dwell on this very much specialized topic and we confine ourselves to mentioning some series of simple groups that are easily defined.

Assuming the list of simple groups as given one can study the possible *extensions* which allow the construction of new groups having the simple groups as building blocks. We will give some general ideas about that mathematical construction, which is one of the most sophisticated areas of development in modern group theory, in Chapter 8.

Also for groups of infinite order there are some general results in the line of a classification, mainly regarding Abelian groups. We shall touch briefly upon this issue and an interesting explicit example will be studied in Chapter 9 when considering the notion of *monodromy group* of a linear differential equation.

For Lie groups the quest of classification follows a pattern very similar to the case of finite groups, involving the definition of *simple Lie groups* to be classified first. We will discuss this in full detail in Chapters 6 and 11.

https://doi.org/10.1515/9783111201535-004

In order to address the issue of group classification we need to introduce several concepts and general theorems related to the inner structure of a given group; for instance, the concepts of *conjugacy classes* and *invariant subgroups* are essential.

These concepts and the theorems that allow us to discuss in much finer detail the structure of groups are the subject of Section 4.2.

The issue of representations of finite groups is addressed in detail in Section 4.3.

As we are going to see, the intrinsic structure of the abstract G determines the possible linear representations of G.

4.2 Basic notions and structural theorems for finite groups

Let us start with a theorem that strongly delimits the operational ground for finite group classification.

4.2.1 Cayley's theorem

Theorem 4.2.1. *Any group G of* finite *order $|G|$ is* isomorphic to a subgroup of the permutation group *on $|G|$ objects, $S_{|G|}$.*

This theorem was not stated in this way by Cayley, but it is rightly named after him since it streams from Cayley's paper of 1854, which is extensively reviewed in [28]. An obvious consequence of Cayley's theorem is that the number of distinct groups of order G is finite, as the number of subgroups of $S_{|G|}$ certainly is. It can also be used to determine the possible group structures of low orders. However, though the permutation groups are easily defined, the structure of their subgroups is far from obvious, and the problem of classifying finite groups is far from being solved by this simple token.

Proof of Theorem 4.2.1. The proof of Cayley's theorem relies on the observation that each row of the group multiplication table defines a distinct permutation of the elements of the group. Thus, with any element $g \in G$ we can associate the permutation $\pi \in S_{|G|}$ that acts as $g_k \overset{\pi_g}{\longmapsto} (gg_k)$ $(k = 1, \ldots, |G|)$. Distinct group elements are associated with distinct permutations. The identity e is mapped into the identical permutation π_e. The product is preserved by the mapping; indeed, for any $c, b \in G$, we have

$$\pi_c \pi_b = \begin{pmatrix} bg_1 & \cdots & bg_n \\ cbg_1 & \cdots & cbg_n \end{pmatrix} \begin{pmatrix} g_1 & \cdots & g_n \\ bg_1 & \cdots & bg_n \end{pmatrix} = \begin{pmatrix} g_1 & \cdots & g_n \\ cbg_1 & \cdots & cbg_n \end{pmatrix} = \pi_{cb}, \tag{4.1}$$

where, for convenience, we have described π_c by its action on the elements bg_i; this amounts just to a relabeling of the elements g_i. All in all, the set $\{\pi_b : b \in G\}$ is a subgroup of $S_{|G|}$ isomorphic to G. □

4.2.1.1 Regular permutations

The permutations π_g associated with the elements $g \in G$ in the previously described isomorphism can be read directly from the multiplication table of the group. Such permutations are called *regular* permutations, and the subgroups of S_n isomorphic to groups G of order n are subgroups of regular permutations. Let us summarize the properties of such subgroups.

(i) Apart from the identical permutation, all other π_g do not leave any *symbol* (any of the objects on which the permutation acts) invariant; this corresponds to the property of the rows of the group multiplication table.[1]
(ii) Any of the n permutations π_g maps a given symbol into a different symbol; this corresponds to the property of the columns of the multiplication table.
(iii) All the cycles in the cycle decomposition of a regular permutation have the same length. Indeed, if a regular permutation π_g had two cycles of lengths $l_1 < l_2$, then $(\pi_g)^{l_1}$ would leave the elements of the first cycle invariant, but not those of the second, which however cannot be the case for a regular permutation.

Cayley's theorem is useful in determining the possible group structures of low order. The following result is a corollary of Cayley's theorem.

4.2.1.2 Groups of prime order

Lemma 4.2.1. *The only finite group of order p, where p is a prime, is the cyclic group* $\mathbb{Z}_p$.

Proof of Lemma 4.2.1. By Cayley's theorem, any group is isomorphic to a subgroup of S_p made of regular permutations. Since all cycles of a regular permutation must have the same length ℓ, this latter must be a divisor of p. If p is prime, the only possible regular permutations have either p cycles of length 1 (identical permutation) or one cycle of length p. This is the case of cyclic permutations which form a group isomorphic to $\mathbb{Z}_p$. □

4.2.2 Left and right cosets

Let $H = \{e = h_1, h_2, \dots h_m\}$ be a subgroup of G of order $|H| = m$. Given an element a_1 not in H, $a_1 \in G \setminus H$, define its *left coset*:

$$a_1 H = \{a_1, a_1 h_2, \dots a_1 h_m\}. \tag{4.2}$$

1 Concerning this property one might wonder if there cannot be idempotent group elements such that $\gamma \cdot \gamma = \gamma$ is a true statement. If such an element existed in the group $\gamma \in G$, the corresponding permutation π_γ would leave one symbol invariant, namely, γ itself. Yet from the axioms of a group it follows that $\gamma \cdot \gamma \cdot \gamma^{-1} = e$, which implies $\gamma = e$. Hence, in any group the only idempotent element is the unique neutral element e and all the permutations induced by group elements are regular.

Since $h_i \neq h_j, \forall i, j = 1, \dots m$, we also have $a_1 h_i \neq a_1 h_j$ (otherwise $a_1 = e$, but then a_1 would belong to H). Moreover, $\forall i$, $a_1 h_i \notin H$; otherwise, $a_1 h_i = h_j$ for some j, and therefore $a_1 = h_j (h_i)^{-1}$ would belong to H. We can now take another element a_2 of $G \setminus H$ not contained in $a_1 H$. The m elements of its left coset $a_2 H$ are again all distinct, for the same reasons as above. Moreover, $\forall i$, $a_2 h_i \notin H$, as above, but also $a_2 h_i \notin a_1 H$; otherwise, we would have $a_2 h_i = a_1 h_j$, for some j, so that $a_2 = a_1 h_j (h_i)^{-1}$ would belong to $a_1 H$. We can iterate the reasoning until we exhaust all the elements of G.

Thus G decomposes into a *disjoint union* of left cosets with respect to any subgroup H:

$$G = H \cup a_1 H \cup a_2 H \cup \cdots \cup a_l H. \tag{4.3}$$

Another way to reach the same result relies on the concept of equivalence classes. First note that we can introduce the following relation:

$$\forall g_1, g_2 \in G : g_1 \sim g_2 \quad \Leftrightarrow \quad \exists h \in H \subset G / g_1 = g_2 h, \tag{4.4}$$

which satisfies the axioms of an equivalence relation since it is both reflexive and transitive. Hence, the entire set G decomposes into a finite set of disjoint equivalence classes and these are the cosets mentioned in equation (4.3).

In complete analogy, given a subgroup H, we can introduce the *right cosets*

$$H a_1 = \{a_1, h_2 a_1, \dots h_m a_1\}, \tag{4.5}$$

and we can repeat the entire reasoning for the right cosets.

4.2.3 Lagrange's theorem

An immediate consequence of the above reasoning is the following.

Theorem 4.2.2. *The order of a subgroup H of a finite group G is a divisor of the order of G:*

$$\exists l \in \mathbb{N} \quad \textit{such that} \quad |G| = l|H|. \tag{4.6}$$

The integer $l \equiv [G : H]$ is named the index *of H in G.*

The above theorem, named after Lagrange, is very important for the classification of the possible subgroups of given groups. For instance, if $|G| = p$ is a prime, G does not admit any proper subgroup. This is the case for $G = \mathbb{Z}_p$ (the only group of order p, as we saw before).

Corollary 4.2.1. *The* order *of any element of a finite group G is a* divisor *of the order of G.*

Proof of Corollary 4.2.1. Indeed, if the order of an element $a \in G$ is h, then a generates a cyclic subgroup of order h $\{e, a, a^2, \dots, a^{h-1}\}$. This being a subgroup of G, its order h must be a divisor of $|G|$. □

4.2.4 Conjugacy classes

We have already introduced the conjugacy relation between elements of a group G in Section 2.3.4 ($g' \sim g \Leftrightarrow \exists h \in G$ such that $g' = h^{-1}gh$) and we have noted that it is an equivalence relation. Therefore, we can consider the quotient of G (as a set) by means of this equivalence relation. The elements of the quotient set are named the *conjugacy classes*. Any group element g defines a conjugacy class $[g]$:

$$[g] \equiv \{g' \in G \text{ such that } g' \sim g\} = \{h^{-1}gh, \text{ for } h \in G\}. \tag{4.7}$$

Basically, conjugation is the implementation of an inner automorphism of the group; we may think of it as a *change of basis* in the group (it is indeed so for matrix groups); in many instances one is interested in those properties and those quantities that are independent of conjugation. Such properties pertain to the conjugacy classes rather than to the individual elements.

4.2.4.1 Example: conjugacy classes of the symmetric groups

The key feature that allows for an efficient description of the conjugacy classes of the symmetric group S_n is the invariance under conjugation of the *cycle decomposition* of any permutation $P \in S_n$.

Indeed, suppose that P contains a cycle $(p_1, p_2, \dots, p_k)$ of length k. Then a conjugate permutation $Q^{-1}PQ$ contains a cycle of the same length, namely,

$$(Q^{-1}(p_1), Q^{-1}(p_2), \dots, Q^{-1}(p_k)).$$

Indeed, we have

$$\begin{aligned}(Q^{-1}(p_1), Q^{-1}(p_2), \dots, Q^{-1}(p_k)) &\xrightarrow{Q} (p_1, p_2, \dots, p_k) \xrightarrow{P} (p_2, p_3, \dots, p_1) \\ &\xrightarrow{Q^{-1}} (Q^{-1}(p_2), Q^{-1}(p_3), \dots, Q^{-1}(p_1)).\end{aligned} \tag{4.8}$$

Thus, *conjugacy classes* of S_n are in one-to-one correspondence with the possible *structures of cycle decompositions*. Let a permutation P be decomposed into cycles and name r_l the number of cycles of length l ($l = 1, \dots n$); then we have the sum rule

$$\sum_{l=1}^{n} r_l l = n. \tag{4.9}$$

Hence the conjugacy class which the permutation P belongs to is determined by the set of integers $\{r_l\}$ describing how many cycles of each length l appear in the decomposition. Thus the possible conjugacy classes are in one-to-one correspondence with the set of solutions of equation (4.9). These solutions, in turn, correspond to the set of *partitions* of the integer n into integers. A partition[2] of n is a set of integers $\{\lambda_i\}$, with

$$\sum_i \lambda_i = n, \quad \lambda_1 \geq \lambda_2 \geq \cdots \geq \lambda_n \geq 0. \tag{4.11}$$

Indeed, a set $\{r_l\}$ of integers satisfying (4.9) is obtained from a partition $\{\lambda_i\}$ by setting

$$r_1 = \lambda_1 - \lambda_2, \quad r_2 = \lambda_2 - \lambda_3, \ldots, r_{n-1} = \lambda_{n-1} - \lambda_n, \quad r_n = \lambda_n. \tag{4.12}$$

Thus, conjugacy classes of S_n are in one-to-one correspondence with partitions of n, which in turn can be graphically represented by means of *Young tableaux* with n boxes. In a Young tableau, the boxes are distributed in rows of non-increasing length. The length of the i-th row is λ_i; the label r_l (the number of cycles of length l) instead corresponds to the difference between the length of the l-th and the $(l+1)$-th row. As an example we display below the Young tableau $(8, 6, 6, 5, 3, 2)$ corresponding to a partition of 30.

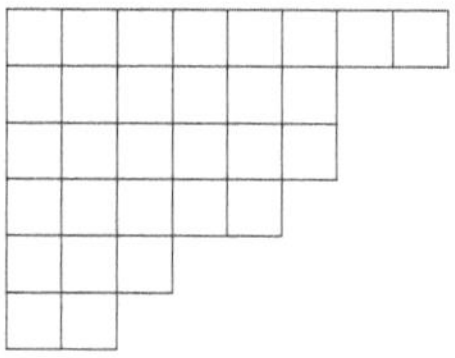

(4.13)

The corresponding cycle structure is $r_1 = 2, r_2 = 0, r_3 = 1, r_4 = 2, r_5 = 1, r_6 = 2$.

The number of elements in a given conjugacy class $\{r_l\}$, which we call the order of the class and denote as $|\{r_l\}|$, is obtained as follows. The n elements $1, \ldots, n$ must be distributed in the collection $\{r_l\}$ of cycles, ordered as follows:

$$\underbrace{(\cdot)\ldots(\cdot)}_{r_1}\ \underbrace{(\cdot\cdot)\ldots(\cdot\cdot)}_{r_2}\ \underbrace{(\cdot\cdot\cdot)\ldots(\cdot\cdot\cdot)}_{r_3}\ \ldots. \tag{4.14}$$

There are n possible positions, so there are $n!$ possibilities. However, distributions differing for a permutation between cycles of the same length correspond to the same element (of course, (12)(45) is the same as (45)(12)); thus, we must divide by $r_1!r_2!\ldots$. Moreover,

---

2 Recall that the number of partitions of n, $p(n)$, is expressed through the generating function

$$P(q) \equiv \sum_{n=0}^{\infty} p(n)q^n = \prod_{k=1}^{\infty} \frac{1}{1-q^k}. \tag{4.10}$$

The coefficient of q^n in the expansion of the infinite product gives the number of partitions of n.

in each cycle of length l we can make l periodic shifts (by 1, by 2, ... by $l-1$) that leave the cycle invariant. Thus we must divide by $1^{r_1}2^{r_2}3^{r_3}\dots$. Altogether we have obtained

$$|\{r_l\}| = \frac{n!}{r_1!\,2^{r_2}r_2!\,3^{r_3}r_3!\dots}. \tag{4.15}$$

4.2.5 Conjugate subgroups

Let H be a subgroup of a group G. Let us consider

$$H_g \equiv \{h_g \in G : h_g = g^{-1}hg,\ \text{for } h \in H\}, \tag{4.16}$$

which we simply write as $H_g = g^{-1}Hg$. It is easy to see that H_g is a subgroup. The subgroups H_g are called *conjugate subgroups* to H.

A subgroup H of a group G is called an *invariant* (or *normal*) subgroup if it coincides with all its conjugate subgroups: $\forall g \in G$, $H_g = H$. Given any subgroup H of G, we can define two equivalence relations in G:

$$\begin{aligned} g_1 \sim_L g_2 \quad &\Leftrightarrow \quad \exists h \in H : g_1 = hg_2 \quad \text{(left equivalence)},\\ g_1 \sim_R g_2 \quad &\Leftrightarrow \quad \exists h \in H : g_1 = g_2h \quad \text{(right equivalence)}. \end{aligned} \tag{4.17}$$

Hence we can consider the set of equivalence classes with respect to either the left or the right equivalence, namely, the *left (right) cosets*. The left coset $H\backslash G$ contains the left classes already introduced, which we write simply as gH, and the right coset contains the right classes Hg.

If H is a normal subgroup, then the two equivalence relations of (4.17) coincide:

$$g_1 \sim_L g_2 \quad \Leftrightarrow \quad g_2 \sim_R g_1. \tag{4.18}$$

It follows that the left and the right coset coincide: $H\backslash G = G/H$. This is the same as saying that the two equivalence relations are *compatible* with the group structure:

$$\begin{cases} g_1 \sim g_2, \\ g_3 \sim g_4 \end{cases} \Leftrightarrow \quad g_1g_3 \sim g_2g_4, \tag{4.19}$$

where $\sim$ stands for $\sim_L$ (or $\sim_R$).

If H is a normal subgroup of G, then G/H (= $H\backslash G$) is a group with respect to the product of classes defined as follows:

$$(g_1H)(g_2H) = g_1g_2H. \tag{4.20}$$

This product is well-defined. Indeed, since $g_2H = Hg_2$ for H invariant, we have $g_1Hg_2H = g_1g_2H$. The subgroup H is the identity of G/H, and the inverse of an element gH is $g^{-1}H$.

The converse of the above statement is also true: if $\phi : G \to G'$ is an homomorphism, then there exists a normal subgroup $H \subset G$ such that $G' = G/H$.

4.2.5.1 Example

Consider the group $n\mathbb{Z}$, namely, the set of multiples of n, $\{\dots, -2n, -n, 0, n, 2n, \dots\}$, with addition as the group law. The factor group $n\mathbb{Z}/\mathbb{Z}$ is isomorphic to the cyclic group $\mathbb{Z}_n$.

4.2.6 Center, centralizers, and normalizers

The center $Z(G)$ of a group G is the set of all those elements of G that commute (in the group sense) with all the elements of G:

$$Z(G) = \{f \in G : g^{-1}fg = f, \ \forall g \in G\}. \tag{4.21}$$

$Z(G)$ is an Abelian subgroup of G.

4.2.6.1 The centralizer of a subset

The *centralizer* $C(A)$ of a subset $A \subset G$ is the subset of G containing all those elements that commute with all the elements of A:

$$C(A) = \{g \in G : \ \forall a \in A, \ g^{-1}ag = a\}. \tag{4.22}$$

If A contains a single element a, then $C(A)$ is simply called the centralizer of a and it is denoted $C(a)$.

For any fixed element g, the product of the order of the conjugacy class of g and its centralizer equals the order of G:

$$\big|[g]\big|\big|C(g)\big| = |G|. \tag{4.23}$$

Indeed, let $g' = u^{-1}gu$ be an element of $[g]$ different from g. Also the conjugation of g by $w = ut$, where $t \in C(g)$, gives g'; in fact, $w^{-1}gw = u^{-1}t^{-1}gtu = u^{-1}gu = g'$. Thus, constructing $[g]$ as the set $\{u^{-1}gu : u \in G\}$ we obtain $|C(g)|$ times each distinct element.

4.2.6.2 Example: centralizers of permutations

Let a permutation $P \in S_n$ be decomposed into a set of $\{r_l\}$ cycles. It is easy to convince oneself that any permutation that (i) permutes between themselves the cycles of equal length in P or (ii) effects arbitrary periodic shifts within any cycle *commutes* with P. Thus, the number of permutations commuting with P is given by

$$\big|C(P)\big| = \prod_{l=1}^{n} r_l!\, l^{r_l}. \tag{4.24}$$

We see that this expression, together with (4.15) giving the order of the conjugacy class of P, is consistent with (4.23).

4.2.6.3 The normalizer of a subset

The *normalizer* $N(A)$ of a subset $A \subset G$ is the subgroup of elements of G with respect to which A is invariant:

$$N(A) = \{g \in G : g^{-1}Ag = A\}. \tag{4.25}$$

If A contains a single element a, then $N(A)$ is simply called the normalizer of a and it is denoted as $N(a)$.

4.2.7 The derived group

The commutator subgroup or the *derived group* $\mathscr{D}(G)$ of a group G is the group *generated* by the set of all group commutators in G (that is, it contains all group commutators and products thereof).

The derived group $\mathscr{D}(G)$ is normal in G, i. e., it is a *normal subgroup*. Indeed, take an element of $\mathscr{D}(G)$ which is a commutator, say $ghg^{-1}h^{-1}$. Then any conjugate of it by an element $f \in G, f^{-1}ghg^{-1}h^{-1}f$, is still a commutator, that of $f^{-1}gf$ and $f^{-1}hf$. The same reasoning applies with little modification to an element of $\mathscr{D}(G)$ that is a product of commutators.

The factor group $G/\mathscr{D}(G)$ is *Abelian*: it is the group obtained from G by *pretending* it is Abelian. Another property is that any subgroup $H \subset G$ that contains $\mathscr{D}(G)$ is *normal*, since $\forall g \in G$ and $\forall h \in H$ we have $g^{-1}hg = [g^{-1}, h]h$.

4.2.8 Simple, semi-simple, and solvable groups

In general a group G admits a chain of invariant subgroups, called its *subnormal series*:[3]

$$G = G_r \triangleright G_{r-1} \triangleright G_{r-2} \triangleright \cdots \triangleright G_1 \triangleright \{e\}, \tag{4.26}$$

where every G_i is a normal subgroup.

3 Following a convention widely utilized in finite group theory we make a distinction between subgroups and normal subgroups. The notation $G \supset H$ simply means that H is a subgroup of G, not necessarily an invariant one. On the other hand, $G \triangleright N$ means that N is a normal (invariant) subgroup of G.

Definition 4.2.1. G is a *simple* group if it has *no proper normal subgroup*. For simple groups, the subnormal series is minimal:

$$G \triangleright \{e\}. \tag{4.27}$$

Simple groups are the *hard core* of possible group structures. There is no factor group G/H smaller than G out of which the group G could be obtained by some *extension*, because there is no normal subgroup H other than the trivial one $\{e\}$ or G itself.

Definition 4.2.2. G is a *semi-simple* group if it has *no proper normal subgroup* which is *Abelian*.

Definition 4.2.3. A group G is *solvable* if it admits a subnormal series (4.26) such that all the factor groups $G/G_1, G_1/G_2, \ldots, G_{k-1}/G_k, \ldots$ are *Abelian*.

4.2.9 Examples of simple groups

4.2.9.1 Cyclic groups of prime order

Cyclic groups $\mathbb{Z}_p$ with p a prime are simple. Indeed, they are Abelian, so every subgroup would be a normal subgroup. However, by Lagrange's theorem, the order of any subgroup of $\mathbb{Z}_p$ should be a divisor of p, which leaves only the improper subgroups $\{e\}$ and $\mathbb{Z}_p$ itself. What is absolutely non-trivial is that the cyclic groups of prime order are the *only simple groups of odd order*.

4.2.9.2 The alternating groups A_n

One can show that the alternating groups A_n with $n \geq 5$ are simple. Here we omit the proof.

4.2.10 Homomorphism theorems

Next we collect a few simple but fundamental theorems concerning homomorphisms and isomorphisms of groups that are extremely useful to understand the possible group structures.

Theorem 4.2.3 (First isomorphism theorem). *Given a homomorphism ϕ of G onto G':*
(i) *The kernel of the homomorphism,* $\ker\phi$, *is an* invariant subgroup, *namely,* $\ker\phi \triangleleft G$.
(ii) *The map ϕ gives rise to an* isomorphism *between the factor group $G/\ker\phi$ and G'.*

Proof of Theorem 4.2.3. By definition the kernel of ϕ is the subset of G that is mapped onto the identity element e' of G':

$$\ker\phi = \{g \in G : \phi(g) = \mathbf{e}'\}. \tag{4.28}$$

It is immediate to see that $\ker\phi$ is a subgroup. It is also normal, because if $g \in \ker\phi$, then any of its conjugates belongs to it: $\phi(u^{-1}gu) = \phi(u^{-1})\phi(g)\phi(u) = [\phi(u)]^{-1}e'\phi(u) = e'$; this proves item (i). In case of finite groups, if $\ker\phi$ has order m, then ϕ is an m-to-one mapping. Indeed, if k_i ($i = 1, \dots m$) are the elements of $\ker\phi$, then the image $\phi(g)$ of a given element coincides with that of the elements $\phi(k_i g)$: the kernel being a normal subgroup, we can define the factor group $G/\ker\phi$. Since the kernel of the map $\phi : G/\ker\phi \to G'$ contains now only the identity of the factor group, this map is an isomorphism. This proves (ii). □

There are other theorems concerning homomorphisms that we mention without proof.

The *correspondence theorem* states that if $\phi : G \to G'$ is an homomorphism, then:

(i) The *preimage* $H = \phi^{-1}(H')$ of any subgroup H' of G' is a subgroup of G containing $\ker\phi$ (this generalizes the property of $\ker\phi = \phi^{-1}(e')$ being a subgroup). If H' is normal in G', then so is H in G.
(ii) If there is any other subgroup H_1 of G, containing $\ker\phi$ that is mapped onto H' by ϕ then $H_1 = H$.

The above statements can be rephrased (via the first isomorphism theorem) in terms of factor groups:

(i) Let L be a subgroup of a factor group G/N. Then $L = H/N$ for H a subgroup of G (containing N). If L is normal in G/N, then H is normal in G.
(ii) If $H/N = H_1/N$, with H and H_1 being subgroups of G containing N, then $H = H_1$.

The *factor of a factor* theorem states that if in the factor group G/N there is a normal subgroup of the form M/N, with $M \supseteq N$, then M is a normal subgroup of G and

$$G/M \sim (G/N)/(M/N). \tag{4.29}$$

4.2.11 Direct products

As a set, the direct product $G \otimes F$ of two groups G and F is the Cartesian product of G and F:

$$G \times F = \{(g,f) : g \in G, f \in F\}. \tag{4.30}$$

Elements of $G\otimes F$ are pairs, and $|G\otimes F| = |G||F|$. The group operation is defined as follows. Elements of $G \otimes F$ have to be multiplied "independently" in each entry, in the first entry with the product law of G and in the second with the product law of F:

$$(g,f)(g',f') = (gg',ff'). \tag{4.31}$$

Conversely, given a group G, we say that it is the direct product of certain subgroups,

$$G = H_1 \otimes H_2 \otimes \cdots \otimes H_n, \tag{4.32}$$

if and only if:
(i) Elements belonging to different subgroups H_i commute.
(ii) The only element common to the various subgroups H_i is the identity.
(iii) Any element $g \in G$ can be expressed as a product,

$$g = h_1 h_2 \dots h_n, \quad (h_1 \in H_1, \dots h_n \in H_n). \tag{4.33}$$

From conditions (ii) and (iii) it follows that the decomposition in equation (4.33) is uniquely defined. Condition (i) is equivalent to all the subgroups H_i being normal.

4.2.11.1 Example

The orthogonal group in three dimensions, O(3), is the direct product of the special orthogonal group SO(3) and of the matrix group (isomorphic to $\mathbb{Z}_2$) given by the two 3×3 matrices $\{\mathbf{1}, -\mathbf{1}\}$.

The direct product is the simplest way to build larger groups out of smaller building blocks. We can start from, say, two simple groups G and F to obtain a larger group $G \otimes F$, which is no longer simple as it admits G and F as normal subgroups. These normal subgroups are embedded into $G \otimes F$ in the simplest way; the homomorphism $\phi : G \otimes F \to G$ corresponds simply to neglecting the F component: $\phi : (g, f) \mapsto g$.

4.2.12 Action of a group on a set

Before proceeding it is convenient to summarize a few notions and basic definitions related to the action of a group G as a group of transformations on a set $\mathscr{S}$ that can be finite, countably infinite, or continuously infinite.

Definition 4.2.4. Let G be a group and let $\mathscr{S}$ be a set. We say that G acts as a transformation group on a set $\mathscr{S}$ if with each group element $\gamma \in G$ we can associate a map $\mu(\gamma)$ of the set into itself:

$$\forall \gamma \in G : \quad \mu(\gamma) : \mathscr{S} \to \mathscr{S}. \tag{4.34}$$

Furthermore, given two group elements $\gamma_{1,2} \in G$ we must have

$$\mu(\gamma_1 \cdot \gamma_2) = \mu(\gamma_1) \circ \mu(\gamma_2), \tag{4.35}$$

where $\circ$ denotes the composition of two maps and also

$$\mu(\mathbf{e}) = \mathbf{id}, \tag{4.36}$$

where **e** is the neutral element of G and **id** denotes the identity map which associates with each element $s \in \mathscr{S}$ the same element s.

Next we introduce the following notions.

Definition 4.2.5. Let G be a group that acts on the set $\mathscr{S}$ as a transformation group. For simplicity we denote by γ the map $\mu(\gamma) : \mathscr{S} \to \mathscr{S}$ associated with the group element $\gamma \in G$. Let $s \in S$ be an element in the set. We call the collection of all those transformations of G which map s into itself a *stability subgroup* of s ($H_s \subset G$):

$$H_s = \{h \in G \mid h(s) = s\}. \tag{4.37}$$

It is immediately evident that H_s defined as above is a subgroup.

Conversely, we have the following.

Definition 4.2.6. Let G be a group that acts on the set $\mathscr{S}$ as a transformation group and consider an element $s \in \mathscr{S}$. We call the collection of all those elements of $\mathscr{S}$ which are reached from s by a suitable element of G the *orbit* of s with respect to G, denoted $\text{Orbit}_G(s)$:

$$\text{Orbit}_G(s) = \{p \in \mathscr{S} \mid \exists \gamma \in G \backslash \gamma(s) = p\}. \tag{4.38}$$

Furthermore, we have the following.

Definition 4.2.7. Let G be a group that acts on the set $\mathscr{S}$ as a transformation group. We say that the action of G on $\mathscr{S}$ is *transitive* if given any two elements $s_{1,2} \in \mathscr{S}$ there exists a suitable group element that maps the first into the second:

$$\forall s_1, s_2 \in \mathscr{S} : \quad \exists \gamma_{12} \in G \backslash \gamma_{12}(s_1) = s_2, \tag{4.39}$$

When a group G has a transitive action on a set $\mathscr{S}$ we can identify it with the orbit of any of its elements.

Lemma 4.2.2. *Let G have a transitive action on the set $\mathscr{S}$. Then*

$$\forall s \in \mathscr{S} : \quad \text{Orbit}_G(s) = \mathscr{S}. \tag{4.40}$$

The proof of the above lemma is obvious from the very definitions of orbit and transitivity.

Consider now the case of finite groups.

Let the finite group G act on the set $\mathscr{S}$ (which can be finite or infinite, continuous or discrete). Consider an element $s \in \mathscr{S}$ and its orbit $\mathscr{O}_s \equiv \text{Orbit}_G(s)$. By definition $\mathscr{O}_s \subset \mathscr{S}$

is a finite subset and G has a transitive action on it. The natural question is: What is the cardinality of this subset? The immediate answer is

$$|\mathcal{O}_s| = \frac{|G|}{|H_s|}, \tag{4.41}$$

where H_s is the stability subgroup of the element s. Indeed, only the elements of G that are not in H_s map s to new elements of the orbit. Yet any two elements $\gamma_1, \gamma_2 \in G$ that differ by right multiplication by any element of H_s map s to the same element of the orbit. Hence, we can conclude that

$$\mathcal{O}_s \sim G/H_s, \tag{4.42}$$

and equation (4.41) immediately follows. Furthermore, any element $s' \in \mathcal{O}_s$ has a stability subgroup $H_{s'}$ which is isomorphic to H_s being conjugate to it by means of any group element $\gamma \in G$ that maps s into s'. We conclude that the possible orbits of G are in one-to-one correspondence with the possible cosets G/H. Whether the orbit corresponding to a given subgroup H exists or not depends on the explicit form of the action of G on $\mathcal{S}$, yet there is no orbit $\mathcal{O}$ of a certain cardinality $n = |\mathcal{O}|$ if there is no subgroup H of the required cardinality $|H| = |G|/n$.

Examples of stability subgroups are the normalizer and centralizer considered above.

4.2.13 Semi-direct products

A slightly more complicated construction to obtain a larger group from two building blocks is the *semi-direct product*. Let G and K be two groups and assume that G acts as a *group of transformations* on K:

$$\forall g \in G, \quad g : k \in K \mapsto g(k) \in K. \tag{4.43}$$

Let us use the symbols $k_1 \circ k_2$ and $g_1 \cdot g_2$ for the group products in K and G, respectively. The *semi-direct product* of G and K, denoted as $G \ltimes K$, is the Cartesian product of G and K as sets,

$$G \ltimes K = \{(g, k) : g \in G,\ k \in K\}, \tag{4.44}$$

but the product in $G \ltimes K$ is defined as follows:

$$(g_1, k_1)(g_2, k_2) = (g_1 \cdot g_2, k_1 \circ g_1(k_2)). \tag{4.45}$$

That is, before being multiplied by k_1, the element k_2 is acted upon by g_1. The inverse of an element of $G \ltimes K$ is then given by

$$(g,k)^{-1} = (g^{-1}, [g^{-1}(k)]^{-1}), \tag{4.46}$$

where the inverse g^{-1} is such with respect to the product in G, while the "external" inverse in $[g^{-1}(k)]^{-1}$ is with respect to the product in K.

The semi-direct product $G \ltimes K$ possesses a normal subgroup $\tilde{K}$, isomorphic to K, given by the elements of the form (e,k), with e being the identity of G and $k \in K$. Indeed, any conjugate of such an element is again in the subgroup $\tilde{K}$:

$$(g,k)^{-1}(e,h)(g,k) = (g^{-1}, [g^{-1}(k)]^{-1})(g, h\circ k) = (e, [g^{-1}(k)]^{-1} \circ g^{-1}(h \circ k)). \tag{4.47}$$

Instead, the subgroup isomorphic to G containing elements of the form (g,e), where e is the identity in K and $g \in G$, is *not* a normal subgroup. The above discussion is summarized as follows.

Definition 4.2.8. Conversely, given a group G, we say that it is the semi-direct product $G = G_1 \ltimes G_2$ of two subgroups G_1 and G_2 iff:
(i) G_2 is a normal subgroup of G.
(ii) G_1 and G_2 have only the identity in common.
(iii) Every element of G can be written as a product of an element of G_1 and one of G_2.

From (ii) and (iii) it follows that the decomposition (iii) is unique.

4.2.13.1 The Euclidean groups

The Euclidean groups Eucl_d are semi-direct products of the orthogonal group $\mathrm{O}(d)$ and of the Abelian group of d-dimensional translations. Their structure as semi-direct product has been discussed in equations (3.12) and (3.13) for the inhomogeneous rotation group ISO(2), the extension to larger Euclidean groups being immediate.

4.2.13.2 The Poincaré group

Consider a Poincaré transformation of a 4-vector x^μ:

$$x^\mu \to \Lambda^\mu{}_\nu x^\nu + c^\mu. \tag{4.48}$$

Here Λ is a pseudo-orthogonal matrix, $\Lambda \in \mathrm{O}(1,3)$, namely, it is a matrix such that $\Lambda^T \eta \Lambda = \eta$, where $\eta = \mathrm{diag}(-1,1,1,1)$ is the Minkowski metric. These matrices encode all the Lorentz transformations. Instead, the four-vector c^μ is a translation parameter. Poincaré transformations are the *isometries* of Minkowski space $\mathbb{R}^{1,3}$; they are the analog of the transformations of the Euclidean groups, the metric that they preserve, $\eta_{\mu\nu}$, being non-positive definite. Notice that the translation parameters c^μ are four-vectors and as such they are acted on by Lorentz transformations: $c^\mu \to \Lambda^\mu{}_\nu c^\nu$. The composition of two Poincaré transformations is the following one:

$$x^{\mu} \xrightarrow{(2)} \Lambda^{\mu}_{(2)\nu}x^{\nu} + c^{\nu}_{(2)} \xrightarrow{(1)} \Lambda^{\mu}_{(1)\nu}(\Lambda^{\nu}_{(2)\rho}x^{\rho} + c^{\nu}_{(2)}) + c^{\mu}_{(1)} = (\Lambda_{(1)}\Lambda_{(2)})^{\mu}{}_{\rho}x^{\rho} + (\Lambda_{(1)}c_{(2)})^{\mu} + c^{\mu}_{(1)}. \tag{4.49}$$

We see that the product law for the Poincaré group is

$$(\Lambda_{(1)}, c_{(1)})(\Lambda_{(2)}, c_{(2)}) = (\Lambda_{(1)}\Lambda_{(2)}, \Lambda_{(1)}c_{(2)} + c_{(1)}) \tag{4.50}$$

and the Poincaré group is the *semi-direct product* of the Lorentz group $O(1,3)$ with the translation group.

4.3 Linear representations of finite groups

We next turn to the group theory issue which is most relevant in most applications in any branch of science, namely, that of *linear representations*. In pure mathematics the notion of linear representations is usually replaced by the equivalent one of *module* (see Section 2.9).

Definition 4.3.1. Let G be a group and let V be a vector space of dimension n. Any homomorphism

$$D : G \to \mathrm{Hom}(V, V) \tag{4.51}$$

is named a linear representation of dimension n of the group G.

If the vectors $\{\mathbf{e}_i\}$ form a basis of the vector space V, then each group element $\gamma \in G$ is mapped into an $n \times n$ matrix $D_{ij}(\gamma)$ such that

$$D(\gamma).\mathbf{e}_i = D_{ij}(\gamma)\mathbf{e}_j \tag{4.52}$$

and we see that the homomorphism D can also be rephrased as the following one:

$$D : G \to \mathrm{GL}(n, \mathbb{F}), \quad \mathbb{F} = \begin{cases} \mathbb{R}, \\ \mathbb{C}, \end{cases} \tag{4.53}$$

where the field $\mathbb{F}$ is that of the real or complex numbers, depending on whether V is a real or complex vector space. Correspondingly we say that D is a *real* or *complex representation.*

Definition 4.3.1 applies in the same way to finite, countably infinite, and continuously infinite groups. The same is true for the concept of irreducible representations introduced by the following two definitions.

Definition 4.3.2. Let $D : G \to \mathrm{Hom}(V, V)$ be a linear representation of a group G. A vector subspace $W \subset V$ is said to be *invariant* iff

$$\forall \gamma \in G, \quad \forall \mathbf{w} \in W : \quad D(\gamma).\mathbf{w} \in W. \tag{4.54}$$

Definition 4.3.3. A linear representation $D : G \to \mathrm{Hom}(V, V)$ of a group G is named *irreducible* iff the only invariant subspaces of V are $\mathbf{0}$ and V itself.

In other words, a representation is irreducible if it does not admit any proper invariant subspace.

Definition 4.3.4. A linear representation $D : G \to \mathrm{Hom}(V, V)$ that admits at least one proper invariant subspace $W \subset V$ is named *reducible*. A reducible representation $D : G \to \mathrm{Hom}(V, V)$ is named *fully reducible* iff the orthogonal complement $W^\perp$ of any invariant subspace W is also invariant.

Let D be a fully reducible representation and let $W_i \subset V$ be a sequence of invariant subspaces such that:

(1)

$$V = \bigoplus_{i=1}^{r} W_i,$$

(2) none of the W_i contains invariant subspaces.

This situation can always be achieved by further splitting any invariant subspace W into smaller ones if it happens to admit further invariant subspaces. When we have reached (1) and (2) we see that the restriction to W_i of the representation D defines an irreducible representation D_i and we can write

$$D = \bigoplus_{i=1}^{r} D_i; \quad D_i : G \to \mathrm{Hom}(W_i, W_i). \tag{4.55}$$

Matrix-wise, we have

$$\forall \gamma \in G, \quad D(\gamma) = \left(\begin{array}{c|c|c|c|c|c} D_1(\gamma) & 0 & 0 & \cdots & 0 & 0 \\ \hline 0 & D_2(\gamma) & 0 & \cdots & \cdots & 0 \\ \hline \vdots & \vdots & \vdots & \vdots & \vdots & \vdots \\ \hline 0 & \cdots & \cdots & 0 & D_{r-1}(\gamma) & 0 \\ \hline 0 & \cdots & \cdots & \cdots & 0 & D_r(\gamma) \end{array}\right). \tag{4.56}$$

The above is described as the decomposition of the considered reducible representation D into irreducible ones.

We see from the above discussion that irreducible representations are the building blocks for any representation and the main issue becomes the classification of irreducible representations, frequently called *irreps*. Is the set of irreps an infinite set or a finite one? Here comes the main difference between continuous groups (in particular

Lie groups) and finite ones. While the set of irreducible representations of Lie groups is infinite (nevertheless, as we show in later chapters, they can be tamed and classified within the constructive framework of roots and weights), the irreps of a finite group G constitute a finite set whose order is just equal to $c(G)$, this number being the number of conjugacy classes into which the elements of G are distributed. As we are going to see, with each irrep D_i we can associate a $c(G)$-dimensional vector $\chi[D_i]$ named its character. The theory of characters is a very simple and elegant piece of mathematics with profound implications for geometry, physics, chemistry, and crystallography. We devote this section to its development, starting from two lemmas due to Schur that provide the foundation of the whole story.

4.3.1 Schur's lemmas

Lemma 4.3.1. *Let V, W be two vector spaces of dimension n and m, respectively, with $n > m$. Let G be a finite group and let $D_1 : G \to \mathrm{Hom}(V, V)$ and $D_2 : G \to \mathrm{Hom}(W, W)$ be two irreducible representations of dimension n and m, respectively. Consider a linear map $\mathscr{A} : W \to V$ and impose the constraint*

$$\forall \gamma \in G \quad \forall \mathbf{w} \in W, \quad D_1(\gamma)\mathscr{A}.\mathbf{w} = \mathscr{A}.D_2(\gamma).\mathbf{w}. \tag{4.57}$$

The only element $\mathscr{A} \in \mathrm{Hom}(W, V)$ that satisfies (4.57) is $\mathscr{A} = 0$.

Proof of Lemma 4.3.1. Let $\mathrm{Im}[\mathscr{A}]$ be the image of the linear map $\mathscr{A}$, namely, the subspace of V spanned by all vectors $\mathbf{v}$ that have a preimage in W:

$$\mathbf{v} \in \mathrm{Im}[\mathscr{A}] \quad \Leftrightarrow \quad \exists \mathbf{w} \in W \backslash \mathscr{A}.\mathbf{w} = \mathbf{v}. \tag{4.58}$$

The vector subspace $\mathrm{Im}[\mathscr{A}]$ is a proper subspace of V of dimension $m > 0$. Equation (4.57) states that this subspace should be invariant under the action of G in the representation D_1. This contradicts the hypothesis that D_1 is an irreducible representation. Ergo, $\mathrm{Im}[\mathscr{A}] = \mathbf{0} \in V$, namely, $\mathscr{A} = 0$. □

Lemma 4.3.2. *Let $D : G \to \mathrm{Hom}(V, V)$ be an n-dimensional irreducible representation of a finite group G. Let $C \in \mathrm{Hom}(V, V)$ be such that*

$$\forall \gamma \in G, \quad CD(\gamma) = D(\gamma)C. \tag{4.59}$$

Then $C = \lambda \mathbf{1}$, where $\lambda \in \mathbb{C}$ and $\mathbf{1}$ is the identity map of the n-dimensional vector space V into itself.

Proof of Lemma 4.3.2. Consider the eigenvalue equation

$$C\mathbf{v} = \lambda \mathbf{v}; \quad \mathbf{v} \in V. \tag{4.60}$$

The eigenspace $V_\lambda \subset V$ belonging to the eigenvalue λ is invariant under the action of G by the hypothesis of the lemma, namely, $\forall \mathbf{v} \in V_\lambda$, $\forall \gamma \in G$, we have $D(\gamma).\mathbf{v} \in V_\lambda$. As is well known, given any homomorphism $C \in \mathrm{Hom}(V, V)$ the vector space splits into the direct sum of the eigenspaces pertaining to different eigenvalues:

$$V = \bigoplus_{\lambda \in \mathrm{Spec}(C)} V_\lambda. \tag{4.61}$$

Hence, if the spectrum of C is composed of several different eigenvalues we have as many invariant subspaces of V and this contradicts the hypothesis that D is an irreducible representation. Ergo, there is only one eigenvalue $\lambda \in C$ and the map C is proportional to the identity map $\mathbf{1}$. □

4.3.1.1 Orthogonality relations

Relying on Schur's lemmas we can now derive some very useful orthogonality relations that lead to the main result, namely, the classification of irreducible representations for finite groups.

Let $X : V \to V$ be an arbitrary homomorphism of a finite-dimensional vector space into itself, $X \in \mathrm{Hom}(V, V)$. Consider

$$C \equiv \sum_{\gamma \in G} D(\gamma).X.D(\gamma^{-1}), \tag{4.62}$$

where D is a linear representation of a finite group G. We can easily prove that

$$\forall g \in G : \quad D(g).C = C.D(g). \tag{4.63}$$

To this effect it suffices to name $\tilde{\gamma} = g.\gamma$ and observe that the sum over all group elements γ is the same as the sum over all group elements $\tilde{\gamma}$, so that

$$D(g).C = \sum_{\gamma \in G} D(g).D(\gamma).X.D(\gamma^{-1}) = \sum_{\gamma \in G} D(g.\gamma).X.D(\gamma^{-1}) \tag{4.64}$$

$$= \sum_{\tilde{\gamma} \in G} D(\tilde{\gamma}).X.D(\tilde{\gamma}^{-1}).D(g) = C.D(g). \tag{4.65}$$

If $D = D^\mu$ is an irreducible representation it follows from the above discussion and Schur's lemmas that

$$X \in \mathrm{Hom}(V, V) : \quad \sum_{\gamma \in G} D^\mu(\gamma).X.D^\mu(\gamma^{-1}) = \lambda \mathbf{1}, \tag{4.66}$$

where $\mathbf{1}$ denotes the identity map. Let us specialize the result (4.66) to the case where, in some given basis, the homomorphism X is represented by a matrix with all vanishing entries except one in the crossing of the $\mathfrak{m}$-th row with the $\mathfrak{l}$-th column:

$$X = \begin{pmatrix} 0 & 0 & \cdots & \cdots & \cdots & 0 & 0 \\ 0 & \cdots & \cdots & \cdots & \cdots & \cdots & 0 \\ \vdots & \cdots & \cdots & 0 & \cdots & \cdots & \vdots \\ \vdots & \cdots & 0 & X_{\mathfrak{m}\mathfrak{l}} & 0 & \cdots & \vdots \\ 0 & \cdots & \cdots & 0 & \cdots & \cdots & 0 \\ 0 & 0 & \cdots & \cdots & \cdots & 0 & 0 \end{pmatrix}. \tag{4.67}$$

Choosing $X_{\mathfrak{m}\mathfrak{l}} = 1$, from equation (4.66) we obtain

$$\sum_{\gamma \in G} D^{\mu}_{i\mathfrak{m}}(\gamma) D^{\mu}_{\mathfrak{l}j}(\gamma^{-1}) = \lambda_{\mathfrak{m}\mathfrak{l}} \delta_{ij}. \tag{4.68}$$

The number $\lambda_{\mathfrak{m}\mathfrak{l}}$ can be evaluated from equation (4.68) by setting $i = j$ and summing over i. We obtain the relation

$$|G| \times \delta_{\mathfrak{m}\mathfrak{l}} = n_{\mu} \times \lambda_{\mathfrak{m}\mathfrak{l}}, \tag{4.69}$$

where n_{μ} is the dimension of the irreducible representation under consideration and $|G|$ is the order of the group. Putting together the above results we obtain the following orthogonality relations for the matrix elements of an irreducible representation of a finite group:

$$\sum_{\gamma \in G} D^{\mu}_{i\mathfrak{m}}(\gamma) D^{\mu}_{\mathfrak{l}j}(\gamma^{-1}) = \frac{|G|}{n_{\mu}} \times \delta_{\mathfrak{m}\mathfrak{l}} \delta_{ij}. \tag{4.70}$$

As one sees, the above result is just a consequence of Schur's second lemma. If we utilize also the first lemma, by considering two inequivalent irreducible representations μ and ν and going through the very same steps we arrive at

$$\sum_{\gamma \in G} D^{\mu}_{ij}(\gamma) D^{\nu}_{IJ}(\gamma^{-1}) = \frac{|G|}{n_{\mu}} \times \delta^{\mu\nu} \times \delta_{iJ} \delta_{jI}. \tag{4.71}$$

Hence, given an irreducible representation D^{μ} of dimension n_{μ} the matrix elements give rise to n_{μ}^2 vectors with $|G|$ components that are orthogonal to each other and to the n_{ν}^2 vectors provided by another irreducible representation. Since in a G-dimensional vector space the maximal number of orthogonal vectors is $|G|$, it follows that

$$\sum_{\mu} n_{\mu}^2 \leq |G|. \tag{4.72}$$

This is already a very strong result since it implies that the number of irreducible representations is finite and that their dimensionality is upper-bounded. Actually, refining our arguments we can show that in equation (4.72) the inequality sign $\leq$ can be substituted with the equality sign $=$, and this will provide an even stronger result. Provisionally let us denote by κ the number of inequivalent irreducible representations of the finite group G and let us denote by r the number of conjugacy classes of the same. We shall shortly prove that $\kappa = r$.

4.3.2 Characters

Let us now introduce the following definition.

Definition 4.3.5. Let $D : G \to \mathrm{Hom}(V, V)$ be a linear representation of a finite group G of dimension $n = \dim V$. Let r be the number of conjugacy classes $\mathscr{C}_i$ into which the whole group is split:

$$G = \bigcup_{i=1}^{r} \mathscr{C}_i; \quad \mathscr{C}_i \cap \mathscr{C}_j = \delta_{ij}\mathscr{C}_j \quad \forall \gamma, \tilde{\gamma} \in \mathscr{C}_i \quad \exists g \in G / \tilde{\gamma} = g\gamma g^{-1}. \tag{4.73}$$

By *character* we mean the following representation D of an r-dimensional vector:

$$\chi[D] = \{\mathrm{Tr}[D(\gamma_1)], \mathrm{Tr}[D(\gamma_2)], \ldots, \mathrm{Tr}[D(\gamma_r)]\}, \tag{4.74}$$

where

$$\gamma_i \in \mathscr{C}_i \tag{4.75}$$

is any set of representatives of the r conjugacy classes.

The above definition of characters makes sense because of the following obvious result.

Lemma 4.3.3. *In any linear representation $D : G \to \mathrm{Hom}(V, V)$ of a finite group the matrices $D(\gamma)$ and $D(\tilde{\gamma})$ representing two conjugate elements $\gamma = g\gamma g^{-1}$ have the same trace:* $\mathrm{Tr}[D(\gamma)] = \mathrm{Tr}[D(\tilde{\gamma})]$.

Proof of Lemma 4.3.3. Indeed, from the fundamental cyclic property of the trace we have

$$\begin{aligned} \mathrm{Tr}[D(\tilde{\gamma})] &= \mathrm{Tr}[D(g\gamma g^{-1})] = \mathrm{Tr}[D(g)D(\gamma)D(g^{-1})] \\ &= \mathrm{Tr}[D(\gamma)D(g^{-1})D(g)] = \mathrm{Tr}[D(\gamma)]. \end{aligned} \tag{4.76}$$

□

Keeping in mind the definition of characters, let us reconsider equation (4.71) and put $i = j$ and $I = J$. We obtain

$$\sum_{\gamma \in G} D^{\mu}_{ii}(\gamma) D^{\nu}_{JJ}(\gamma^{-1}) = \frac{|G|}{n_\mu} \times \delta^{\mu\nu} \times \delta_{iJ}. \tag{4.77}$$

Summing over all i and J in equation (4.77) we get

$$\sum_{\gamma \in G} \mathrm{Tr}[D^{\mu}(\gamma)]\, \mathrm{Tr}[D^{\nu}(\gamma^{-1})] = |G| \times \delta^{\mu\nu}. \tag{4.78}$$

Let us now assume that on top of being irreducible the representations $D^{\mu} : G \to \mathrm{Hom}(V, V)$ are unitary according to the following standard.

Definition 4.3.6. A linear representation $D : G \to \mathrm{Hom}(V, V)$ of a finite group G of dimension $n = \dim V$ is called *unitary* iff:

$$\gamma \in G : \quad D(\gamma^{-1}) = D^{\dagger}(\gamma). \tag{4.79}$$

The above definition requires a sesquilinear scalar product to be defined on the vector space V according to the definitions in Sections 2.6.2 and 3.5.6.

If the considered irreducible representations are all unitary, then equation (4.78) becomes

$$\sum_{\gamma \in G} \mathrm{Tr}[D^{\mu}(\gamma)]\, \mathrm{Tr}[D^{\nu\dagger}(\gamma)] = |G| \times \delta^{\mu\nu}, \tag{4.80}$$

$$\Downarrow$$

$$\sum_{i=1}^{r} g_i \bar{\chi}_i^{\nu} \chi_i^{\mu} = |G| \delta^{\nu\mu}, \tag{4.81}$$

where we have introduced the following standard notation:

$$g_i \equiv |\mathscr{C}_i| = \#\text{ of elements in the conj. class } i. \tag{4.82}$$

4.3.3 Decomposition of a representation into irreducible representations

Given any representation D we can decompose it into irreducible ones setting

$$\forall \gamma \in G : \quad D(\gamma) = \bigoplus_{\mu=1}^{\kappa} a_{\mu} D^{\mu}(\gamma), \tag{4.83}$$

where a_{ν} are necessarily all integers (the multiplicity of occurrence of the ν-th representation). Correspondingly, taking the traces of the matrices on the left and on the right of the relation (4.83), it follows that

$$\chi_i[D] = \sum_{\mu=1}^{\kappa} a_{\mu} \chi_i^{\mu}, \tag{4.84}$$

namely, the character of the representation D is decomposed in a linear combination with integer coefficients of the characters of the irreducible representations, which are called *simple characters.*

Utilizing the orthogonality relation (4.81) we immediately obtain a formula which expresses the multiplicity of each irreducible representation contained in a given D in terms of the character of such a representation:

$$a_{\mu} = \frac{1}{|G|} \sum_{i=1}^{r} g_i \bar{\chi}_i^{\mu} \chi[D]. \tag{4.85}$$

4.3.4 The regular representation

In order to complete our proof of the main theorem concerning group characters we have to introduce the so-called regular permutation. Let us label the elements of the finite group G as follows:

$$\underbrace{\gamma_1}_{=\mathbf{e}}, \gamma_2, \dots, \gamma_{g-1}, \gamma_g; \quad g = |G|, \tag{4.86}$$

where the first is the neutral element and all the others are listed in some arbitrarily chosen order. Multiplying the entire list on the left by any element γ_a reproduces the same list in a permuted order:

$$\gamma_a \cdot \{\gamma_1, \gamma_2, \dots, \gamma_{g-1}, \gamma_g\} = \{\gamma_{\pi_a(1)}, \gamma_{\pi_a(2)}, \dots, \gamma_{\pi_a(g-1)}, \gamma_{\pi_a(g)}\}. \tag{4.87}$$

The permutation π_a is represented by a $g \times g$ matrix as already specified in equation (3.7), namely,

$$D^R_{i\ell}(\gamma_a) = \delta_{i,\pi_a(\ell)}, \tag{4.88}$$

and the set of these matrices for $a = 1, \dots, g$ forms the *regular representation* of the group.

The character of the regular representation is very particular. The matrices $D^R(\gamma)$ have no non-vanishing entry on the diagonal except in the case of the identity element $\gamma = e$. This follows from the discussion of Section 4.2.1.1, which states that no permutation π_a leaves any symbol unchanged unless $a = 1$. From this simple observation it follows that the character of the regular permutation is the following one:

$$\chi[D^R] = \{g, 0, 0, \dots, 0\}. \tag{4.89}$$

Relying on (4.89) and applying equation (4.84) to the case of the regular representation, we find

$$a_\nu[D^R] = \frac{1}{g} \sum_{i=1}^{r} g_i \overline{\chi}_i^\nu \chi_i[D^R] = \overline{\chi}_1^\nu = n_\nu. \tag{4.90}$$

On the other hand, by the definition of decomposition of the regular representation into irreducible ones we must have

$$g = \sum_{\nu=1}^{\kappa} a_\nu[D^R] n_\nu. \tag{4.91}$$

Combining the two above results we conclude that

$$\sum_{\nu=1}^{\kappa} n_\nu^2 = g \equiv |G|. \tag{4.92}$$

This is the promised refinement of equation (4.72). It remains only to prove that $\kappa = r$, namely, that the number of irreducible inequivalent representations is equal to the number of conjugacy classes of the group.

The proof of this last statement, which we omit because its details are slightly boring, is provided for instance in [69]. It relies on showing the complementary orthogonality relations,

$$\sum_{\nu=1}^{\kappa} \overline{\chi}_i^{\nu} \chi_j^{\nu} = \frac{g_i}{g} \delta_{ij}, \tag{4.93}$$

from which it follows that $r \leq \kappa$, since in a space there cannot be a larger number of orthonormal vectors than the number of its dimensions. From equation (4.81) it follows instead that for the same reason $\kappa \leq r$. Hence, $r = \kappa$ and the characters form a square matrix.

4.4 Strategy to construct the irreducible representations of a solvable group

In general, the derivation of the irreps and the ensuing character table of a finite group G is a quite hard task. Yet a definite constructive algorithm can be devised if G is solvable and one can establish a chain of normal subgroups ending with an Abelian one, whose index is, at each step, a prime number q_i, namely, if we have the following situation:

$$\begin{aligned} &G = \mathrm{G}_{\mathrm{N_p}} \rhd \mathrm{G}_{\mathrm{N_{p-1}}} \rhd \cdots \rhd \mathrm{G}_{\mathrm{N_1}} \rhd \mathrm{G}_{\mathrm{N_0}} = \text{Abelian group}, \\ &\left| \frac{\mathrm{G}_{\mathrm{N_i}}}{\mathrm{G}_{\mathrm{N_{i-1}}}} \right| = \frac{\mathrm{N_i}}{\mathrm{N_{i-1}}} \equiv q_i = \text{prime integer number}. \end{aligned} \tag{4.94}$$

The algorithm for the construction of the irreducible representations is based on an inductive procedure that allows to derive the irreps of the group $\mathrm{G}_{\mathrm{N_i}}$ if we know those of the group $\mathrm{G}_{\mathrm{N_{i-1}}}$ and the index q_i is a prime number. The first step of the induction is immediately solved because any Abelian finite group is necessarily a direct product of cyclic groups $\mathbb{Z}_k$, whose irreps are all one-dimensional and obtained by assigning to their generator one of the k-th roots of unity. When one considers crystallographic groups in three dimensions the index q_i is always either 2 or 3. Hence, we sketch the inductive algorithms with particular reference to the two cases of $q = 2$ and $q = 3$.

4.4.1 The inductive algorithm for irreps

To simplify notation we denote $\mathcal{G} = \mathrm{G}_{\mathrm{N_i}}$ and $\mathcal{H} = \mathrm{G}_{\mathrm{N_{i-1}}}$. By hypothesis $\mathcal{H} \lhd \mathcal{G}$ is a normal subgroup. Furthermore, $q \equiv |\frac{\mathcal{G}}{\mathcal{H}}| = \textit{prime number}$ (in particular $q = 2$ or 3). Let us denote by $D_\alpha[\mathcal{H}, d_\alpha]$ the irreducible representations of the subgroup. The index α (with $\alpha = 1, \dots, r_H \equiv \#$ of conj. classes of $\mathcal{H}$) enumerates them. In each case d_α denotes the dimension of the corresponding carrying vector space or, in mathematical jargon, of the corresponding module.

The first step to be taken is to distribute the $\mathcal{H}$ irreps into conjugation classes with respect to the bigger group. Conjugation classes of irreps are defined as follows. First one observes that, given an irreducible representation $D_\alpha[\mathcal{H}, d_\alpha]$, for every $g \in \mathcal{G}$ we can create another irreducible representation $D_\alpha^{(g)}[\mathcal{H}, d_\alpha]$, named the conjugate of $D_\alpha[\mathcal{H}, d_\alpha]$ with respect to g. The new representation is as follows:

$$\forall h \in \mathcal{H}: \quad D_\alpha^{(g)}[\mathcal{H}, d_\alpha](h) = D_\alpha[\mathcal{H}, d_\alpha](g^{-1}hg). \tag{4.95}$$

That the one defined above is a homomorphism of $\mathcal{H}$ onto $\mathrm{GL}(d_\alpha, \mathbb{F})$ is obvious and, as a consequence, it is also obvious that the new representation has the same dimension as the first. Secondly, if $g = \tilde{h} \in \mathcal{H}$ is an element of the subgroup we get

$$D_\alpha^{(\tilde{h})}[\mathcal{H}, d_\alpha](h) = A^{-1} D_\alpha[\mathcal{H}, d_\alpha](h) A, \quad \text{where } A = D_\alpha[\mathcal{H}, d_\alpha](\tilde{h}), \tag{4.96}$$

so that conjugation amounts simply to a change of basis (a similarity transformation) inside the same representation. This does not alter the character vector and the new representation is equivalent to the old one. Hence, the only non-trivial conjugations to be considered are those with respect to representatives of the different equivalence classes in $\frac{\mathcal{G}}{\mathcal{H}}$. Let us denote by γ_i ($i = 0, \dots, q-1$) a set of representatives of such equivalence classes and define the orbit of each irrep $D_\alpha[\mathcal{H}, d_\alpha]$ as follows:

$$\mathrm{Orbit}_\alpha \equiv \{D_\alpha^{(\gamma_0)}[\mathcal{H}, d_\alpha], D_\alpha^{(\gamma_1)}[\mathcal{H}, d_\alpha], \dots, D_\alpha^{(\gamma_{q-1})}[\mathcal{H}, d_\alpha]\}. \tag{4.97}$$

Since the available irreducible representations are a finite set, every $D_\alpha^{(\gamma_i)}[\mathcal{H}, d_\alpha]$ necessarily is identified with one of the existing $D_\beta[\mathcal{H}, d_\beta]$. Furthermore, since conjugation preserves the dimension, it follows that $d_\alpha = d_\beta$. It follows that $\mathcal{H}$-irreps of the same dimensions d arrange themselves into $\mathcal{G}$-orbits:

$$\mathrm{Orbit}_\alpha[d] = \{D_{\alpha_1}[\mathcal{H}, d], D_{\alpha_2}[\mathcal{H}, d], \dots, D_{\alpha_q}[\mathcal{H}, d]\}, \tag{4.98}$$

and there are only two possibilities: either all $\alpha_i = \alpha$ are equal (self-conjugate representations) or they are all different (non-conjugate representations).

Once the irreps of $\mathcal{H}$ have been organized into conjugation orbits, we can proceed to promote them to irreps of the big group $\mathcal{G}$ according to the following scheme:

(A) Each self-conjugate $\mathcal{H}$-irrep $D_\alpha[\mathcal{H}, d]$ is uplifted to q distinct irreducible $\mathcal{G}$-representations of the same dimension d, namely, $D_{\alpha_i}[\mathcal{G}, d]$, where $i = 1, \dots, q$.

(B) From each orbit β of q distinct but conjugate $\mathcal{H}$-irreps $\{D_{\alpha_1}[\mathcal{H}, d], D_{\alpha_2}[\mathcal{H}, d], \dots, D_{\alpha_q}[\mathcal{H}, d]\}$ one extracts a single $(q \times d)$-dimensional $\mathcal{G}$-representation.

4.4.1.1 (A) Uplifting of self-conjugate representations

Let $D_\alpha[\mathcal{H}, d]$ be a self-conjugate irrep. If the index q of the normal subgroup is a prime number, this means that $\frac{\mathcal{G}}{\mathcal{H}} \simeq \mathbb{Z}_q$. In this case the representatives γ_j of the q equivalence classes that form the quotient group can be chosen in the following way:

$$\gamma_1 = \mathbf{e}, \quad \gamma_2 = g, \quad \gamma_3 = g^2, \dots, \gamma_q = g^{q-1}, \tag{4.99}$$

where $g \in \mathcal{G}, g \notin \mathcal{H}$, is a single group element satisfying $g^q = \mathbf{e}$. The key point in uplifting the representation $D_\alpha[\mathcal{H}, d]$ to the bigger group resides in the determination of a $d \times d$ matrix U that should satisfy the following constraints:

$$U^q = \mathbf{1}, \tag{4.100}$$

$$\forall h \in \mathcal{H} : \quad D_\alpha[\mathcal{H}, d](g^{-1}hg) = U^{-1} D_\alpha[\mathcal{H}, d](h) U. \tag{4.101}$$

These algebraic equations have exactly q distinct solutions $U_{[j]}$ and each of the solutions leads to one of the irreducible $\mathcal{G}$-representations induced by $D_\alpha[\mathcal{H}, d]$. Any element $\gamma \in \mathcal{G}$ can be written as $\gamma = g^p h$ with $p = 0, 1, \dots, q-1$ and $h \in \mathcal{H}$. Then it suffices to write

$$D_{\alpha_j}[\mathcal{G}, d](\gamma) = D_{\alpha_j}[\mathcal{G}, d](g^p h) = U_{[j]}^p D_\alpha[\mathcal{H}, d](h) \tag{4.102}$$

4.4.1.2 (B) Uplifting of non-self-conjugate representations

In the case of non-self-conjugate representations the induced representation of dimensions $q \times d$ is constructed relying once again on the possibility to write all group elements in the form $\gamma = g^p h$ with $p = 0, 1, \dots, q-1$ and $h \in \mathcal{H}$. Furthermore, chosen one representation $D_\alpha[\mathcal{H}, d]$ in the q-orbit (4.97), the other members of the orbit can be represented as $D_\alpha^{(g^j)}[\mathcal{H}, d_\alpha]$ with $j = 1, \dots, q-1$. In view of this one writes

$$\forall h \in \mathcal{H} : \quad D_\alpha[\mathcal{G}, d](h) = \left(\begin{array}{c|c|c|c|c} D_\alpha[\mathcal{H}, d](h) & 0 & 0 & \cdots & 0 \\ \hline 0 & D_\alpha^{(g)}[\mathcal{H}, d](h) & 0 & \cdots & 0 \\ \hline 0 & 0 & D_\alpha^{(g^2)}[\mathcal{H}, d](h) & \cdots & 0 \\ \hline \vdots & \vdots & \vdots & \vdots & \vdots \\ \hline 0 & 0 & \cdots & 0 & D_\alpha^{(g^{q-1})}[\mathcal{H}, d](h) \end{array}\right),$$

$$g : \quad D_\alpha[\mathcal{G}, d](g) = \left(\begin{array}{c|c|c|c|c} 0 & \mathbf{1} & 0 & \cdots & 0 \\ \hline 0 & 0 & \mathbf{1} & \cdots & 0 \\ \hline \vdots & \vdots & \vdots & \vdots & \vdots \\ \hline 0 & 0 & \cdots & 0 & \mathbf{1} \\ \hline \mathbf{1} & 0 & \cdots & 0 & 0 \end{array}\right),$$

$$\gamma = g^p h, \quad D_\alpha[\mathcal{G}, d](g) = \left(D_\alpha[\mathcal{G}, d](g)\right)^p D_\alpha[\mathcal{G}, d](h). \tag{4.103}$$

4.4.2 The octahedral group $O_{24} \sim S_4$ and its irreps

In Chapter 8, when discussing crystallographic lattices and their point groups, we will meet the octahedral group and derive its irreducible representations using its interpretation as a transformation group in three-dimensional Euclidean space. In the present section, as an application of the methods described in the previous section, we want to derive the irreducible representations of O_{24} simply from the knowledge of its multiplication table and from the fact that it is solvable. The MATHEMATICA NoteBook mentioned in Appendix A is devoted to the octahedral group.

Abstractly, the octahedral group $O_{24} \sim S_4$ is isomorphic to the symmetric group of permutations of four objects. It is defined by the following generators and relations:

$$\mathrm{A},\mathrm{B}: \quad \mathrm{A}^3 = \mathbf{e}; \quad \mathrm{B}^2 = \mathbf{e}; \quad (\mathrm{BA})^4 = \mathbf{e}. \tag{4.104}$$

Since O_{24} is a finite, discrete subgroup of the three-dimensional rotation group, any $\gamma \in O_{24} \subset SO(3)$ of its 24 elements can be uniquely identified by its action on the coordinates x, y, z, as displayed below:

e	1_1	=	$\{x,y,z\}$		4_1	=	$\{-x,-z,-y\}$
	2_1	=	$\{-y,-z,x\}$		4_2	=	$\{-x,z,y\}$
	2_2	=	$\{-y,z,-x\}$	C_2	4_3	=	$\{-y,-x,-z\}$
	2_3	=	$\{-z,-x,y\}$		4_4	=	$\{-z,-y,-x\}$
C_3	2_4	=	$\{-z,x,-y\}$		4_5	=	$\{z,-y,x\}$
	2_5	=	$\{z,-x,-y\}$		4_6	=	$\{y,x,-z\}$
	2_6	=	$\{z,x,y\}$		5_1	=	$\{-y,x,z\}$
	2_7	=	$\{y,-z,-x\}$		5_2	=	$\{-z,y,x\}$
	2_8	=	$\{y,z,x\}$	C_4	5_3	=	$\{z,y,-x\}$
	3_1	=	$\{-x,-y,z\}$		5_4	=	$\{y,-x,z\}$
C_4^2	3_2	=	$\{-x,y,-z\}$		5_5	=	$\{x,-z,y\}$
	3_3	=	$\{x,-y,-z\}$		5_6	=	$\{x,z,-y\}$

(4.105)

As one sees from the above list, the 24 elements are distributed into five conjugacy classes mentioned in the first column of the table, according to a nomenclature which is standard in the chemical literature on crystallography. The relation between the abstract and concrete presentations of the octahedral group is obtained by identifying in the list (4.105) the generators A and B mentioned in equation (4.104). Explicitly, we have

$$\mathrm{A} = 2_8 = \begin{pmatrix} 0 & 1 & 0 \\ 0 & 0 & 1 \\ 1 & 0 & 0 \end{pmatrix}; \quad \mathrm{B} = 4_6 = \begin{pmatrix} 0 & 1 & 0 \\ 1 & 0 & 0 \\ 0 & 0 & -1 \end{pmatrix}. \tag{4.106}$$

All other elements are reconstructed from the above two using the following multiplication table of the group:

	1_1	2_1	2_2	2_3	2_4	2_5	2_6	2_7	2_8	3_1	3_2	3_3	4_1	4_2	4_3	4_4	4_5	4_6	5_1	5_2	5_3	5_4	5_5	5_6
1_1	1_1	2_1	2_2	2_3	2_4	2_5	2_6	2_7	2_8	3_1	3_2	3_3	4_1	4_2	4_3	4_4	4_5	4_6	5_1	5_2	5_3	5_4	5_5	5_6
2_1	2_1	2_5	2_4	3_3	3_2	1_1	3_1	2_6	2_3	2_7	2_2	2_8	5_3	4_4	5_6	4_6	5_4	4_2	4_1	4_3	5_1	5_5	4_5	5_2
2_2	2_2	2_6	2_3	1_1	3_1	3_3	3_2	2_5	2_4	2_8	2_1	2_7	4_5	5_2	5_5	5_4	4_6	4_1	4_2	5_1	4_3	5_6	5_3	4_4
2_3	2_3	3_2	1_1	2_2	2_8	2_7	2_1	3_3	3_1	2_4	2_6	2_5	4_6	5_1	5_3	5_6	4_1	4_5	5_2	4_2	5_5	4_4	4_3	5_4
2_4	2_4	3_1	3_3	2_1	2_7	2_8	2_2	1_1	3_2	2_3	2_5	2_6	5_4	4_3	4_5	5_5	4_2	5_3	4_4	4_1	5_6	5_2	5_1	4_6
2_5	2_5	1_1	3_2	2_8	2_2	2_1	2_7	3_1	3_3	2_6	2_4	2_3	5_1	4_6	5_2	4_2	5_5	4_4	5_3	5_6	4_1	4_5	5_4	4_3
2_6	2_6	3_3	3_1	2_7	2_1	2_2	2_8	3_2	1_1	2_5	2_3	2_4	4_3	5_4	4_4	4_1	5_6	5_2	4_5	5_5	4_2	5_3	4_6	5_1
2_7	2_7	2_3	2_6	3_1	1_1	3_2	3_3	2_4	2_5	2_1	2_8	2_2	5_2	4_5	4_2	5_1	4_3	5_6	5_5	5_4	4_6	4_1	4_4	5_3
2_8	2_8	2_4	2_5	3_2	3_3	3_1	1_1	2_3	2_6	2_2	2_7	2_1	4_4	5_3	4_1	4_3	5_1	5_5	5_6	4_6	5_4	4_2	5_2	4_5
3_1	3_1	2_8	2_7	2_6	2_5	2_4	2_3	2_2	2_1	1_1	3_3	3_2	5_6	5_5	4_6	5_3	5_2	4_3	5_4	4_5	4_4	5_1	4_2	4_1
3_2	3_2	2_7	2_8	2_5	2_6	2_3	2_4	2_1	2_2	3_3	1_1	3_1	5_5	5_6	5_4	4_5	4_4	5_1	4_6	5_3	5_2	4_3	4_1	4_2
3_3	3_3	2_2	2_1	2_4	2_3	2_6	2_5	2_8	2_7	3_2	3_1	1_1	4_2	4_1	5_1	5_2	5_3	5_4	4_3	4_4	4_5	4_6	5_6	5_5
4_1	4_1	5_4	4_6	4_5	5_3	5_2	4_4	5_1	4_3	5_5	5_6	4_2	1_1	3_3	2_8	2_6	2_3	2_2	2_7	2_5	2_4	2_1	3_1	3_2
4_2	4_2	4_6	5_4	5_3	4_5	4_4	5_2	4_3	5_1	5_6	5_5	4_1	3_3	1_1	2_7	2_5	2_4	2_1	2_8	2_6	2_3	2_2	3_2	3_1
4_3	4_3	5_3	5_2	5_6	4_2	5_5	4_1	4_5	4_4	4_6	5_1	5_4	2_6	2_4	1_1	2_8	2_7	3_1	3_2	2_2	2_1	3_3	2_5	2_3
4_4	4_4	4_2	5_5	5_1	5_4	4_6	4_3	5_6	4_1	5_2	4_5	5_3	2_8	2_1	2_6	1_1	3_2	2_5	2_3	3_1	3_3	2_4	2_2	2_7
4_5	4_5	5_6	4_1	4_6	4_3	5_1	5_4	4_2	5_5	5_3	4_4	5_2	2_2	2_7	2_4	3_2	1_1	2_3	2_5	3_3	3_1	2_6	2_8	2_1
4_6	4_6	4_4	4_5	4_1	5_5	4_2	5_6	5_2	5_3	4_3	5_4	5_1	2_3	2_5	3_1	2_1	2_2	1_1	3_3	2_7	2_8	3_2	2_4	2_6
5_1	5_1	4_5	4_4	5_5	4_1	5_6	4_2	5_3	5_2	5_4	4_3	4_6	2_5	2_3	3_3	2_7	2_8	3_2	3_1	2_1	2_2	1_1	2_6	2_4
5_2	5_2	4_1	5_6	4_3	4_6	5_4	5_1	5_5	4_2	4_4	5_3	4_5	2_7	2_2	2_5	3_3	3_1	2_6	2_4	3_2	1_1	2_3	2_1	2_8
5_3	5_3	5_5	4_2	5_4	5_1	4_3	4_6	4_1	5_6	4_5	5_2	4_4	2_1	2_8	2_3	3_1	3_3	2_4	2_6	1_1	3_2	2_5	2_7	2_2
5_4	5_4	5_2	5_3	4_2	5_6	4_1	5_5	4_4	4_5	5_1	4_6	4_3	2_4	2_6	3_2	2_2	2_1	3_3	1_1	2_8	2_7	3_1	2_3	2_5
5_5	5_5	4_3	5_1	4_4	5_2	5_3	4_5	4_6	5_4	4_1	4_2	5_6	3_2	3_1	2_2	2_4	2_5	2_8	2_1	2_3	2_6	2_7	3_3	1_1
5_6	5_6	5_1	4_3	5_2	4_4	4_5	5_3	5_4	4_6	4_2	4_1	5_5	3_1	3_2	2_1	2_3	2_6	2_7	2_2	2_4	2_5	2_8	1_1	3_3

(4.107)

This observation is important in relation with representation theory. Any linear representation of the group is uniquely specified by giving the matrix representation of the two generators $A = 2_8$ and $B = 4_6$.

4.4.2.1 The solvable structure of O_{24}

The group O_{24} is solvable since the following chain of normal subgroups exists:

$$O_{24} \triangleright N_{12} \triangleright N_4, \tag{4.108}$$

where the mentioned normal subgroups are given by the following lists of elements:

$$N_{12} \equiv \{1_1, 2_1, 2_2, \dots, 2_8, 3_1, 3_2, 3_3\}, \tag{4.109}$$

$$N_4 \equiv \{1_1, 3_1, 3_2, 3_3\}. \tag{4.110}$$

The group N_4 is Abelian and we have

$$N_4 \sim \mathbb{Z}_2 \times \mathbb{Z}_2 \tag{4.111}$$

since all of its elements are of order 2. On the other hand, the indices of the two normal subgroups are respectively $q = 2$ and $q = 3$ and we have

$$\frac{O_{24}}{N_{12}} \sim \mathbb{Z}_2; \quad \frac{N_{12}}{N_4} \sim \mathbb{Z}_3, \tag{4.112}$$

so that we can apply the strategy outlined in the previous section for the construction of the irreducible representations.

4.4.2.2 Irreps of N_4 (Klein group)

Since N_4 is Abelian, its irreducible representations are all one-dimensional and they are immediately determined from its $\mathbb{Z}_2 \times \mathbb{Z}_2$ structure. Denoting $3_1 = a$, $3_2 = b$, we obtain $3_3 = ab$ and the multiplication table of this normal subgroup is

	1	a	b	ab
1	1	a	b	ab
a	a	1	ab	b
b	b	ab	1	a
ab	ab	b	a	1

(4.113)

The representations are obtained by assigning to a, b the values ± 1 in all possible ways. We have the following representation table, which is also the character table:

	1	3_1	3_2	3_3
Δ_1	1	1	1	1
Δ_2	1	−1	1	−1
Δ_3	1	1	−1	−1
Δ_4	1	−1	−1	1

(4.114)

4.4.2.3 The irreps of N_{12} by induction

Next we construct the irreducible representation of the subgroup N_{12} by induction from those of its normal subgroup N_4. First we analyze the conjugacy class structure of N_{12}. Looking at the multiplication table (4.107) we determine four conjugacy classes:

$$\begin{aligned} \mathscr{C}_1 &= \{1\}, \\ \mathscr{C}_2 &= \{3_1, 3_2, 3_3\}, \\ \mathscr{C}_3 &= \{2_1, 2_2, 2_7, 2_8\}, \\ \mathscr{C}_4 &= \{2_3, 2_4, 2_5, 2_6\}. \end{aligned} \tag{4.115}$$

Hence, we expect four irreducible representations. In order to determine them by induction we consider also the three equivalence classes in the quotient N_{12}/N_4. We find

$$\begin{aligned}
\mathfrak{C}_1 &= \mathscr{C}_1 \cup \mathscr{C}_2 \sim N_4, \\
\mathfrak{C}_2 &= \mathscr{C}_3 \sim 2_8 \cdot N_4, \\
\mathfrak{C}_2 &= \mathscr{C}_4 \sim 2_6 \cdot N_4 = (2_8)^2 \cdot N_4.
\end{aligned} \tag{4.116}$$

According to the strategy outlined in Section 4.4 we analyze the orbits of the four irreducible representations of N_4 with respect to the action of N_{12}. According to equation (4.116) it suffices to consider conjugation with respect to 2_8 and its powers. We find two orbits:

$$\mathscr{O}_1 = \{\Delta_1, \Delta_1, \Delta_1\}; \quad \mathscr{O}_2 = \{\Delta_2, \Delta_3, \Delta_4\}. \tag{4.117}$$

Hence there is one self-conjugate representation (the identity representation) of the normal subgroup N_4 which gives rise to three inequivalent one-dimensional representations ($\Pi_{1,2,3}$) of the big group N_{12} and there is an orbit of three conjugate representations of the normal subgroup which gives rise to a three-dimensional representation (Π_4) of the big group. The summation rule is easily verified: $12 = 1 + 1 + 1 + 3^2$.

Let us construct these representations according to the algorithm outlined in Section 4.4. We begin with the one-dimensional representations.

4.4.2.4 Representation Π_1 = identity

In this representation all 12 elements of the group are represented by 1.

4.4.2.5 Representation Π_2

In this representation the elements of the group in $\mathfrak{C}_1$ are represented by 1, those in $\mathfrak{C}_2$ are represented by $\exp[\frac{2\pi}{3}\mathrm{i}]$, and those in $\mathfrak{C}_3$ are represented by $\exp[\frac{4\pi}{3}i]$.

4.4.2.6 Representation Π_3

In this representation the elements of the group in $\mathfrak{C}_1$ are represented by 1, those in $\mathfrak{C}_2$ are represented by $\exp[\frac{4\pi}{3}\mathrm{i}]$, and those in $\mathfrak{C}_3$ are represented by $\exp[\frac{2\pi}{3}i]$.

4.4.2.7 The three-dimensional representation Π_4

In order to construct this representation we first of all write the block-diagonal matrices representing the elements of the normal subgroup, namely,

$$
\begin{aligned}
1_1 &= \left(\begin{array}{c|c|c} 1 & 0 & 0 \\ \hline 0 & 1 & 0 \\ \hline 0 & 0 & 1 \end{array}\right), \\
3_1 &= \left(\begin{array}{c|c|c} -1 & 0 & 0 \\ \hline 0 & 1 & 0 \\ \hline 0 & 0 & -1 \end{array}\right), \\
3_2 &= \left(\begin{array}{c|c|c} 1 & 0 & 0 \\ \hline 0 & -1 & 0 \\ \hline 0 & 0 & -1 \end{array}\right), \\
3_3 &= \left(\begin{array}{c|c|c} -1 & 0 & 0 \\ \hline 0 & -1 & 0 \\ \hline 0 & 0 & 1 \end{array}\right).
\end{aligned}
\tag{4.118}
$$

Furthermore, we introduce the representation of the element 2_8 according to the recipe provided in equation (4.103):

$$
2_8 = \left(\begin{array}{c|c|c} 0 & 1 & 0 \\ \hline 0 & 0 & 1 \\ \hline 1 & 0 & 0 \end{array}\right). \tag{4.119}
$$

All the other elements of the order 12 group N_{12} are obtained by multiplication of the above five matrices according to the multiplication table, schematically summarized in the structure of the equivalence classes in (4.116).

4.4.2.8 The irreps of the octahedral group

As already anticipated, the irreducible representations of the octahedral group will be directly constructed in Section 8.2.2. What we did above should have sufficiently illustrated the method of construction by induction which is available when the group is solvable. We briefly outline the next step, namely, the construction of the irreps of O_{24} from those of N_{12} if we were to do it by induction instead that directly as we will do.

The group O_{24} has five conjugacy classes, so we expect five irreducible representations. Since there is only one three-dimensional representation of N_{12}, it is necessarily self-conjugate. As the index of N_{12} in O_{24} is 2, it follows that the self-conjugate three-dimensional representation gives rise to two three-dimensional irreps of O_{24}; let us denote them D_4 and D_5, respectively. The identity representation is also self-conjugate for any group. Hence, by the same argument as before, there must be two one-dimensional representations D_1 and D_2, of which the first is necessarily the identity representation. Only one irrep remains at our disposal, D_3 ($\dim D_3 = x$), and this is necessarily two-dimensional ($x = 2$), since, by numerology, we must have

$$
1 + 1 + x^2 + 3^2 + 3^2 = 24. \tag{4.120}
$$

Indeed, we can verify that (Π_2, Π_3) form an orbit of conjugate representation and they are combined together in the unique two-dimensional representation D_3 of the octahedral group.

This discussion clarifies how the method of induction allows to determine all the irreducible representations in the case of solvable groups.

4.5 Bibliographical note

The elements of finite group theory contained in this chapter are covered in different combinations in several textbooks and monographs. Our treatment of representation theory closely follows the classical exposition of Hamermesh [69] and, in some instances, the mathematical textbook by James and Liebeck [70], but is modernized in terms of both notation and presentation. Furthermore, the induction method for the construction of solvable group representations presented here was systematized by one of us in collaboration with Alexander Sorin in a paper on Arnold–Beltrami flows [71].

Suggested further references are the same as those already mentioned in Section 3.6.

5 Finite subgroups of SO(3): the ADE classification

χαλεπὰ, τὰ καλά
Nothing beautiful without struggle

Plato

5.1 Introduction

The theory of symmetric and, more generally, homogeneous spaces, interlaced with Lie algebra theory, is ubiquitous in many scientific contexts related to dynamical systems and non-linear differential equations and is applied in natural sciences, engineering, socioeconomic sciences, and recently even artificial intelligence and deep learning (see for instance the review [29]). In the case of theoretical physics, the prominent fields of supergravity and string theory bring to the forefront not only symmetric spaces associated with classical groups, but also those associated with exceptional ones, which turn out to be of utmost relevance in that context (see for instance the review book [72]). It is predicted that the exceptional instances of symmetric spaces will soon play relevant roles also in other applications.

At various stages of these geometric constructions also the finite groups play an important role, including those that are crystallographic in certain dimensions. This is not very surprising since there exists a profound relation between the classification of simple, simply laced, complex Lie algebras and the classification of finite subgroups of the three-dimensional rotation group, the so-called ADE classification.[1]

This ADE correspondence, which has been known for a long time, finds a deeper and fertile interpretation in the McKay correspondence, which is crucial for the Kronheimer construction of gravitational instantons as hyper-Kähler quotients and in several different fields of algebraic geometry. The McKay correspondence admits a generalization to finite subgroups $\Gamma \subset \mathrm{SU(n)}$, in particular for the case $n = 3$, which plays a significant role in the context of the $\mathrm{AdS}_4/\mathrm{CFT}_3$ correspondence [73, 74, 75].

We introduce here the discussion of the joint ADE classification of binary extensions $\Gamma_b \subset \mathrm{SU}(2)$ of finite subgroups $\Gamma \subset \mathrm{SO}(3)$ and simple, simply laced, complex Lie algebras.

1 The ADE classification system, which is frequently utilized in the physical–mathematical literature, is based on a diophantine inequality that we present in the sequel of the present chapter. It encompasses in just one scheme the classification of several different types of mathematical objects:

1. The finite rotation groups.
2. The simple, simply laced, complex Lie algebras.
3. The locally Euclidean gravitational instantons.
4. The singularities $\mathbb{C}^2/\Gamma$.
5. The modular invariant partition functions of 2D conformal field theories.

https://doi.org/10.1515/9783111201535-005

After obtaining the classification of these finite subgroups of SU(2) we will use them as main examples in Chapter 8, where we illustrate the theorems and the constructions of finite group theory discussed in the previous chapter. We do this for a quite precise reason. Among the finite subgroups of SO(3) we find all the possible *point groups* of crystallographic lattices, and in this way we come in touch with some of the basic applications of discrete groups to *chemistry*, *crystallography*, and *molecular physics*. In that context the issue of *space groups* and tessellations will be addressed in general mathematical terms related to the issue of group extensions.

5.2 ADE classification of the finite subgroups of SU(2)

Let us start by considering the homomorphism

$$\omega : \mathrm{SU}(2) \longrightarrow \mathrm{SO}(3) \tag{5.1}$$

between the group SU(2) of unitary 2×2 matrices, each of which can be written as

$$\mathrm{SU}(2) \ni \mathcal{U} = \begin{pmatrix} \alpha & i\beta \\ \overline{i\beta} & \overline{\alpha} \end{pmatrix} \tag{5.2}$$

in terms of two complex numbers α, β satisfying the constraint

$$|\alpha|^2 + |\beta|^2 = 1, \tag{5.3}$$

and the group SO(3) of 3×3 of orthogonal matrices with unit determinant

$$\mathcal{O} \in \mathrm{SO}(3) \Leftrightarrow \mathcal{O}^T \mathcal{O} = \mathbf{1} \quad \text{and} \quad \det \mathcal{O} = 1. \tag{5.4}$$

The homomorphism ω can be explicitly constructed utilizing the so-called triplet σ^x of Hermitian Pauli matrices:

$$\sigma^1 = \begin{pmatrix} 0 & 1 \\ 1 & 0 \end{pmatrix}; \quad \sigma^2 = \begin{pmatrix} 0 & -i \\ i & 0 \end{pmatrix}; \quad \sigma^3 = \begin{pmatrix} 1 & 0 \\ 0 & -1 \end{pmatrix}. \tag{5.5}$$

Using the above we can define

$$\mathcal{H} = \sum_{x=1}^{3} h_x \sigma^x, \tag{5.6}$$

where h_x is a 3-vector with real components. The matrix $\mathcal{H} = \mathcal{H}^\dagger$ is Hermitian by construction and we have

$$\frac{1}{2} \mathrm{Tr}[\mathcal{H}^2] = \sum_{x=1}^{3} h_x^2. \tag{5.7}$$

Consider next the following matrix transformed by means of an SU(2) element:

$$\begin{aligned} \tilde{\mathscr{H}} &= \mathscr{U}^\dagger \mathscr{H} \mathscr{U} = \tilde{h}_x \sigma^x, \\ \tilde{h}_x &= \mathscr{O}_x{}^y h_y. \end{aligned} \tag{5.8}$$

The first line of equation (5.8) can be written since the Pauli matrices form a complete basis for the space of 2×2 Hermitian traceless matrices. The second line can be written since the matrix $\tilde{\mathscr{H}}$ depends linearly on the matrix $\mathscr{H}$. Next we observe that because of its definition the matrix $\tilde{\mathscr{H}}$ has the following property:

$$\frac{1}{2}\operatorname{Tr}[\tilde{\mathscr{H}}^2] = \sum_{x=1}^{3} \tilde{h}_x^2 = \sum_{x=1}^{3} h_x^2. \tag{5.9}$$

This implies that the matrix $\mathscr{O}_x{}^y$ is orthogonal and by definition it is the image of $\mathscr{U}$ through the homomorphism ω. We can write an explicit formula for the matrix elements $\mathscr{O}_x{}^y$ in terms of $\mathscr{U}$:

$$\forall \mathscr{U} \in \mathrm{SU}(2) : \omega[\mathscr{U}] = \mathscr{O} \in \mathrm{SO}(3) / \mathscr{O}_x{}^y = \frac{1}{2}\operatorname{Tr}[\mathscr{U}^\dagger \sigma_x \mathscr{U} \sigma^y], \tag{5.10}$$

which follows from the trace orthogonality of the Pauli matrices $\frac{1}{2}\operatorname{Tr}[\sigma^y \sigma_x] = \delta_x^y$.

We named the map defined above a homomorphism rather than an isomorphism since it has a non-trivial kernel of order two. Indeed, the following two SU(2) matrices constitute the kernel of ω since they are both mapped into the identity element of SO(3):

$$\begin{aligned} \ker \omega &= \left\{ \mathbf{e} = \begin{pmatrix} 1 & 0 \\ 0 & 1 \end{pmatrix}, \mathscr{Z} = \begin{pmatrix} -1 & 0 \\ 0 & -1 \end{pmatrix} \right\}, \\ \mathbf{1} &= \omega[\mathbf{e}] = \omega[\mathscr{Z}]. \end{aligned} \tag{5.11}$$

We will now obtain the classification of all finite subgroups of SU(2), which we collectively name G_{2n}^b, denoting by $2n$ their necessarily even order. Through the isomorphism ω each of them maps into a finite subgroup $\mathrm{G}_\mathrm{n} \subset \mathrm{SO}(3)$, whose order is just n because of the two-dimensional kernel mentioned above:

$$\omega[\mathrm{G}_{2n}^b] = \mathrm{G}_\mathrm{n}. \tag{5.12}$$

The groups G_{2n}^b are named the binary extensions of the finite rotation groups G_n.

5.2.1 The argument leading to the Diophantine equation

We begin by considering one-parameter subgroups of SO(3). These are singled out by a rotation axis, namely, by a point on the 2-sphere $\mathbb{S}^2$ (see Figure 5.1). Explicitly, let us consider a solution (ℓ, m, n) to the sphere equation

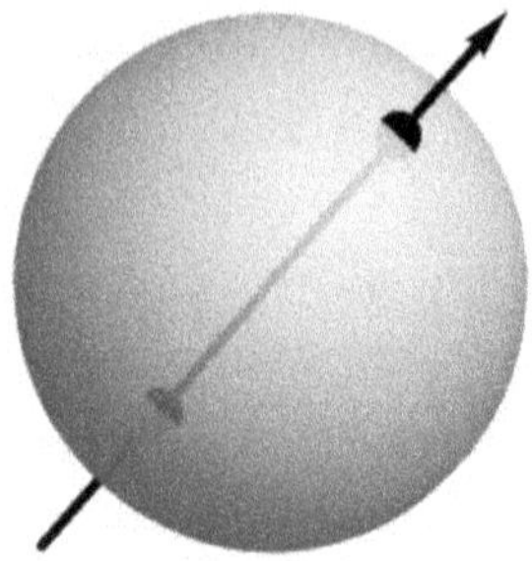

Figure 5.1: Every element of the rotation group $\mathcal{O}_{(\ell,m,n)} \in \mathrm{SO}(3)$ corresponds to a rotation around some axis $\mathbf{a} = \{\ell, m, n\}$. On the surface of the 2-sphere $\mathbb{S}^2$ this rotation has two fixed points, a North Pole and a South Pole, that do not rotate around any other point. The rotation $\mathcal{O}_{(\ell,m,n)}$ is the image, under the homomorphism ω of either one of the 2×2 matrices $\mathcal{U}^{\pm}_{\ell,m,n}$ that, acting on the space $\mathbb{C}^2$, admit two eigenvectors $\mathbf{z}_1$ and $\mathbf{z}_2$. The one-dimensional complex spaces $p_{1,2} \equiv \lambda_{1,2}\mathbf{z}_{1,2}$ are called the two poles of the unitary rotation.

$$\ell^2 + m^2 + n^2 = 1. \tag{5.13}$$

The triplet of real numbers (ℓ, m, n) parameterizes the direction of a possible rotation angle. The generator of infinitesimal rotations around such an axis is given by the following matrix:

$$A_{\ell,m,n} = \begin{pmatrix} 0 & -n & m \\ n & 0 & -\ell \\ -m & \ell & 0 \end{pmatrix} = -A^T_{\ell,m,n}, \tag{5.14}$$

which being antisymmetric belongs to the SO(3) Lie algebra. The matrix A has the property that $A^3 = -A$ and explicitly we have

$$A^2_{\ell,m,n} = \begin{pmatrix} -1+\ell^2 & \ell m & \ell n \\ \ell m & -1+m^2 & mn \\ \ell n & mn & -1+n^2 \end{pmatrix}. \tag{5.15}$$

Hence a finite element of the group SO(3) corresponding to a rotation of an angle θ around this axis is given by

$$\mathcal{O}_{(\ell,m,n)} = \exp[\theta A_{\ell,m,n}] = \mathbf{1} + \sin\theta A_{\ell,m,n} + (1-\cos\theta)A^2_{\ell,m,n}. \tag{5.16}$$

Setting

$$\lambda = \ell \sin\frac{\theta}{2}; \quad \mu = m \sin\frac{\theta}{2}; \quad \nu = n \sin\frac{\theta}{2}; \quad \rho = \cos\frac{\theta}{2}, \tag{5.17}$$

the corresponding SU(2) finite group elements realizing the double covering are

$$\mathscr{U}^{\pm}_{\ell,m,n} = \pm \begin{pmatrix} \rho + \mathrm{i}\nu & \mu - \mathrm{i}\lambda \\ -\mu - \mathrm{i}\lambda & \rho - \mathrm{i}\nu \end{pmatrix}, \tag{5.18}$$

namely, we have

$$\omega[\mathscr{U}^{\pm}_{\ell,m,n}] = \mathscr{O}_{(\ell,m,n)}. \tag{5.19}$$

We can now consider the argument that leads to the ADE classification of the finite subgroups of SU(2). Let us consider the action of the SU(2) matrices on $\mathbb{C}^2$. A generic $\mathscr{U} \in \mathrm{SU}(2)$ acts on a $\mathbb{C}^2$-vector $\mathbf{z} = \binom{z_1}{z_2}$ by usual matrix multiplication $\mathscr{U}\mathbf{z}$. Each element $\mathscr{U} \in \mathrm{SU}(2)$ has two eigenvectors $\mathbf{z}_1$ and $\mathbf{z}_2$ such that

$$\begin{aligned} \mathscr{U}\mathbf{z}_1 &= \exp[\mathrm{i}\theta]\mathbf{z}_1, \\ \mathscr{U}\mathbf{z}_2 &= \exp[-\mathrm{i}\theta]\mathbf{z}_2, \end{aligned} \tag{5.20}$$

where θ is some (half-)rotation angle. Namely, for each $\mathscr{U} \in \mathrm{SU}(2)$ we can find an orthogonal basis where $\mathscr{U}$ is diagonal and given by

$$\mathscr{U} = \begin{pmatrix} \exp[\mathrm{i}\theta] & 0 \\ 0 & \exp[-\mathrm{i}\theta] \end{pmatrix} \tag{5.21}$$

for some angle θ. Then let us consider the rays $\{\lambda\mathbf{z}_1\}$ and $\{\mu\mathbf{z}_2\}$, where $\lambda, \mu \in \mathbb{C}$ are arbitrary complex numbers. Since $\mathbf{z}_1 \cdot \mathbf{z}_2 = \mathbf{z}_1^{\dagger}\mathbf{z}_2 = 0$ it follows that each element of SU(2) singles out two rays, hereafter named *poles*, that are determined one from the other by the orthogonality relation. This concept of pole is a key tool in the argument leading to the classification of finite rotation groups.

Let $H \subset \mathrm{SO}(3)$ be a finite, discrete subgroup of the rotation group and let $\hat{H} \subset \mathrm{SU}(2)$ be its preimage in SU(2) with respect to the homomorphism ω. Then the order of H is some positive integer number:

$$|H| = n \in \mathbb{N}. \tag{5.22}$$

The total number of poles associated with H is

$$\#\text{ of poles} = 2n - 2, \tag{5.23}$$

since $n - 1$ is the number of elements in H that are different from the identity. Let us then adopt the notation

$$p_i \equiv \{\lambda\mathbf{z}_i\} \tag{5.24}$$

for the pole or ray singled out by the eigenvector $\mathbf{z}_i$. We say that two poles are equivalent if there exists an element of the group H that maps one into the other:

$$p_i \sim p_j \quad \text{iff} \quad \exists\gamma \in H/\gamma p_i = p_j. \tag{5.25}$$

Let us distribute the poles p_i into orbits under the action of the group H,

$$\mathcal{Q}_\alpha = \{p_1^\alpha, \dots, p_{m_\alpha}^\alpha\}; \quad \alpha = 1, \dots, r, \tag{5.26}$$

and name m_α the cardinality of the orbit class $\mathcal{Q}_\alpha$, namely, the number of poles it contains. Hence we have assumed that there are r orbits and that each orbit $\mathcal{Q}_\alpha$ contains m_α elements.

Each pole $p \in \mathcal{Q}_\alpha$ has a stability subgroup $K_p \subset H$,

$$\forall h \in K_p : \quad hp = p, \tag{5.27}$$

that is finite, Abelian, and cyclic of order k_α. Indeed, it must be finite since it is a subgroup of a finite group, it must be Abelian since in the basis $\mathbf{z}_1, \mathbf{z}_2$ the SU(2) matrices that preserve the poles $\lambda \mathbf{z}_1$ and $\mu \mathbf{z}_2$ are of the form (5.21), and therefore it is cyclic of some order. The H group can be decomposed into cosets according to the subgroup K_p:

$$H = K_p \cup v_1 K_p \cup \dots \cup v_{m_\alpha} K_p, \quad m_\alpha \in \mathbb{N}. \tag{5.28}$$

Consider now an element $x_i \in v_i K_p$ belonging to one of the cosets and define the group conjugate to K_p through x_i:

$$K_{(xp)_i} = x_i K_p x_i^{-1}. \tag{5.29}$$

Each element $h \in K_{(xp)_i}$ admits a pole p_x,

$$hp_x = p_x, \tag{5.30}$$

that is given by

$$p_x = x_i p, \tag{5.31}$$

since

$$hp_x = xh_p x x^{-1} p = xh_p p = xp = p_x. \tag{5.32}$$

Hence the set of poles $\{p, v_1 p, v_2 p, \dots v_{m_\alpha} p\}$ are equivalent forming an orbit. Each of them has a stability group K_{p_i} conjugate to K_p, which implies that all K_{p_i} are finite of the same order:

$$\forall v_i p \quad |K_{p_i}| = k_\alpha. \tag{5.33}$$

By this token we have proven that in each orbit $\mathcal{Q}_\alpha$ the stability subgroups of each element are isomorphic and cyclic of the same order k_α, which is a property of the orbit. Hence, we must have

$$\forall \mathcal{Q}_\alpha; \quad k_\alpha m_\alpha = n. \tag{5.34}$$

The total number of poles we have in the orbit $\mathcal{Q}_\alpha$ (counting coincidences) is

$$\#\text{ of poles in the orbit } \mathcal{Q}_\alpha = m_\alpha(k_\alpha - 1), \tag{5.35}$$

since the number of elements in K_p different from the identity is $k_\alpha - 1$. Hence, we find

$$2n - 2 = \sum_{\alpha=1}^{r} m_\alpha(k_\alpha - 1). \tag{5.36}$$

Dividing by n we obtain

$$2\left(1 - \frac{1}{n}\right) = \sum_{\alpha=1}^{r}\left(1 - \frac{1}{k_\alpha}\right). \tag{5.37}$$

We consider next the possible solutions to the Diophantine equation (5.37), and to this effect we rewrite it as follows:

$$r + \frac{2}{n} - 2 = \sum_{\alpha=1}^{r} \frac{1}{k_\alpha}. \tag{5.38}$$

We observe that $k_\alpha \geq 2$. Indeed, each pole admits at least two group elements that keep it fixed, the identity and the non-trivial group element that defines it by diagonalization. Hence, we have the bound

$$r + \frac{2}{n} - 2 \leq \frac{r}{2}, \tag{5.39}$$

which implies

$$r \leq 4 - \frac{4}{n} \quad \Rightarrow \quad r = 1, 2, 3. \tag{5.40}$$

On the other hand, we also have $k_\alpha \leq n$ so that

$$r + \frac{2}{n} - 2 \geq \frac{r}{n} \quad \Rightarrow \quad r\left(1 - \frac{1}{n}\right) \geq 2\left(1 - \frac{1}{n}\right) \quad \Rightarrow \quad r \geq 2. \tag{5.41}$$

Therefore, there are only two possible cases:

$$r = 2 \quad \text{or} \quad r = 3. \tag{5.42}$$

Let us now consider the solutions of the Diophantine equation (5.39) and identify the finite rotation groups and their binary extensions.

Taking into account the conclusion (5.42) we have two cases.

5.2.2 Case $r = 2$: the infinite series of cyclic groups $\mathbb{A}_n$

Choosing $r = 2$, the Diophantine equation (5.38) reduces to

$$\frac{2}{n} = \frac{1}{k_1} + \frac{1}{k_2}. \tag{5.43}$$

Since we have $k_{1,2} \leq n$, the only solution of (5.43) is $k_1 = k_2 = n$, with n arbitrary. Since the order of the cyclic stability subgroup of the two poles coincides with the order of the full group H, it follows that H itself is a cyclic subgroup of SU(2) of order n. We name it $\Gamma_b[n, n, 1]$. The two orbits are given by the two eigenvectors of the unique cyclic group generator:

$$\mathscr{A} \in \mathrm{SU}(2): \quad \mathscr{Z} \equiv \mathscr{A}^n. \tag{5.44}$$

The finite subgroup of SU(2) isomorphic to the abstract group $\mathbb{Z}_{2n}$ is composed of the following $2n$ elements:

$$\mathbb{Z}_{2n} \sim \Gamma_b[n, n, 1] = \{\mathbf{1}, \mathscr{A}, \mathscr{A}^2, \dots, \mathscr{A}^{n-1}, \mathscr{Z}, \mathscr{Z}\mathscr{A}, \mathscr{Z}\mathscr{A}^2, \dots, \mathscr{Z}\mathscr{A}^{n-1}\}. \tag{5.45}$$

Under the homomorphism ω, the SU(2)-element $\mathscr{Z}$ maps into the identity and both $\mathscr{A}$ and $\mathscr{Z}\mathscr{A}$ map into the same 3×3 orthogonal matrix $\mathrm{A} \in \mathrm{SO}(3)$ with the property $\mathrm{A}^n = \mathbf{1}$. Hence, we have

$$\omega[\Gamma_b[n, n, 1]] = \Gamma[n, n, 1] \sim \mathbb{Z}_n. \tag{5.46}$$

In conclusion, we can define the cyclic subgroups of SO(3) and their binary extensions in SU(2) by means of the following presentation in terms of generators and relations:

$$\mathbb{A}_n \quad \Leftrightarrow \quad \begin{cases} \Gamma_b[n, n, 1] = (\mathscr{A}, \mathscr{Z} \mid \mathscr{A}^n = \mathscr{Z};\ \mathscr{Z}^2 = \mathbf{1}), \\ \Gamma[n, n, 1] = (\mathrm{A} \mid \mathrm{A}^n = \mathbf{1}), \end{cases} \tag{5.47}$$

where $\mathscr{Z}$, being by definition a central extension, commutes with the other generators and henceforth with all the group elements. The nomenclature $\mathbb{A}_n$ introduced in the above equation is just for future comparison. As we will see, in the ADE classification of simply laced Lie algebras the case of cyclic groups corresponds to that of $\mathfrak{a}_n$ algebras.

5.2.3 Case $r = 3$ and its solutions

In the $r = 3$ case the Diophantine equation becomes

$$\frac{1}{k_1} + \frac{1}{k_2} + \frac{1}{k_3} = 1 + \frac{2}{n}. \tag{5.48}$$

In order to analyze its solutions in a unified way, inspired by the above case, it is convenient to introduce the notation

$$\mathscr{R} = 1 + \sum_{a}^{r} k_a \tag{5.49}$$

and consider the abstract groups that turn out to be of finite order associated with each triple of integers $\{k_1, k_2, k_3\}$ satisfying (5.48) and defined by the following presentation:

$$\begin{aligned} \Gamma_b[k_1, k_2, k_3] &= (\mathscr{A}, \mathscr{B}, \mathscr{Z} \mid (\mathscr{A}\mathscr{B})^{k_1} = \mathscr{A}^{k_2} = \mathscr{B}^{k_3} = \mathscr{Z}; \mathscr{Z}^2 = \mathbf{1}), \\ \Gamma[k_1, k_2, k_3] &= (\mathrm{A}, \mathrm{B} \mid (\mathrm{AB})^{k_1} = \mathrm{A}^{k_2} = \mathrm{B}^{k_3} = \mathbf{1}). \end{aligned} \tag{5.50}$$

We will see that the finite subgroups of SU(2) are indeed isomorphic to the above defined abstract groups $\Gamma_b[k_1, k_2, k_3]$ and that their images under the homomorphism ω are isomorphic to $\Gamma[k_1, k_2, k_3]$.

5.2.3.1 The solution $(k, 2, 2)$ and the dihedral groups Dih_k

One infinite class of solutions of the Diophantine equation (5.48) is given by

$$\{k_1, k_2, k_3\} = \{k, 2, 2\}; \quad 2 < k \in \mathbb{Z}. \tag{5.51}$$

The corresponding subgroups of SU(2) and SO(3) are

$$\mathrm{Dih}_k \quad \Leftrightarrow \quad \begin{cases} \Gamma_b[k, 2, 2] = (\mathscr{A}, \mathscr{B}, \mathscr{Z} \mid (\mathscr{A}\mathscr{B})^{k} = \mathscr{A}^{2} = \mathscr{B}^{2} = \mathscr{Z}; \mathscr{Z}^2 = \mathbf{1}), \\ \Gamma[k, 2, 2] = (\mathrm{A}, \mathrm{B} \mid (\mathrm{AB})^{k} = \mathrm{A}^{2} = \mathrm{B}^{2} = \mathbf{1}), \end{cases} \tag{5.52}$$

whose structure we illustrate next.

$\Gamma_b[k, 2, 2] \simeq \mathrm{Dih}_k^b$ is the binary dihedral subgroup. Its order is

$$|\mathrm{Dih}_k^b| = 4k \tag{5.53}$$

and it contains a cyclic subgroup of order k that we name K. Its index in Dih_k^b is 2. The elements of Dih_k^b that are not in K are of period equal to 2 since $k_2 = k_3 = 2$. Altogether, the elements of the dihedral group are the following matrices:

$$F_l = \begin{pmatrix} e^{il\pi/k} & 0 \\ 0 & e^{-il\pi/k} \end{pmatrix} \quad (l = 0, 1, 2, \ldots, 2k-1),$$

$$G_l = \begin{pmatrix} 0 & i\,e^{-il\pi/k} \\ i\,e^{il\pi/k} & 0 \end{pmatrix} \quad (l = 0, 1, 2, \ldots, 2k-1).$$

In terms of them the generators are identified as follows:

$$F_0 = 1; \quad F_1 G_0 = \mathscr{A}; \quad F_k = \mathscr{Z}; \quad G_0 = \mathscr{B}. \tag{5.54}$$

There are exactly $\mathscr{R} = k + 3$ conjugacy classes:
1. K_e contains only the identity F_0.
2. K_Z contains the central extension $\mathscr{Z}$.
3. $K_{G\,\mathrm{even}}$ contains the elements $G_{2\nu}$ ($\nu = 1, \ldots, k-1$).
4. $K_{G\,\mathrm{odd}}$ contains the elements $G_{2\nu+1}$ ($\nu = 1, \ldots, k-1$).
5. The $k-1$ classes K_{F_μ}; each of these classes contains the pair of elements F_μ and $F_{2k-\mu}$ for ($\mu = 1, \ldots, k-1$).

Correspondingly, the group Dih_k^b admits $k+3$ irreducible representations, four of which are one-dimensional, while $k-1$ are two-dimensional. We name them as follows:

$$\begin{cases} D_e; D_Z; D_{G\,\mathrm{even}}; D_{G\,\mathrm{odd}}; & \text{one-dimensional,} \\ D_{F_1}; \ldots; D_{F_{k-1}}; & \text{two-dimensional.} \end{cases} \tag{5.55}$$

The combinations of the $\mathbb{C}^2$-vector components (z_1, z_2) that transform in the four one-dimensional representations are easily listed:

$$\begin{aligned} D_e &\longrightarrow |z_1|^2 + |z_2|^2, \\ D_Z &\longrightarrow z_1 z_2, \\ D_{G\,\mathrm{even}} &\longrightarrow z_1^k + z_2^k, \\ D_{G\,\mathrm{odd}} &\longrightarrow z_1^k - z_2^k. \end{aligned} \tag{5.56}$$

The matrices of the $k-1$ two-dimensional representations are obtained in the following way. In the DF_s representation, $s = 1, \ldots, k-1$, the generator $\mathscr{A}$, namely, the group element F_1, is represented by the matrix F_s. The generator $\mathscr{B}$ is instead represented by $(i)^{s-1} G_0$ and the generator $\mathscr{Z}$ is given by F_{sk}, so that

$$\begin{aligned} DF_s(F_j) &= F_{sj}, \\ DF_s(G_j) &= (i)^{s-1} G_{sj}. \end{aligned} \tag{5.57}$$

The character table is immediately obtained; it is displayed in Table 5.1. This concludes the discussion of the binary dihedral groups.

5.2.3.2 The three isolated solutions corresponding to the tetrahedral, octahedral, and icosahedral groups

There remain three isolated solutions of the Diophantine equation (5.48), namely,

$$\{k_1, k_2, k_3\} = \{3, 3, 2\}, \tag{5.58}$$

$$\{k_1, k_2, k_3\} = \{4, 3, 2\}, \tag{5.59}$$

$$\{k_1, k_2, k_3\} = \{5, 3, 2\}. \tag{5.60}$$

Table 5.1: Character table of the group Dih_k^b.

$\cdot$	KE	KZ	KG_e	KG_o	KF_1	$\cdots$	KF_{k-1}
DE	1	1	1	1	1	$\cdots$	1
DZ	1	1	-1	-1	1	$\cdots$	1
DG_e	1	$(-1)^k$	i^k	$-i^k$	$(-1)^1$	$\cdots$	$(-1)^{k-1}$
DG_o	1	$(-1)^k$	$-i^k$	i^k	$(-1)^1$	$\cdots$	$(-1)^{k-1}$
DF_1	2	$(-2)^1$	0	0	$2\cos\frac{\pi}{k}$	$\cdots$	$2\cos\frac{(k-1)\pi}{k}$
$\vdots$	$\vdots$	$\vdots$	$\vdots$	$\vdots$	$\vdots$	$\ddots$	$\vdots$
DF_{k-1}	2	$(-2)^{k-1}$	0	0	$2\cos\frac{(k-1)\pi}{k}$	$\cdots$	$2\cos\frac{(k-1)^2\pi}{k}$

They respectively correspond to the tetrahedral T_{12}, octahedral O_{24}, and icosahedral I_{60} groups and to their binary extensions, namely,

$$\Gamma[3,3,2] \simeq T_{12}, \tag{5.61}$$

$$\Gamma[4,3,2] \simeq O_{24}, \tag{5.62}$$

$$\Gamma[5,3,2] \simeq I_{60}. \tag{5.63}$$

As their name reveals, these three groups have 12, 24, and 60 elements, respectively. The corresponding binary extensions have 24, 48, and 120 elements, respectively. With a procedure completely analogous to the one utilized in the case of the dihedral groups we might reconstruct all these elements and organize them into conjugacy classes. We do not do this explicitly; in Chapter 8, while discussing crystallographic groups, we will rather study in full detail the examples of the octahedral O_{24} and tetrahedral T_{12} groups and we will do that starting from the three-dimensional realization in SO(3).

5.2.4 Summary of the ADE classification of finite rotation groups

Here we prepare the stage for the illustration of the deep and surprising relation, several times already anticipated, between the platonic classification of finite rotation groups and that of semi-simple Lie algebras. To this effect let us consider Figure 5.2 and diagrams of the sort there displayed. Such diagrams are named Dynkin diagrams and will obtain a well-defined interpretation while studying root spaces and the classification of simple Lie algebras (see Chapter 11). For the time being let us note that Dynkin diagrams such as that in Figure 5.2 are characterized by three integer numbers $\{k_1, k_2, k_3\}$, denoting the lengths of three chains of dots, linked to one another and departing from a central node which belongs to each of the three chains. In the case one of the numbers k_a is equal to one (say k_3), the corresponding chain disappears and we are left with a simple chain of length $k_1 + k_2 - 1$. In Chapter 11 we will see that the admissible Dynkin diagrams with one node are those and only those where the numbers $\{k_1, k_2, k_3\}$ satisfy

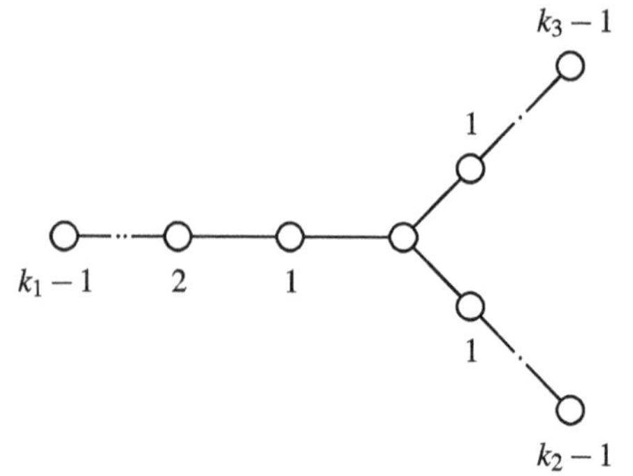

	Simple Lie algebras	Finite subgroups of $\Gamma_b \subset \mathrm{SU}(2)$
r	number of simple chains in the Dynkin diagram	# of different group element orders in $\Gamma \equiv \omega[\Gamma_b]$
k_a	$k_a - 1$ = lengths of the simple chains in the Dynkin diagram	group element orders in $\Gamma \equiv (\mathrm{A}, \mathrm{B} \mid (\mathrm{AB})^{k_1} = \mathrm{A}^{k_2} = \mathrm{B}^{k_3} = \mathbf{1})$
$\mathscr{R} - 1 \equiv \sum_{a=1}^{r}(k_a - 1)$	$\mathscr{R}$ = rank of Lie algebra	$\mathscr{R} + 1$ = # of conj. classes in Γ_b

Figure 5.2: Interpretation of the solutions of the same Diophantine equation in the case of finite subgroups of $\Gamma_b \subset \mathrm{SU}(2)$ and of simply laced Lie algebras.

the diophantine equation (5.48). Hence each solution of that equation has a double interpretation: it singles out a finite rotation group and labels a simple Lie algebra. The anticipated correspondence is the following one:

$$\Gamma[\ell, \ell, 1] \simeq \mathbb{Z}_\ell \Leftrightarrow \mathfrak{a}_\ell, \tag{5.64}$$

$$\Gamma[\ell, 2, 2] \simeq \mathrm{Dih}_\ell \Leftrightarrow \mathfrak{d}_\ell, \tag{5.65}$$

$$\Gamma[3, 3, 2] \simeq \mathrm{T}_{12} \Leftrightarrow \mathfrak{e}_6, \tag{5.66}$$

$$\Gamma[4, 3, 2] \simeq \mathrm{O}_{24} \Leftrightarrow \mathfrak{e}_7, \tag{5.67}$$

$$\Gamma[5, 3, 2] \simeq \mathrm{I}_{60} \Leftrightarrow \mathfrak{e}_8, \tag{5.68}$$

where $\mathfrak{a}_\ell$ is the Lie algebra associated with the classical Lie group $\mathrm{SL}(\ell+1, \mathbb{C})$, $\mathfrak{d}_\ell$ is the Lie algebra associated with the classical Lie group $\mathrm{SO}(2\ell, \mathbb{C})$, and $\mathfrak{e}_{6,7,8}$ are the Lie algebras of three exceptional Lie groups of dimensions 78, 133, and 248, respectively. As we will see in Chapter 10, a very important concept in Lie algebra theory is that of *rank*, that is, the maximal number of mutually commuting and diagonalizable elements of the algebra. As we see from Figure 5.2, the rank has a counterpart in the binary extension of the corresponding finite rotation group: it is the number of non-trivial conjugacy classes of the group, except the class of the identity element. The property of Lie algebras that in Dynkin diagrams there are no nodes with more than three converging lines corresponds on the finite rotation group side to the property that in such groups there are at most three different types of group element orders.

A further challenging reinterpretation of the ADE classification regards the construction of the so-called ALE manifolds, which are four-dimensional spaces with a self-dual curvature and asymptotic flatness.[2]

5.3 Bibliographical note

ADE classification is a widely discussed topic in contemporary algebraic and differential geometry and it is the subject of a vast body of specialized literature. An in depth review of its application to ALE manifolds and to advanced issues in supersymmetric field theories is presented in the book [72] by the same author.

2 For this topic we refer the reader to [72].

6 Manifolds and Lie groups

6.1 Introduction

> The analysts try in vain to conceal the fact that they do not deduce: they combine, they compose ... when they do arrive at the truth they stumble over it after groping their way along.
>
> Évariste Galois

In Chapter 2 we focused on algebraic structures and we reviewed the basic algebraic concepts that apply both to discrete and to continuous groups. In the present chapter we turn to basic concepts of differential geometry preparing the stage for the study of Lie groups. These latter, which constitute one of the main subjects of this text, arise from the consistent merging of two structures:

1. An algebraic structure, since the elements of a Lie group G can be composed via an internal binary operation, generically called product, that obeys the axioms of a group.
2. A differential geometric structure, since G is an analytic differentiable manifold and the group operations are infinitely differentiable in such a topology.

General relativity is founded on the concept of *differentiable manifolds*. The adopted mathematical model of *space-time* is given by a pair $(\mathcal{M}, g)$ where $\mathcal{M}$ is a differentiable manifold of dimension $D = 4$ and g is a *metric*, that is, a rule to calculate the length of curves connecting points of $\mathcal{M}$.

However, as already stated several times, discrete and Lie groups and differential geometry have a much larger spectrum of applications than fundamental physics. Differentiable manifolds can be used as mathematical models for a much larger variety of sets of data, pertaining to disparate fields of knowledge, and this is the perspective adopted in the present book. The notions of distance, of curvature, of geodesics, and first of all of *symmetry* pertain to a large spectrum of problems from different sciences yet their mathematical formalization is always the same.

The central notions are those which fix the geometric environment,

- *differentiable manifolds*
- *fiber bundles*

and those which endow such environment with structures accounting for the measure of lengths and for the rules of parallel transport, namely,

- *metrics*
- *connections*

Once the geometric environments are properly mathematically defined, the metrics and connections one can introduce over them turn out to be the structures which, on the side of physics, encode the fundamental forces of Nature, and in applications to other

https://doi.org/10.1515/9783111201535-006

sciences have different interpretations, yet they are equally relevant since they provide the most efficient framework for the analysis of the phenomena associated with the considered space.

Summarizing, *differential geometry* and *Lie group theory*:
- are *intimately and inextricably related*, and;
- have a *much wider range of applications* in all branches of pure and applied science.

Indeed, that of a manifold is the appropriate mathematical model of any continuous space whose points can have the most disparate interpretations and that of a group (discrete or continuous) is the appropriate mathematical tool to deal with symmetry operations acting on that space.

In the following sections we introduce first differentiable manifolds and then Lie groups.

6.2 Differentiable manifolds

First and most fundamental in the list of geometrical concepts that we need to introduce is that of a *manifold*, which corresponds, as already explained, to our intuitive idea of a *continuous space*. In mathematical terms this is, to begin with, a *topological space*, namely, a set of elements where one can define the notions of *neighborhood* and *limit*. This is the correct mathematical description of our intuitive ideas of vicinity and close-by points. Secondly the characterizing feature that distinguishes a manifold from a simple topological space is the possibility of labeling its points with a set of coordinates. Coordinates are a set of real numbers $x_1(p), \dots, x_D(p) \in \mathbb{R}$ associated with each point $p \in \mathcal{M}$ that tell us *where* we are. Actually, in general relativity each point is an event so that coordinates specify not only its *where* but also its *when*. In other applications the coordinates of a point can be the most disparate parameters specifying the state of some complex system of the most general kind (dynamical, biological, economical, or whatever).

In classical physics the laws of motion are formulated as a set of differential equations of the second order where the unknown functions are the three Cartesian coordinates x, y, z of a particle and the variable t is time. Solving the dynamical problem amounts to determining the continuous functions $x(t), y(t), z(t)$ that yield a parametric description of a curve in $\mathbb{R}^3$, or better, defining a curve in $\mathbb{R}^4$, having included the time t in the list of coordinates of each event. Coordinates, however, are not uniquely defined. Each observer has his own way of labeling space points and the laws of motion take a different form if expressed in the coordinate frame of different observers. There is however a privileged class of observers in whose frames the laws of motion have always the same form: these are the inertial frames, which are in rectilinear relative motion with constant velocity. The existence of a privileged class of inertial frames is common to classical Newtonian physics and to special relativity; the only difference is the form of coordinate transformations connecting them, Galilei transformations in the first case and Lorentz transformations in the second. This goes hand in hand with the fact that the space-time manifold is the *flat affine*[1] *manifold* $\mathbb{R}^4$ in both cases. By definition all points of $\mathbb{R}^N$ can be covered by one coordinate frame $\{x^i\}$ and all frames with such a property are related to each other by general linear transformations, that is, by the elements of the general linear group $\mathrm{GL}(\mathrm{N}, \mathbb{R})$:

$$x^{i'} = A^{i}_{\ j} x^j; \quad A^{i}_{\ j} \in \mathrm{GL}(\mathrm{N}, \mathbb{R}). \tag{6.1}$$

The restriction to the Galilei or Lorentz subgroups of $\mathrm{GL}(4, \mathbb{R})$ is a consequence of the different *scalar products* on $\mathbb{R}^4$-vectors one wants to preserve in the two cases, but the relevant common feature is the fact that the space-time manifold has a vector space structure. The privileged coordinate frames are those that use the corresponding vectors as labels of each point.

In application of differential geometry to different disciplines where the considered manifold is *not space-time*, but rather its points represent completely different entities, Galilei and Lorentz groups are replaced by other groups more pertinent to the specific case, yet the following argumentations apply as well. Whatever its interpretation might be, a different situation arises when the manifold is not flat and affine, like, for instance, the surface of an N-dimensional sphere $\mathbb{S}^N$. As cartographers know very well, there is no way of representing all points of a curved surface in a single coordinate frame, namely, in a single *chart*. However we can succeed in representing all points of a curved surface by means of an *atlas*, namely, with a collection of charts, each of which maps one open region of the surface and such that the union of all these regions covers the entire surface. Knowing the transition rule from one chart to the next one, in the regions where they overlap, we obtain a complete coordinate description of the curved surface by means of our atlas.

The intuitive idea of an *atlas* of *open charts*, suitably reformulated in mathematical terms, provides the very definition of a differentiable manifold, the geometrical concept that generalizes our notion of a continuous space, from $\mathbb{R}^N$ to more complicated non-flat situations.

There are many possible *atlases* that describe the same manifold $\mathcal{M}$, related to each other by more or less complicated transformations. For a generic $\mathcal{M}$ no privileged choice of the atlas is available differently from the case of $\mathbb{R}^N$: here the inertial frames are singled out by the additional *vector space* structure of the manifold, which allows to label each point with the corresponding vector. Therefore, if the laws of physics have to be universal and have to accommodate non-flat space-times, then they must be formulated in such a way that they have the same form in any atlas. This is the principle of *general covariance* at the basis of general relativity.

1 A manifold (defined in this section) is named *affine* when it is also a vector space.

Similarly, in the wider perspective here adopted, the choice of a particular set of parameters to describe the state of a complex system should not be privileged with respect to any other choice. The laws that govern the dynamics of a system of whatever type should be intrinsic and should not depend on the set of variables chosen to describe it.

6.2.1 Homeomorphisms and the definition of manifolds

A fundamental ingredient in formulating the notion of differential manifolds is that of homeomorphism.[2]

Definition 6.2.1. Let X and Y be two topological spaces and let h be a map:

$$h : X \to Y. \tag{6.2}$$

If h is one-to-one and if both h and its inverse h^{-1} are continuous, then we say that h is a *homeomorphism*.

As a consequence of the theorems proved in all textbooks about elementary topology and calculus, homeomorphisms preserve all topological properties. Indeed, let h be a homeomorphism mapping X onto Y and let $A \subset X$ be an open subset. Its image through h, namely, $h(A) \subset Y$, is also an open subset in the topology of Y. Similarly, the image $h(C) \subset Y$ of a closed subset $C \subset X$ is a closed subset. Furthermore, for all $A \subset X$ we have

$$h(\overline{A}) = \overline{h(A)}, \tag{6.3}$$

namely, the closure of the image of a set A coincides with the image of the closure.

Definition 6.2.2. Let X and Y be two topological spaces. If there exists a homeomorphism $h : X \to Y$, then we say that X and Y are homeomorphic.

It is easy to see that given a topological space X, the set of all homeomorphisms $h : X \to X$ constitutes a group, usually denoted $\mathrm{Hom}(X)$. Indeed, if $h \in \mathrm{Hom}(X)$ is a homeomorphism, then also $h^{-1} \in \mathrm{Hom}(X)$ is a homeomorphism. Furthermore, if $h \in \mathrm{Hom}(X)$ and $h' \in \mathrm{Hom}(X)$, then also $h \circ h' \in \mathrm{Hom}(X)$. Finally, the identity map

$$\mathbf{1} : X \to X \tag{6.4}$$

2 We assume that the reader is familiar with the basic notions of general topology concerning the notions of bases of neighborhoods, open and close subsets, boundary, and limit (see for instance [42, 76, 77, 41, 78]).

is certainly one-to-one and continuous and it coincides with its own inverse. Hence, $\mathbf{1} \in \mathrm{Hom}(X)$. As we discuss later on, for any manifold X the group $\mathrm{Hom}(X)$ is an example of an infinite and continuous group.

Let now $\mathcal{M}$ be a topological Hausdorff space. An *open chart* of $\mathcal{M}$ is a pair (U, φ), where $U \subset \mathcal{M}$ is an open subset of $\mathcal{M}$ and φ is a homeomorphism of U on an open subset $\mathbb{R}^m$ (m being a positive integer). The concept of open chart allows to introduce the notion of coordinates for all points $p \in U$. Indeed, the coordinates of p are the m real numbers that identify the point $\varphi(p) \in \varphi(U) \subset \mathbb{R}^m$.

Using the notion of open chart we can finally introduce the notion of differentiable structure.

Definition 6.2.3. Let $\mathcal{M}$ be a topological Hausdorff space. A differentiable structure of dimension m on $\mathcal{M}$ is an *atlas* $\mathcal{A} = \bigcup_{i \in A}(U_i, \varphi_i)$ of open charts (U_i, φ_i), where $\forall i \in A$, $U_i \subset \mathcal{M}$ is an open subset and

$$\varphi_i : U_i \to \varphi_i(U_i) \subset \mathbb{R}^m \tag{6.5}$$

is a homeomorphism of U_i in $\mathbb{R}^m$, namely, a continuous, invertible map onto an open subset of $\mathbb{R}^m$ such that the inverse map

$$\varphi_i^{-1} : \varphi_i(U_i) \to U_i \subset \mathcal{M} \tag{6.6}$$

is also continuous (see Figure 6.1). The atlas must fulfill the following axioms:

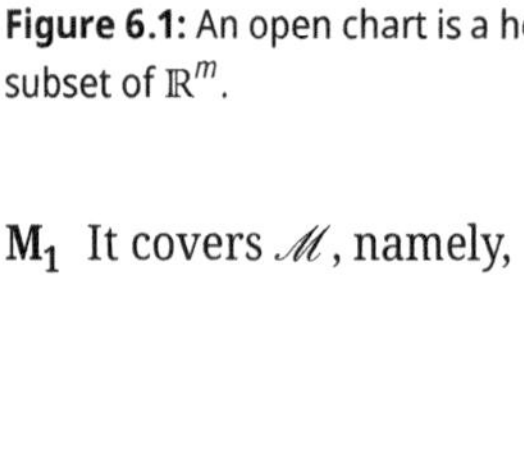

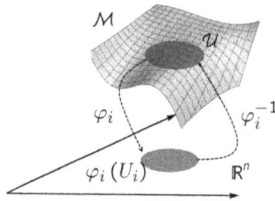

Figure 6.1: An open chart is a homeomorphism of an open subset U_i of the manifold $\mathcal{M}$ onto an open subset of $\mathbb{R}^m$.

M$_1$ It covers $\mathcal{M}$, namely,

$$\bigcup_i U_i = \mathcal{M}, \tag{6.7}$$

so that each point of $\mathcal{M}$ is contained at least in one chart and generically in more than one: $\forall p \in \mathcal{M} \mapsto \exists (U_i, \varphi_i) / p \in U_i$.

M$_2$ Chosen any two charts (U_i, φ_i), (U_j, φ_j) such that $U_i \cap U_j \neq \emptyset$, on the intersection

$$U_{ij} \stackrel{\text{def}}{=} U_i \cap U_j \tag{6.8}$$

there exist two homeomorphisms,

$$\begin{aligned} \varphi_i|_{U_{ij}} &: U_{ij} \to \varphi_i(U_{ij}) \subset \mathbb{R}^m, \\ \varphi_j|_{U_{ij}} &: U_{ij} \to \varphi_j(U_{ij}) \subset \mathbb{R}^m, \end{aligned} \tag{6.9}$$

and the composite map

$$\begin{aligned} &\psi_{ij} \stackrel{\text{def}}{=} \varphi_j \circ \varphi_i^{-1} \\ &\psi_{ij} : \varphi_i(U_{ij}) \subset \mathbb{R}^m \to \varphi_j(U_{ij}) \subset \mathbb{R}^m \end{aligned} \tag{6.10}$$

named the *transition function*, which is actually an m-tuple of m real functions of m real variables that is required to be *differentiable* (see Figure 6.2).

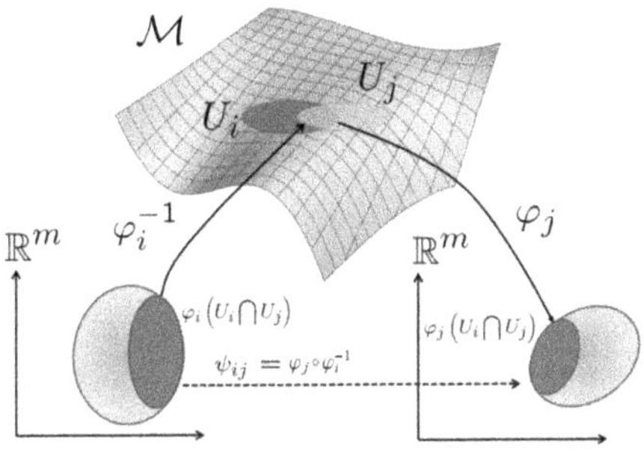

Figure 6.2: A transition function between two open charts is a differentiable map from an open subset of $\mathbb{R}^m$ to another open subset of the same.

M_3 The collection $(U_i, \varphi_i)_{i\in A}$ is the maximal family of open charts for which both M_1 and M_2 hold true.

Now we can finally introduce the definition of differentiable manifold.

Definition 6.2.4. A differentiable manifold of dimension m is a topological space $\mathcal{M}$ that admits at least one differentiable structure $(U_i, \varphi_i)_{i\in A}$ of dimension m.

The definition of a differentiable manifold is constructive in the sense that it provides a way to construct it explicitly. What one has to do is to give an atlas of open charts (U_i, φ_i) and the corresponding transition functions ψ_{ij} which should satisfy the necessary consistency conditions:

$$\forall i,j \quad \psi_{ij} = \psi_{ji}^{-1}, \tag{6.11}$$

$$\forall i,j,k \quad \psi_{ij} \circ \psi_{jk} \circ \psi_{ki} = \mathbf{1}. \tag{6.12}$$

In other words, a general recipe to construct a manifold is to specify the open charts and how they are *glued* together. The properties assigned to a manifold are the properties fulfilled by its transition functions. In particular, we have the following definition.

Definition 6.2.5. A differentiable manifold $\mathcal{M}$ is said to be *smooth* if the transition functions (6.10) are *infinitely differentiable*:

$$\mathcal{M} \text{ is smooth} \quad \Leftrightarrow \quad \psi_{ij} \in \mathbb{C}^{\infty}(\mathbb{R}^m). \tag{6.13}$$

Similarly, we have the following definition of a complex manifold.

Definition 6.2.6. A real manifold of even dimension $m = 2\nu$ is *complex* of dimension ν if the 2ν real coordinates in each open chart U_i can be arranged into ν complex numbers so that equation (6.5) can be replaced by

$$\varphi_i : U_i \to \varphi_i(U_i) \subset \mathbb{C}^{\nu} \tag{6.14}$$

and the transition functions ψ_{ij} are *holomorphic maps*:

$$\psi_{ij} : \varphi_i(U_{ij}) \subset \mathbb{C}^{\nu} \to \varphi_j(U_{ij}) \subset \mathbb{C}^{\nu}. \tag{6.15}$$

Although the constructive definition of a differentiable manifold is always in terms of an atlas, in many occurrences we can have other intrinsic global definitions of what $\mathcal{M}$ is and the construction of an atlas of coordinate patches is an a posteriori operation. Typically this happens when the manifold admits a description as an algebraic locus. The prototype example is provided by the $\mathbb{S}^N$ sphere, which can be defined as the locus in $\mathbb{R}^{N+1}$ of points with distance r from the origin:

$$\{X_i\} \in \mathbb{S}^N \quad \Leftrightarrow \quad \sum_{i=1}^{N+1} X_i^2 = r^2. \tag{6.16}$$

In particular, for $N = 2$ we have the familiar $\mathbb{S}^2$, which is diffeomorphic to the compactified complex plane $\mathbb{C} \cup \{\infty\}$. Let us briefly discuss this issue in some detail. A 2-sphere of radius r, named $\mathbb{S}^2_r$, can be defined as the locus in $\mathbb{R}^3$ of all the points with distance r from the origin:

$$\{x, y, u\} \in \mathbb{S}^2_r \quad \Leftrightarrow \quad x^2 + y^2 + u^2 = r^2. \tag{6.17}$$

Since all spheres are topologically equivalent, we can choose as representatives of all the spheres that of radius $r = 1$, which we will simply name $\mathbb{S}^2$. We want to show that $\mathbb{S}^2$ is homeomorphic to the compactified complex plane $\mathbb{C} \cup \{\infty\}$. Indeed, we can easily verify that $\mathbb{S}^2$ is a one-dimensional complex manifold considering the atlas of holomorphic open charts suggested by the geometrical construction named *the stereographic projection*. To this effect consider the picture in Figure 6.3, where we have drawn the 2-sphere

$\mathbb{S}^2$ of radius $r = 1$ centered in the origin of $\mathbb{R}^3$. Given a generic point $P \in \mathbb{S}^2$ we can construct its image on the equatorial plane $\mathbb{R}^2 \sim \mathbb{C}$ drawing the straight line in $\mathbb{R}^3$ that goes through P and through the *North Pole* of the sphere N. Such a line will intersect the equatorial plane in the point P_N whose value z_N, regarded as a complex number, we can identify with the complex coordinate of P in the open chart under consideration. In terms of the Cartesian coordinates $\{x, y, u\}$ of P we have

$$\varphi_N(P) = \frac{x + iy}{1 - u} \equiv z_N \in \mathbb{C}. \tag{6.18}$$

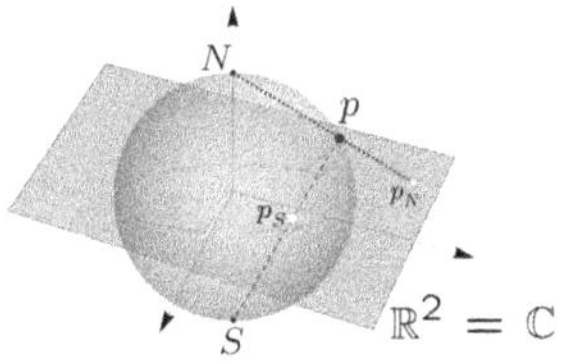

Figure 6.3: Stereographic projection of the 2-sphere.

Alternatively we can draw the straight line through P and the *South Pole S*. This intersects the equatorial plane in another point P_S whose value as a complex number, named z_S, is just the reciprocal of z_N: $z_S = 1/z_N$. We can take z_S as the complex coordinate of the same point P. In other words, we have another open chart:

$$\varphi_S(P) = z_S \in \mathbb{C}. \tag{6.19}$$

What is the domain of these two charts? That is, what are the open subsets U_N and U_S? This is rather easily established considering that the North Pole projection yields a finite result $z_N < \infty$ for all points P except the North Pole itself. Hence, $U_N \subset \mathbb{S}^2$ is the open set obtained by subtracting one point (the North Pole) from the sphere. Similarly, the South Pole projection yields a finite result for all points P except the South Pole itself and U_S is $\mathbb{S}^2$ minus the south pole. More definitely we can choose for U_N and U_S any two open neighborhoods of the South and North Pole, respectively, with non-vanishing intersection (see Figure 6.4). In this case the intersection $U_N \cap U_S$ is a band wrapped around the equator of the sphere and its image in the complex equatorial plane is a circular corona that excludes both a circular neighborhood of the origin and a circular neighborhood of infinity. On such an intersection we have the transition function

$$\psi_{NS} : z_N = \frac{1}{z_S}, \tag{6.20}$$

which is clearly holomorphic and satisfies the consistency conditions in equations (6.11) and (6.12). Hence we see that $\mathbb{S}^2$ is a complex 1-manifold that can be constructed with an atlas composed of two open charts related by the transition function (6.20). Obviously, a complex 1-manifold is a fortiori a *smooth real 2-manifold.* Manifolds with infinitely differentiable transition functions are named smooth not without a reason. Indeed, they correspond to our intuitive notion of smooth hypersurfaces without conical points or edges. The presence of such defects manifests itself through the lack of differentiability in some regions.

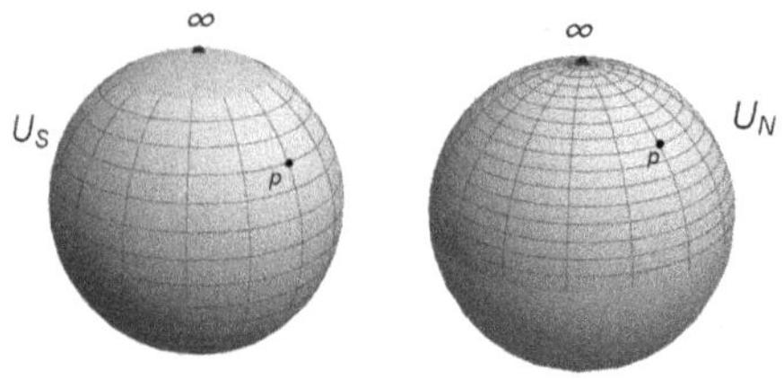

Figure 6.4: In this picture we display the two open charts $U_N \subset \mathbb{S}^2$ and $U_S \subset \mathbb{S}^2$, which are respectively a neighborhood of $S = 0$ and of $N = \infty$. The same point P that is neither of the two poles has a double description, since it belongs to $U_N \cap U_S$. In the North Pole stereographic projection it corresponds to the point $z_N \in \mathbb{C}$, while in the South Pole stereographic projection it corresponds to the point $z_S \in \mathbb{C}$.

6.2.2 Functions on manifolds

Being the mathematical model of the space of available configurations of whatever system we consider, manifolds are the geometrical support of all the processes pertaining to that system. They are the arenas where such processes take place and where the quantities, pertinent to case-by-case relevant branches of science, take their values. Mathematically, this implies that calculus, originally introduced on $\mathbb{R}^N$, must be extended to manifolds. The entities defined over manifolds with which we have to deal are mathematically characterized as *scalar functions, vector fields, tensor fields, differential forms,* and *sections of more general fiber bundles.* We introduce such basic geometrical notions slowly, beginning with the simplest concept of a *scalar function.*

Definition 6.2.7. A real scalar function on a differentiable manifold $\mathcal{M}$ is a map

$$f : \mathcal{M} \to \mathbb{R} \tag{6.21}$$

that assigns a real number $f(p)$ to every point $p \in \mathcal{M}$ of the manifold.

The properties of a scalar function, for instance its differentiability, are the properties characterizing its local description in the various open charts of an atlas. For each open chart (U_i, φ_i) let us define

$$f_i \stackrel{\text{def}}{=} f \circ \varphi_i^{-1}. \tag{6.22}$$

By construction

$$f_i : \mathbb{R}^m \supset \varphi_i(U_i) \to \mathbb{R} \tag{6.23}$$

is a map of an open subset of $\mathbb{R}^m$ into the real line $\mathbb{R}$, namely, a real function of m real variables (see Figure 6.5). The collection of the real functions $f_i(x_1^{(i)}, \ldots, x_m^{(i)})$ constitute the local description of the scalar function f. The function is said to be *continuous, differentiable, infinitely differentiable* if the real functions f_i have such properties. From the definition (6.22) of the local description and from the definition (6.10) of the transition functions it follows that we must have

$$\forall U_i, U_j : \quad f_j|_{U_i \cap U_j} = f_i|_{U_i \cap U_j} \circ \psi_{ij}. \tag{6.24}$$

Let $x^\mu_{(i)}$ be the coordinates in the patch U_i and let $x^\mu_{(j)}$ be the coordinates in the patch U_j. For points p that belong to the intersection $U_i \cap U_j$ we have

$$x^\mu_{(j)}(p) = \psi^\mu_{ji}(x^1_{(i)}(p), \ldots x^m_{(i)}(p)) \tag{6.25}$$

and the gluing rule (6.24) takes the form

$$f(p) = f_j(x_{(j)}) = f_j(\psi_{ji}(x_{(i)})) = f_i(x_{(i)}). \tag{6.26}$$

The practical way of assigning a function on a manifold is therefore that of writing its local description in the open charts of an atlas, taking care that the various f_i glue together correctly, namely, through equation (6.24). Although the number of continuous and differentiable functions one can write on any open region of $\mathbb{R}^m$ is infinite, the smooth functions globally defined on a non-trivial manifold can be very few. Indeed, it is only occasionally that we can consistently glue together various local functions $f_i \in \mathbb{C}^\infty(U_i)$ into a global f. When this happens we say that $f \in \mathbb{C}^\infty(\mathscr{M})$.

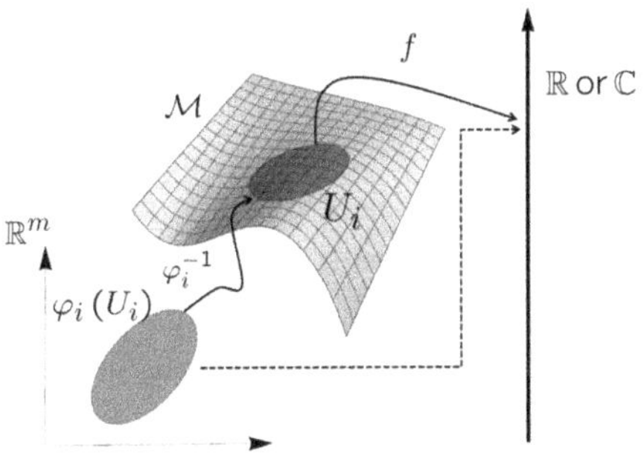

Figure 6.5: Local description of a scalar function on a manifold.

All we said about real functions can be trivially repeated for complex functions. It suffices to replace $\mathbb{R}$ by $\mathbb{C}$ in equation (6.21).

6.2.3 Germs of smooth functions

The local geometry of a manifold is studied by considering operations not on the space of smooth functions $\mathbb{C}^\infty(\mathcal{M})$, which, as just explained, can be very small, but on the space of germs of functions defined at each point $p \in \mathcal{M}$, which is always an infinite-dimensional space.

Definition 6.2.8. Given a point $p \in \mathcal{M}$, the space of germs of smooth functions at p, denoted $\mathbb{C}_p^\infty$, is defined as follows. Consider all the open neighborhoods of p, namely, all the open subsets $U_p \subset \mathcal{M}$ such that $p \in U_p$. Consider the space of smooth functions $\mathbb{C}^\infty(U_p)$ on each U_p. Two functions $f \in \mathbb{C}^\infty(U_p)$ and $g \in \mathbb{C}^\infty(U_p')$ are said to be equivalent if they coincide on the intersection $U_p \cap U_p'$ (see Figure 6.6):

$$f \sim g \quad \Leftrightarrow \quad f|_{U_p \cap U_p'} = g|_{U_p \cap U_p'}. \tag{6.27}$$

The union of all the spaces $\mathbb{C}^\infty(U_p)$ modded by the equivalence relation (6.27) is the space of germs of smooth functions at p:

$$\mathbb{C}_p^\infty \equiv \frac{\bigcup_{U_p} \mathbb{C}^\infty(U_p)}{\sim}. \tag{6.28}$$

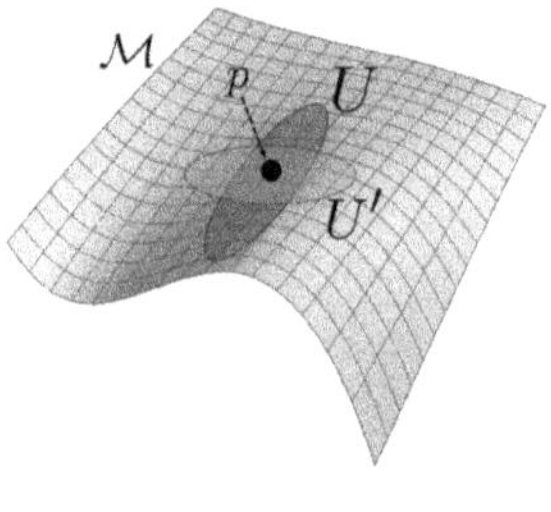

Figure 6.6: A germ of a smooth function is the equivalence class of all locally defined functions that coincide in some neighborhood of a point p.

What underlies the above definition of germs is the familiar principle of analytic continuation. Of the same function we can have different definitions that have different domains of validity; apparently we have different functions, but if they coincide on some open region, then we consider them just as different representations of a single function. Given any germ in some open neighborhood U_p we try to extend it to a larger domain by suitably changing its representation. In general there is a limit to such extension and only very special germs extend to globally defined functions on the whole manifold $\mathcal{M}$. For instance, the power series $\sum_{k \in \mathbb{N}} z^k$ defines a holomorphic function within its radius of convergence $|z| < 1$. As is well known, within the convergence radius the sum of this series coincides with $1/(1-z)$, which is a holomorphic function defined on a much larger neighborhood of $z = 0$. According to our definition the two functions are equivalent and correspond to two different representatives of the same *germ*. The germ, however, does not extend to a holomorphic function on the whole Riemann sphere $\mathbb{C} \cup \infty$ since it has a singularity in $z = 1$. Indeed, as stated by the Liouville theorem, the space of global holomorphic functions on the Riemann sphere contains only the constant function.

6.3 Holomorphic functions revisited and the conformal automorphism group of $\mathbb{C} \cup \{\infty\}$

Before continuing the development of differential geometry by introducing tangent and cotangent spaces, it is convenient to stop for a while in order to revisit some aspects of holomorphic function theory in view of the more precise differential geometric notions that we have so far introduced. This both provides an illustration of the abstract definitions and gives the opportunity to define and study an important and simple instance of Lie group already touched upon earlier, namely, the Möbius group of conformal automorphisms of the extended complex plane $\mathbb{C} \cup \{\infty\}$. In the following discussion we occasionally refer to concepts that are familiar to science undergraduate students, like integrals of holomorphic functions along paths in the complex plane, yet they are known in a traditional old form that obscures their true meaning in the framework of modern differential geometry and algebraic topology. This revisitation is quite pedagogical since it gradually brings the reader to absorb the concrete content of the abstract definitions that we expose systematically in the present chapter.

6.3.1 The annulus and its preimage on the Riemann sphere

Thinking of the complex plane $\mathbb{C}$ in terms of the stereographic projection, one realizes that the point at infinity is indeed unique since it just corresponds to the North Pole. Furthermore, one discovers a canonical way to deal with ∞ which, from the Riemann sphere viewpoint, is just a standard point as any other. Simply it is not covered by the open chart (U_N, ϕ_N) so that to investigate what happens in its neighborhood we have to change open chart. This is done by means of the transition function in equation (6.20).

The behavior of any holomorphic function $f(z_N)$ for $z_N \to \infty$ is studied by looking at its behavior in the neighborhood of $z_S = 0$. Since we are mainly interested in the differential 1-form[3] $\omega[f] = f(z_N)\, dz_N$, which is what is supposed to be integrated along

3 Here in order to make a connection with familiar issues, as the topic of holomorphic functions should be for most students of any science major, we anticipate the concept of differential forms that will be explained later in this very chapter so as to give to $f(z)dz$ its proper name in the context of differential geometry.

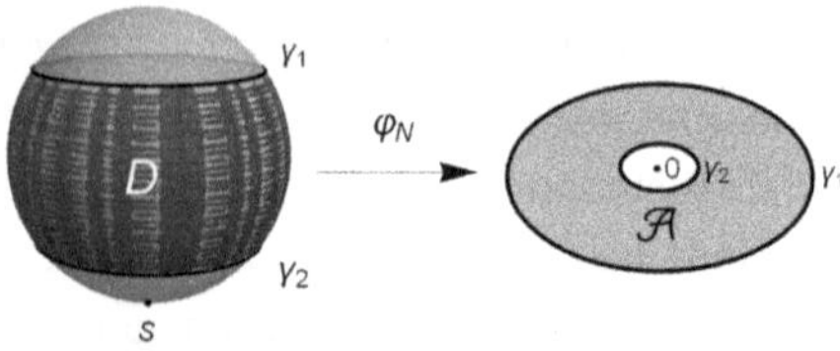

Figure 6.7: The preimage D on the Riemann 2-sphere of the annulus $\mathscr{A}$ around $z_0 = 0$.

cycles, we just ought to rewrite the same differential form in terms of the South Pole canonical coordinate. We immediately obtain

$$\omega[f] = f(z_N)\,dz_N = \underbrace{-\frac{1}{z_S^2} f\left(\frac{1}{z_S}\right)}_{=g(z_S)} dz_S = g(z_S)\,dz_S. \tag{6.29}$$

Hence, the behavior of $f(z)$ near infinity is dictated by the behavior of $g(z) = -z^{-2} f(1/z)$ for z close to zero.

A function $f(z)$ that is holomorphic in the interior of an annulus[4] $\mathscr{A}$ around some point $z_0 \in \mathbb{C}$ admits there a development in Laurent series. Since all points of the plane are equivalent, we can assume, without loss of generality, that $z_0 = 0$. It is very much instructive to consider the preimage on the Riemann sphere $\mathbb{S}^2$ of the annulus $\mathscr{A}$. We utilize the stereographic projection of the North Pole and we want to determine the domain $D \subset \mathbb{S}^2$ which is mapped into $\mathscr{A}$ by φ_N:

$$\varphi_N(D) = \mathscr{A} \equiv \{z \in \mathbb{C} \parallel \rho_2 < |z| < \rho_1\}. \tag{6.30}$$

Such a domain D is shown in Figure 6.7. The domain D consists of a band wrapped around the sphere that excludes both the arctic and the antarctic region and that is bounded by two parallels at different latitudes. The main conceptual point is that D is the *intersection of two simply connected regions* of the sphere (see Figure 6.8), namely,

$$D = D_1 \cap D_2. \tag{6.31}$$

By definition this property is preserved by the homeomorphism φ_N and we have

$$\mathscr{A} = \varphi_N(D_1 \cap D_2) = \varphi_N(D_1) \cap \varphi_N(D_2) = A_1 \cap A_2. \tag{6.32}$$

This result reproduces

4 Here we refer to an object which is very familiar and can be seen in Figure 6.8. We warn the reader that the concepts mentioned in the corresponding figure caption about homotopy and homology will be illustrated later in this very chapter.

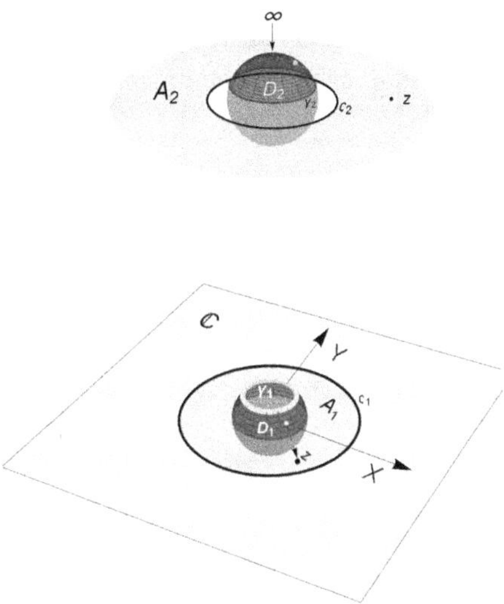

Figure 6.8: This figure displays the preimages $D_{1,2}$ on the compact Riemann sphere of the two domains $A_{1,2}$ of which the annulus is the intersection. The key point is that the preimages $D_{1,2}$ are simply connected and the two curves $\gamma_{1,2}$ are boundaries on $\mathbb{S}^2$. The same is obviously true of their homeomorphic images in $\mathbb{C}$: $\mathfrak{c}_{1,2} = \partial A_{1,2}$.

$$\mathscr{A} \equiv A_1 \cap A_2 = \{z \in C \parallel \rho_2 \leq |z - z_0| \leq \rho_1\}, \tag{6.33}$$

where

$$A_1 = \Delta_1 \equiv \{z \in \mathbb{C} \parallel |z - z_0| < \rho_1\} \tag{6.34}$$

and

$$A_2 = \mathbb{C} - \Delta_2; \quad \Delta_2 \equiv \{z \in \mathbb{C} \parallel |z - z_0| > \rho_2\}. \tag{6.35}$$

The new quality resides in the following, which refers to the basic notions of homotopy and homology, developed in general in Section 6.9. We state that looking at the preimages of $A_{1,2}$ on the Riemann sphere we clearly see that A_2 is simply connected just as A_1 (see Section 6.9 for the rigorous definition of simply connected). Moreover, both paths $\gamma_{1,2}$ are homotopically and homologically trivial since they are the homeomorphic images of the paths $\gamma_{1,2}$ on $\mathbb{S}^2$ (see the quoted Section 6.9 for the concepts of homology and homotopy and also for what we say next about the fundamental group). The fundamental group of the sphere is trivial $\pi_1(\mathbb{S}^2) = \mathbb{e}$ and every closed path is the boundary of some region. In our case $\gamma_{1,2} = \partial D_{1,2}$. The domains $D_{1,2}$ and their homeomorphic images $A_{1,2}$ in the complex plane are illustrated in Figure 6.8.

6.3.1.1 The automorphism group of the Riemann sphere and PSL(2, ℂ)

Every holomorphic function $f(z)$ can be seen as a map of the Riemann sphere $\mathbb{S}^2 = \mathbb{C} \cup \{\infty\}$ into itself,

$$f : \mathbb{S}^2 \longrightarrow \mathbb{S}^2, \tag{6.36}$$

which does not need to be surjective or one-to-one. For instance, for a polynomial function

$$\mathscr{P}(z) \equiv \sum_{s=0}^{n} a_s z^s \tag{6.37}$$

the preimages of every point $w \in \mathbb{C}$ are the n roots z_i of the degree n algebraic equation $\mathscr{P}(z) = w$. Under the action of $f(z)$ an open domain $D \subset \mathbb{C} \cup \{\infty\}$ is mapped into a new open domain $D \subset \mathbb{C} \cup \{\infty\}$. For instance, look at Figure 6.9 to see how the triangle D with vertices in $0, 1 + i, 2$ is mapped into new curvilinear triangles by the two functions

$$\begin{aligned} f(z) &= 1 - \frac{1}{7}z^3, \\ f(z) &= \frac{1}{1 + \frac{1}{5}z^7}, \end{aligned} \tag{6.38}$$

which have been arbitrarily chosen just for illustrative purposes. The maps induced by a holomorphic function are named *conformal maps* because of the following property: they change the shape and the area of figures, yet they conserve all the angles. Precisely, we have the following.

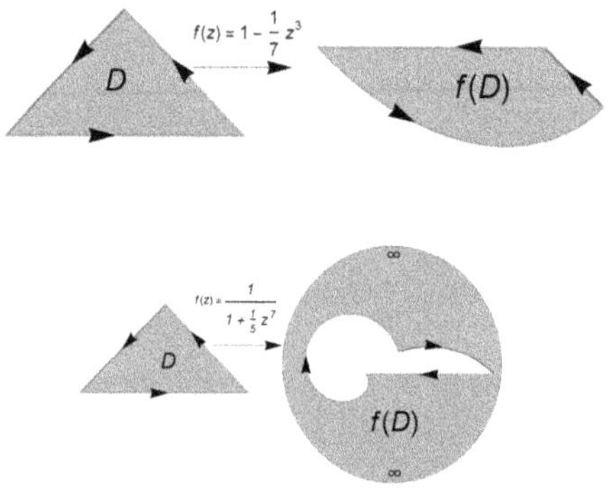

Figure 6.9: Two conformal maps of a standard triangle. Note that in the second map the direction of the boundary is reversed so that the actual image of the domain is the portion of the plane complementary to the interior of the figure. This means that in the second case $\infty \in f(D)$.

Lemma 6.3.1. *Let*

$$\gamma_{1,2} : \mathbb{R} \supset [0,1] \longrightarrow D \subset \mathbb{C} \cup \{\infty\}$$

be two differentiable paths $\gamma_{1,2}(t)$ contained in D that intersect each other at the point $z_0 \in D$. This means that $\exists t_{1,2} \in [0,1]$ such that $\gamma_1(t_1) = \gamma_2(t_2) = z_0 \in D$. Let θ be the angle between the tangent vector to γ_1 and that to γ_2 at the crossing point. Let $f(z)$ be any holomorphic function in the domain D and let

$$g_{1,2}(t) = f(\gamma_{1,2}(t))$$

be their images in $f(D)$. These images cross each other at $f(z_0) \in D$ and the angle between their tangent vectors is the same θ.

Proof of Lemma 6.3.1. In order to prove the lemma let us calculate the angle between the tangents before and after the conformal transformation $f(z)$. Before the transformation we have

$$\mathbf{t}_{1,2} \equiv \left\{ \mathrm{Re}\left[\frac{\mathrm{d}}{\mathrm{d}t}\gamma_{1,2}(t)\bigg|_{t=t_{1,2}} \right], \mathrm{Im}\left[\frac{\mathrm{d}}{\mathrm{d}t}\gamma_{1,2}(t)\bigg|_{t=t_{1,2}} \right] \right\} \tag{6.39}$$

and the angle is defined by the standard identity of two-dimensional Euclidean geometry:

$$(\mathbf{t}_1, \mathbf{t}_2) = |\mathbf{t}_1||\mathbf{t}_2| \cos\theta. \tag{6.40}$$

After the transformation we obtain

$$\tilde{\mathbf{t}}_{1,2} \equiv \left\{ \mathrm{Re}\left[\frac{df}{dz}(z_0)\frac{\mathrm{d}}{\mathrm{d}t}\gamma_{1,2}(t)\bigg|_{t=t_{1,2}} \right], \mathrm{Im}\left[\frac{df}{dz}(z_0)\frac{\mathrm{d}}{\mathrm{d}t}\gamma_{1,2}(t)\bigg|_{t=t_{1,2}} \right] \right\}, \tag{6.41}$$

so that

$$(\tilde{\mathbf{t}}_1, \tilde{\mathbf{t}}_2) = \left| \frac{df}{dz}(z_0) \right|^2 \times (\mathbf{t}_1, \mathbf{t}_2). \tag{6.42}$$

Analogously we have

$$|\tilde{\mathbf{t}}_{1,2}| = \left| \frac{df}{dz}(z_0) \right| |\mathbf{t}_{1,2}|. \tag{6.43}$$

Combining these results we finally get

$$\cos\tilde{\theta} = \frac{(\tilde{\mathbf{t}}_1, \tilde{\mathbf{t}}_2)}{|\tilde{\mathbf{t}}_1||\tilde{\mathbf{t}}_2|} = \frac{(\mathbf{t}_1, \mathbf{t}_2)}{|\mathbf{t}_1||\mathbf{t}_2|} = \cos\theta, \tag{6.44}$$

which proves the lemma. □

Having established that a holomorphic function $f(z)$ describes a conformal mapping of a region D of the Riemann sphere into another one $f(D)$, the question arises naturally whether holomorphic invertible maps can exist that map a given $D \subset \mathbb{C} \cup \{\infty\}$ into itself. If such a holomorphic function exists, it is named an *automorphism* of D.

Lemma 6.3.2. *Given an open subset $D \subset \mathbb{C} \cup \{\infty\}$, the set of automorphisms of D forms a group under the composition of maps. This group is named the automorphism group of D and it is denoted* $\mathrm{Aut}[D]$.

Proof of Lemma 6.3.2. It is evident that if $f(z)$ is a holomorphic, invertible function that maps D onto itself and $g(z)$ is another holomorphic function with the same properties, also $h(z) = g \circ f = g(f(z))$ will map D in D and will be equally invertible. The identity map $f(z) = z$ is certainly an automorphism of D. By definition every automorphism is invertible. This shows that all the axioms of a group are satisfied and the lemma is proved. □

6.3.1.2 The automorphism group of the complex plane $\mathbb{C}$

We would like to determine the automorphism group the complex plane $\mathrm{Aut}[\mathbb{C}]$. First of all we observe that any element $f(z) \in \mathrm{Aut}[\mathbb{C}]$ cannot have singular points anywhere in $\mathbb{C}$, because no point $z \in \mathbb{C}$ has to be mapped into ∞. Hence we have only two possibilities:

(1) $f(z)$ is holomorphic everywhere in $\mathbb{C}$ and it has an essential singularity for $z \to \infty$.
(2) $f(z)$ is a polynomial $\mathscr{P}^{(n)}(z) = \sum_{s=0}^{n} a_s z^s$ of some degree n.

Case (1) is excluded by the following argument. By definition $f(z)$ is one-to-one. Hence the f image of the open ball $\mathscr{B}^{(1)}[\infty] = \{z \in \mathbb{C} \mid |z| > 1$ cannot intersect the f image of the open ball $\mathscr{B}^{(1)}[0] = \{z \in \mathbb{C} \mid |z| < 1\}$:

$$f(\mathscr{B}^{(1)}[\infty]) \cap f(\mathscr{B}^{(1)}[0]) = \emptyset. \tag{6.45}$$

Equation (6.45) contradicts the Weierstrass theorem,[5] since the image of $\mathscr{B}^{(1)}[\infty]$, which is a neighborhood of the supposedly essential singular point $z = \infty$, is not dense in the whole plane $\mathbb{C}$. Hence, an automorphism of $\mathbb{C}$ cannot have an essential singularity at the point at infinity.

We are left with case (2). As we know from the fundamental theorem of algebra, the equation $\mathscr{P}^{(n)}(z) = w$ has n roots. It follows that, in order to be one-to-one, the

5 In the theory of holomorphic functions the quite famous Weierstrass theorem states the following. *If z_0 is an isolated **essential singular point** of a function $f(z)$ which is holomorphic in the punctured disk*

$$\Delta_{z_0}^{(\rho)} = \left\{ z \in \mathbb{C} \parallel 0 < |z - z_0| < \rho \right\},$$

then for any $\epsilon > 0$ the image through f of the punctured disk $\Delta_{z_0}^{(\epsilon)}$ is dense in the complex plane $\mathbb{C}$.

polynomial describing an automorphism of the plane $\mathbb{C}$ must be of degree $n = 1$. This means

$$f(z) \in \text{Aut}[\mathbb{C}] \quad \Rightarrow \quad f(z) = az + b; \quad a, b \in \mathbb{C}, \; a \neq 0. \tag{6.46}$$

As we see from the above formula, Aut[$\mathbb{C}$] is a continuous group with two parameters a, b. We will explain its structure in more detail in the next section.

6.3.1.3 The group PSL(2, $\mathbb{C}$) and the automorphisms of the Riemann sphere

Let us consider the set of all 2×2 matrices whose entries are arbitrary complex numbers apart from the constraint that their determinant should be equal to 1. This set is named SL(2, $\mathbb{C}$) according to the nomenclature introduced in Chapter 3,

$$\mathfrak{A} \in \text{SL}(2, \mathbb{C}) \Rightarrow \mathfrak{A} = \begin{pmatrix} a & b \\ c & d \end{pmatrix}; \quad a, b, c, d \in \mathbb{C}; \quad ad - bc = 1, \tag{6.47}$$

and it corresponds to the $n = 2$ instance in the first of the infinite series of classical complex Lie groups, namely, the series of groups SL(n, $\mathbb{C}$) described in Chapter 3. In the specific case of SL(2, $\mathbb{C}$) we note two facts. Firstly, as we already stressed the one under consideration is not a discrete group whose elements can be enumerated; rather it is a continuous group whose elements fill up a complex manifold of complex dimension three, namely, the locus in $\mathbb{C}^4$ singled out by the quadric constraint $ad - bc = 1$. Secondly, we note that SL(2, $\mathbb{C}$) has a normal subgroup which we presently describe. The set composed of the two matrices

$$\text{N} = \left\{ \begin{pmatrix} 1 & 0 \\ 0 & 1 \end{pmatrix}, \begin{pmatrix} -1 & 0 \\ 0 & -1 \end{pmatrix} \right\} \tag{6.48}$$

constitutes a normal subgroup of SL(2, $\mathbb{C}$):

$$\text{N} \lhd \text{SL}(2, \mathbb{C}). \tag{6.49}$$

As a useful exercise the reader can easily verify that all the properties advocated by the relevant definitions are satisfied. The multiplication table of N shows that it is isomorphic with $\mathbb{Z}_2$, which, in any case, is the only group of order two. Having singled out a normal subgroup, we can introduce the factor group, which in the mathematical literature is given the standard name of PSL(2, $\mathbb{C}$):

$$\text{PSL}(2, \mathbb{C}) \equiv \frac{\text{SL}(2, \mathbb{C})}{\text{N}}. \tag{6.50}$$

Any time we construct a factor group $\frac{\text{G}}{\text{N}}$ we have a homomorphism,

$$h; \quad \text{G} \longrightarrow \frac{\text{G}}{\text{N}}, \tag{6.51}$$

and the kernel of the map h is isomorphic to N:

$$\ker h \sim \mathrm{N}. \tag{6.52}$$

In the case under consideration the homomorphism (6.51) can be realized in an analytic way and this realization is the very reason to consider PSL(2, ℂ) in the theory of analytic functions on the Riemann sphere.

Let us describe the homomorphism h in the following way. With every element of SL(2, ℂ) the homomorphism h associates an element of the set $\mathfrak{M}(\mathbb{C})$ of all meromorphic functions, defined as follows:

$$\forall \mathfrak{A} = \begin{pmatrix} a & b \\ c & d \end{pmatrix} \in \mathrm{SL}(2,\mathbb{C}); \quad h(\mathfrak{A}) = f_{\mathfrak{A}}(z) \equiv \frac{az+b}{cz+d}. \tag{6.53}$$

This is a homomorphism since, by explicit calculation, one can easily verify the following properties:

1. The matrix product is translated into the composition law of holomorphic maps, namely,

$$f_{\mathfrak{A}} \circ f_{\mathfrak{B}}(z) \equiv f_{\mathfrak{A}}(f_{\mathfrak{B}}(z)) = f_{\mathfrak{A}\mathfrak{B}}(z). \tag{6.54}$$

2. Both the map of the identity element and of its negative, namely, the elements of the normal subgroup N, are the identity holomorphic transformation:

$$\forall \mathfrak{N} \in \mathrm{N}: \quad f_{\mathfrak{N}}(z) = z. \tag{6.55}$$

3. The image of the inverse is the inverse function, namely,

$$\begin{aligned} &\forall \mathfrak{A} = \begin{pmatrix} a & b \\ c & d \end{pmatrix}: \quad f_{\mathfrak{A}^{-1}}(z) = \frac{dz-b}{-cz+a}, \\ &f_{\mathfrak{A}^{-1}}(w) = z \quad \mathit{iff} \quad f_{\mathfrak{A}}(z) = w. \end{aligned} \tag{6.56}$$

Because of item (2) in the above list, we see that the kernel of the homomorphism h is N, so that the Möbius group of the homeographic transformations

$$\frac{az+b}{cz+d} \tag{6.57}$$

is isomorphic to PSL(2, ℂ). From now on we will identify the elements of PSL(2, ℂ) with the corresponding homeographic transformation (6.57).

Properties of the PSL(2, ℂ) transformations

Considering now the homeographic transformations of PSL(2, ℂ), we observe the properties enumerated in the following list:

(a) Let $\mathscr{T} \subset \mathrm{PSL}(2, \mathbb{C})$ be the Abelian subgroup of homeographic transformations of the following form:

$$\tau_b(z) = z + b; \quad b \in \mathbb{C}. \tag{6.58}$$

The product law inside $\mathscr{T}$ reduces to

$$\tau_c(\tau_b(z)) = \tau_{b+c}(z), \tag{6.59}$$

which shows that $\mathscr{T} \sim \mathbb{C}$, named the *translation group*, is isomorphic to the additive group of complex numbers. The translation group is the homeomorphic image in $\mathrm{PSL}(2, \mathbb{C})$ of the following subgroup $\mathrm{T} \subset \mathrm{SL}(2, \mathbb{C})$:

$$T \ni \mathrm{t} = \begin{pmatrix} 1 & b \\ 0 & 1 \end{pmatrix}. \tag{6.60}$$

The action of the group $\mathscr{T}$ is *transitive* on the complex plane $\mathbb{C}$ since any point $z \in \mathbb{C}$ can be mapped by a suitable element of $\mathscr{T}$ into any other point $w \in \mathbb{C}$. Indeed, it suffices to choose $b = w - z$ and we have $\tau_{w-z}(z) = w$. In particular the transformation $\tau_{-w}(z)$ maps the arbitrary point $w \in \mathbb{C}$ into 0.

(b) Every point $w \in \mathbb{C}$ has a two-parameter *isotropy subgroup* $\mathscr{I}_w \subset \mathrm{PSL}(2, \mathbb{C})$ that leaves the point invariant. This group is the homeomorphic image $h(\mathrm{I}_w)$ in $\mathrm{PSL}(2, \mathbb{C})$ of the following subgroup of $\mathrm{SL}(2, \mathbb{C})$:

$$\mathrm{SL}(2, \mathbb{C}) \supset \mathrm{I}_w \ni \mathbf{i}_w[\lambda, \mu] \equiv \begin{pmatrix} \lambda + \mu w & , & \frac{w}{\lambda} - w(\lambda + \mu w) \\ \mu & , & \frac{1}{\lambda} - \mu w \end{pmatrix}, \quad \lambda, \mu \in \mathbb{C},\ \lambda \neq 0. \tag{6.61}$$

Indeed, we can easily verify that

$$\iota_w[\lambda, \mu](z) = f_{\mathbf{i}_w[\lambda,\mu]}(z) = \frac{w(-(\lambda + \mu w)) + \frac{w}{\lambda} + z(\lambda + \mu w)}{\frac{1}{\lambda} + \mu(z - w)} \tag{6.62}$$

has the property

$$\iota_w[\lambda, \mu](w) = w \quad \forall \lambda, \mu \in \mathbb{C}. \tag{6.63}$$

This result can be understood considering the isotropy subgroup of the special point $z = 0$. This is easily derived. We see that

$$\mathrm{SL}(2, \mathbb{C}) \supset \mathrm{I}_0 \ni \mathbf{i}_0[\lambda, \mu] \equiv \begin{pmatrix} \lambda & 0 \\ \mu & \frac{1}{\lambda} \end{pmatrix}, \tag{6.64}$$

whose image in $\mathrm{PSL}(2, \mathbb{C})$ is provided by the homeomorphic transformations

$$\iota_0[\lambda,\mu](z) = f_{\mathbf{i}_0[\lambda,\mu]}(z) = \frac{\lambda^2 z}{\lambda\mu z + 1}. \tag{6.65}$$

By multiplication of the 2×2 matrices in equation (6.64) we easily see that

$$\mathbf{i}_0[\lambda,\mu] \cdot \mathbf{i}_0[\rho,\sigma] = \mathbf{i}_0\left[\lambda\rho, \frac{\sigma}{\lambda} + \mu\rho\right]. \tag{6.66}$$

Obviously, the same composition law of the parameters is respected by the homomorphic image in $\mathrm{PSL}(2,\mathbb{C})$:

$$\iota_0[\lambda,\mu] \circ \iota_0[\rho,\sigma] = \iota_0\left[\lambda\rho, \frac{\sigma}{\lambda} + \mu\rho\right]. \tag{6.67}$$

Actually, all the groups I_w are isomorphic to I_0 because they are conjugate to it in $\mathrm{SL}(2,\mathbb{C})$ through the transformation of the translation group which brings the point w to 0. We can write

$$\mathrm{I}_w = \begin{pmatrix} 1 & w \\ 0 & 1 \end{pmatrix} \mathrm{I}_0 \begin{pmatrix} 1 & -w \\ 0 & 1 \end{pmatrix}, \tag{6.68}$$

which is indeed verified performing explicitly the following matrix multiplication:

$$\begin{pmatrix} 1 & w \\ 0 & 1 \end{pmatrix} \begin{pmatrix} \lambda & 0 \\ \mu & \frac{1}{\lambda} \end{pmatrix} \begin{pmatrix} 1 & -w \\ 0 & 1 \end{pmatrix} = \begin{pmatrix} \lambda + \mu w & , & \frac{w}{\lambda} - w(\lambda + \mu w) \\ \mu & , & \frac{1}{\lambda} - \mu w \end{pmatrix}. \tag{6.69}$$

Obviously, the same conjugation structure is preserved by the homeomorphic images in $\mathrm{PSL}(2,\mathbb{C})$ and we have

$$\mathscr{I}_w = \tau_w \circ \mathscr{I}_0 \circ \tau_{-w}. \tag{6.70}$$

The above procedure to find the isotropy subgroup of a point is standard whenever a group G has a transitive action on some manifold $\mathscr{M}$. Since any point $p \in \mathscr{M}$ can be mapped to some reference point $p_0 \in \mathscr{M}$, by a suitable element g_p of the group, it suffices to study the isotropy subgroup $\mathscr{I}_{p_0}$ of the reference point p_0. The isotropy subgroups of all the other points will be conjugate to $\mathscr{I}_{p_0}$ and therefore isomorphic to it. Indeed, we have

$$\mathscr{I}_p = g_p^{-1} \cdot \mathscr{I}_{p_0} \cdot g_p. \tag{6.71}$$

In Chapter 16 we discuss the general notion of homogeneous spaces G/H and we extensively come back to the above notions that constitute a pedagogical anticipation.

(c) The $\mathrm{PSL}(2,\mathbb{C})$ group has a transitive action not only on the complex plane, but rather on the whole Riemann sphere $\mathbb{S}^2 \simeq \mathbb{C} \cup \{\infty\}$. This is easily verified. We already saw that the translation subgroup $\mathscr{T} \subset \mathrm{PSL}(2,\mathbb{C})$ can map any finite point $z \in \mathbb{C}$ into any other finite point w. It suffices to show that there always exists at least one group element $\pm\mathfrak{A} = \pm\left(\begin{smallmatrix} a & b \\ c & d \end{smallmatrix}\right) \in \mathrm{PSL}(2,\mathbb{C})$ able to map the arbitrary finite point $w \in \mathbb{C}$ in ∞. Nothing is simpler to find. Considering the corresponding homeographic transformation

$$f_{\mathfrak{A}}(z) = \frac{az+b}{cz+d}, \tag{6.72}$$

it suffices that $f_{\mathfrak{A}}(w) = \infty$. This happens if $c = -\frac{d}{w}$. Hence, a particular element that does the required job is provided by

$$\pm\,\mathfrak{A} = \pm\begin{pmatrix} 1 & 0 \\ -\frac{1}{w} & 1 \end{pmatrix}. \tag{6.73}$$

All the other elements that do the same job are obtained from this one by multiplying it on the right by an element of the isotropy subgroup of w. In this way we get the two-parameter set of transformations

$$f_{w\to\infty}(z) = \frac{w(w(-(\lambda+\mu w)) + \frac{w}{\lambda} + z(\lambda+\mu w))}{\lambda w - \lambda z} \tag{6.74}$$

that for all λ, μ have the property $\lim_{z\to w} f_{w\to\infty}(z) = \infty$. To complete the proof of our statement about transitivity we also have to show that infinity can be mapped into any point w of the complex plane by a suitable homeographic transformation. To this effect it suffices that

$$\lim_{z\to\infty}\frac{az+b}{cz+d} = \frac{a}{c} = w, \tag{6.75}$$

which can be satisfied in many different ways. Hence, $\mathrm{PSL}(2,\mathbb{C})$ has a transitive action of the Riemann sphere $\mathbb{S}^2 = \mathbb{C} \cup \{\infty\}$.

(d) The group $\mathrm{PSL}(2,\mathbb{C})$ is the conformal automorphism group on the Riemann sphere. This follows from the fact that the translation subgroup $T \subset \mathrm{PSL}(2,\mathbb{C})$ is the automorphism group of the complex plane and hence the isotropy group of the point at infinity. Indeed, this fact realizes the conditions of a general lemma that we prove next.

Lemma 6.3.3. *Let D be an open subset of the Riemann sphere $\mathbb{S}^2$ and let G be a subgroup of the group $\Gamma(D)$ of all automorphisms of D. Let us assume that the following two conditions are verified:*

1. *G is transitive on D.*
2. *There exists at least one point $p \in D$ such that its isotropy group $\mathscr{I}_p$ is contained as a subgroup in G: $\mathscr{I}_p \subset G$.*

Then G is the group of all automorphisms of D: $G = \mathrm{Aut}[D]$.

Proof of Lemma 6.3.3. Let $\mathfrak{s} \in \mathrm{Aut}[D]$ and let $p \in D$ be a point of D whose isotropy group $\mathcal{I}_p$ is contained in G. Since G is transitive on D there exists a $\tau \in G$ such that $\mathfrak{s}(p) = \tau(p)$. Hence, the transformation $\mathfrak{K} = \tau^{-1} \circ S$ has the property that $\mathfrak{K}(p) = p$ so that it belongs to the isotropy group of p, $K \in \mathcal{I}_p \subset G$. Since $\tau \in G$ and $\mathfrak{K} \in G$, we have $G \ni \tau \circ \mathfrak{K} = \mathfrak{s}$. In conclusion, all the elements of $\mathrm{Aut}[D]$ are in G and $G = \mathrm{Aut}[D]$. The lemma is proved. □

For the application of this lemma to our purposes we choose $D = \mathbb{S}^2$, $G = \mathrm{PSL}(2,\mathbb{C})$, $p = \infty$, $\mathcal{I}_\infty = \mathrm{T} \subset \mathrm{PSL}(2,\mathbb{C})$. All the hypotheses are verified, so we conclude

$$\mathrm{PSL}(2,\mathbb{C}) = \mathrm{Aut}[\mathbb{S}^2]. \tag{6.76}$$

What we have seen in the present section is an illustration of the previously discussed group theoretical structures within the framework of a familiar contest, namely, that of the theory of complex analytic functions. In particular the study of the automorphism group PSL(2, ℂ) has given us the opportunity to inspect the case of one of the smaller Lie groups in its capacity of transformation group on a manifold. Furthermore, the considered example provided an illustration of a case of transitive action of the group G on the manifold $\mathcal{M}$. In such cases the manifold $\mathcal{M}$ is homeomorphic to the coset manifold G/H, where H is the isotropy subgroup of any point of $\mathcal{M}$. This is a very important lesson to be learned, since coset manifolds constitute a distinguished very important class of manifolds.

After this intermission let us come back to the development of basic notions of differential geometry.

6.4 Tangent and cotangent spaces

In elementary geometry the notion of a *tangent line* is associated with the notion of a curve. Hence, to introduce tangent vectors we have to begin with the notion of *curves in a manifold.*

Definition 6.4.1. A curve $\mathcal{C}$ in a manifold $\mathcal{M}$ is a continuous and differentiable map of an interval of the real line (say $[0,1] \subset \mathbb{R}$) into $\mathcal{M}$:

$$\mathcal{C} : [0,1] \to \mathcal{M}. \tag{6.77}$$

In other words, a curve is a one-dimensional submanifold $\mathcal{C} \subset \mathcal{M}$ (see Figure 6.10).

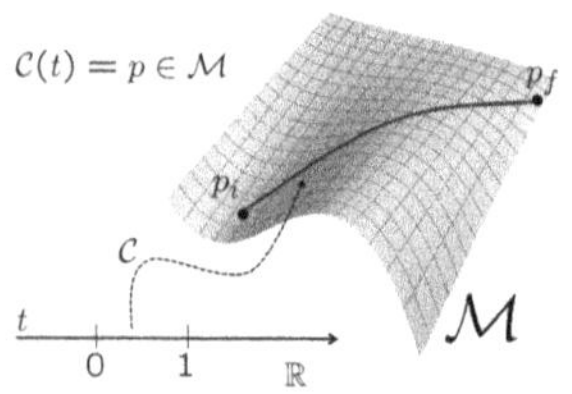

Figure 6.10: A curve in a manifold is a continuous map of an interval of the real line into the manifold.

There are curves with a *boundary*, namely, $\mathscr{C}(0) \cup \mathscr{C}(1)$, and open curves that do not contain a boundary. This happens if in equation (6.77) we replace the closed interval $[0,1]$ with the open interval $]0,1[$. *Closed curves* or *loops* correspond to the case where the initial and the final point coincide, that is, when $p_i \equiv \mathscr{C}(0) = \mathscr{C}(1) \equiv p_f$. Differently said, we have the following definition.

Definition 6.4.2. A closed curve is a continuous differentiable map of a circle into the manifold:

$$\mathscr{C} : \mathbb{S}^1 \to \mathscr{M}. \tag{6.78}$$

Indeed, identifying the initial and the final point means considering the points of the curve as being in one-to-one correspondence with the equivalence classes

$$\mathbb{R}/\mathbb{Z} \equiv \mathbb{S}^1, \tag{6.79}$$

which constitutes the mathematical definition of the circle. Explicitly equation (6.79) means that two real numbers r and r' are declared to be equivalent if their difference $r' - r = n$ is an integer number $n \in \mathbb{Z}$. As representatives of these equivalence classes we have the real numbers contained in the interval $[0,1]$ with the proviso that $0 \sim 1$.

We can also consider *semi-open curves* corresponding to maps of the semi-open interval $[0,1[$ into $\mathscr{M}$. In particular, in order to define tangent vectors we are interested in open branches of curves defined in the neighborhood of a point.

6.4.1 Tangent vectors at a point $p \in \mathscr{M}$

For each point $p \in \mathscr{M}$ let us fix an open neighborhood $U_p \subset \mathscr{M}$ and let us consider the semi-open curves of the following type:

$$\begin{cases} \mathscr{C}_p : [0,1[\to U_p, \\ \mathscr{C}_p(0) = p. \end{cases} \tag{6.80}$$

In other words, for each point p let us consider all possible curves $\mathscr{C}_p(t)$ that go through p (see Figure 6.11). Intuitively the tangent in p to a curve that starts from p is the vector that specifies the curve's *initial* direction. The basic idea is that in an m-dimensional manifold there are as many directions in which the curve can depart as there are vectors in $\mathbb{R}^m$; furthermore, for sufficiently small neighborhoods of p we cannot tell the difference between the manifold $\mathscr{M}$ and the flat vector space $\mathbb{R}^m$. Hence, to each point $p \in \mathscr{M}$ of a manifold we can attach an m-dimensional real vector space

$$\forall p \in \mathscr{M} : p \mapsto T_p\mathscr{M} \quad \dim T_p\mathscr{M} = m \tag{6.81}$$

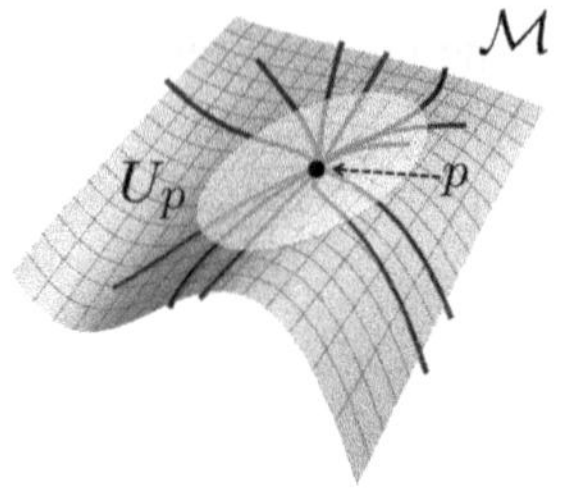

Figure 6.11: In a neighborhood U_p of each point $p \in \mathscr{M}$ we consider the curves that go through p.

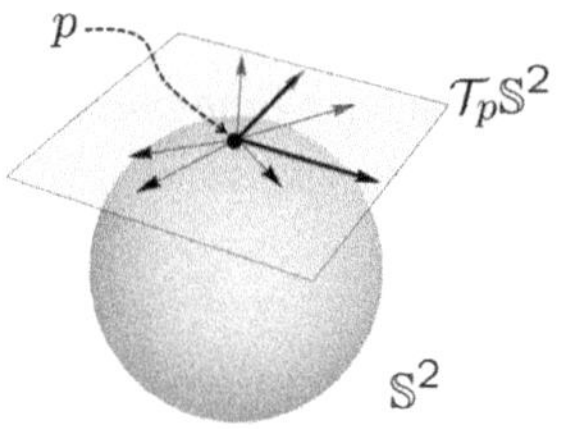

Figure 6.12: The tangent space in a generic point of an $\mathbb{S}^2$ sphere.

which parameterizes the possible directions in which a curve starting at p can depart. This vector space is named the tangent space to $\mathscr{M}$ at the point p and is, by definition, isomorphic to $\mathbb{R}^m$, namely, $T_p\mathscr{M} \sim \mathbb{R}^m$. For instance, to each point of an $\mathbb{S}^2$ sphere we attach a tangent plane $\mathbb{R}^2$ (see Figure 6.12) Let us now make this intuitive notion mathematically precise. Consider a point $p \in \mathscr{M}$ and a germ of a smooth function $f_p \in C_p^\infty(\mathscr{M})$. In any open chart $(U_\alpha, \varphi_\alpha)$ that contains the point p, the germ f_p is represented by an infinitely differentiable function of m variables:

$$f_p(x^1_{(\alpha)}, \ldots, x^m_{(\alpha)}). \tag{6.82}$$

Let us now choose an open curve $\mathscr{C}_p(t)$ that lies in U_α and starts at p,

$$\mathscr{C}_p(t) : \begin{cases} \mathscr{C}_p : [0,1[\rightarrow U_\alpha, \\ \mathscr{C}_p(0) = p, \end{cases} \tag{6.83}$$

and consider the composed map

$$f_p \circ \mathscr{C}_p : [0,1[\subset \mathbb{R} \rightarrow \mathbb{R}, \tag{6.84}$$

which is a real function

$$f_p(\mathscr{C}_p(t)) \equiv g_p(t) \tag{6.85}$$

of one real variable (see Figure 6.13).

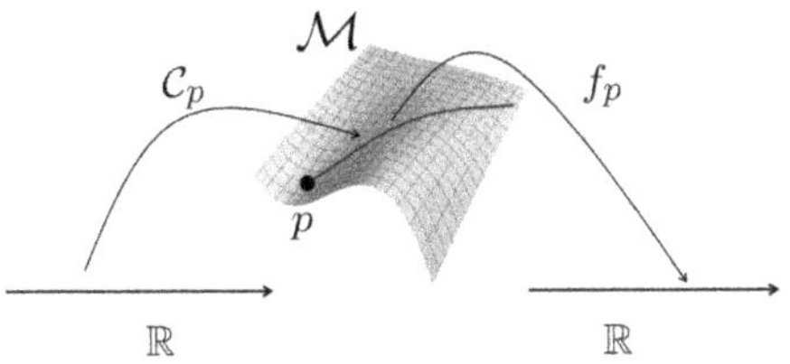

Figure 6.13: The composed map $f_p \circ \mathscr{C}_p$, where f_p is a germ of a smooth function in p and $\mathscr{C}_p$ is a curve departing from $p \in \mathscr{M}$.

We can calculate its derivative with respect to t in $t = 0$ which, in the open chart $(U_\alpha, \varphi_\alpha)$, reads as follows:

$$\frac{d}{dt} g_p(t)\bigg|_{t=0} = \frac{\partial f_p}{\partial x^\mu} \cdot \frac{dx^\mu}{dt}\bigg|_{t=0}. \tag{6.86}$$

We see from the above formula that the increment of any germ $f_p \in \mathbb{C}_p^\infty(\mathscr{M})$ along a curve $\mathscr{C}_p(t)$ is defined by means of the following m real coefficients:

$$c^\mu \equiv \frac{dx^\mu}{dt}\bigg|_{t=0} \in \mathbb{R}, \tag{6.87}$$

which can be calculated whenever the parametric form of the curve is given: $x^\mu = x^\mu(t)$. Explicitly we have

$$\frac{df_p}{dt} = c^\mu \frac{\partial f_p}{\partial x^\mu}. \tag{6.88}$$

Equation (6.88) can be interpreted as the action of a differential operator on the space of germs of smooth functions, namely,

$$\mathbf{t}_p \equiv c^\mu \frac{\partial}{\partial x^\mu} \quad \Rightarrow \quad \mathbf{t}_p : \mathbb{C}_p^\infty(\mathscr{M}) \mapsto \mathbb{C}_p^\infty(\mathscr{M}). \tag{6.89}$$

Indeed, for any germ f and for any curve

$$\mathbf{t}_p f = \frac{dx^\mu}{dt}\bigg|_{t=0} \frac{\partial f}{\partial x^\mu} \in \mathbb{C}_p^\infty(\mathscr{M}) \tag{6.90}$$

is a new germ of a smooth function in the point p. This discussion justifies the mathematical definition of the tangent space.

Definition 6.4.3. The tangent space $T_p\mathscr{M}$ to the manifold $\mathscr{M}$ in the point p is the vector space of *first order differential operators* on the germs of smooth functions $\mathbb{C}_p^\infty(\mathscr{M})$.

Next let us observe that the space of germs $\mathbb{C}_p^\infty(\mathscr{M})$ is an *algebra* with respect to linear combinations with real coefficients $(\alpha f + \beta g)(p) = \alpha f(p) + \beta g(p)$ and point-wise multiplication $f \cdot g(p) \equiv f(p)g(p)$,

$$\begin{aligned} \forall \alpha,\beta \in \mathbb{R} \quad \forall f,g \in \mathbb{C}_p^{\infty}(\mathcal{M}) \qquad & \alpha f + \beta g \in \mathbb{C}_p^{\infty}(\mathcal{M}), \\ \forall f,g \in \mathbb{C}_p^{\infty}(\mathcal{M}) \qquad & f \cdot g \in \mathbb{C}_p^{\infty}(\mathcal{M}), \\ & (\alpha f + \beta g)\cdot h = \alpha f \cdot h + \beta g \cdot h, \end{aligned} \tag{6.91}$$

and a tangent vector $\mathbf{t}_p$ is a *derivation* of this algebra.

Definition 6.4.4. A *derivation* $\mathcal{D}$ of an algebra $\mathcal{A}$ is a map

$$\mathcal{D} : \mathcal{A} \to \mathcal{A} \tag{6.92}$$

that:
1. is linear

$$\forall \alpha,\beta \in \mathbb{R} \quad \forall f,g \in \mathcal{A}: \quad \mathcal{D}(\alpha f + \beta g) = \alpha \mathcal{D} f + \beta \mathcal{D} g; \tag{6.93}$$

2. obeys the Leibnitz rule

$$\forall f,g \in \mathcal{A}: \quad \mathcal{D}(f\cdot g) = \mathcal{D} f \cdot g + f \cdot \mathcal{D} g. \tag{6.94}$$

That tangent vectors fit into Definition 6.4.4 is clear from their explicit realization as differential operators (equations (6.89) and (6.90)). It is also clear that the set of *derivations* $D[\mathcal{A}]$ of an algebra constitutes a real vector space. Indeed, a linear combination of derivations is still a derivation, having set

$$\forall \alpha,\beta \in \mathbb{R}, \forall \mathcal{D}_1,\mathcal{D}_2 \in D[\mathcal{A}], \forall f \in \mathcal{A}: \quad (\alpha\mathcal{D}_1 + \beta\mathcal{D}_2) f = \alpha \mathcal{D}_1 f + \beta \mathcal{D}_2 f. \tag{6.95}$$

Hence, an equivalent and more abstract definition of the tangent space is the following.

Definition 6.4.5. The tangent space to a manifold $\mathcal{M}$ at the point p is the vector space of derivations of the algebra of germs of smooth functions in p:

$$T_p\mathcal{M} \equiv D[\mathbb{C}_p^{\infty}(\mathcal{M})]. \tag{6.96}$$

Indeed, for any tangent vector (6.89) and any pair of germs $f,g \in \mathbb{C}_p^{\infty}(\mathcal{M})$ we have

$$\begin{aligned} \mathbf{t}_p(\alpha f + \beta g) &= \alpha \mathbf{t}_p(f) + \beta \mathbf{t}_p(g), \\ \mathbf{t}_p(f\cdot g) &= \mathbf{t}_p(f)\cdot g + f \cdot \mathbf{t}_p(g). \end{aligned} \tag{6.97}$$

In each coordinate patch a tangent vector is, as we have seen, a first order differential operator singled out by its *components*, namely, by the coefficients c^{μ}. In the language of tensor calculus the tangent vector *is identified* with the m-tuple of real numbers c^{μ}. The relevant point, however, is that such m-tuple representing the *same tangent vector* is different in different coordinate patches. Consider two coordinate patches (U,φ) and (V,ψ) with non-vanishing intersection. Denote by x^{μ} the coordinate of a point $p \in U \cap V$

in the patch (U, φ) and by y^α the coordinate of the same point in the patch (V, ψ). The transition function and its inverse are expressed by setting

$$x^\mu = x^\mu(y); \quad y^\nu = y^\nu(x). \tag{6.98}$$

Then the same first order differential operator can be alternatively written as

$$\mathbf{t}_p = c^\mu \frac{\partial}{\partial x^\mu} \quad \text{or} \quad \mathbf{t}_p = c^\mu \left(\frac{\partial y^\nu}{\partial x^\mu} \right) \frac{\partial}{\partial y^\nu} = c^\nu \frac{\partial}{\partial y^\nu}, \tag{6.99}$$

having defined

$$c^\nu \equiv c^\mu \left(\frac{\partial y^\nu}{\partial x^\mu} \right). \tag{6.100}$$

Equation (6.100) expresses the transformation rule for the components of a tangent vector from one coordinate patch to another one (see Figure 6.14). Such a transformation is *linear* and the matrix that realizes it is the *inverse of the Jacobian matrix* $(\partial y/\partial x) = (\partial x/\partial y)^{-1}$. For this reason we say that the components of a tangent vector constitute a *contravariant world vector*. By definition a *covariant world vector* transforms instead with the *Jacobian matrix*. We will see that covariant world vectors are the components of a differential form.

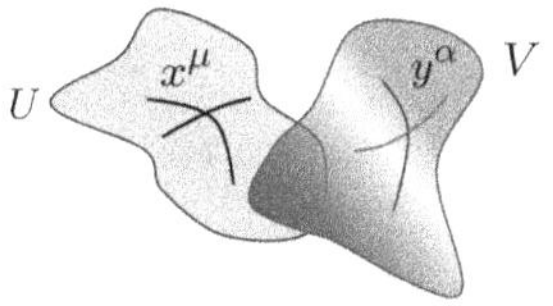

Figure 6.14: Two coordinate patches.

6.4.2 Differential forms at a point $p \in \mathcal{M}$

Let us now consider the total differential of a function (better said: of the germ of a smooth function) when we evaluate it along a curve. For all $f \in \mathbb{C}_p^\infty(\mathcal{M})$ and for each curve $c(t)$ starting at p we have

$$\left. \frac{d}{dt} f(c(t)) \right|_{t=0} = c^\mu \frac{\partial}{\partial x^\mu} f \equiv \mathbf{t}_p f, \tag{6.101}$$

where we have denoted by $\mathbf{t}_p = \frac{dc^\mu}{dt}|_{t=0} \frac{\partial}{\partial x^\mu}$ the tangent vector to the curve in its initial point p. So, fixing a tangent vector means that for any germ f we know its total differential along the curve that admits such a vector as tangent in p. Let us now reverse our viewpoint. Rather than keeping the tangent vector fixed and letting the germ f vary, let us keep the germ f fixed and consider all possible curves that depart from the point p. We would like to evaluate the total derivative of the germ $\frac{df}{dt}$ along each curve. The solution of such a problem is easily obtained: given the tangent vector $\mathbf{t}_p$ to the curve in p we have $df/dt = \mathbf{t}_p f$. The moral of this tale is the following: the concept of *total differential of a germ* is the *dual* of the concept of *tangent vector*. Indeed, we recall from linear algebra that the dual of a vector space is the space of linear functionals on that vector space and our discussion shows that the total differential of a germ is precisely a linear functional on the tangent space $T_p\mathcal{M}$.

Definition 6.4.6. The total differential df_p of a smooth germ $f \in \mathbb{C}_p^\infty(\mathcal{M})$ is a *linear functional* on $T_p\mathcal{M}$ such that

$$\begin{aligned} &\forall \mathbf{t}_p \in T_p\mathcal{M} && df_p(\mathbf{t}_p) = \mathbf{t}_p f, \\ &\forall \mathbf{t}_p, \mathbf{k}_p \in T_p\mathcal{M},\ \forall \alpha, \beta \in \mathbb{R} && df_p(\alpha \mathbf{t}_p + \beta \mathbf{k}_p) = \alpha df_p(\mathbf{t}_p) + \beta df_p(\mathbf{k}_p). \end{aligned} \tag{6.102}$$

The linear functionals on a finite-dimensional vector space $\mathcal{V}$ constitute a vector space $\mathcal{V}^\star$ (the dual) with the same dimension. This justifies the following.

Definition 6.4.7. We call the vector space $T_p^*\mathcal{M}$ of linear functionals (or 1-forms in p) on the tangent space $T_p\mathcal{M}$ the *cotangent space* to the manifold $\mathcal{M}$ in the point p:

$$T_p^*\mathcal{M} \equiv \mathrm{Hom}(T_p\mathcal{M}, \mathbb{R}) = (T_p\mathcal{M})^\star. \tag{6.103}$$

So we call the elements of the cotangent space differential 1-forms in p, and $\forall \omega_p \in T_p^*\mathcal{M}$ we have

$$\begin{aligned} &(1) && \forall \mathbf{t}_p \in T_p\mathcal{M}: && \omega_p(\mathbf{t}_p) \in \mathbb{R}, \\ &(2) && \forall \alpha, \beta \in \mathbb{R}, \forall \mathbf{t}_p, \mathbf{k}_p \in T_p\mathcal{M}: && \omega_p(\alpha \mathbf{t}_p + \beta \mathbf{k}_p) = \alpha \omega_p(\mathbf{t}_p) + \beta \omega_p(\mathbf{k}_p). \end{aligned} \tag{6.104}$$

The reason why the above linear functionals are named differential 1-forms is that in every coordinate patch $\{x^\mu\}$ they can be expressed as linear combinations of the coordinate differentials,

$$\omega_p = \omega_\mu\, dx^\mu, \tag{6.105}$$

and their action on the tangent vectors is expressed as follows:

$$\mathbf{t}_p = c^\mu \frac{\partial}{\partial x^\mu} \quad \Rightarrow \quad \omega_p(\mathbf{t}_p) = \omega_\mu c^\mu \in \mathbb{R}. \tag{6.106}$$

Indeed, in the particular case where the 1-form is exact (namely, it is the differential of a germ) $\omega_p = df_p$ we can write $\omega_p = \partial f/\partial x^\mu\, dx^\mu$ and we have $df_p(\mathbf{t}_p) \equiv \mathbf{t}_p f = c^\mu \partial f/\partial x^\mu$.

Hence, when we extend our definition to differential forms that are not exact we continue to make the same statement, namely, that the value of the 1-form on a tangent vector is given by equation (6.106).

Summarizing, in each coordinate patch, a differential 1-form in a point $p \in \mathscr{M}$ has the representation (6.105) and its coefficients ω_μ constitute a *contravariant vector*. Indeed, in complete analogy to equation (6.99), we have

$$\omega_p = \omega_\mu \, dx^\mu \quad \text{or} \quad \omega_p = \omega_\mu \left(\frac{\partial x^\mu}{\partial y^\nu} \right) dy^\nu = \omega_\nu \, dy^\nu, \tag{6.107}$$

having defined

$$\omega_\nu \equiv \omega_\mu \left(\frac{\partial x^\mu}{\partial y^\nu} \right). \tag{6.108}$$

Finally, the duality relation between 1-forms and tangent vectors can be summarized writing the rule

$$dx^\mu \left(\frac{\partial}{\partial x^\nu} \right) = \delta^\mu_\nu. \tag{6.109}$$

6.5 About the concept of fiber bundle

The next step we have to take is *gluing together* all the tangent spaces $T_p\mathscr{M}$ and cotangent spaces $T_p^*\mathscr{M}$ we have discussed in the previous sections. The result of such a gluing procedure is not a vector space; rather it is a vector bundle. Vector bundles are specific instances of the more general notion of *fiber bundles*.

The concept of *fiber bundle* is absolutely central in contemporary physics and provides the appropriate mathematical framework to formulate modern field theory since all the fields one can consider are either *sections* of *associated bundles* or *connections* on *principal bundles*. There are two kinds of fiber bundles:

1. principal bundles
2. associated bundles

The notion of a principal fiber bundle is the appropriate mathematical concept underlying the formulation of *gauge theories* that provide the general framework to describe the dynamics of all non-gravitational interactions. The concept of a connection on such principal bundles codifies the physical notion of the bosonic particles mediating the interaction, namely, the gauge bosons, like the photon, the gluon, or the graviton. Indeed, gravity itself is a gauge theory although of a very special type. On the other hand, the notion of associated fiber bundles is the appropriate mathematical framework to describe *matter fields* that interact through the exchange of the *gauge bosons*.

Also from a more general viewpoint and in relation with all sorts of applications the notion of fiber bundles is absolutely fundamental. As we already emphasized, the points of a manifold can be identified with the possible states of a complex system specified by an m-tuple of parameters $x_1, \dots x_m$. Real or complex functions of such parameters are the natural objects one expects to deal with in any scientific theory that explains the phenomena observed in such a system. Yet, as we already anticipated, calculus on manifolds that are not trivial as the flat $\mathbb{R}^m$ cannot be confined to functions, which is a too restrictive notion. The appropriate generalization of functions is provided by the *sections* of fiber bundles. Locally, namely in each coordinate patch, functions and sections are just the same thing. Globally, however, there are essential differences. A section is obtained by gluing together many local functions by means of non-trivial transition functions that reflect the geometric structure of the fiber bundle.

6.6 The notion of Lie group

To introduce the mathematical definition of a fiber bundle we need to introduce the definition of a Lie group, which will be a central topic in the sequel.

Definition 6.6.1. A Lie group G is:

- a group from the algebraic point of view, namely, a set with an internal composition law, the product

$$\forall g_1 g_2 \in G, \quad g_1 \cdot g_2 \in G, \tag{6.110}$$

  which is associative, admits a unique neutral element e, and yields an inverse for each group element;
- a smooth manifold of finite dimension $\dim G = n < \infty$ whose transition functions are not only infinitely differentiable but also real analytic, namely, they admit an expansion in power series;
- in the topology defined by the manifold structure the two algebraic operations of taking the inverse of an element and performing the product of two elements are real analytic (admit a power series expansion).

The last point in Definition 6.6.1 deserves a more extended explanation. With each group element the product operation associates two maps of the group into itself:

$$\begin{aligned} \forall g \in G: \quad L_g: \quad G \to G: \quad g' \to L_g(g') \equiv g' \cdot g, \\ \forall g \in G: \quad R_g: \quad G \to G: \quad g' \to R_g(g') \equiv g \cdot g', \end{aligned} \tag{6.111}$$

respectively named the *left translation* and the *right translation*. Both maps are required to be real analytic for each choice of $g \in G$. Similarly, the group structure induces a map,

$$(\cdot)^{-1} : G \to G : g \to g^{-1}, \tag{6.112}$$

which is also required to be real analytic.

6.7 Developing the notion of fiber bundle

Coming now to fiber bundles let us begin by recalling that a pedagogical and pictorial example of such spaces is provided by the celebrated case of a Möbius strip (see Figure 6.15). The basic idea is that if we consider a piece of the bundle this cannot be distinguished from a trivial direct product of two spaces, an open subset of the base manifold and the fiber. In Figure 6.15 the base manifold is a circle and the fiber is a segment $I \equiv [-1,1]$. Locally the space is the direct product of an open interval of $U =]a,b[\subset \mathbb{R}$ with the standard fiber I, as is evident from Figure 6.16. However, the relevant point is that, *globally*, the bundle *is not a direct product of spaces*.

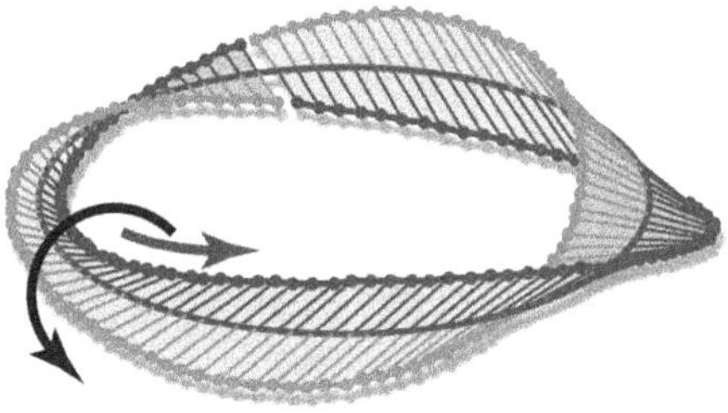

Figure 6.15: The Möbius strip provides a pedagogical example of a fiber bundle.

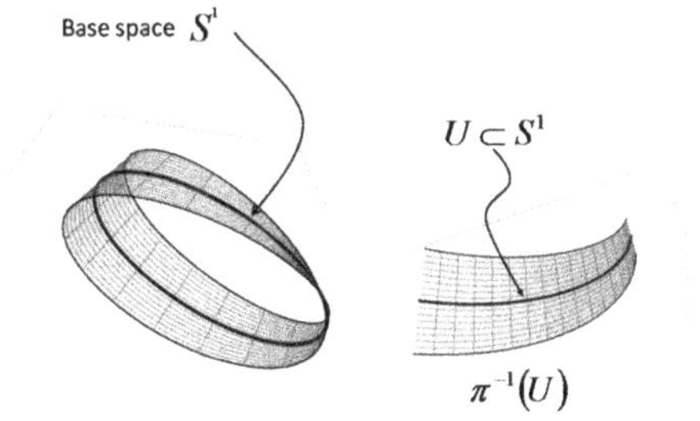

Figure 6.16: Local triviality of an open piece of the Möbius strip.

Hence the notion of fiber bundle corresponds to that of a differentiable manifold P with dimension $\dim P = m + n$ that locally *looks like* the direct product $U \times F$ of an open manifold U of dimension $\dim U = m$ with another manifold F (the standard fiber) of dimension $\dim F = n$. Essential in the definition is the existence of a map

$$\pi : P \rightarrow \mathcal{M} \tag{6.113}$$

named *the projection* from the *total manifold* P of dimension $m + n$ to a manifold $\mathcal{M}$ of dimension m, named the *base manifold.* Such a map is required to be continuous. Due to the difference in dimensions the projection cannot be invertible. Indeed, with every point $\forall p \in \mathcal{M}$ of the base manifold the projection associates a submanifold $\pi^{-1}(p) \subset P$ of dimension $\dim \pi^{-1}(p) = n$ composed of those points of $x \in P$ whose projection on $\mathcal{M}$ is the chosen point p: $\pi(x) = p$. The submanifold $\pi^{-1}(p)$ is called the *fiber over p* and the basic idea is that each fiber is homeomorphic to the *standard fiber F*. More precisely, for each open subset $U_\alpha \subset \mathcal{M}$ of the base manifold the submanifold

$$\pi^{-1}(U_\alpha)$$

must be homeomorphic to the direct product

$$U_\alpha \times F.$$

This is the precise meaning of the statement that, locally, the bundle looks like a direct product (see Figure 6.17). Explicitly what we require is the following: there should be a family of pairs (U_α, ϕ_α) where U_α are open charts covering the base manifold $\bigcup_\alpha U_\alpha = \mathcal{M}$ and ϕ_α are maps,

$$\phi_\alpha : \pi^{-1}(U_\alpha) \subset P \rightarrow U_\alpha \otimes F, \tag{6.114}$$

that are required to be one-to-one and bicontinuous (= continuous, together with its inverse) and to satisfy the property that

$$\pi \circ \phi_\alpha^{-1}(p,f) = p. \tag{6.115}$$

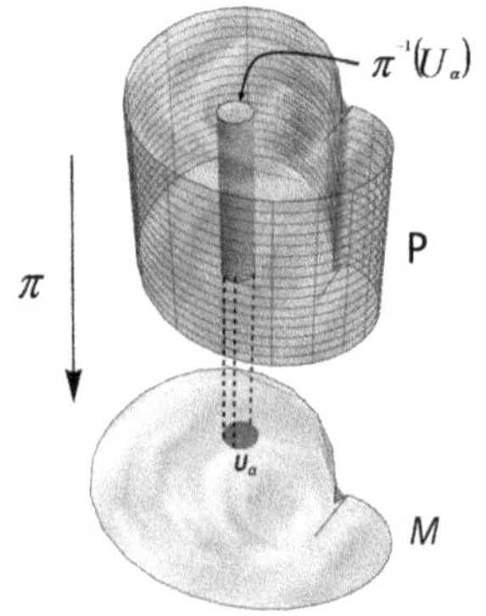

Figure 6.17: A fiber bundle is locally trivial.

Namely, the projection of the image in P of a base manifold point p *times* some fiber point f is p itself. Each pair (U_α, ϕ_α) is named a *local trivialization*. As for the case of manifolds, the interesting question is what happens in the intersection of two different local trivializations. Indeed, if $U_\alpha \cap U_\beta \neq \emptyset$, then we also have $\pi^{-1}(U_\alpha) \cap \pi^{-1}(U_\beta) \neq \emptyset$. Hence, each point $x \in \pi^{-1}(U_\alpha \cap U_\beta)$ is mapped by ϕ_α and ϕ_β in two different pairs $(p, f_\alpha) \in U_\alpha \otimes F$ and $(p, f_\beta) \in U_\alpha \otimes F$ with the property, however, that the first entry p is the same in both pairs. This follows from property (6.115). It implies that there must exist a map

$$t_{\alpha\beta} \equiv \phi_\beta^{-1} \circ \phi_\alpha : (U_\alpha \cap U_\beta) \otimes F \to (U_\alpha \cap U_\beta) \otimes F \tag{6.116}$$

named *transition function* which acts exclusively on the fiber points in the sense that

$$\forall p \in U_\alpha \cap U_\beta, \ \forall f \in F \quad t_{\alpha\beta}(p,f) = (p, t_{\alpha\beta}(p).f), \tag{6.117}$$

where for each choice of the point $p \in U_\alpha \cap U_\beta$,

$$t_{\alpha\beta}(p) : F \mapsto F \tag{6.118}$$

is a continuous and invertible map of the standard fiber F into itself (see Figure 6.18). The last bit of information contained in the notion of fiber bundle is related to the *structural group*. This has to do with answering the following question: Where are the transition functions chosen from? Indeed, the set of all possible continuous invertible maps of the standard fiber F into itself constitutes a group, so that it is no restriction to say that the transition functions $t_{\alpha\beta}(p)$ are group elements. Yet the group of all homeomorphisms Hom(F, F) is very large and it makes sense to include into the definition of fiber bundle the requirement that the transition functions should be chosen within a smaller hunting ground, namely, inside some finite-dimensional Lie group G that has a well-defined action on the standard fiber F.

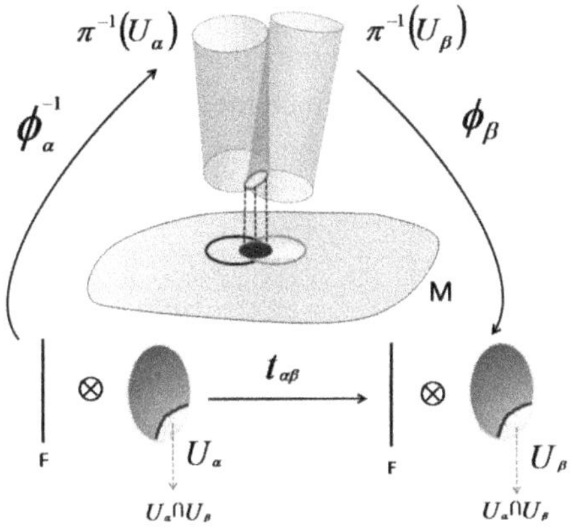

Figure 6.18: Transition function between two local trivializations of a fiber bundle.

The above discussion can be summarized into the following technical definition of fiber bundles.

Definition 6.7.1. A fiber bundle $(P,\pi,\mathcal{M},F,G)$ is a geometrical structure that consists of the following list of elements:

1. A differentiable manifold P named the *total space.*
2. A differentiable manifold $\mathcal{M}$ named the *base space.*
3. A differentiable manifold F named the *standard fiber.*
4. A Lie group G, named the *structure group*, which acts as a transformation group on the standard fiber:

$$\forall g \in G; \quad g : F \longrightarrow F \quad \{\text{i. e., } \forall f \in F \ g.f \in F\}. \tag{6.119}$$

5. A surjection map $\pi : P \longrightarrow \mathcal{M}$, named the *projection*. If $n = \dim \mathcal{M}$, $m = \dim F$, then we have $\dim P = n+m$ and $\forall p \in \mathcal{M}, F_p = \pi^{-1}(p)$ is an m-dimensional manifold diffeomorphic to the standard fiber F. The manifold F_p is named the *fiber at the point p*.
6. A covering of the base space $\bigcup_{(\alpha\in A)} U_\alpha = \mathcal{M}$, realized by a collection $\{U_\alpha\}$ of open subsets ($\forall \alpha \in A\ U_\alpha \subset \mathcal{M}$), equipped with a homeomorphism

$$\phi_\alpha^{-1} : U_\alpha \times F \longrightarrow \pi^{-1}(U_\alpha) \tag{6.120}$$

such that

$$\forall p \in U_\alpha, \forall f \in F : \quad \pi \cdot \phi_\alpha^{-1}(p,f) = p. \tag{6.121}$$

The map ϕ_α^{-1} is named a *local trivialization* of the bundle, since its inverse ϕ_α maps the open subset $\pi^{-1}(U_\alpha) \subset P$ of the total space into the direct product $U_\alpha \times F$.

7. If we write $\phi_\alpha^{-1}(p,f) = \phi_{\alpha,p}^{-1}(f)$, the map $\phi_{\alpha,p}^{-1} : F \longrightarrow F_p$ is the homeomorphism required by point (6) of the present definition. For all points $p \in U_\alpha \cap U_\beta$ in the intersection of two different local trivialization domains, the composite map $t_{\alpha\beta}(p) = \phi_{\alpha,p} \cdot \phi_{\beta,p}^{-1} F \longrightarrow F$ is an element of the structure group $t_{\alpha\beta} \in G$, named the *transition function*. Furthermore, the transition function realizes a smooth map $t_{\alpha\beta} : U_\alpha \cap U_\beta \longrightarrow G$. We have

$$\phi_\beta^{-1}(p,f) = \phi_\alpha^{-1}(p, t_{\alpha\beta}(p).f). \tag{6.122}$$

Just as manifolds can be constructed by gluing together open charts, fiber bundles can be obtained by gluing together local trivializations. Explicitly one proceeds as follows.

1. First choose a base manifold $\mathcal{M}$, a typical fiber F, and a structural Lie group G whose action on F must be well-defined.

2. Then choose an atlas of open neighborhoods $U_\alpha \subset \mathcal{M}$ covering the base manifold $\mathcal{M}$.
3. Next to each non-vanishing intersection $U_\alpha \cap U_\beta \neq \emptyset$ assign a transition function, namely, a smooth map

$$\psi_{\alpha\beta} : U_\alpha \cap U_\beta \mapsto G, \tag{6.123}$$

from the open subset $U_\alpha \cap U_\beta \subset \mathcal{M}$ of the base manifold to the structural Lie group. For consistency the transition functions must satisfy the following two conditions:

$$\begin{aligned} \forall U_\alpha, U_\beta / U_\alpha \cap U_\beta \neq \emptyset : \quad & \psi_{\beta\alpha} = \psi_{\alpha\beta}^{-1}, \\ \forall U_\alpha, U_\beta, U_\gamma / U_\alpha \cap U_\beta \cap U_\gamma \neq \emptyset : \quad & \psi_{\alpha\beta} \cdot \psi_{\beta\gamma} \cdot \psi_{\gamma\alpha} = \mathbf{1}_G. \end{aligned} \tag{6.124}$$

Whenever a set of local trivializations with consistent transition functions satisfying equation (6.124) has been given, a fiber bundle is defined. A different and much more difficult task to complete is determining whether two sets of local trivializations define the same fiber bundle or not. We do not address such a problem whose proper treatment is beyond the scope of this textbook. We just point out that the classification of inequivalent fiber bundles one can construct on a given base manifold $\mathcal{M}$ is a problem of global geometry which can also be addressed with the techniques of algebraic topology and algebraic geometry.

Typically inequivalent bundles are characterized by topological invariants that receive the name of *characteristic classes*.

In physical language the transition functions in (6.123) from one local trivialization to another one are the *gauge transformations*, namely, group transformations depending on the position in space-time (i. e., the point on the base manifold).

Definition 6.7.2. A principal bundle $P(\mathcal{M}, G)$ is a fiber bundle where the standard fiber coincides with the structural Lie group $F = G$ and the action of G on the fiber is the left (or right) multiplication (see equation (6.111)):

$$\forall g \in G \quad \Rightarrow \quad L_g : G \mapsto G. \tag{6.125}$$

The name principal is given to the fiber bundle in Definition 6.7.2 since it is a *"father"* bundle which, once given, generates an infinity of *associated vector bundles*, one for each linear representation of the Lie group G.

Let us anticipate the notion of linear representations of a Lie group, which will be extensively discussed in Chapter 12.

Definition 6.7.3. Let V be a vector space of finite dimension $\dim V = m$ and let $\mathrm{Hom}(V, V)$ be the group of all linear homomorphisms of the vector space into itself:

$$\begin{aligned} & f \in \mathrm{Hom}(\mathrm{V}, \mathrm{V}) / f : V \to V, \\ \forall \alpha, \beta \in \mathbb{R}, \ \forall v_1, v_2 \in V : \quad & f(\alpha \mathbf{v}_1 + \beta \mathbf{v}_2) = \alpha f(\mathbf{v}_1) + \beta f(\mathbf{v}_2). \end{aligned} \tag{6.126}$$

A linear representation of the Lie group G of dimension n is a *group homomorphism*:

$$\begin{cases} \forall g \in G & g \mapsto D(g) \in \mathrm{Hom}(\mathrm{V},\mathrm{V}), \\ \forall g_1 g_2 \in G & D(g_1 \cdot g_2) = D(g_1) \cdot D(g_2), \\ & D(e) = \mathbf{1}, \\ \forall g \in G & D(g^{-1}) = [D(g)]^{-1}. \end{cases} \tag{6.127}$$

Whenever we choose a basis $\mathbf{e}_1, \mathbf{e}_2, \ldots, \mathbf{e}_n$ of the vector space V every element $f \in \mathrm{Hom}(V, V)$ is represented by a matrix $f^j_{\ i}$ defined by

$$f(\mathbf{e}_i) = \mathbf{e}_j f^j_{\ i}. \tag{6.128}$$

Therefore, a linear representation of a Lie group associates with each abstract group element g an $n \times n$ matrix $D(g)^j_{\ i}$. As should be known to the reader, linear representations are said to be *irreducible* if the vector space V admits *no* non-trivial vector subspace $W \subset V$ that is *invariant* with respect to the action of the group: $\forall g \in G / D(g) W \subset W$ (see Section 12.1). For simple Lie groups reducible representations can always be decomposed into a direct sum of irreducible representations, namely, $V = V_1 \oplus V_2 \oplus \cdots \oplus V_r$ (with V_i irreducible), and irreducible representations are completely defined by the structure of the group. These notions that we have recalled from group theory motivate the following definition.

Definition 6.7.4. An *associated vector bundle* is a fiber bundle where the standard fiber $F = V$ is a vector space and the action of the structural group on the standard fiber is a linear representation of G on V.

The reason why the bundles in Definition 6.7.4 are called associated is almost obvious. Given a principal bundle and a linear representation of G we can immediately construct a corresponding vector bundle. It suffices to use as transition functions the linear representation of the transition functions of the principal bundle:

$$\psi^{(V)}_{\alpha\beta} \equiv D(\psi^{(G)}_{\alpha\beta}) \in \mathrm{Hom}(V, V). \tag{6.129}$$

For any vector bundle the dimension of the standard fiber is called the *rank* of the bundle.

Whenever the base manifold of a fiber bundle is complex and the transition functions are holomorphic maps, we say that the bundle is *holomorphic*.

A very important and simple class of holomorphic bundles are the *line bundles*. By definition these are principal bundles on a complex base manifold $\mathcal{M}$ with structural group $\mathbb{C}^\star \equiv \mathbb{C} \backslash 0$, namely, the multiplicative group of non-zero complex numbers. Let $z_\alpha(p) \in \mathbb{C}^\star$ be an element of the standard fiber above the point $p \in U_\alpha \cap U_\beta \subset \mathcal{M}$ in the local trivialization α and let $z_\beta(p) \in \mathbb{C}^\star$ be the corresponding fiber point in the local trivialization β. The transition function between the two trivializations is expressed by (see Figure 6.19):

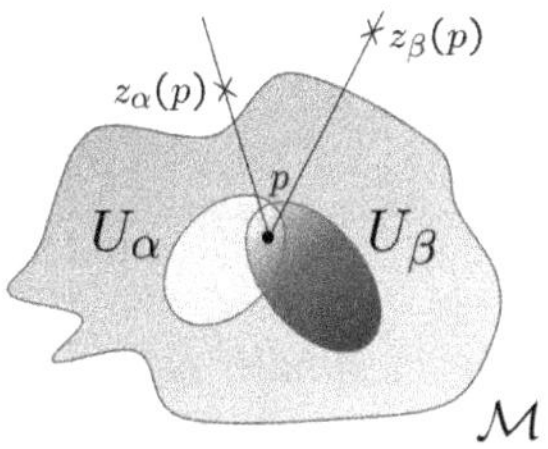

Figure 6.19: The intersection of two local trivializations of a line bundle.

$$z_\alpha(p) = \underbrace{f_{\alpha\beta}(p)}_{\in \mathbb{C}^\star} \cdot z_\beta(p) \quad \Rightarrow \quad f_{\alpha\beta}(p) = \frac{z_\alpha(p)}{z_\beta(p)} \neq 0. \tag{6.130}$$

6.8 Tangent and cotangent bundles

Let $\mathcal{M}$ be a differentiable manifold of dimension $\dim \mathcal{M} = m$: in Section 6.4 we have seen how to construct the tangent spaces $T_p\mathcal{M}$ associated with each point $p \in \mathcal{M}$ of the manifold. We have also seen that each $T_p\mathcal{M}$ is a real vector space isomorphic to $\mathbb{R}^m$. Considering the definition of fiber bundles discussed in the previous section we now realize that what we actually did in Section 6.4 was constructing a vector bundle, the *tangent bundle* $T\mathcal{M}$ (see Figure 6.20). In the tangent bundle $T\mathcal{M}$ the *base manifold* is the differentiable manifold $\mathcal{M}$, the *standard fiber* is $F = \mathbb{R}^m$, and the structural group is $\mathrm{GL}(m, \mathbb{R})$, namely, the group of real $m \times m$ matrices. The main point is that the transition functions are not newly introduced to construct the bundle; rather they are completely determined from the transition functions relating open charts of the base manifold. In other words, whenever we define a manifold $\mathcal{M}$, associated with it there is a unique vector bundle $T\mathcal{M} \to \mathcal{M}$ which encodes many intrinsic properties of $\mathcal{M}$. Let us see how.

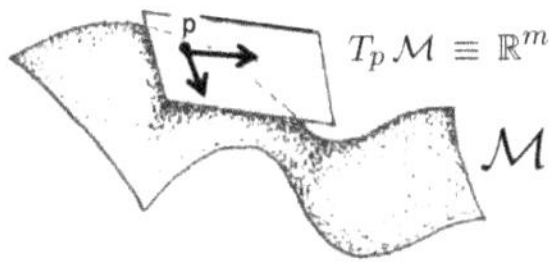

Figure 6.20: The tangent bundle is obtained by gluing together all the tangent spaces.

Consider two intersecting local charts (U_α, ϕ_α) and (U_β, ϕ_β) of our manifold. A tangent vector in a point $p \in \mathcal{M}$ was written as

$$\mathbf{t}_p = c^\mu(p) \frac{\partial}{\partial x^\mu}\bigg|_p. \tag{6.131}$$

Now we can consider choosing smoothly a tangent vector for each point $p \in \mathscr{M}$, namely, introducing a map:

$$p \in \mathscr{M} \mapsto \mathbf{t}_p \in T_p\mathscr{M}. \tag{6.132}$$

Mathematically what we have obtained is a *section of the tangent bundle*, namely, a smooth choice of a point in the fiber for each point of the base. Explicitly this just means that the components $c^\mu(p)$ of the tangent vector are smooth functions of the base point coordinates x^μ. Since we use coordinates, we need an extra label denoting in which local patch the vector components are given:

$$\begin{cases} \mathbf{t} = c^\mu_{(\alpha)}(x) \left.\dfrac{\partial}{\partial x^\mu}\right|_p & \Rightarrow \quad \text{in chart } \alpha, \\ \mathbf{t} = c^\nu_{(\beta)}(y) \left.\dfrac{\partial}{\partial y^\nu}\right|_p & \Rightarrow \quad \text{in chart } \beta, \end{cases} \tag{6.133}$$

having denoted by x^μ and y^ν the local coordinates in patches α and β, respectively. Since the tangent vector is the same, irrespectively of the coordinates used to describe it, we have

$$c^\nu_{(\beta)}(y)\frac{\partial}{\partial y^\nu} = c^\mu_{(\alpha)}(x)\frac{\partial y^\nu}{\partial x^\mu}\frac{\partial}{\partial y^\nu}, \tag{6.134}$$

that is,

$$c^\nu_{(\beta)}(p) = c^\mu_{(\alpha)}(p)\left(\frac{\partial y^\nu}{\partial x^\mu}\right)(p). \tag{6.135}$$

In formula (6.135) we see the explicit form of the transition function between two local trivializations of the tangent bundle: it is simply the *inverse Jacobian matrix* associated with the transition functions between two local charts of the base manifold $\mathscr{M}$. On the intersection $U_\alpha \cap U_\beta$ we have

$$\forall p \in U_\alpha \cap U_\beta : \quad p \to \psi_{\beta\alpha}(p) = \left(\frac{\partial y}{\partial x}\right)(p) \in \mathrm{GL}(\mathrm{m}, \mathbb{R}), \tag{6.136}$$

as pictorially described in Figure 6.21.

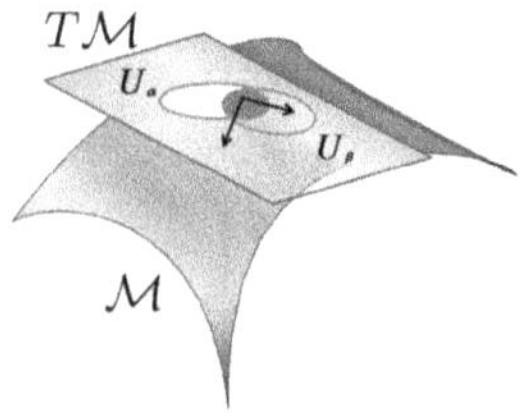

Figure 6.21: Two local charts of the base manifold $\mathscr{M}$ yield two local trivializations of the tangent bundle $T\mathscr{M}$. The transition function maps a vector into another one.

6.8.1 Sections of a bundle

It is now the appropriate time to associate a precise definition with the notion of bundle section that we have implicitly advocated in equation (6.132).

Definition 6.8.1. Consider a generic fiber bundle $E \xrightarrow{\pi} \mathcal{M}$ with generic fiber F. We give the name of *section of the bundle* to any rule s that with each point $p \in \mathcal{M}$ of the base manifold associates a point $s(p) \in F_p$ in the fiber above p, namely, a map

$$s : \mathcal{M} \mapsto E \tag{6.137}$$

such that

$$\forall p \in \mathcal{M} : s(p) \in \pi^{-1}(p). \tag{6.138}$$

The above definition is illustrated in Figure 6.22, which also clarifies the intuitive idea behind the name chosen for such a concept.

It is clear that sections of the bundle can be chosen to be *continuous*, *differentiable*, *smooth* or, in the case of complex manifolds, even *holomorphic*, depending on the properties of the map s in each local trivialization of the bundle. Indeed, given a local trivialization and given open charts for both the base manifold $\mathcal{M}$ and the fiber F, the local description of the section reduces to a map:

$$\mathbb{R}^m \supset U \mapsto F_U \subset \mathbb{R}^n, \tag{6.139}$$

where m and n are the dimensions of the base manifold and the fiber, respectively. We

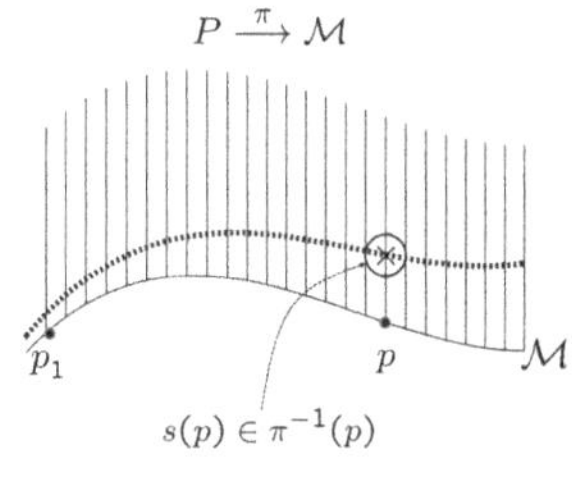

Figure 6.22: A section of a fiber bundle.

are specifically interested in smooth sections, namely, in sections that are infinitely differentiable. Given a bundle $E \xrightarrow{\pi} \mathcal{M}$, the set of all such sections is denoted by

$$\Gamma(E, \mathcal{M}). \tag{6.140}$$

Of particular relevance are the smooth sections of vector bundles. In this case, with each point of the base manifold p we associate a vector $\mathbf{v}(p)$ in the vector space above the point p. In particular we can consider sections of the tangent bundle $T\mathcal{M}$ associated with a smooth manifold M. Such sections correspond to the notion of *vector fields*.

Definition 6.8.2. Given a smooth manifold $\mathcal{M}$, we give the name of *vector field* on $\mathcal{M}$ to any smooth section $\mathbf{t} \in \Gamma(T\mathcal{M}, \mathcal{M})$ of the tangent bundle . The local expression of such vector field in any open chart (U, ϕ) is

$$\mathbf{t} = t^{\mu}(x)\frac{\vec{\partial}}{\partial x^{\mu}} \qquad \forall x \in U \subset \mathcal{M}. \tag{6.141}$$

6.8.1.1 Example: holomorphic vector fields on $\mathbb{S}^2$

As we have seen above, the 2-sphere $\mathbb{S}^2$ is a complex manifold of complex dimension one covered by an atlas composed of two charts, that of the North Pole and that of the South Pole (see Figure 6.3), and the transition function between the local complex coordinates in the two patches is the following one:

$$z_N = \frac{1}{z_S}. \tag{6.142}$$

Correspondingly, in the two patches, the local description of a holomorphic vector field $\mathbf{t}$ is given by

$$\begin{aligned} \mathbf{t} &= v_N(z_N)\frac{d}{dz_N}, \\ \mathbf{t} &= v_S(z_S)\frac{d}{dz_S}, \end{aligned} \tag{6.143}$$

where the two functions $v_N(z_N)$ and $v_S(z_S)$ are supposed to be holomorphic functions of their argument, namely, to admit a Taylor power series expansion:

$$\begin{aligned} v_N(z_N) &= \sum_{k=0}^{\infty} c_k z_N^k, \\ v_S(z_S) &= \sum_{k=0}^{\infty} d_k z_S^k. \end{aligned} \tag{6.144}$$

However, from the transition function (6.142) we obtain the relations

$$\frac{d}{dz_N} = -z_S^2\frac{d}{dz_S}; \qquad \frac{d}{dz_S} = -z_N^2\frac{d}{dz_N}, \tag{6.145}$$

and hence

$$\mathbf{t} = -\sum_{k=0}^{\infty} c_k z_S^{2-k}\frac{d}{dz_S} = \sum_{k=0}^{\infty} d_k z_S^k\frac{d}{dz_S} = -\sum_{k=0}^{\infty} d_k z_N^{2-k}\frac{d}{dz_N} = \sum_{k=0}^{\infty} c_k z_N^k\frac{d}{dz_N}. \tag{6.146}$$

The only way for equation (6.146) to be self-consistent is to have

$$\forall k > 2 \quad c_k = d_k = 0; \quad c_0 = -d_2, \quad c_1 = -d_1, \quad c_2 = -d_0. \tag{6.147}$$

This shows that the space of holomorphic sections of the tangent bundle $T\mathbb{S}^2$ is a *finite-dimensional* vector space of dimension *three* spanned by the three differential operators:

$$\begin{aligned} \mathbf{L}_0 &= -z\frac{d}{dz}, \\ \mathbf{L}_1 &= -\frac{d}{dz}, \\ \mathbf{L}_{-1} &= -z^2\frac{d}{dz}. \end{aligned} \tag{6.148}$$

What we have so far discussed can be summarized by stating the transformation rule of vector field components when we change coordinate patch from x^μ to $x^{\mu'}$:

$$t^{\mu'}(x') = t^\nu(x)\frac{\partial x^{\mu'}}{\partial x^\nu}. \tag{6.149}$$

Indeed a convenient way of defining a fiber bundle is provided by specifying the way its sections transform from one local trivialization to another one, which amounts to giving all the transition functions. This method can be used to discuss the construction of the cotangent bundle.

6.8.2 The Lie algebra of vector fields

In Section 6.4 we saw that the tangent space $T_p\mathcal{M}$ at point $p \in \mathcal{M}$ of a manifold can be identified with the vector space of derivations of the algebra of germs (see Definition 6.4.5). After gluing together all tangent spaces into the tangent bundle $T\mathcal{M}$ such an identification of tangent vectors with the derivations of an algebra can be extended from the local to the global level. The crucial observation is that the set of smooth functions on a manifold $\mathbb{C}^\infty(\mathcal{M})$ constitutes an algebra with respect to point-wise multiplication just as the set of germs at point p. The vector fields, namely, the sections of the tangent bundle, are derivations of this algebra. Indeed, each vector field $\mathbf{X} \in \Gamma(T\mathcal{M}, \mathcal{M})$ is a linear map of the algebra $\mathbb{C}^\infty(\mathcal{M})$ into itself,

$$\mathbf{X} : \mathbb{C}^\infty(\mathcal{M}) \to \mathbb{C}^\infty(\mathcal{M}), \tag{6.150}$$

that satisfies the analogous properties of those mentioned in equations (6.97) for tangent vectors, namely,

$$\begin{aligned} \mathbf{X}(\alpha f + \beta g) &= \alpha \mathbf{X}(f) + \beta \mathbf{X}(g), \\ \mathbf{X}(f \cdot g) &= \mathbf{X}(f) \cdot g + f \cdot \mathbf{X}(g) \\ &\quad [\forall \alpha, \beta \in \mathbb{R} \text{ (or } \mathbb{C}); \ \forall f, g \in \mathbb{C}^{\infty}(\mathcal{M})]. \end{aligned} \tag{6.151}$$

On the other hand, the set of vector fields, renamed for this reason,

$$\mathbb{D}\mathrm{iff}(\mathcal{M}) \equiv \Gamma(T\mathcal{M}, \mathcal{M}), \tag{6.152}$$

forms a Lie algebra with respect to the following Lie bracket operation:

$$[\mathbf{X}, \mathbf{Y}]f = \mathbf{X}(\mathbf{Y}(f)) - \mathbf{Y}(\mathbf{X}(f)). \tag{6.153}$$

Indeed, the set of vector fields is a vector space with respect to the scalar numbers ($\mathbb{R}$ or $\mathbb{C}$, depending on the type of manifold, real or complex), namely, we can take linear combinations of the form

$$\forall \lambda, \mu \in \mathbb{R} \text{ or } \mathbb{C} \quad \forall \mathbf{X}, \mathbf{Y} \in \mathbb{D}\mathrm{iff}(\mathcal{M}) : \quad \lambda \mathbf{X} + \mu \mathbf{Y} \in \mathbb{D}\mathrm{iff}(\mathcal{M}), \tag{6.154}$$

having defined

$$[\lambda \mathbf{X} + \mu \mathbf{Y}](f) = \lambda[\mathbf{X}(f)] + \mu[\mathbf{Y}(f)], \quad \forall f \in \mathbb{C}^{\infty}(\mathcal{M}). \tag{6.155}$$

Furthermore, the operation (6.153) is the commutator of two maps and as such it is antisymmetric and satisfies the Jacobi identity.

The Lie algebra of vector fields is named $\mathbb{D}\mathrm{iff}(\mathcal{M})$ since each of its elements can be interpreted as the generator of an infinitesimal diffeomorphism of the manifold onto itself. As we are going to see, $\mathbb{D}\mathrm{iff}(\mathcal{M})$ is a Lie algebra of *infinite dimension*, but it can contain finite-dimensional subalgebras generated by particular vector fields. The typical example will be the case of the Lie algebra of a Lie group: this is the finite-dimensional subalgebra $\mathbb{G} \subset \mathbb{D}\mathrm{iff}(G)$ spanned by those vector fields defined on the Lie group manifold that have an additional property of invariance with respect to either left or right translations (see Chapter 15).

6.8.3 The cotangent bundle and differential forms

Let us recall that a differential 1-form in the point $p \in \mathcal{M}$ of a manifold $\mathcal{M}$, namely, an element $\omega_p \in T_p^*\mathcal{M}$ of the cotangent space over such a point, was defined as a real valued linear functional over the tangent space at p, namely,

$$\omega_p \in \mathrm{Hom}(T_p\mathcal{M}, \mathbb{R}), \tag{6.156}$$

which implies

$$\forall \mathbf{t}_p \in T_p\mathscr{M} \quad \omega_p : \mathbf{t}_p \mapsto \omega_p(\mathbf{t}_p) \in \mathbb{R}. \tag{6.157}$$

The expression of ω_p in a coordinate patch around p is

$$\omega_p = \omega_\mu(p)\, dx^\mu, \tag{6.158}$$

where $dx^\mu(p)$ are the differentials of the coordinates and $\omega_\mu(p)$ are real numbers. We can glue together all the cotangent spaces and construct the cotangent bundle by stating that a *generic smooth section* of such a bundle is of the form (6.158), where $\omega_\mu(p)$ are now smooth functions of the base manifold point p. Clearly, if we change coordinate system, an argument completely similar to that employed in the case of the tangent bundle tells us that the coefficients $\omega_\mu(x)$ transform as

$$\omega'_\mu(x') = \omega_\nu(x)\frac{\partial x^\nu}{\partial x^{\mu'}}, \tag{6.159}$$

and equation (6.159) can be taken as a definition of the *cotangent bundle* $T^*\mathscr{M}$, whose sections transform with the Jacobian matrix rather than with the inverse Jacobian matrix as the sections of the tangent bundle do (see equation (6.149)). So we can write the following definition.

Definition 6.8.3. A differential 1-form ω on a manifold $\mathscr{M}$ is a section of the cotangent bundle, namely, $\omega \in \Gamma(T^*\mathscr{M}, \mathscr{M})$.

This means that a differential 1-form is a map

$$\omega : \Gamma(T\mathscr{M}, \mathscr{M}) \mapsto \mathbb{C}^\infty(\mathscr{M}) \tag{6.160}$$

from the space of vector fields (i. e., the sections of the tangent bundle) to smooth functions. Locally we can write

$$\begin{aligned} \Gamma(T\mathscr{M}, \mathscr{M}) \ni \omega &= \omega_\mu(x)dx^\mu, \\ \Gamma(T^*\mathscr{M}, \mathscr{M}) \ni \mathbf{t} &= t^\mu(x)\frac{\partial}{\partial x^\mu}, \end{aligned} \tag{6.161}$$

and we obtain

$$\omega(\mathbf{t}) = \omega_\mu(x)t^\nu(x)dx^\mu\left(\frac{\partial}{\partial x^\nu}\right) = \omega_\mu(x)t^\mu(x) \tag{6.162}$$

using

$$dx^\mu\left(\frac{\partial}{\partial x^\nu}\right) = \delta^\mu_\nu, \tag{6.163}$$

which is the statement that coordinate differentials and partial derivatives are dual bases for 1-forms and tangent vectors, respectively.

Since $T\mathcal{M}$ is a vector bundle it is meaningful to consider the addition of its sections, namely, the addition of vector fields and also their point-wise multiplication by smooth functions. Taking this into account we see that the map (6.160) used to define sections of the cotangent bundles, namely, 1-forms, is actually an F-linear map. This means the following. Considering any F-linear combination of two vector fields, namely,

$$f_1\mathbf{t}_1 + f_2\mathbf{t}_2, \quad f_1, f_2 \in \mathbb{C}^\infty(\mathcal{M}), \quad \mathbf{t}_1, \mathbf{t}_2 \in \Gamma(T\mathcal{M}, \mathcal{M}), \tag{6.164}$$

for any 1-form $\omega \in \Gamma(T^*\mathcal{M}, \mathcal{M})$ we have

$$\omega(f_1\mathbf{t}_1 + f_2\mathbf{t}_2) = f_1(p)\omega(\mathbf{t}_1)(p) + f_2(p)\omega(\mathbf{t}_2)(p), \tag{6.165}$$

where $p \in \mathcal{M}$ is any point of the manifold $\mathcal{M}$.

It is now clear that the definition of differential 1-form generalizes the concept of *total differential* of the germ of a smooth function. Indeed, in an open neighborhood $U \subset \mathcal{M}$ of a point p we have

$$\forall f \in \mathbb{C}_p^\infty(\mathcal{M}) \quad df = \partial_\mu f \, dx^\mu \tag{6.166}$$

and the value of df at p on any tangent vector $\mathbf{t}_p \in T_p\mathcal{M}$ is defined to be

$$df_p(\mathbf{t}_p) \equiv \mathbf{t}_p(f) = t^\mu \partial_\mu f, \tag{6.167}$$

which is the directional derivative of the local function f along $\mathbf{t}_p$ in the point p. If rather than the germ of a function we take a global function $f \in \mathbb{C}^\infty(\mathcal{M})$, we realize that the concept of 1-form generalizes the concept of total differential of such a function. Indeed, the total differential df fits into the definition of a 1-form, since for any vector field $\mathbf{t} \in \Gamma(T\mathcal{M}, \mathcal{M})$ we have

$$df(\mathbf{t}) = t^\mu(x)\partial_\mu f(x) \equiv \mathbf{t}f \in \mathbb{C}^\infty(\mathcal{M}). \tag{6.168}$$

A first obvious question is the following. Is any 1-form $\omega = \omega_\mu(x)dx^\mu$ the differential of some function? The answer is clearly no, and in any coordinate patch there is a simple test to see whether this is the case or not. Indeed, if $\omega_\mu^{(1)} = \partial_\mu f$ for some germ $f \in \mathbb{C}_p^\infty(\mathcal{M})$, then we must have

$$\frac{1}{2}(\partial_\mu \omega_\nu^{(1)} - \partial_\nu \omega_\mu^{(1)}) = \frac{1}{2}[\partial_\mu, \partial_\nu]f = 0. \tag{6.169}$$

The left hand side of equation (6.169) are the components of what we will call a differential 2-form,

$$\omega^{(2)} = \omega_{\mu\nu}^{(2)} \, dx^\mu \wedge dx^\nu, \tag{6.170}$$

and in particular the 2-form of equation (6.169) will be identified with the exterior differential of the 1-form $\omega^{(1)}$, namely, $\omega^{(2)} = d\omega^{(1)}$. In simple words the exterior differential operator d is the generalization on any manifold and to differential forms of any degree of the concept of *curl*, familiar from ordinary tensor calculus in $\mathbb{R}^3$. Forms whose exterior differential vanishes will be named *closed forms*. All these concepts need appropriate explanations that will be provided shortly from now. Yet, already at this intuitive level, we can formulate the next basic question. We saw that in order to be the total differential of a function, a 1-form must be necessarily closed. Is such a condition also sufficient? In other words, are all closed forms the differential of something? Locally the correct answer is yes, but globally it may be no. Indeed, in any open neighborhood a closed form can be represented as the differential of another differential form, but the forms that do the job in the various open patches may not glue together nicely into a globally defined one. This problem and its solution constitute an important chapter of geometry, named cohomology. Actually, cohomology is a central issue in algebraic topology, the art of characterizing the topological properties of manifolds through appropriate algebraic structures.

6.8.4 Differential k-forms

Next we introduce differential forms of degree k and the exterior differential d. In a later section, after the discussion of homology we show how this relates to the important construction of cohomology. For the time being our approach is simpler and down to earth.

We have seen that the 1-forms at a point $p \in \mathcal{M}$ of a manifold are linear functionals on the tangent space $T_p\mathcal{M}$. First of all we recall the construction of exterior k-forms on any vector space W defined to be the k-th linear antisymmetric functionals on such a space (see Section 3.5.4).

6.8.4.1 Exterior differential forms

It follows that on $T_p\mathcal{M}$ we can construct not only the 1-forms but also all the higher degree k-forms. They span the vector space $\Lambda_k(T_p\mathcal{M})$. By gluing together all such vector spaces, as we did in the case of 1-forms, we obtain the vector bundles of k-forms. More explicitly, we can set the following.

Definition 6.8.4. A differential k-form $\omega^{(k)}$ is a smooth assignment

$$\omega^{(k)} : p \mapsto \omega_p^{(k)} \in \Lambda_k(T_p\mathcal{M}) \tag{6.171}$$

of an exterior k-form on the tangent space at p for each point $p \in \mathcal{M}$ of a manifold.

Let now (U, φ) be a local chart and let $\{dx_p^1, \dots, dx_p^m\}$ be the usual natural basis of the cotangent space $T_p^*\mathcal{M}$. Then in the same local chart the differential form $\omega^{(k)}$ is written as

$$\omega^{(k)} = \omega_{i_1,\dots,i_k}(x_1, \dots, x_m)\, dx^{i_1} \wedge \cdots \wedge dx^{i_k}, \tag{6.172}$$

where $\omega_{i_1,\dots,i_k}(x_1, \dots, x_m) \in \mathbb{C}^\infty(U)$ are smooth functions on the open neighborhood U, completely antisymmetric in the indices $i_1, \dots, i_k$.

At this point it is obvious that the operation of exterior product, defined on exterior forms, can be extended to *exterior differential forms*. In particular, if $\omega^{(k)}$ and $\omega^{(k')}$ are a k-form and a k'-form, respectively, then $\omega^{(k)} \wedge \omega^{(k')}$ is a $(k + k')$-form. As a consequence of equation (3.29) we have

$$\omega^{(k)} \wedge \omega^{(k')} = (-)^{kk'} \omega^{(k')} \wedge \omega^{(k)} \tag{6.173}$$

and in local coordinates we find

$$\omega^{(k)} \wedge \omega^{(k')} = \omega^{(k)}_{[i_1 \dots i_k} \omega^{(k')}_{i_{k+1} \dots i_{k+k'}]}\, dx^1 \wedge \cdots dx^{k+k'}, \tag{6.174}$$

where $[\dots]$ denotes the complete antisymmetrization on the indices.

Let $\mathcal{A}_0(\mathcal{M}) = \mathbb{C}^\infty(\mathcal{M})$ and let $\mathcal{A}_k(\mathcal{M})$ be the $\mathbb{C}^\infty(\mathcal{M})$-module of differential k-forms. To justify the naming module, observe that we can construct the product of a smooth function $f \in \mathbb{C}^\infty(\mathcal{M})$ with a differential form $\omega^{(k)}$ setting

$$[f\omega^{(k)}](\mathbf{Z}_1, \dots \mathbf{Z}_k) = f \cdot \omega^{(k)}(\mathbf{Z}_1, \dots \mathbf{Z}_k) \tag{6.175}$$

for each k-tuple of vector fields $\mathbf{Z}_1, \dots \mathbf{Z}_k \in \Gamma(T\mathcal{M}, \mathcal{M})$.

Furthermore, let

$$\mathcal{A}(\mathcal{M}) = \bigoplus_{k=0}^{m} \mathcal{A}_k(\mathcal{M}), \quad \text{where } m = \dim \mathcal{M}. \tag{6.176}$$

Then $\mathcal{A}$ is an algebra over $\mathbb{C}^\infty(\mathcal{M})$ with respect to the exterior wedge product $\bigwedge$.

To introduce the exterior differential d we proceed as follows. Let $f \in \mathbb{C}^\infty(\mathcal{M})$ be a smooth function: for each vector field $\mathbf{Z} \in \mathbb{D}\mathrm{iff}(\mathcal{M})$, we have $\mathbf{Z}(f) \in \mathbb{C}^\infty(\mathcal{M})$ and therefore there is a unique differential 1-form, denoted df, such that $df(\mathbf{Z}) = \mathbf{Z}(f)$. This differential form is named the total differential of the function f. In a local chart U with local coordinates $x^1, \dots, x^m$ we have

$$df = \frac{\partial f}{\partial x^j}\, dx^j. \tag{6.177}$$

More generally, we can see that there exists an endomorphism d ($\omega \mapsto d\omega$) of $\mathcal{A}(\mathcal{M})$ onto itself with the following properties:

$$\begin{aligned}
(i) &\quad \forall\omega \in \mathscr{A}_k(\mathcal{M}) \quad d\omega \in \mathscr{A}_{k+1}(\mathcal{M}), \\
(ii) &\quad \forall\omega \in \mathscr{A}(\mathcal{M}) \quad dd\omega = 0, \\
(iii) &\quad \forall\omega^{(k)} \in \mathscr{A}_k(\mathcal{M}) \quad \forall\omega^{(k')} \in \mathscr{A}_{k'}(\mathcal{M}) \\
&\quad d(\omega^{(k)} \wedge \omega^{(k')}) = d\omega^{(k)} \wedge \omega^{(k')} + (-1)^k \omega^{(k)} \wedge d\omega^{(k')}, \\
(iv) &\quad \text{if } f \in \mathscr{A}_0(\mathcal{M}) \quad df = \text{total differential.}
\end{aligned} \tag{6.178}$$

In each local coordinate patch the above intrinsic definition of the exterior differential leads to the following explicit representation:

$$d\omega^{(k)} = \partial_{[i_1}\omega_{i_2\ldots i_{k+1}]}\, dx^{i_1} \wedge \cdots \wedge dx^{i_{k+1}}. \tag{6.179}$$

As already stressed, the exterior differential is the generalization of the concept of curl, well known in elementary vector calculus.

6.9 Homotopy, homology, and cohomology

Differential 1-forms can be integrated along differentiable paths on manifolds. The higher differential p-forms can be integrated on p-dimensional submanifolds. An appropriate discussion of such integrals and their properties requires the fundamental concepts of algebraic topology *homotopy* and *homology*. Also the global properties of Lie groups and their many-to-one relation with Lie algebras can be understood only in terms of homotopy. For this reason we devote the present section to an introductory discussion of homotopy, homology, and the dual of the latter, namely, cohomology.

The kind of problems we are going to consider can be intuitively grasped if we consider Figure 6.23, displaying a closed two-dimensional surface with two handles (actually an oriented, closed Riemann surface of genus $g = 2$) on which we have drawn several different closed one-dimensional paths $\gamma_1, \ldots, \gamma_6$. Consider first the path γ_5. It is an intuitive fact that γ_5 can be continuously deformed to just a point on the surface. Paths with such a property are named *homotopically trivial* or *homotopic to zero*. It is also an intuitive fact that neither γ_2, nor γ_3, nor γ_1, nor γ_4 is homotopically trivial. Paths of such a type are *homotopically non-trivial*. Furthermore, we say that two paths are homotopic if one can be continuously deformed into the other. This is for instance the case for γ_6, which is clearly homotopic to γ_3.

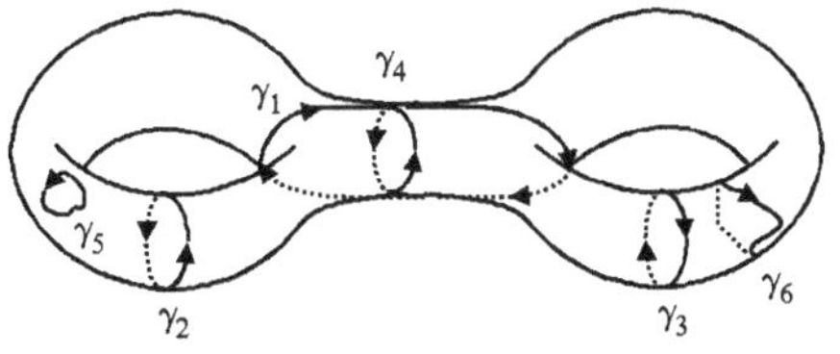

Figure 6.23: A closed surface with two handles marked by several different closed one-dimensional paths.

Let us now consider the difference between path γ_4 and path γ_1 from another viewpoint. Imagine the result of cutting the surface along the path γ_4. After the cut the surface splits into two separate parts, R_1 and R_2, as shown in Figure 6.24. Such a splitting does not occur if we cut the original surface along the path γ_1. The reason for this different behavior resides in the following. The path γ_4 is the boundary of a region on the surface (the region R_1 or, equivalently, its complement R_2), while γ_1 is not the boundary of any region. A similar statement is true for γ_2 or γ_3. We say that γ_4 is *homologically trivial*, while γ_1, γ_2, and γ_3 are *homologically non-trivial*.

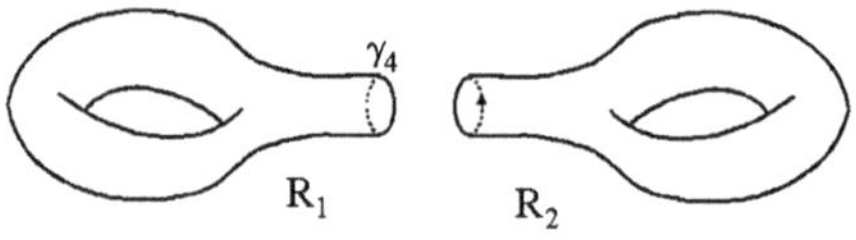

Figure 6.24: When we cut a surface along a path that is a boundary, namely, it is homologically trivial, the surface splits into two separate parts.

Next let us observe that if we simultaneously cut the original surface along γ_1, γ_2, γ_3, the surface splits once again into two separate parts, as shown in Figure 6.25. This is due to the fact that the sum of the three paths is the boundary of a region: either R_1 or R_2 of Figure 6.25. In this case we say that $\gamma_2 + \gamma_3$ is *homologous* to $-\gamma_1$, since the difference $\gamma_2 + \gamma_3 - (-\gamma_3)$ is a boundary.

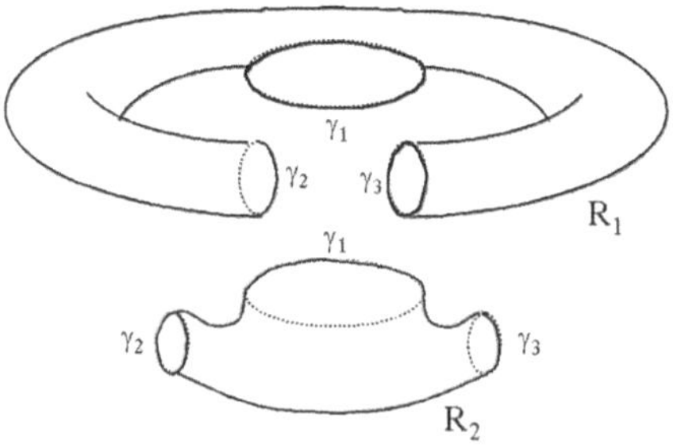

Figure 6.25: The sum of the three paths $\gamma_1, \gamma_2, \gamma_3$ is homologically trivial, namely, $\gamma_2 + \gamma_3$ is homologous to $-\gamma_1$.

In order to give a rigorous formulation to these intuitive concepts, which can be extended also to higher-dimensional submanifolds of any manifold, we proceed as follows.

6.9.1 Homotopy

Let us come back to Definition 6.4.1 of a curve (or path) in a manifold and slightly generalize it.

Definition 6.9.1. Let $[a, b]$ be a closed interval of the real line $\mathbb{R}$ parameterized by the parameter t and subdivide it into a finite number of closed, partial intervals:

$$[a, t_1], [t_1, t_2], \ldots, [t_{n-1}, t_n], [t_n, b]. \tag{6.180}$$

By *piece-wise differentiable path* we mean a continuous map

$$\gamma : [a, b] \to \mathcal{M} \tag{6.181}$$

of the interval $[a, b]$ into a differentiable manifold $\mathcal{M}$ such that there exists a splitting of $[a, b]$ into a finite set of closed subintervals as in equation (6.180) with the property that on each of these intervals the map γ is not only continuous but also infinitely differentiable.

Since we have parametric invariance we can always rescale the interval $[a, b]$ and reduce it to

$$[0, 1] \equiv I. \tag{6.182}$$

Let

$$\begin{aligned} \sigma &: I \to \mathcal{M}, \\ \tau &: I \to \mathcal{M} \end{aligned} \tag{6.183}$$

be two piece-wise differentiable paths with coinciding extrema, namely, such that (see Figure 6.26)

$$\begin{aligned} \sigma(0) &= \tau(0) = x_0 \in \mathcal{M}, \\ \sigma(1) &= \tau(1) = x_1 \in \mathcal{M}. \end{aligned} \tag{6.184}$$

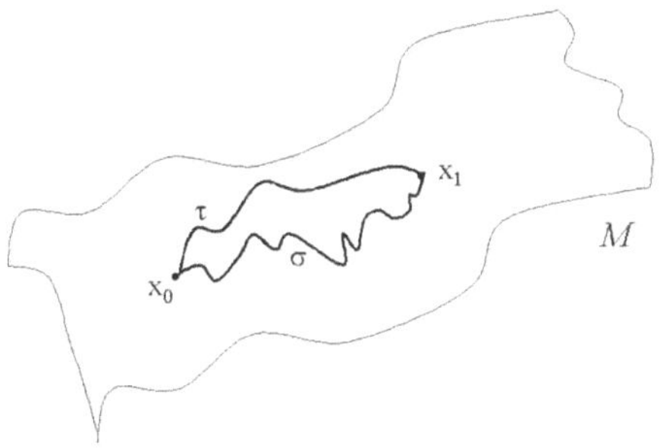

Figure 6.26: Two paths with coinciding extrema.

Definition 6.9.2. We say that σ is homotopic to τ and we write $\sigma \simeq \tau$ if there exists a continuous map

$$F : I \times I \to \mathcal{M} \tag{6.185}$$

such that

$$\begin{aligned} F(s,0) &= \sigma(s) \quad \forall s \in I, \\ F(s,1) &= \tau(s) \quad \forall s \in I, \\ F(0,t) &= x_0 \quad \forall t \in I, \\ F(1,t) &= x_1 \quad \forall t \in I. \end{aligned} \tag{6.186}$$

In particular, if σ is a closed path, namely, it forms a loop at x_0, that is, $x_0 = x_1$, and τ homotopic to σ is a *constant loop*, that is,

$$\forall s \in I : \quad \tau(s) = x_0, \tag{6.187}$$

then we say that σ is *homotopically trivial* and that it can be contracted to a point.

It is quite obvious that the homotopy relation $\sigma \simeq \tau$ is an equivalence relation. Hence, we shall consider the homotopy classes $[\sigma]$ of paths from x_0 to x_1.

Next we can define a binary product operation on the space of paths in the following way. If σ is a path from x_0 to x_1 and τ is a path from x_1 to x_2, we can define a path from x_0 to x_2 traveling first along σ and then along τ. More precisely, we set

$$\sigma\tau(t) = \begin{cases} \sigma(2t) & 0 \le t \le \frac{1}{2}, \\ \tau(2t-1) & \frac{1}{2} \le t \le 1. \end{cases} \tag{6.188}$$

What we can immediately verify from this definition is that if $\sigma \simeq \sigma'$ and $\tau \simeq \tau'$, then $\sigma\tau \simeq \sigma'\tau'$. The proof is immediate and it is left to the reader. Hence, without any ambiguity we can multiply the equivalence class of σ with the equivalence class of τ always assuming that the final point of σ coincides with the initial point of τ. Relying on these definitions we have a theorem which is very easy to prove but has an outstanding relevance.

Theorem 6.9.1. *Let $\pi_1(\mathcal{M}, x_0)$ be the set of homotopy classes of loops in the manifold $\mathcal{M}$ with base in the point $x_0 \in \mathcal{M}$. If the product law of paths is defined as we just explained, then with respect to this operation $\pi_1(\mathcal{M}, x_0)$ is a group whose identity element is provided by the homotopy class of the constant loop at x_0 and the inverse of the homotopy class $[\sigma]$ is the homotopy class of the loop σ^{-1} defined by*

$$\sigma^{-1}(t) = \sigma(1-t), \quad 0 \le t \le 1. \tag{6.189}$$

In other words, σ^{-1} is the same path followed backward.

Proof of Theorem 6.9.1. Clearly, the composition of a loop σ with the constant loop (from now on denoted as x_0) yields σ. Hence, x_0 is effectively the identity element of the group. We still have to show that $\sigma\sigma^{-1} \simeq x_0$. The explicit realization of the required homotopy is provided by the following function:

$$F(s,t) = \begin{cases} \sigma(2s) & 0 \le 2s \le t, \\ \sigma(t) & t \le 2s \le 2-t, \\ \sigma^{-1}(2s-1) & 2-t \le 2s \le 2. \end{cases} \tag{6.190}$$

Let us observe that having defined F as above we have

$$\begin{aligned} F(s,0) &= \sigma(0) = x_0 \quad \forall s \in I, \\ F(s,1) &= \begin{cases} \sigma(2s) & 0 \le s \le \frac{1}{2}, \\ \sigma^{-1}(2s-1) & \frac{1}{2} \le s \le 1. \end{cases} \end{aligned} \tag{6.191}$$

Furthermore,

$$\begin{aligned} F(0,t) &= \sigma(0) = x_0 \quad \forall t \in I, \\ F(1,t) &= \sigma^{-1}(1) = x_0 \quad \forall t \in I. \end{aligned} \tag{6.192}$$

Therefore, it is sufficient to check that $F(s,t)$ is continuous. Dividing the square $[0,1] \times [0,1]$ into three triangles as in Figure 6.27 we see that $F(s,t)$ is continuous in each of the triangles and that it is consistently glued on the sides of the triangles. Hence F as defined in equation (6.190) is continuous. This concludes the proof of the theorem. □

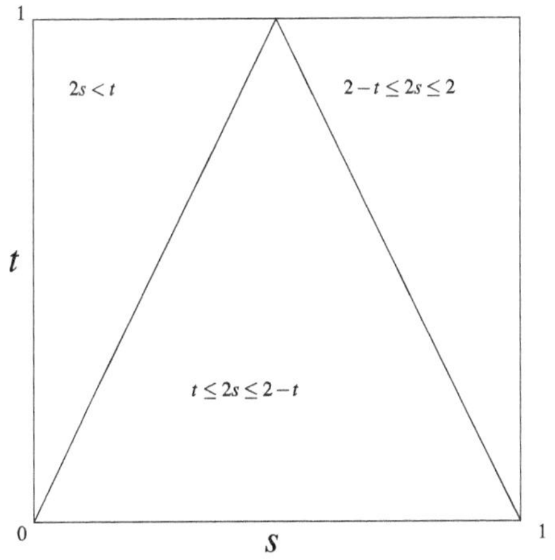

Figure 6.27: Description of the homotopy map.

Theorem 6.9.2. *Let α be a path from x_0 to x_1. Then*

$$[\sigma] \xrightarrow{\alpha} [\alpha^{-1}\sigma\alpha] \tag{6.193}$$

is an isomorphism of $\pi_1(\mathscr{M}, x_0)$ into $\pi_1(\mathscr{M}, x_1)$.

Proof of Theorem 6.9.2. Indeed, since

$$[\sigma\tau] \longrightarrow [\alpha^{-1}\sigma\alpha][\alpha^{-1}\tau\alpha] = [\alpha^{-1}\sigma\tau\alpha] \tag{6.194}$$

we see that $\xrightarrow{\alpha}$ is a homomorphism. Since also the inverse $\xrightarrow{\alpha^{-1}}$ exists, the homomorphism is actually an isomorphism. □

From this theorem it follows that in an arc-wise connected manifold, namely, in a manifold where every point is connected to any other by at least one piece-wise differentiable path, the group $\pi_1(\mathscr{M}, x_0)$ is independent of the choice of the base point x_0 and we can call it simply $\pi_1(\mathscr{M})$. The group $\pi_1(\mathscr{M})$ is named the first homotopy group of the manifold or simply the *fundamental group* of $\mathscr{M}$.

Definition 6.9.3. A differentiable manifold $\mathscr{M}$ which is arc-wise connected is called *simply connected* if its fundamental group $\pi_1(\mathscr{M})$ is the trivial group composed only of the identity element,

$$\pi_1(\mathscr{M}) = \text{id} \quad \Leftrightarrow \quad \mathscr{M} = \text{simply connected}. \tag{6.195}$$

6.9.2 Homology

The notion of homotopy led us to introduce an internal composition group for paths, the fundamental group $\pi_1(\mathscr{M})$, whose structure is a topological invariant of the manifold $\mathscr{M}$, since it does not change under continuous deformations of the latter. For this group we have used multiplicative notations since nothing guarantees a priori that it should be Abelian. Generically the fundamental homotopy group of a manifold is non-Abelian. There are higher homotopy groups $\pi_n(\mathscr{M})$ whose elements are the homotopy classes of spheres $\mathbb{S}^n$ drawn on the manifold.

In this section we turn our attention to another series of groups that also codify topological properties of the manifold and are on the contrary all Abelian. These are the homology groups

$$H_k(\mathscr{M}); \quad k = 0, 1, 2, \ldots, \dim(\mathscr{M}). \tag{6.196}$$

We can grasp the notion of *homology* if we persuade ourselves that it makes sense to consider linear combinations of submanifolds or regions of dimension p of a manifold $\mathscr{M}$, with coefficients in a ring $\mathscr{R}$ that can be either $\mathbb{Z}$, or $\mathbb{R}$, or sometimes $\mathbb{Z}_n$. The reason is that the submanifolds of dimension p are just fit to integrate p-differential forms over them. This fact allows to give a meaning to an expression of the following form:

$$\mathscr{C}^{(p)} = m_1 S_1^{(p)} + m_2 S_2^{(p)} + \cdots + m_k S_k^{(p)}, \tag{6.197}$$

where $S_i^{(p)} \subset \mathscr{M}$ are suitable p-dimensional submanifolds of the manifold $\mathscr{M}$, later on called *simplexes*, and $m_i \in \mathscr{R}$ are elements of the chosen ring of coefficients. What we systematically do is the following. For each differential p-form $\omega^{(p)} \in \Lambda_p(\mathscr{M})$ we set

$$\int_{\mathscr{C}^{(p)}} \omega^{(p)} = \int_{m_1 S_1^{(p)} + m_2 S_2^{(p)} + \cdots + m_k S_k^{(p)} \mathscr{C}^{(p)}} \omega^{(p)} = \sum_{i=1}^{k} m_i \int_{S_i^{(p)}} \omega^{(p)}, \tag{6.198}$$

and in this way we define the integral of $\omega^{(p)}$ on the region $\mathscr{C}^{(p)}$. Next let us give the precise definition of the p-simplexes of which we want to take linear combinations.

Definition 6.9.4. Let us consider the Euclidean space $\mathbb{R}^{p+1}$. The standard p-simplex Δ^p is the set of all points $\{t_0, t_1, \ldots, t_p\} \in \mathbb{R}^{p+1}$ such that the following conditions are satisfied:

$$t_i \geq 0; \quad t_0 + t_1 + \cdots + t_p = 1. \tag{6.199}$$

It is easy to see that the standard 0-simplex is a point, namely, $t_0 = 1$, the standard 1-simplex is a line segment, the standard 2-simplex is a triangle, the standard 3-simplex is a tetrahedron, and so on (see Figure 6.28). Let us now consider the standard $(p-1)$-simplex $\Delta^{(p-1)}$ and let us observe that there are $(p+1)$ canonical maps ϕ_i that map $\Delta^{(p-1)}$ into Δ^p:

$$\phi_i : \Delta^{(p-1)} \mapsto \Delta^p. \tag{6.200}$$

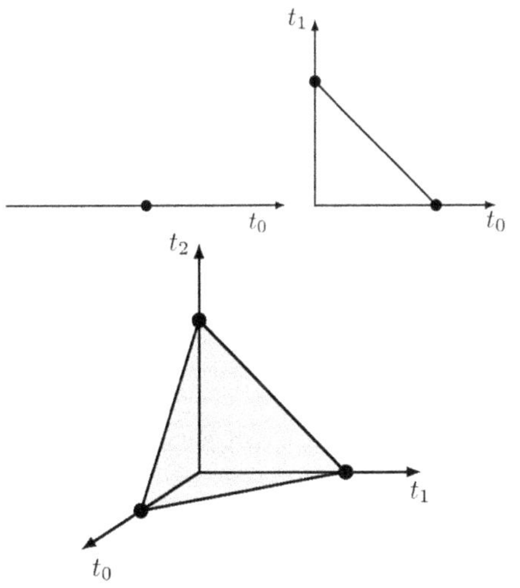

Figure 6.28: The standard p-simplexes for $p = 0, 1, 2$.

These maps are defined as follows:

$$\phi_i^{(p)}(t_0, \dots, t_{i-1}, t_{i+1}, \dots, t_p) = (t_0, \dots, t_{i-1}, 0, t_{i+1}, \dots, t_p). \tag{6.201}$$

Definition 6.9.5. The $p+1$ standard simplexes Δ^{p-1} immersed in the standard p-simplex Δ^p by means of the $p+1$ maps of equation (6.201) are named the *faces* of Δ^p and the index i enumerates them. Hence, the map $\phi_i^{(p)}$ yields, as a result, the *i-th face* of the standard p-simplex.

For instance, the two faces of the standard 1-simplex are the two points ($t_0 = 0$, $t_1 = 1$) and ($t_0 = 1$, $t_1 = 0$), as shown in Figure 6.29. Similarly the three segments ($t_0 = 0$, $t_1 = t$, $t_2 = 1 - t$), ($t_0 = t$, $t_1 = 0$, $t_2 = 1 - t$), and ($t_0 = t$, $t_1 = 1 - t$, $t_2 = 0$) are the three faces of the standard 2-simplex (see Figure 6.30).

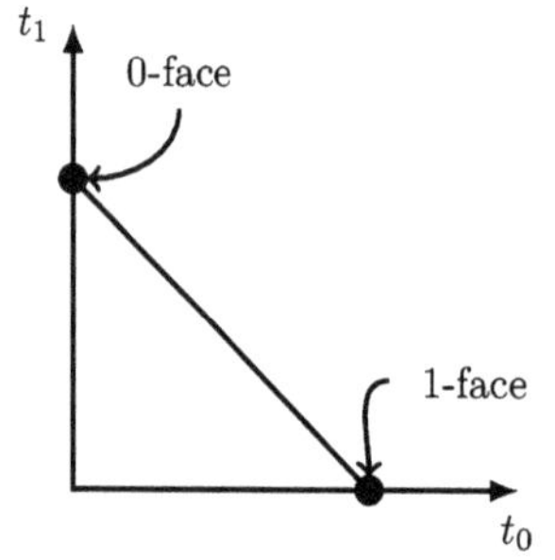

Figure 6.29: The faces of the standard 1-simplex.

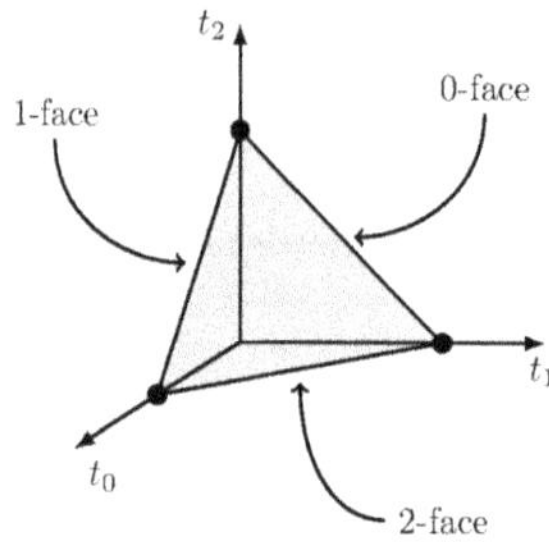

Figure 6.30: The faces of the standard 2-simplex.

Definition 6.9.6. Let $\mathscr{M}$ be a differentiable manifold of dimension m. A continuous map

$$\sigma^{(p)} : \Delta^{(p)} \mapsto \mathscr{M} \tag{6.202}$$

of the standard p-simplex into the manifold is called a *singular p-simplex* or simply a *simplex* of $\mathscr{M}$.

Clearly, a 1-simplex is a continuous path in $\mathcal{M}$, a 2-simplex is a portion of a surface immersed in $\mathcal{M}$, and so on. The i-th face of the simplex $\sigma^{(p)}$ is given by the $(p-1)$-simplex obtained by composing $\sigma^{(p)}$ with ϕ_i:

$$\sigma^{(p)} \circ \phi_i : \Delta^{(p-1)} \mapsto \mathcal{M}. \tag{6.203}$$

Let $\mathcal{R}$ be a commutative ring.

Definition 6.9.7. Let $\mathcal{M}$ be a manifold of dimension m. For each $0 \leq n \leq m$ the group of n-chains with coefficients in $\mathcal{R}$, named $C(\mathcal{M}, \mathcal{R})$, is defined as the *free $\mathcal{R}$-module* having a generator for each n-simplex in $\mathcal{M}$.

In simple words, Definition 6.9.7 states that $C_p(\mathcal{M}, \mathcal{R})$ is the set of all possible linear combinations of p-simplexes with coefficients in $\mathcal{R}$:

$$\mathcal{C}^{(p)} = m_1 S_1^{(p)} + m_2 S_2^{(p)} + \cdots m_k S_k^{(p)}, \tag{6.204}$$

where $m_i \in \mathcal{R}$. The elements of $C_p(\mathcal{M}, \mathcal{R})$ are named *p-chains*.

The concept of p-chains gives a rigorous meaning to the intuitive idea that any p-dimensional region of a manifold can be constructed by gluing together a certain number of simplexes. For instance, a path γ can be constructed gluing together a finite number of segments (or better their homeomorphic images). In the case $p = 2$, the construction of a two-dimensional region by means of 2-simplexes corresponds to a triangulation of a surface.

As an example consider the case where the manifold we deal with is just the complex plane $\mathcal{M} = \mathbb{C}$ and let us focus on the 2-simplexes drawn in Figure 6.31. The chain

$$\mathcal{C}^{(2)} = S_1^{(2)} + S_2^{(2)} \tag{6.205}$$

denotes the region of the complex plane depicted in Figure 6.32, with the proviso that when we compute the integral of any 2-form on $\mathcal{C}^{(2)}$ the contribution from the simplex

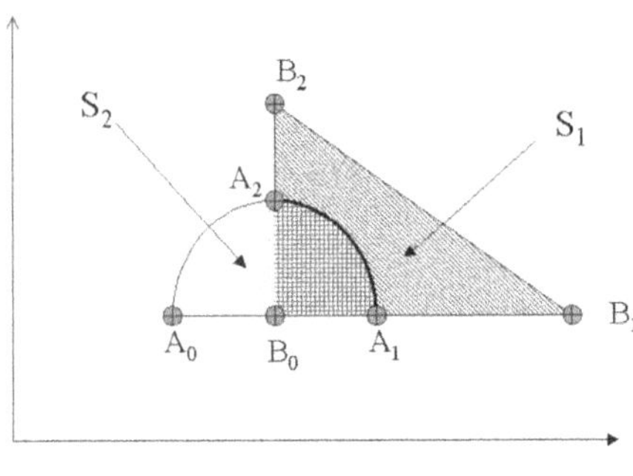

Figure 6.31: $S_1^{(2)}$ and $S_2^{(2)}$ are two distinct 2-simplexes, namely, two triangles with vertices respectively given by (A_0, A_1, A_2) and B_0, B_1, B_2. The 2-simplex $S_3^{(2)}$ with vertices B_0, A_1, A_2 is the intersection of the other two, $S_3^{(2)} = S_1^{(2)} \cap S_2^{(2)}$.

$S_3^{(2)} = S_1^{(2)} \cap S_2^{(2)}$ (the shadowed area in Figure 6.32) has to be counted twice since it belongs to both $S_1^{(2)}$ and $S_2^{(2)}$. Relying on these notions we can introduce the boundary operator.

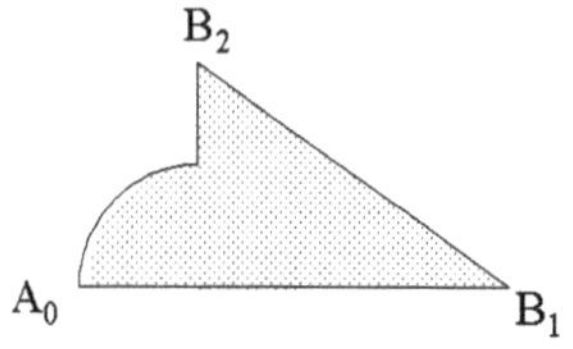

Figure 6.32: Geometrically the chain $S_1^{(2)} + S_2^{(2)}$ is the union of the two simplexes $S_1^{(2)} \cup S_2^{(2)}$.

Definition 6.9.8. The boundary operator ∂ is the map

$$\partial : C_n(\mathcal{M}, \mathcal{R}) \to C_{n-1}(\mathcal{M}, \mathcal{R}) \tag{6.206}$$

defined by the following properties:

1. *$\mathcal{R}$-linearity:*

$$\begin{aligned} &\forall \mathcal{C}_1^{(p)}, \mathcal{C}_2^{(p)} \in C_p(\mathcal{M}, \mathcal{R}), \quad \forall m_1, m_2 \in \mathcal{R}, \\ &\partial(m_1 \mathcal{C}_1^{(p)} + m_2 \mathcal{C}_2^{(p)}) = m_1 \partial \mathcal{C}_1^{(p)} + m_2 \partial \mathcal{C}_2^{(p)}. \end{aligned} \tag{6.207}$$

2. *Action on the simplexes:*

$$\begin{aligned} \partial \sigma &\equiv \sigma \circ \phi_0 - \sigma \circ \phi_1 + \sigma \circ \phi_2 - \cdots \\ &= \sum_{i=1}^{p} (-)^i \sigma \circ \phi_i. \end{aligned} \tag{6.208}$$

The image of a chain $\mathcal{C}$ through ∂, namely, $\partial\mathcal{C}$, is called the *boundary* of the chain.

As an exercise we can compute the boundary of the 2-chain $\mathcal{C}^{(2)} = \mathcal{S}_1^{(2)} + \mathcal{S}_2^{(2)}$ of Figure 6.31, with the understanding that the relevant ring is in this case $\mathbb{Z}$. We have

$$\begin{aligned} \partial C^{(2)} &= \partial S_1^{(2)} + \partial S_2^{(2)} \\ &= \overrightarrow{A_1A_2} - \overrightarrow{A_0A_2} + \overrightarrow{A_0A_1} + \overrightarrow{B_1B_2} - \overrightarrow{B_0B_2} + \overrightarrow{B_1B_2}, \end{aligned} \tag{6.209}$$

where $\mathbf{A_1A_2}, \ldots$ denote the oriented segments from A_1 to A_2, and so on. As one sees, the change in sign is interpreted as a change of orientation (which is the correct interpretation if one thinks of the chain and of its boundary as the support of an integral). With this convention the 1-chain

$$\overrightarrow{A_1A_2} - \overrightarrow{A_0A_2} + \overrightarrow{A_0A_1} = \overrightarrow{A_1A_2} + \overrightarrow{A_2A_0} + \overrightarrow{A_0A_1} \tag{6.210}$$

is just the oriented boundary of the $S_1^{(2)}$ simplex as shown in Figure 6.33.

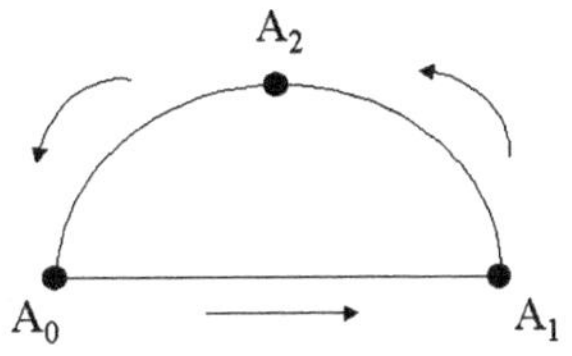

Figure 6.33: The oriented boundary of the $S^{(2)}$ simplex.

Theorem 6.9.3. *The boundary operator ∂ is nilpotent, namely, it is true that*

$$\partial^2 \equiv \partial \circ \partial = 0. \tag{6.211}$$

Proof of Theorem 6.9.3. It is sufficient to observe that, as a consequence of their own definition, the maps ϕ_i defined in equation (6.201) have the following property:

$$\phi_i^{(p)} \circ \phi_j^{(p-1)} = \phi_j^{(p)} \circ \phi_{i-1}^{(p-1)}. \tag{6.212}$$

Then for the p-simplex σ we have

$$\begin{aligned}
\partial\partial\sigma &= \sum_{i=0}^{p} (-)^i \partial[\sigma \circ \phi_i] \\
&= \sum_{i=0}^{p}\sum_{j=0}^{p-1} (-)^i (-)^j \sigma \circ (\phi_i^{(p)} \circ \phi_j^{(p-1)}) \\
&= \sum_{j<i=1}^{p} (-)^{i+j} \sigma \circ (\phi_j^{(p)} \circ \phi_{i-1}^{(p-1)}) + \sum_{0=i\leq j}^{p-1} \sigma(\phi_i^{(p)} \circ \phi_j^{(p-1)}).
\end{aligned} \tag{6.213}$$

We can verify that everything in the last line of equation (6.213) cancels identically and this proves the theorem. □

As an illustration we can calculate $\partial\partial S_1^{(2)}$ for the 2-simplex $S_1^{(2)}$ described in Figure 6.31. We obtain

$$\partial\partial S_1^{(2)} = A_2 - A_1 - A_2 + A_0 + A_1 - A_0 = 0. \tag{6.214}$$

The nilpotency of the boundary operator ∂ that acts on the *chains* is the counterpart of the nilpotency of the exterior derivative d that acts on differential forms, as explained in Section 6.8.4. Consider Figure 6.34. As one sees, the sequence of the vector spaces C_m of m-chains can be put into correspondence with the sequence of vector spaces Λ_m of differential m-forms. The operator

$$\partial : C_k \to C_{k-1} \tag{6.215}$$

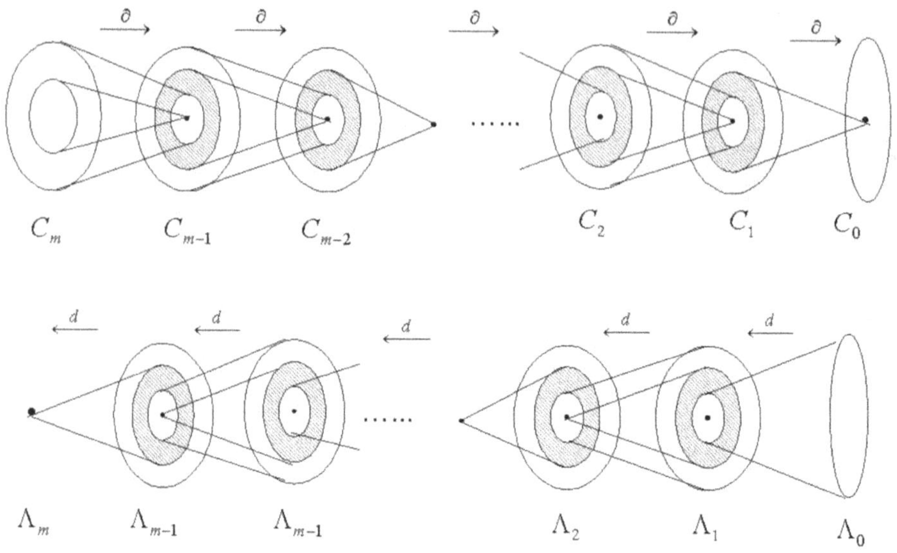

Figure 6.34: Homology versus cohomology groups.

makes you travel on the sequence from left to right, while the exterior derivative operator

$$d : \Lambda_k \to \Lambda_{k-1} \tag{6.216}$$

causes you to travel along the same sequence in the opposite direction from right to left. Both ∂ and d are nilpotent maps.

6.9.3 Homology and cohomology groups: general construction

Let $\pi : X \to Y$ be a linear map between vector spaces. We define the *kernel* of π (denoted $\ker \pi$) as the subspace of X whose elements have the property of being mapped into $0 \in Y$ by π:

$$\ker \pi = \{x \in X/\pi(x) = 0 \in Y\}. \tag{6.217}$$

We call the image of π (denoted $\operatorname{Im} \pi$) the subspace of Y whose elements have the property that they are the image through π of some element of X:

$$\operatorname{Im} \pi = \{y \in Y/\exists x \in X/\pi(x) = y\}. \tag{6.218}$$

A nilpotent operator that acts on a sequence of vector spaces X_i defines a sequence of linear maps π_i

$$X_1 \xrightarrow{\pi_1} X_2 \xrightarrow{\pi_2} X_3 \longrightarrow \cdots X_i \xrightarrow{\pi_i} X_{i+1} \tag{6.219}$$

that have the following property:

$$\operatorname{Im} \pi_i \subset \ker \pi_{i+1}. \tag{6.220}$$

The inclusion of $\operatorname{Im} \pi_i$ in $\ker \pi_{i+1}$ is what is pictorially described in Figure 6.34 and applies both to the boundary and to the exterior derivative operator. This situation suggests the following terminology.

Definition 6.9.9. In every space $C_k(\mathscr{M}, \mathbb{R})$ we call the elements of $\ker \partial$, namely, the chains $\mathscr{C}$ whose boundaries vanish, $\partial\mathscr{C} = 0$, *cycles*. Similarly, in every space $\Lambda_k(\mathscr{M})$ we call the elements of $\ker d$, namely, the differential forms ω such that $d\omega = 0$, closed forms or *cocycles*.

At the same time, we have the following.

Definition 6.9.10. In every space $C_k(\mathscr{M}, R)$ we call all k-chains that are the boundary of a $(k+1)$-chain *boundaries*:

$$\mathscr{C}^{(k)} = \text{boundary} \quad \Leftrightarrow \quad \exists \mathscr{C}^{(k+1)} / \mathscr{C}^{(k)} = \partial \mathscr{C}^{(k+1)}. \tag{6.221}$$

Similarly, in every space $\Lambda_k(\mathscr{M})$ we call all differential forms $\omega^{(k)}$ such that they can be written as the exterior derivative of some $(k-1)$-form exact forms or *coboundaries*: $\omega^{(k)} = d\omega^{(k-1)}$.

Clearly, equation (6.220) can be translated by saying that every boundary is a cycle and every coboundary is a cocycle. The reverse statement, however, is not true in general. There are cycles that are not boundaries and there are cocycles that are not coboundaries.

The concept of homology (or cohomology) previously discussed in an intuitive way can be formalized in the following way.

Definition 6.9.11. Consider the k-cycles. We say that two cycles $C_1^{(k)}$ and $C_2^{(k)}$ are *homologous* and we write $C_1^{(k)} \sim C_2^{(k)}$ if their difference is a boundary:

$$C_1^{(k)} \sim C_2^{(k)} \quad \Rightarrow \quad \exists C_3^{(k+1)} / C_1^{(k)} - C_2^{(k)} = \partial C_3^{(k+1)}. \tag{6.222}$$

Clearly, homology is an equivalence relation, since

$$\left.\begin{array}{l} C_1^{(k)} - C_2^{(k)} = \partial C_a^{(k+1)} \\ C_2^{(k)} - C_3^{(k)} = \partial C_b^{(k+1)} \end{array}\right\} \quad \Rightarrow \quad C_1^{(k)} - C_3^{(k)} = \partial\left[C_a^{(k+1)} + C_b^{(k+1)}\right]. \tag{6.223}$$

Definition 6.9.12. We call the group of equivalence classes of the k-th cycles with respect to the k-boundaries the *k-th homology group* (denoted $H_k(\mathscr{M}, \mathbb{R})$).

Similarly, we define the *k-th cohomology group* (denoted $H^k(\mathcal{M}, \mathbb{R})$) as the group of equivalence classes of the k-cocycles with respect to the k-th coboundaries. Indeed, we say that two closed forms ω and ω' are cohomologous if their difference is an exact form: $\omega \sim \omega' \Rightarrow \exists \phi / \omega - \omega' = d\phi$.

More generally, when we have a sequence of vector spaces X_i as in equation (6.219) and a sequence of linear maps π_i satisfying equation (6.220) we define the *cohomology groups* relative to the operator π as

$$H^i_{(\pi)} \equiv \frac{\ker \pi_i}{\operatorname{Im} \pi_{i-1}}. \tag{6.224}$$

The relation existing between homology and cohomology is fully contained in the following formula, which generalizes to an arbitrary smooth manifold and to differential forms of any degree the familiar Gauss lemma or Stokes lemma:

$$\int_{\partial \mathcal{C}^{(k+1)}} \omega^{(k)} = \int_{\mathcal{C}^{(k+1)}} d\omega^{(k)}. \tag{6.225}$$

Equation (6.225), whose general proof we omit, implies that in the case $\mathcal{C}^{(k)}$ is a cycle, we have

$$\int_{\mathcal{C}^{(k)}} [\omega^{(k)} + d\phi^{(k-1)}] = \int_{\mathcal{C}^{(k)}} \omega^{(k)}, \tag{6.226}$$

namely, the integral of a closed differential form along a cycle depends only on the cohomology class and not on the choice of the representative. Similarly, if $\omega^{(k)}$ is a closed form,

$$\int_{\mathcal{C}^{(k)} + \partial \mathcal{C}^{(k+1)}} \omega^{(k)} = \int_{\mathcal{C}^{(k)}} \omega^{(k)}, \tag{6.227}$$

namely, the integral of a cocycle along a cycle depends on the homology class of the class and not on the choice of the representative inside the class.

6.9.4 Relation between homotopy and homology

The relation between homotopy and homology groups of a manifold is provided by a fundamental theorem of algebraic geometry that we state without proof.

Theorem 6.9.4. *Let $\mathcal{M}$ be a smooth manifold. Then there exists a homomorphism:*

$$\chi : \pi_1(\mathcal{M}) \to H_1(\mathcal{M}, \mathbb{Z}) \tag{6.228}$$

that sends the homotopy class of each loop γ into the 1-simplex γ. If $\mathcal{M}$ is arc-wise connected, then the map χ is surjective and the kernel of χ is the subgroup of commutators in $\pi_1(\mathcal{M})$.

We recall that the subgroup of commutators of a discrete group G is the group G' generated by all elements of the form $x^{-1}y^{-1}xy$ for some $x, y \in G$.

From this theorem we have two consequences.

Corollary 6.9.1. *If $\pi_1(\mathcal{M})$ is Abelian, then $H_1(\mathcal{M}) \simeq \pi_1(\mathcal{M})$, namely, the homotopy and cohomology groups coincide.*

Corollary 6.9.2. *If a manifold $\mathcal{M}$ is simply connected ($\pi_1(\mathcal{M}) = 1$), then also the first homology group is trivial, $H_1(\mathcal{M}) = 0$.*

The second of the above corollaries implies that in a simply connected manifold every closed loop is homologous to zero, namely, it is the boundary of some region.

6.10 Holomorphic functions and their integrals in view of homotopy, homology, and cohomology

In this section we revisit once again the theory of holomorphic functions, showing that the concepts of homotopy, homology, and cohomology are at the basis of all classical results in this field of analysis and provide an excellent and simple illustration of what we presented in Section 6.9.

In order to appreciate how much the theory of complex analytic functions is an illustration of the concepts and structures systematically exposed in Section 6.9, we just review how one arrives at these notions starting from the problem of curvilinear integrals in the $\mathbb{R}^2$ plane, which is the origin of the whole mathematical setup of complex function theory.

6.10.1 Curvilinear integrals in the $\mathbb{R}^2$ plane

Let us label the points of the infinite $\mathbb{R}^2$ plane with the coordinates $\{x, y\}$.

Definition 6.10.1. By *differentiable path* we mean a map of the interval $[a, b] \subset \mathbb{R}$ to the plane

$$[a, b] \ni t \to \gamma(t) \in \mathbb{R}^2 \tag{6.229}$$

such that the coordinates $x(t)$, $y(t)$ of the point $\gamma(t)$ are continuous infinitely differentiable functions.

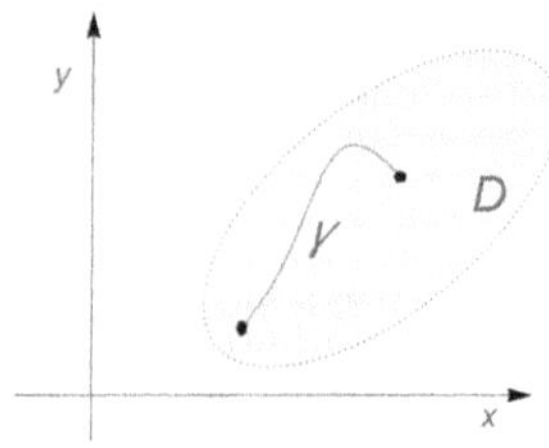

Figure 6.35: A continuous differentiable path in the plane.

The origin of γ is the point $\gamma(a)$, and the endpoint of γ is the point $\gamma(b)$. If $D \subset \mathbb{R}^2$ is an open subset of the plane we say that γ is a differentiable path of D in the case $\gamma(t)$ takes values in D (see Figure 6.35). Next we consider the following *differentiable 1-form* defined over an open domain $D \subset \mathbb{R}^2$:

$$\omega = P(x,y)\,dx + Q(x,y)\,dy, \tag{6.230}$$

where the coefficients $P(x,y)$ and $Q(x,y)$ are real or complex valued continuous functions in the domain D. If γ is a differentiable path of D and ω is a 1-form in D, we can define the integral $\int_\gamma \omega$ by means of the following formula:

$$\int_\gamma \omega = \int_a^b \gamma^\star[\omega], \tag{6.231}$$

where $\gamma^\star[\omega]$ is the *pull-back* of the differential 1-form ω on the path γ given by

$$\gamma^\star[\omega] \equiv f(t)\,dt; \quad f(t) \equiv P(x(t),y(t))x'(t) + Q(x(t),y(t))y'(t). \tag{6.232}$$

Let us now consider a *continuous, differentiable* function $\tau = \tau(u)$ where the new variable belongs to another closed interval of the real line $u \in [\alpha,\beta] \subset \mathbb{R}$. We assume that $\tau(u)$ is monotonic increasing, i. e., $\tau'(u) > 0$, and that $\tau(\alpha) = a$, $\tau(\beta) = b$. The composed map $\gamma_1 \equiv \gamma \circ \tau$ is a new differentiable path of the domain D:

$$\gamma_1 : [\alpha,\beta] \longrightarrow D. \tag{6.233}$$

By means of a standard manipulation of integration variables we obtain

$$\gamma_1^\star[\omega] = f_1(u)du; \quad f_1(u) = f(\tau(u))\tau'(u), \tag{6.234}$$

so we conclude

$$\int_\gamma \omega = \int_{\gamma_1} \omega. \tag{6.235}$$

This means that the value of the integral of a 1-form along a differentiable path does not depend on the way we parameterize such a path; rather, it is an intrinsic property of the path. Such a property is called parametric invariance.

Consider now a function $\tau = \tau(u)$ that is monotonically decreasing, $\tau'(u) < 0$, with $\tau(\alpha) = b$, $\tau(\beta) = a$. This corresponds to inversion of the orientation of the path. One easily verifies that in this case $\int_\gamma \omega = -\int_{\gamma_1} \omega$.

Next let us split the interval $[a, b]$ of the variable t into a finite number of partial intervals,

$$[a, t_1], [t_1, t_2], \ldots, [t_{n-1}, t_n], [t_n, b], \tag{6.236}$$

and let us assume that $a < t_1 < t_2 < \cdots < t_n < b$. Let us denote by γ_i the restriction of the map γ to the i-th subinterval. Clearly, we have

$$\int_\gamma \omega = \sum_{i=1}^{n} \int_{\gamma_i} \omega. \tag{6.237}$$

This property suggests the generalization provided by the following definition.

Definition 6.10.2. By *piece-wise differentiable path* we mean a map

$$\gamma : [a, b] \longrightarrow \mathbb{R}^2 \tag{6.238}$$

such that:
- there exists a finite subdivision of the interval $[a, b]$ in a finite number of subintervals as in equation (6.236);
- the restrictions γ_i of γ to such subintervals are infinitely differentiable.

The integral of any 1-form along a piece-wise differentiable path can be unambiguously defined by setting

$$\int_\gamma \omega \equiv \sum_{i=1}^{n} \int_{\gamma_i} \omega. \tag{6.239}$$

An example of a piece-wise differentiable path is displayed in Figure 6.36.

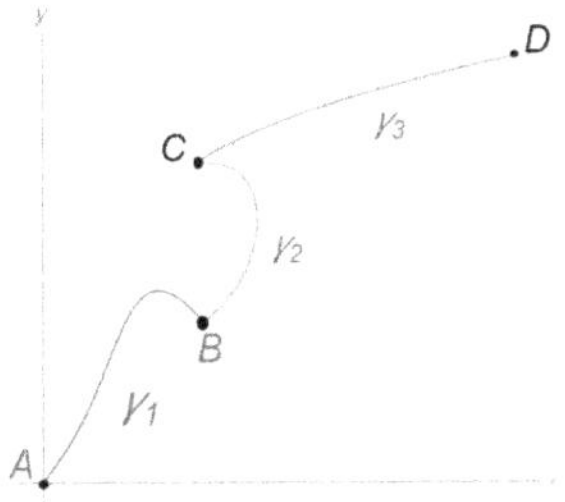

Figure 6.36: A piece-wise differentiable path from the point A to the point D of the plane.

6.10.2 The primitive of a differential 1-form

In what follows we utilize a lemma that can be formally proved, yet is intuitively evident.

Lemma 6.10.1. *Let $D \subset \mathbb{R}^2$ be an open and connected subset of the plane. Given any two points $\mathbf{a}, \mathbf{b} \in D$ there exists a continuous piece-wise differentiable path γ that starts in $\mathbf{a}$, ends in $\mathbf{b}$, and is contained in D.*

Next let us consider exact 1-forms according to the general definitions introduced in Section 6.9. In the present case the 1-form $\omega = P\,dx + Q\,dy$ defined over the open domain D is an *exact form* if there exists a function $F(x,y)$ defined over the same domain D such that $\omega = dF$ is the total differential of the function F. This is equivalent to posing

$$P(x,y) = \frac{\partial F}{\partial x}; \quad Q(x,y) = \frac{\partial F}{\partial y}. \tag{6.240}$$

In complex function theory, such a function F, if it exists, is named according to nineteenth-century traditional nomenclature the *primitive* of ω. From a general cohomological point of view, F is just a 0-form. When a form ω is exact and admits a primitive F, for any path γ we have the evident relation

$$\int_\gamma \omega = \int_\gamma dF = F(\mathbf{b}) - F(\mathbf{a}), \tag{6.241}$$

having denoted by $\mathbf{a}$ and $\mathbf{b}$ the initial and final points of the path, respectively.

According to the general concepts of Section 6.9, a *cycle* $\mathfrak{c}$ is a path whose initial and final points coincide, namely, $\mathfrak{c}(a) = \mathfrak{c}(b) = \mathbf{p} \in \mathbb{R}^2$. Looking at the relation (6.241) we see that whenever the 1-form ω is exact, its integral on any cycle $\mathfrak{c}$ contained in D vanishes, $\int_\mathfrak{c} \omega = 0$. The converse of this statement is a theorem which is just the tip of an iceberg, namely, it happens to be the first simple instance of a proposition in the cohomology/homology theory discussed in Section 6.9. We present both the theorem and its elementary proof as a pedagogical illustration of the general cohomology/homology theory.

Theorem 6.10.1. *Let $D \subset \mathbb{R}^2$ be open and connected. In order for a differential 1-form ω to be exact and admit a primitive F it is necessary and sufficient that*

$$\int_\mathfrak{c} \omega = 0 \tag{6.242}$$

for all cycles $\mathfrak{c}$ contained in D.

Proof of Theorem 6.10.1. Let us split the proof in the usual two parts:

(1) The condition is necessary because of what we have already observed above in the case $\omega = dF$.

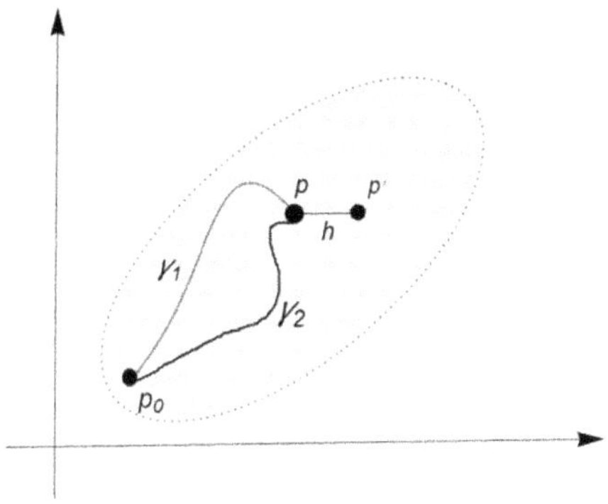

Figure 6.37: If $\int_c \omega = 0$ for all cycles in D, then the integrals $\int_{\gamma_1} \omega$ and $\int_{\gamma_2} \omega$ along two paths that have the same initial and final points do not depend on the choice of the path.

(2) The condition is sufficient because of the following argument. First let us observe Figure 6.37. Given a reference point $\mathbf{p}_0 \in D$ and any other point $\mathbf{p} \in D$, the integral of ω along any path γ going from $\mathbf{p}_0$ to $\mathbf{p}$ does not depend on the choice of γ. Indeed, given a path γ_1 and a path γ_2 as in the mentioned figure, the path $c_{12} \equiv \gamma_1 - \gamma_2$ obtained by following the first in the direct direction and then the second in the reversed direction is just a closed path, namely, a cycle. Hence, by hypothesis of the theorem we have $\int_{c_{12}} \omega = 0$, which implies

$$\int_{\gamma_1} \omega = \int_{\gamma_2} \omega. \tag{6.243}$$

Relying on that, let us define a function $F(\mathbf{p})$ on D given by the value of the integral $\int_{\gamma_{\mathbf{p},\mathbf{p}_0}} \omega$ along any path $\gamma_{\mathbf{p},\mathbf{p}_0}$ which joins a reference fixed point $\mathbf{p}_0$ with a variable point $\mathbf{p}$. Let us consider a small increment h of the coordinate x of the point $\mathbf{p}$. By definition of $F(x,y)$, the difference

$$F(x+h,y) - F(x,y)$$

is given by the integral along any path γ that connects the point $\mathbf{p} = \{x,y\}$ with the point $\mathbf{p}' = \{x+h,y\}$. For very small h, D being open, we can always choose for γ a segment of the straight line parallel to the coordinate axis of the x's, as shown in Figure 6.37. In this way we find

$$\frac{F(x+h,y) - F(x,y)}{h} = \frac{1}{h} \int_{x}^{x+h} P(\xi,y)\, d\xi, \tag{6.244}$$

and in the limit $h \to 0$ we obtain

$$\frac{\partial F(x,y)}{\partial x} = P(x,y). \tag{6.245}$$

In a fully analogous fashion we prove that

$$\frac{\partial F(x,y)}{\partial y} = Q(x,y). \tag{6.246}$$

This concludes the proof of the theorem, since the function F turns out to be a primitive of ω which therefore is exact. Any other primitive differs from F by a constant and this accounts for the arbitrary choice of the reference point $\mathbf{p}_0$. □

6.10.3 The Green–Riemann formula, an instance of the general Stokes theorem

In Section 6.9 we utilized the Gauss–Stokes theorem in equation (6.225) which provides the pairing between homology and cohomology, omitting the proof. As an illustration of the general idea we present here the two-dimensional version of the lemma, traditionally named the Green–Riemann formula, together with its proof obtained in terms of elementary tools.

Theorem 6.10.2. *Let us consider a differential 1-form ω as in equation (6.230) which we assume to be defined over an open, connected domain of the plane D. Let the closed piecewise differentiable path γ be the boundary of such a domain $\gamma = \partial D$. Then we have the relation*

$$\int_\gamma \omega = \int_\gamma (P\,dx + Q\,dy) = \int\int_{\overline{D}} \left(\frac{\partial Q}{\partial x} - \frac{\partial P}{\partial y} \right) dxdy, \tag{6.247}$$

which is called the Green–Riemann formula.

Proof of Theorem 6.247. The proof of this theorem is organized in two steps.

Step one: In the first step let us assume that the bounded domain D is of the type shown in the first picture of Figure 6.38. Namely, we assume that we can separate the boundary γ in two branches α and β, which both project onto the same interval $[a, b]$ of the x-axis. We also assume that each point $t \in [a, b]$ has a unique preimage in either α or β. This means that the entire path α and β can be represented as a one valued function $\alpha(x)$ and $\beta(x)$, respectively. These one valued functions, have, by definition, the property

$$\alpha(a) = \beta(a); \quad \alpha(b) = \beta(b). \tag{6.248}$$

Next consider the following obvious identity:

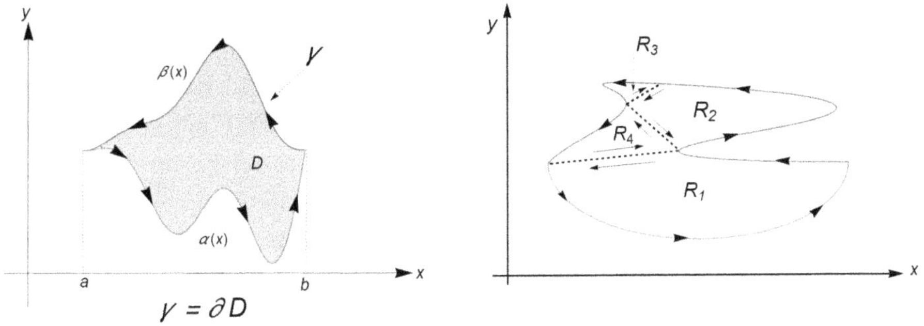

Figure 6.38: Visualization of the paths and subregions utilized in the proof of the Green–Riemann theorem.

$$\begin{aligned}\int\!\!\int_{\overline{D}} \frac{\partial P(x,y)}{\partial y} dx\, dy &= \int_a^b dx \int_{\alpha(x)}^{\beta(x)} \frac{\partial P(x,t)}{\partial t} dt \\ &= \int_a^b dx[P(x,\beta(x)) - P(x,\alpha(x))] \\ &= -\int_\gamma P(x,y)\, dx. \end{aligned} \tag{6.249}$$

This being established, let us introduce a function of three variables defined by the integral

$$F(x,u,v) \equiv \int_u^v Q(x,y)\, dy \tag{6.250}$$

and set

$$\varphi(x) = F(x,\alpha(x),\beta(x)). \tag{6.251}$$

By explicit calculation we get

$$\frac{d}{dx}\varphi(x) = \int_{\alpha(x)}^{\beta(x)} \frac{\partial Q(x,y)}{\partial x} dy + \beta'(x)Q(x,\beta(x)) - \alpha'(x)Q(x,\alpha(x)), \tag{6.252}$$

which implies

$$\begin{aligned}\int_a^b dx \frac{d}{dx}\varphi(x) &= \varphi(b) - \varphi(a) \\ &= \int_a^b dx \int_{\alpha(x)}^{\beta(x)} \frac{\partial Q(x,y)}{\partial x} dy + \int_a^b dx[\beta'(x)Q(x,\beta(x)) - \alpha'(x)Q(x,\alpha(x))]\end{aligned}$$

$$= \int\int_{\bar{D}} \frac{\partial Q(x,y)}{\partial x} dx\, dy - \int_{\gamma} Q(x,y)\, dy. \tag{6.253}$$

Since $\varphi(a) = \varphi(b) = 0$ because of the very definition of the function $\varphi(x)$ and equation (6.248) we get

$$\int\int_{\bar{D}} \frac{\partial Q(x,y)}{\partial x} dx\, dy = \int_{\gamma} Q(x,y)\, dy, \tag{6.254}$$

which combined with equation (6.249) proves the statement of the theorem in the case that the domain D has the properties specified above.

Step two: The second step of the proof is sketched in a more intuitive way. It consists in showing that given an arbitrary bounded domain $D \subset \mathbb{R}^2$, this latter can always be split into a finite number n of regions R_i, each of which satisfies the hypothesis assumed in step one, namely, the boundary of each region ∂R_i can be represented by two functions $\alpha(x)$ and $\beta(x)$ as above. This is intuitively evident looking at the second picture in Figure 6.38. Hence, for each region R_i the proposition of the theorem holds true. On the other hand, taking orientation into consideration we have

$$\begin{aligned} \int\int_{D} \left(\frac{\partial Q}{\partial x} - \frac{\partial P}{\partial y}\right) dx\, dy &= \sum_{i=1}^{n} \int\int_{R_i} \left(\frac{\partial Q}{\partial x} - \frac{\partial P}{\partial y}\right) \\ &= \sum_{i=1}^{n} \int_{\partial R_i} (P\, dx + Q\, dy) \\ &= \int_{\partial D} (P\, dx + Q\, dy). \end{aligned} \tag{6.255}$$

The catch of the proof comes from the fact that

$$\sum_{i=1}^{n} \partial R_i = \partial D, \tag{6.256}$$

which is true since each of the internal paths separating one region R_i from another R_j contributes twice but with opposite orientation to the integral, as is visually evident from the second picture in Figure 6.38.

In this way the proof of the theorem is completed. □

6.10.4 Illustrations of homotopy in $\mathbb{R}^2$

Next we apply the general concept of homotopy defined in Section 6.9.1. We consider piece-wise differentiable paths in a domain $D \subset \mathbb{R}^2$. Due to the previously discussed parametric invariance of the paths we can always assume that all the paths are parameterized by the closed segment $I = [0,1] \subset \mathbb{R}$. According to the general definition, Definition 6.9.2, discussed in Section 6.9.1, we say that two paths in $D \subset \mathbb{R}^2$

$$\gamma_0 : I \longrightarrow D; \quad \gamma_1 : I \longrightarrow D \tag{6.257}$$

that have the same initial and final points, namely, $\gamma_0(0) = \gamma_1(0)$ and $\gamma_0(1) = \gamma_1(1)$, are *homotopic* in D if there exists a continuous map

$$\delta : I \times I \longrightarrow D \tag{6.258}$$

provided by the continuous two-variable function $\delta(t,u)$ such that

$$\begin{cases} \delta(t,0) = \gamma_0(t); & \delta(t,1) = \gamma_1(t), \\ \delta(0,u) = \gamma_0(0) = \gamma_1(0); & \delta(1,u) = \gamma_0(1) = \gamma_1(1). \end{cases} \tag{6.259}$$

At fixed u the map $t \to \delta(t,u)$ is a path γ_u in D which has the same initial and final points as γ_0 and γ_1. Intuitively the path γ_u is a continuous deformation of the path γ_0 which ultimately becomes γ_1 when $u = 1$ (see Figure 6.39). We have a similar definition for closed paths: we say that two closed paths $\gamma_0(t)$ and $\gamma_1(t)$ are *homotopic* in D if there exists a continuous map

$$\delta : I \times I \longrightarrow D \tag{6.260}$$

provided by the continuous two-variable function $\delta(t,u)$ such that

$$\begin{cases} \delta(t,0) = \gamma_0(t); & \delta(t,1) = \gamma_1(t), \\ \delta(0,u) = \delta(1,u) & \forall u \in [0,1]. \end{cases} \tag{6.261}$$

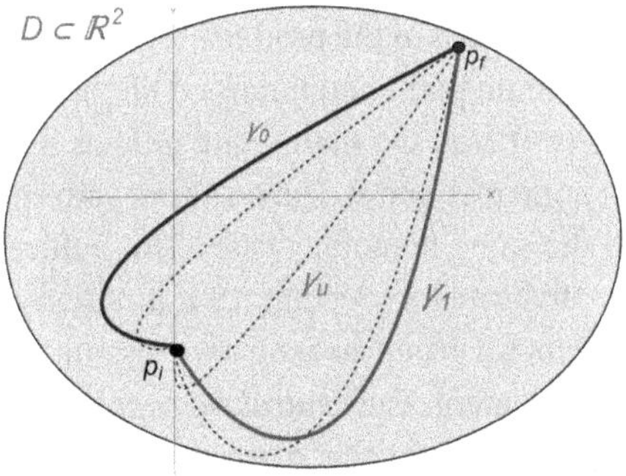

Figure 6.39: Two piece-wise differentiable paths γ_0 and γ_1 in a region $D \subset \mathbb{R}^2$ are homotopic if they can be continuously deformed one into the other while always remaining inside D. The dashed curves are some of the intermediate paths γ_u for $u \neq 0$ and $u \neq 1$.

In this way every intermediate path $\gamma_u = \delta(t, u)$ is also a closed path. In particular we say that a path $\gamma_0(t)$ is homotopic to a point if within the previous setup $\gamma_1(t) = \text{const}$. The definition of homotopy in the case of closed paths is illustrated in Figure 6.40.

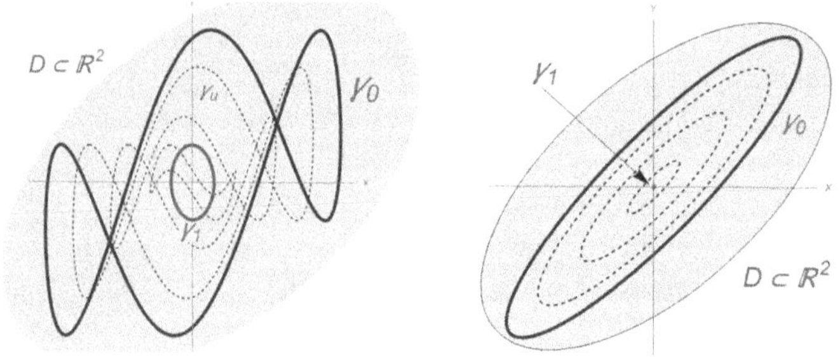

Figure 6.40: Two closed paths γ_0 and γ_1 in a region $D \subset \mathbb{R}^2$ are homotopic if they can be continuously deformed one into the other while always remaining inside D. As in the previous figure, the dashed lines represent some of the intermediate paths γ_u that realize the homotopy. As the reader can appreciate, it is not important whether the number of self-intersections in two paths that are homotopic is different. The only important thing is that they can be continuously deformed into each other, as the figure illustrates.

As we know from the general theory exposed in Section 6.9.1, homotopy establishes an equivalence relation. We say that two closed paths γ_A and γ_B are equivalent if they are homotopic and we write $\gamma_A \overset{h}{\sim} \gamma_B$. Hence, the set of all possible closed paths inside a domain $D \subset \mathbb{R}^2$ can be partitioned into homotopic equivalence classes h_A, whose set can have a finite or infinite cardinality, but it is in any case a discrete set that we name $\pi_1(D)$ and constitutes a group under the product law explained in Section 6.9.1. Explicitly the product of closed loops is taken in the way illustrated in Figure 6.41. Given the homotopy classes $[A]$, $[B]$ of two closed paths A and B, one deforms each of the latter in such a way that the new representatives touch each other in a common point. This is always possible since, by hypothesis, the domain D is connected. Next one defines the product of the two homotopy classes $[A] \cdot [B]$ as the homotopy class of the product $A \cdot B$ obtained in the following way. First one starts from the common point and follows A all the way up to its return to the common point. There one starts on the path B and follows it all the way up to return once again to the common point of A and B. The resulting path can now be deformed at will remaining always in the same homotopy class. This suffices to establish that we have indeed, as generally advocated in Section 6.9.1, a notion of product among homotopy classes. In order to achieve a group structure we also need a neutral element **e** and the notion of inverse of an element. The *neutral element* is quite naturally provided by the homotopy class of the trivial paths that are contractible to a point. As for the inverse of a path, nothing is simpler: it is just the same path followed in the reversed direction; we will denote by A^{-1} the path A in the reverse direction. It is visually evident that the product $A \cdot A^{-1}$ is a homotopically trivial path, as illustrated in

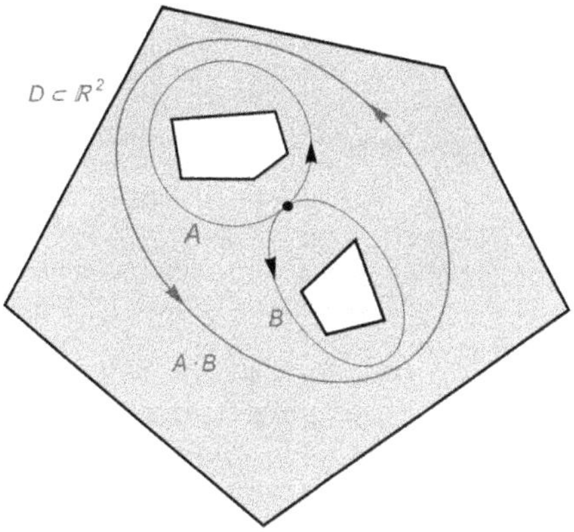

Figure 6.41: This picture illustrates the notion of product in homotopy. The path A winds once around the upper hole of D. The path B winds once around the lower hole. The product $A \cdot B$ winds once around both holes.

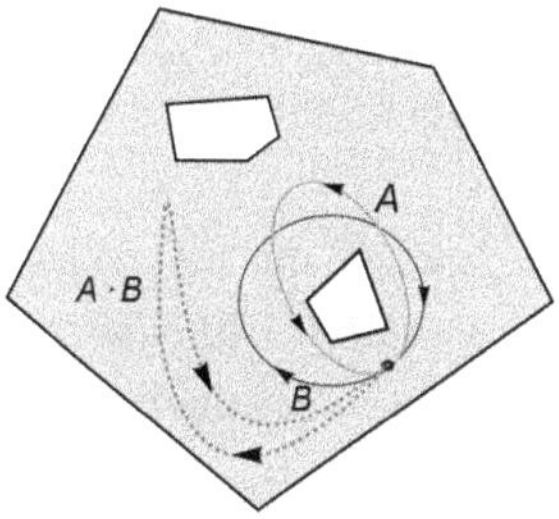

Figure 6.42: This figure illustrates the notion of inverse of a path in homotopy. The cycles A and B both wind around the lower hole in the domain D, yet they do so in opposite directions. The product $A \cdot B$ can be continuously deformed to a point as illustrated in this figure. Hence we can correctly say that $B = A^{-1}$.

Figure 6.42. As we stated, the *trivial class* in $\pi_1(D)$ is composed of all closed paths that are homotopic to a point. If the topology of the considered domain D is *simple* (we will say that D is *simply connected*), the trivial class is actually the only one, namely, all paths can be deformed to a point, as shown in Figure 6.43. If D is connected, yet it has some holes, that is, its boundary is composed of disconnected pieces, like in the second picture of Figure 6.43, then new non-trivial homotopy classes do arise. Essentially the non-trivial homotopy classes are those of the closed paths that wind around a hole of the domain and can do that once, twice, thrice, or in general n times. Indeed, in the simple example provided by the second case of Figure 6.43, taking into account also the orientation of the paths, it follows that all the homotopy classes are in one-to-one correspondence with the integer numbers $\mathbb{Z}$, since a path can wind around the hole n times, or $-n$ times if it is oriented in the opposite direction with respect to a reference one, named the positive direction.

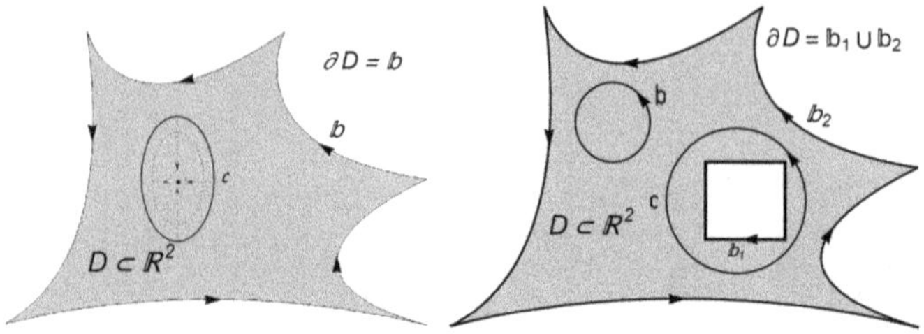

Figure 6.43: The essential difference between the domain D in the first picture of this figure and the domain D in the second picture is that the boundary of the former $\partial D = \mathbf{b}$ consists of only one component, while the boundary of the latter $\partial D = \mathbf{b}_1 \cup \mathbf{b}_2$ consists of two disjoint pieces $\mathbf{b}_1 \cap \mathbf{b}_2 = \emptyset$. As a consequence of this fact there arise non-trivial homotopy classes as that of the path $\mathfrak{c}$ which is not homotopic to a point of the domain D.

6.10.5 Consequences of a fundamental theorem

In Section 6.9 we mentioned without proof the fundamental theorem of algebraic topology (Theorem 6.9.4). Let us mention the immediate consequences of that theorem in the case of the functions defined over regions of the $\mathbb{R}^2$ plane.

First of all let us explain what we mean when we write $H_n(\mathcal{M}, \mathbb{Z})$. It is just a specification of the coefficients that we utilize while defining the *p-chains* as linear combinations of those fundamental p-chains that we have named the *p-simplexes*. Writing $H_n(\mathcal{M}, \mathbb{Z})$, one states that such coefficients are integer numbers. In the case of the first homology group $H_1(\mathcal{M}, \mathbb{Z})$ the fundamental 1-simplex is just a segment, or rather its smooth image in the variety $\mathcal{M}$, namely, what we already defined as a differentiable path γ. Hence, the 1-chains $\mathscr{C}_1 = \sum_{i=1}^{r} c_i \gamma_i$ are just linear combinations with integer coefficients $c_i \in \mathbb{Z}$ of paths γ_i. We already stressed that these entities acquire a precise meaning when we think of the 1-chains as lines along which we integrate the 1-forms $\omega^{[1]}$. In the integral $\int_{C_i} \omega^{[1]}$, each individual subintegral $\int_{\gamma_i} \omega^{[1]}$ contributes c_i times. For instance, in Figure 6.44 we observe that the cycle C, which certainly cannot be continuously deformed to a point, is homologically trivial, being the boundary of a region. According to the above quoted general theorem this happens just because the path C is homotopic to the commutator path $ABA^{-1}B^{-1}$.

From Theorem 6.9.4 it immediately follows that when the homotopy group is trivial $\pi_1(\mathcal{M}) = \mathbf{e}$, also the homology group is trivial $H_1(\mathcal{M}, Z) = \mathbf{e}$, namely, all closed paths are boundaries of some region.

Definition 6.10.3. A manifold $\mathcal{M}$ is called *simply connected* if it is connected and its fundamental group $\pi_1(\mathcal{M})$ is trivial.

In view of the definition of homotopy and Theorem 6.9.4 we have the following corollary.

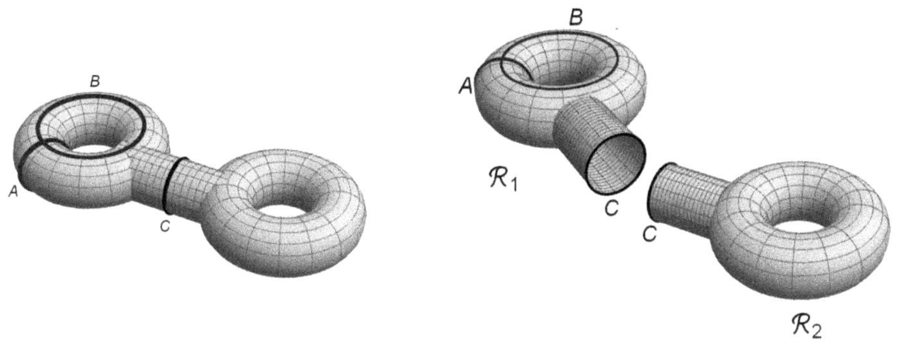

Figure 6.44: In this figure we illustrate the basic concepts of homology. On the bitorus, namely, the surface with two handles, we can draw various closed paths, for instance those marked A, B, C in the first picture. None of these three can be continuously shrunk to zero, yet there is a difference among them. If we perform a cut with scissors along A or B the surface does not split in two pieces. This means that neither A nor B is the boundary of any region of the bitorus. However, if we perform a cut along the path C, what happens is described in the second picture: the surface splits in the two regions $\mathscr{R}_1$ and $\mathscr{R}_2$. This means that the path C is a boundary.

Corollary 6.10.1. *A simply connected open domain of the plane $D \subset \mathbb{R}^2$ is a region where every closed path $\mathfrak{c}$ can be contracted to a point and moreover it is the boundary of some region $\mathscr{R} \subset D$.*

In this way, recalling Theorem 6.10.1 we can easily conclude that the following is true.

Theorem 6.10.3. *Every closed differential 1-form ω defined over a simply connected open domain $D \subset \mathbb{R}^2$ admits there a primitive F.*

Proof of Theorem 6.10.3. Indeed, since any closed path $\mathfrak{c}$ is a boundary, namely, $\exists \mathscr{R} \subset D$ such that $\partial R = \mathfrak{c}$, it follows that $\int_{\mathfrak{c}} \omega = 0$, which is the necessary and sufficient condition for ω to admit a primitive F. □

The difference between simply connected and non-simply connected domains of the plane was exemplified in Figure 6.43.

6.10.6 Holomorphicity

So far we have considered closed 1-forms in $\mathbb{R}^2$, and no explicit mention of complex holomorphic functions was made. Actually, as we are going to show, there is a complete identification among the two mathematical objects. Complex numbers enter the mathematical playing ground at least in two capacities. The first capacity, which is algebraic, corresponds to the need to extend the field of real numbers and make it algebraically complete so that a generic polynomial $\mathscr{P}(z)$ of degree n admits exactly n roots. The second capacity is that of a division algebra over the reals $\mathbb{R}$. From this viewpoint, a complex number is a 2-vector of $\mathbb{R}^2$, which is a standard vector space, yet equipped with a new operation that makes it an algebra, thus giving meaning to the product of two vectors and also to the inverse of a vector. Obviously, at the end of the day $\mathbb{C}$ is not only an algebra, it is even a field.

Nevertheless, if one looks at complex numbers as a two-dimensional vector space, isomorphic to $\mathbb{R}^2$, one can wonder what kind of mathematical beast the imaginary unit i might be. The answer is simple: it is just a *complex structure*, namely, a structure that can be introduced on any even-dimensional vector space.

6.10.6.1 The notion of complex structure

Definition 6.10.4. Let V be the vector space over the reals $\mathbb{R}$ of dimension $2n$. By complex structure over V we mean any linear map

$$\begin{aligned} &\mathbf{J} : V \to V \\ &\mathbf{J}(\alpha\mathbf{v}_1 + \beta\mathbf{v}_2) = \alpha\mathbf{J}(\mathbf{v}_1) + \beta\mathbf{J}(\mathbf{v}_2) \quad \forall\alpha,\beta \in \mathbb{R} \quad \forall\mathbf{v}_1,\mathbf{v}_2 \in V \end{aligned} \tag{6.262}$$

that satisfies the following defining conditions:

(CS1) The map $\mathbf{J}$ is non-degenerate, namely, $\mathbf{J}(\mathbf{v}) = 0$ implies $\mathbf{v} = \mathbf{0}$.

(CS2) The map $\mathbf{J}$ squares to minus the identity map:

$$\mathbf{J} \cdot \mathbf{J} = -\mathbf{Id}. \tag{6.263}$$

It follows from equation (6.263) that the eigenvalues λ of a complex structure $\mathbf{J}$ obey the relation $\lambda^2 = -1$ and hence they are $\pm\sqrt{-1}$. We denote $\sqrt{-1} = i$. In this way we see that the imaginary unit and its negative, $\pm i$, are the possible eigenvalues of a complex structure and they always occur in pairs with opposite signs.

Let us consider the two-dimensional case that is our target. An example of complex structure on $\mathbb{R}^2$ is provided by the 2×2 matrix

$$\mathbf{J} = \begin{pmatrix} 0 & -1 \\ 1 & 0 \end{pmatrix}, \tag{6.264}$$

whose eigenvalues are $\pm i$. We can easily verify that the following unitary transformation diagonalizes $\mathbf{J}$:

$$U = \begin{pmatrix} \frac{1}{\sqrt{2}} & \frac{1}{\sqrt{2}} \\ -\frac{i}{\sqrt{2}} & \frac{i}{\sqrt{2}} \end{pmatrix}; \quad U^\dagger U = \mathbb{1}; \quad U^\dagger \mathbf{J} U = \begin{pmatrix} i & 0 \\ 0 & -i \end{pmatrix}. \tag{6.265}$$

Applying the above change of basis to a generic vector $\mathbf{v} = \{x, y\} \in \mathbb{R}^2$, we obtain

$$U^\dagger \mathbf{v} = \frac{1}{\sqrt{2}} \begin{cases} x + iy \\ x - iy \end{cases} = \frac{1}{\sqrt{2}} \begin{cases} z \\ \overline{z}, \end{cases} \tag{6.266}$$

where we have denoted by z the complex number of which x is the real part and y the imaginary one. In this way we arrive at the conclusion that the complex coordinate $z = x + iy$ is the eigencoordinate of the complex structure $\mathbf{J}$ with eigenvalue i, while the complex conjugate $\overline{z}$ is the eigencoordinate corresponding to $-i$.

Defined on the Cartesian coordinates $\mathbf{X} = \{x, y\}$, the action of the complex structure $\mathbf{J}$ extends to their differentials and to the derivative operators.

If we consider the total differential of a function $f(x, y)$ we get

$$\begin{aligned} df(X) &= \frac{\partial f}{\partial \mathbf{X}} \cdot d\mathbf{X} = \frac{\partial f}{\partial x} dx + \frac{\partial f}{\partial y} dy \\ &= \frac{\partial f}{\partial \mathbf{X}} U^\dagger U d\mathbf{X} = \frac{\partial f}{\partial z} dz + \frac{\partial f}{\partial \overline{z}} d\overline{z}, \end{aligned} \tag{6.267}$$

where

$$\begin{aligned} &dz = dx + i\,dy; \quad d\overline{z} = dx - i\,dy, \\ &\frac{\partial f}{\partial z} = \frac{1}{2}\left(\frac{\partial f}{\partial x} - i\frac{\partial f}{\partial y}\right); \quad \frac{\partial f}{\partial \overline{z}} = \frac{1}{2}\left(\frac{\partial f}{\partial x} + i\frac{\partial f}{\partial y}\right). \end{aligned} \tag{6.268}$$

6.10.7 Holomorphicity condition

Let $D \subset \mathbb{C}$ be an open domain of the complex plane and let

$$f(z) = u(x, y) + iv(x, y) \tag{6.269}$$

be a function of the complex variable $z = x + iy$ defined in D.

Definition 6.10.5. The function $f(z)$ is *holomorphic* at the point $z_0 \in D$ if

$$\lim_{u \to 0} \frac{f(z_0 + u) - f(z_0)}{u} = f'(z_0) \in \mathbb{C} \quad \text{exists and is unique.} \tag{6.270}$$

This amounts to saying that $f(z)$ has a derivative with respect to the complex variable z. We say that $f(z)$ is holomorphic in D if it is holomorphic for all $z \in D$.

The condition (6.270) can be rewritten as

$$f(x_0 + h, y_0 + k) - f(x_0, y_0) = c(h + ik) + \alpha(h, k)\sqrt{h^2 + k^2}, \tag{6.271}$$

where $\alpha(h, k)$ is a function which goes to zero as $\sqrt{h^2 + k^2}$ goes to zero. In this case

$$f'(z_0) \equiv c \tag{6.272}$$

is the derivative of $f(z)$ at z_0. The relation (6.271) shows that a function which is holomorphic at the point z_0 possesses there the partial derivative with respect to x and y. Indeed from equation (6.272) we find

$$\frac{\partial f}{\partial x} = f'(z); \quad \frac{\partial f}{\partial y} = if'(z), \tag{6.273}$$

$$\Downarrow$$

$$\frac{\partial f}{\partial x} + i\frac{\partial f}{\partial y} = 0. \tag{6.274}$$

Comparing equation (6.274) with equation (6.268) we see that the condition of holomorphicity is the same as

$$\frac{\partial f}{\partial \bar{z}} = 0. \tag{6.275}$$

Hence once we replace the coordinates $\{x, y\}$ with the eigencoordinates of the complex structure $\{z, \bar{z}\}$, a holomorphic function $f = f(z)$ depends only on z. In this way we have proved the following.

Statement 6.10.1. In order for a complex function $f(x, y)$ defined in an open domain $D \subset \mathbb{R}^2$ to be holomorphic it suffices that its partial derivatives satisfy the relation (6.274), which, once rewritten in terms of the real and imaginary parts $u(x, y)$ and $v(x, y)$ of the function $f(x, y)$ (see equation (6.269)), reads as follows:

$$\frac{\partial u}{\partial x} = \frac{\partial v}{\partial y}; \quad \frac{\partial u}{\partial y} = -\frac{\partial v}{\partial x}. \tag{6.276}$$

These are named the *Cauchy–Riemann equations.*

6.10.8 Cauchy theorem

One of the fundamental theorems in the theory of analytic functions, which goes under the name of Cauchy, becomes, after the general preparations introduced in the previous sections, almost a triviality whose proof is immediate.

Theorem 6.10.4. *Let $D \subset \mathbb{R}^2 \cong \mathbb{C}$ be an open domain of the complex plane and let $f(z)$ be a holomorphic function defined in D. The differential 1-form $\omega \equiv f(z)dz$ is closed in D.*

Proof of Theorem 6.10.4. The proof is immediate by direct calculation of the exterior differential. For any function $f(x, y) = f(z, \bar{z})$, setting $\omega = f\, dz$ we have

$$d\omega = \underbrace{\frac{\partial f}{\partial z}\, dz \wedge dz}_{=0} + \frac{\partial f}{\partial \bar{z}}\, d\bar{z} \wedge dz. \tag{6.277}$$

Hence, if $f(z)$ is holomorphic, by definition $\frac{\partial f}{\partial \bar{z}} = 0$ and $d\omega = 0$. □

In the proof of the Cauchy theorem, which just follows from the very definition of the exterior differential, we have not made any assumption about the topology of the domain D. Let us consider the case where D is simply connected. Then we obtain an immediate consequence expressed as follows.

Corollary 6.10.2. *Let D be a simply connected open domain of the complex plane, let $f(z)$ be a holomorphic function defined over D, and let $\mathfrak{c}$ be any closed path in D. Then we have*

$$\oint_{\mathfrak{c}} f(z)dz = 0. \tag{6.278}$$

Proof of Corollary 6.10.2. Since D is simply connected, its fundamental group is trivial: $\pi_1(D) = \mathbf{e}$. Hence, also the first homology group $H_1(D) = \mathbf{e}$ is equally trivial. This means that in D every cycle $\mathfrak{c} = \partial\mathscr{R}$ is the boundary of some region $\mathscr{R}$ (see Figure 6.45). Hence, we have

$$\oint_{\mathfrak{c}} f(z)dz = \int\int_{\mathscr{R}} \underbrace{d[f(z)dz]}_{=0} = 0, \tag{6.279}$$

which proves the statement. □

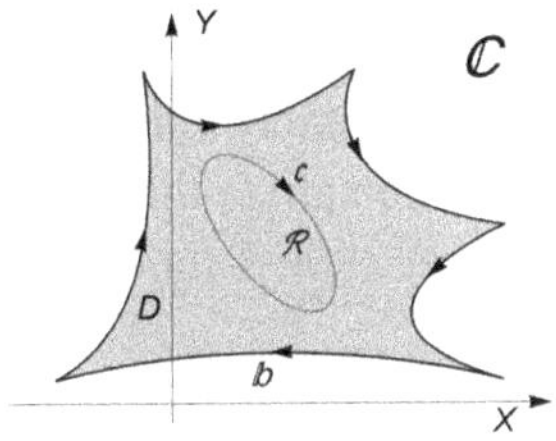

Figure 6.45: In this picture the domain $D \subset \mathbb{C}$ is simply connected. Indeed, its boundary $\partial D = \mathbf{b}$ consists of just one component. Hence, every closed path $\mathfrak{c}$ contained in D is contractible to a point and it is also the boundary of some region $\mathfrak{c} = \partial\mathscr{R} \subset D$.

6.10.8.1 Introducing punctures and the Cauchy integral formula

Our next focus of interest is a punctured domain $D_{\mathfrak{p}}$ of which we provide the following definition.

Definition 6.10.6. Let D be a connected open domain of the plane and let

$$\mathfrak{p} = \bigcup_{i=1,\dots,p} \{z_i\}; \quad \forall i \quad z_i \in D \tag{6.280}$$

be a set of interior points of D. This set is denumerable and can be finite or infinite: the elements of $\mathfrak{p}$ are named the *punctures*.

The punctured domain $D_{\mathfrak{p}}$ is set theoretically defined as the domain D deprived of the punctures (see Figure 6.46):

$$D_{\mathfrak{p}} = D - \mathfrak{p}. \tag{6.281}$$

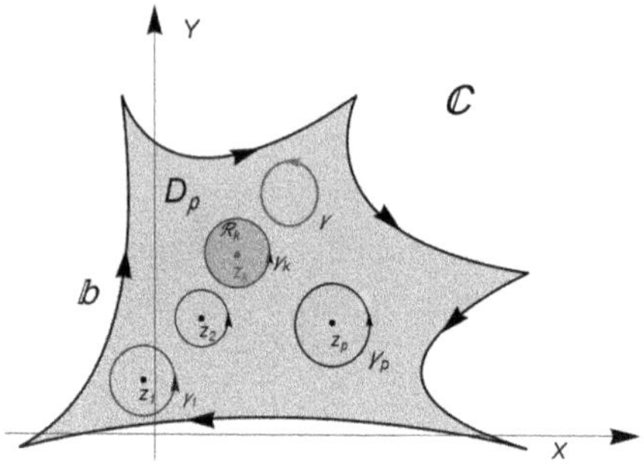

Figure 6.46: In this picture we display a simply connected domain with p punctures. The homotopy group of $D_{\mathfrak{p}}$ is isomorphic to $\mathbb{Z}^p$ since it is the Abelian freely generated group with p generators given by the p circles γ_k that encircle within D one by one all the punctures ($k = 1,\dots,p$). Since D is simply connected, each of these homotopy generators γ_k which is non-trivial in $D_{\mathfrak{p}}$ is trivial in D, both homotopically and homologically. Hence, each $\gamma_k = \partial\mathcal{R}_k$ is the boundary of a region $\mathcal{R}_k \subset D$ and the corresponding puncture $z_k \in \mathcal{R}_k$ belongs to such region.

In Definition 6.10.6 we did not specify whether the domain D is simply connected or not. Yet whatever was the topological character of the original domain D, the corresponding punctured domain is no longer simply connected. Indeed, let the boundary of D be made of a finite number k of disconnected closed paths $\mathbf{b}_\alpha$:

$$\partial D = \bigcup_{\alpha=1,\dots,k} \mathbf{b}_\alpha. \tag{6.282}$$

It is visually evident, looking at Figure 6.46, that the boundary of $D_{\mathfrak{p}}$ is as follows:

$$\partial D_{\mathfrak{p}} = \partial D \cup \mathfrak{p}; \quad \mathfrak{p} \cap \partial D = \emptyset. \tag{6.283}$$

Hence, $\partial D_{\mathfrak{p}}$ is made of several disconnected components and therefore it is not simply connected (see also Figure 6.47). In the case the original domain D was simply connected, i. e., $\pi_1(D) = \mathbf{e}$, we can easily calculate the fundamental group of the punctured domain $D_{\mathfrak{p}}$ with $p < \infty$ punctures:

$$\pi_1(D_{\mathfrak{p}}) \simeq \mathbb{Z}^p. \tag{6.284}$$

Indeed, the homotopy class of any cycle $\mathfrak{c} \subset D_{\mathfrak{p}}$ is specified by how many times $\mathfrak{c}$ winds around each puncture z_k, and that, taking the orientation of the paths into account, is an integer number n_k. This observation is formalized in the following way. Let $\mathfrak{C}$ be the space of cycles in D_p, namely, each element $\mathfrak{c} \in \mathfrak{C}$ is a closed path in D_p. We have the homotopy homomorphism

$$\begin{aligned} h: \quad & \mathfrak{C} \longrightarrow \mathbb{Z}^p, \\ \forall \mathfrak{c} \in \mathfrak{C}: \quad & h(\mathfrak{c}) = \mathbf{n} \equiv \{n_1, n_2, \ldots, n_3\} \quad (n_k \in \mathbb{Z}), \\ & h(\mathfrak{c}_1 \cdot \mathfrak{c}_2) = \mathbf{n}_1 + \mathbf{n}_2. \end{aligned} \tag{6.285}$$

The above formula and the above statements are interpreted in the following way. Given the p punctures, one constructs the small circles γ_k that wind just once around the corresponding puncture z_k (see Figure 6.46). Any closed path c in D_p is homotopic to a product of these paths, that is,

$$c \overset{\text{homotopy}}{\sim} \gamma_1^{n_1} \cdot \gamma_2^{n_2} \cdot \cdots \cdot \gamma_p^{n_p}. \tag{6.286}$$

6.10.8.2 The index of a cycle in a punctured domain D_p

The image $h(\mathfrak{c})$ of a path $\mathfrak{c}$ through the homotopy homomorphism (6.285) is just the ordered collection of the integers n_k, namely, the decomposition of $\mathfrak{c}$ along a basis of generators of the fundamental group $\pi_1(D_\mathrm{p})$: we name $h(\mathfrak{c})$ the index of the considered closed path.

Let us introduce a p-dimensional vector space Res_p over the field $\mathbb{C}$ whose elements are the formal linear combinations of the p punctures,

$$\begin{aligned} \mathfrak{a} &= \sum_{k=1}^{p} a_k z_k \in \mathrm{Res}_\mathrm{p}; \quad (a_i \in \mathbb{C}), \\ \mathfrak{a} &\equiv \{a_1, a_2, \ldots, a_p\} \in \mathbb{C}^p \sim \mathrm{Res}_\mathrm{p}. \end{aligned} \tag{6.287}$$

The space Res_p is called the *space of residues* over the punctures $\mathfrak{p}$. With every residue $\mathfrak{a} \in \mathrm{Res}_\mathrm{p}$ we associate the following holomorphic 1-form:

$$\omega[\mathfrak{a}] \equiv \frac{1}{2\pi i} \sum_{i=1}^{p} a_i \frac{dz}{z - z_i}. \tag{6.288}$$

We have the following basic theorem.

Theorem 6.10.5. *Let D_p be a simply connected domain D with punctures $\mathfrak{p}$. For any closed path $\mathfrak{c}$ in D_p, the integral of the form $\omega[\mathfrak{a}]$ along $\mathfrak{c}$ is equal to the scalar product of the residue with the index of the path:*

$$\oint_{\mathfrak{c}} \omega[\mathfrak{a}] = h(\mathfrak{c}) \cdot \mathfrak{a} \equiv \sum_{k=1}^{p} n_i a_i. \tag{6.289}$$

Proof of Theorem 6.10.5. The proof is obtained through the following chain of arguments:

1. Being holomorphic, the 1-form $\omega[\mathfrak{a}]$ is closed in D_p: $d\omega[\mathfrak{a}] = 0$.
2. Since $\mathfrak{c}$ and $\gamma_2^{n_2} \cdot \cdots \cdot \gamma_p^{n_p}$ are homotopic, the path

$$\xi \equiv \mathfrak{c}^{-1}\gamma_1^{n_1} \cdot \gamma_2^{n_2} \cdot \cdots \cdot \gamma_p^{n_p}$$

   is homotopically and homologically trivial in D_p. Hence, there exists a region $\mathscr{R} \subset D_p$ such that

$$\xi = \partial\mathscr{R}.$$

3. By the Stokes theorem we have

$$\oint_\xi \omega[\mathfrak{a}] = \int\int_R d\omega[\mathfrak{p}] = 0. \tag{6.290}$$

4. From equation (6.290) we conclude

$$\begin{aligned}\oint_c \omega[\mathfrak{a}] &= \sum_{i=1}^p n_i \oint_{\gamma_i} \omega[\mathfrak{a}] \\ &= \sum_{k=1}^p \sum_{j=1}^p n_k \frac{1}{2\pi i} \oint_{\gamma_k} a_j \frac{dz}{z - z_j}.\end{aligned} \tag{6.291}$$

5. With respect to the simply connected D, the cycles γ_k that generate the homotopy group of the punctured domain D_p are both homotopically and homologically trivial. Hence each of them is the boundary of a corresponding region $\mathscr{R}_k \subset D$. Suppose that $u \notin \mathscr{R}_k$. The function $g(z)\frac{1}{z-u}$ is well-defined, holomorphic, and finite over the entire $\mathscr{R}_k$ region. It follows that

$$\oint_{\gamma_k} \frac{dz}{z-u} = \int\int_{\mathscr{R}_k} d\left(\frac{dz}{z-a}\right) = 0; \quad \textit{if } u \notin \mathscr{R}_k. \tag{6.292}$$

6. Consider the integral

$$I_k = \oint_{\gamma_k} \frac{dz}{z - z_k}, \tag{6.293}$$

   where γ_k encircles the puncture z_k. The function $g_k(z)\frac{1}{z-z_k}$ is well-defined, finite, and holomorphic everywhere except at the puncture z_i, where it diverges to infinity. However, we can calculate I_k explicitly. By definition the path γ_k is a circle of small radius ρ with center in z_k. Henceforth the coordinate z along the path can be parameterized as

$$z(t) = z_i + \rho \exp[2\pi i t], \tag{6.294}$$

and substituting (6.294) into equation (6.293) we get

$$I_k = 2\pi i \times \int_0^1 dt = 2\pi i. \tag{6.295}$$

Using (6.295) and (6.292) in equation (6.290) we finally get

$$\begin{aligned} &= \sum_{k=1}^{p}\sum_{j=1}^{p} n_k \frac{1}{2\pi i} \times 2\pi i a_j \times \delta_{k,j} \\ &= \sum_{k=1}^{p} n_k a_k = h(\mathfrak{c}) \cdot \mathfrak{a}. \end{aligned} \tag{6.296}$$

Equation (6.296) concludes the proof of the theorem. □

6.10.8.3 Cauchy's integral formula

The main application of the basic Theorem 6.10.5 is the celebrated integral formula named after Cauchy.

Theorem 6.10.6. *Let $D \subset \mathbb{C}$ be an open domain and let $f(z)$ be a holomorphic function that is well-defined and finite everywhere in D. Introduce a single puncture at the point $z_0 \in D$: $\mathfrak{p}_0 = \{z_0\}$. Consider the punctured domain $D_{\mathfrak{p}_0}$. Let $\mathfrak{c}$ be a path in D which is homotopically trivial and does not touch the puncture. Whatever the fundamental group $\pi_1(D)$ was, the puncture extends it with a new generator that is a circle γ_0 around the puncture. The path $\mathfrak{c}$, homotopically trivial in D, is homotopic in $D_{\mathfrak{p}_0}$ to $\gamma_0^{n_\mathfrak{c}}$, where $n_\mathfrak{c} \in \mathbb{Z}$.*

Under these hypotheses the following integral formula is true:

$$\frac{1}{2\pi i} \oint_{\mathfrak{c}} \frac{f(z)\, dz}{z - z_0} = n_\mathfrak{c} f(z_0). \tag{6.297}$$

Proof of Theorem 6.10.6. Since $f(z)$ is everywhere holomorphic in D, it is in particular holomorphic in the region $\mathscr{R}_0$ of which the contractible path γ_0 is the boundary: $\partial\mathscr{R}_0 = \gamma_0$. Hence, $f(z)$ admits a derivative everywhere in $\mathscr{R}_0$, also at the puncture z_0, and we can write

$$f(z) = f(z_0) + f'(z_0)(z - z_0) + \mathscr{O}((z - z_0)^2). \tag{6.298}$$

Correspondingly,

$$g(z) \equiv \frac{f(z) - f(z_0)}{z - z_0} \tag{6.299}$$

is holomorphic and finite in the whole D and we can apply the Stokes theorem to the integral $\oint_{\gamma_0} g(z)\,dz$:

$$\frac{1}{2\pi i}\oint_{\gamma_0} g(z)\,dz = \frac{1}{2\pi i}\int\int_{\mathscr{R}_0} \underbrace{d(g(z)dz)}_{=0} = 0. \tag{6.300}$$

From equation (6.300) and Theorem 6.10.5 we obtain

$$\frac{1}{2\pi i}\oint_{c} \frac{dz}{z - z_0} dz = \oint_{c} \omega[f(z_0)z_0] = n_c f(z_0). \tag{6.301}$$

Note that in equation (6.301) we have recognized that, according to equation (6.288), $\frac{1}{2\pi i} f(z_0)\frac{dz}{z-z_0}$ is just the definition of the 1-form $\omega[\mathfrak{a}]$ associated with the residue $\mathfrak{a} = f(z_0)\{z_0\}$. In this way the proof of the theorem is completed. □

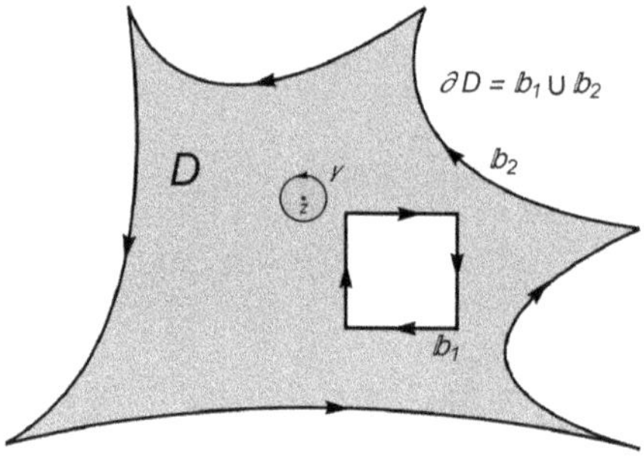

Figure 6.47: We consider the non-simply connected domain D whose boundary is the union of the two disconnected curves $\mathbf{b}_1$ and $\mathbf{b}_2$. We introduce a puncture at the point $\mathfrak{p} = \{z\}$ and we encircle it with the path γ, which is homotopically trivial in D, but not in $D_{\mathfrak{p}} = D - \mathfrak{p}$. In D the path γ is the boundary of a region S, i. e., of the small disk drawn around the puncture z.

6.11 Bibliographical note

The main bibliographical sources for this chapter are the following books:
1. The first volume of [79].
2. The monograph [80].
3. The textbook [81].
4. The textbook [82].

Furthermore, a considerable part of the illustrative material on analytic functions of a complex variable, including all the related original figures, comes from unpublished lecture notes of the present author written for second year students following his innovative course *Mathematical Methods of Physics*.

7 The relation between Lie groups and Lie algebras

> If only I knew how to get mathematicians interested in transformation groups and their applications to differential equations! I am certain, absolutely certain, that these theories will some time in the future be recognized as fundamental. When I wish such a recognition sooner, it is partly because then I could accomplish ten times more.
>
> Sophus Lie 1884 in a letter to Adolf Mayer

After the previous long introductory chapter about manifolds, fiber bundles, homology, and homotopy, we turn to the essential relation that exists among Lie groups and Lie algebras that allows to classify the former by means of a classification of the latter. Furthermore, such a relation is a fundamental tool in order to construct the linear representations of Lie groups since such representations are deduced from the linear representations of the corresponding Lie algebras.

7.1 The Lie algebra of a Lie group

The definition of a Lie group was given in Section 6.6.1. As already sketched there, we recall that, as a consequence of its definition, on a Lie group G one can define two transitive actions of the same group on itself, left and right multiplication, respectively. Indeed, with each element $\gamma \in G$ we can associate two continuous, infinitely differentiable and invertible maps of G in G, named left and right translation, which we introduced in equation (6.111).

Here we introduce the concepts of pull-back and push-forward of any diffeomorphism mapping a differential manifold $\mathcal{M}$ into an open submanifold of another differentiable manifold $\mathcal{N}$. Let ϕ be any such map

$$\phi : \mathcal{M} \to \mathcal{N}. \tag{7.1}$$

The push-forward of ϕ, denoted ϕ_*, is a map from the space of sections of the tangent bundle $T\mathcal{M}$ to the space of sections of the tangent bundle $T\mathcal{N}$:

$$\phi_* : \Gamma(T\mathcal{M}, \mathcal{M}) \to \Gamma(T\mathcal{N}, \mathcal{N}). \tag{7.2}$$

Explicitly if $\mathbf{X} \in \Gamma(T\mathcal{M}, \mathcal{M})$ is a vector field over $\mathcal{M}$, we can use it to define a new vector field $\phi_*\mathbf{X} \in \Gamma(T\mathcal{N}, \mathcal{N})$ over $\mathcal{N}$ using the following procedure. For any $f \in C^\infty(\mathcal{N})$, namely, for any smooth function on $\mathcal{N}$, the action of $\phi_*\mathbf{X}$ on such a function is given by

$$\phi_*\mathbf{X}(f) \equiv [\mathbf{X}(f \circ \phi)] \circ \phi^{-1}. \tag{7.3}$$

To clarify this concept let us describe the push-forward in a pair of open charts for both manifolds. Consider Figure 7.1. On the open neighborhood $U \subset \mathcal{M}$ we have coordi-

https://doi.org/10.1515/9783111201535-007

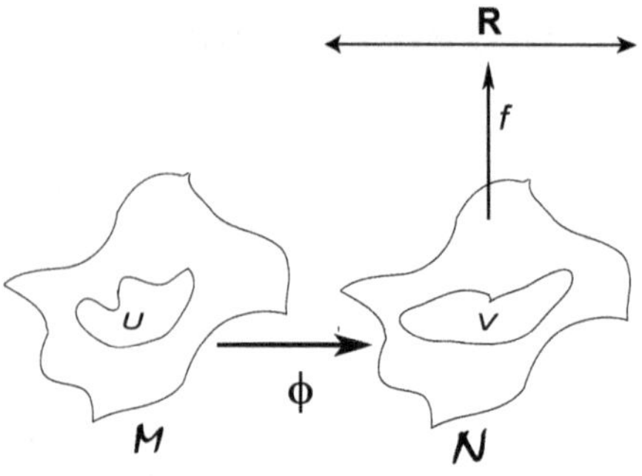

Figure 7.1: Graphical description of the concept of push-forward.

nates x^μ, while on the open neighborhood $V \subset \mathcal{N}$ we have coordinates y^ν. In this pair of local charts $\{U, V\}$ the diffeomorphism (7.1) is described by giving the coordinates y^ν as smooth functions of the coordinates x^μ:

$$y^\nu = y^\nu(x) \equiv \phi^\nu(x). \tag{7.4}$$

On the neighborhood V the smooth function $f \in C^\infty(\mathcal{N})$ is described by a real function $f(y)$ of the n-coordinates y^ν. It follows that $f \circ \phi$ is a smooth function $\tilde{f}(x)$ on the open neighborhood $U \subset \mathcal{M}$ simply given by

$$\tilde{f}(x) = f(y(x)). \tag{7.5}$$

Any tangent vector $\mathbf{X} \in \Gamma(T\mathcal{M}, \mathcal{M})$ is locally described on the open chart U by a first order differential operator of the form

$$\mathbf{X} = X^\mu(x)\frac{\partial}{\partial x^\mu}, \tag{7.6}$$

which therefore can act on $\tilde{f}(x)$:

$$\mathbf{X}\tilde{f}(x) = X^\mu(x)\frac{\partial y^\nu}{\partial x^\mu}\frac{\partial}{\partial y^\nu}f(y(x)). \tag{7.7}$$

Considering now the coordinates x^μ on U as functions of y^ν on V, through the inverse of the diffeomorphism ϕ,

$$x^\mu = x^\mu(y) = (\phi^{-1}(y))^\mu, \tag{7.8}$$

we realize that equation (7.7) defines a new linear first order differential operator acting on any function $f : V \to \mathbb{R}$. This differential operator is the push-forward of $\mathbf{X}$ through the diffeomorphism ϕ:

$$\phi_* \mathbf{X} = X^\nu(y) \frac{\partial}{\partial y^\nu} = X^\mu(x(y)) \frac{\partial y^\nu}{\partial x^\mu} \frac{\partial}{\partial y^\nu}. \tag{7.9}$$

Similarly, the pull-back of ϕ, denoted ϕ^*, is a map from the space of sections of the cotangent bundle $T^*\mathcal{N}$ to the space of sections of the cotangent bundle $T^*\mathcal{M}$:

$$\phi^* : \Gamma(T^*\mathcal{N}, \mathcal{N}) \to \Gamma(T^*\mathcal{M}, \mathcal{M}). \tag{7.10}$$

Explicitly if $\omega \in \Gamma(T^*\mathcal{N}, \mathcal{N})$ is a differential 1-form over $\mathcal{N}$, we can use it to define a differential 1-form $\phi^*\omega$ over $\mathcal{M}$ as follows. We recall that a 1-form is defined if we assign its value on any vector field; hence, we set

$$\forall \mathbf{X} \in \Gamma(T\mathcal{M}, \mathcal{M}); \quad \phi^*\omega(\mathbf{X}) \equiv \omega(\phi_* \mathbf{X}). \tag{7.11}$$

Considering equation (7.9), the local description of the pull-back on a pair of open charts $\{U, V\}$ is easily derived from the definition (7.11). If we name $\omega_\mu(y)$ the local components of the 1-form ω on the coordinate patch V,

$$\omega = \omega_\mu(y)\, dy^\mu, \tag{7.12}$$

the components of the pull-back are immediately deduced:

$$\phi^*\omega = (\phi^*\omega)_\mu(x)\, dx^\mu \equiv \omega_\nu(y(x)) \frac{\partial y^\nu}{\partial x^\mu}\, dx^\mu. \tag{7.13}$$

7.1.1 Left/right-invariant vector fields

Let us now consider the case where the manifold $\mathcal{M}$ coincides with manifold $\mathcal{N}$ and both are equal to a Lie group manifold G. The left and right translations defined in equation (6.111) are diffeomorphisms and for each of them we can consider both the push-forward and the pull-back. This construction allows to introduce the notion of left (respectively right)-invariant vector fields and 1-forms over the Lie group manifolds G.

Definition 7.1.1. A vector field $\mathbf{X} \in \Gamma(TG, G)$ defined over a Lie group manifold G is named left-invariant (respectively right-invariant) if the following condition holds true:

$$\forall \gamma \in G: \quad L_{\gamma *}\mathbf{X} = \mathbf{X} \quad (\text{respectively } R_{\gamma *}\mathbf{X} = \mathbf{X}). \tag{7.14}$$

Similarly, we have the following definition.

Definition 7.1.2. A 1-form $\sigma \in \Gamma(T^*G, G)$ defined over a Lie group manifold G is named left-invariant (respectively right-invariant) if the following condition holds true:

$$\forall \gamma \in G: \quad L_\gamma^*\sigma = \sigma \quad (\text{respectively } R_\gamma^*\sigma = \sigma). \tag{7.15}$$

Let us recall that the space of sections of the tangent bundle has the structure of an infinite-dimensional Lie algebra for any manifold $\mathcal{M}$. Indeed, given any two sections, we can compute their commutator as differential operators and this defines the necessary Lie bracket:

$$\forall \mathbf{X}, \mathbf{Y} \in \Gamma(T\mathcal{M}, \mathcal{M}); \quad [\mathbf{X}, \mathbf{Y}] = \mathbf{Z} \in \Gamma(T\mathcal{M}, \mathcal{M}). \tag{7.16}$$

Viewed as a Lie algebra the space of sections of the tangent bundle is usually denoted $\mathrm{Diff}_0(\mathcal{M})$ since every vector field can be regarded as the generator of a diffeomorphism infinitesimally close to the identity.

In the case of group manifolds we have the following simple but very fundamental theorem.

Theorem 7.1.1. *The two sets of left-invariant and right-invariant vector fields over a Lie group manifold G close two finite-dimensional Lie subalgebras of* $\mathrm{Diff}_0(G)$*, respectively named* $\mathbb{G}_{L/R}$*, which are isomorphic to each other and define the abstract Lie algebra* $\mathbb{G}$ *of the Lie group* G*. Furthermore,* $\mathbb{G}_{L/R}$ *commute with each other.*

This theorem is essentially, reformulated in modern terms, the content of Lie's work of 1874. The proof is obtained through a series of steps and through the proof of some intermediate lemmas. Let us begin with the first.

Lemma 7.1.1. *For any diffeomorphism ϕ the push-forward map ϕ_* has the following property:*

$$\forall \mathbf{X}, \mathbf{Y} \in \mathrm{Diff}_0(\mathcal{M}): \quad \phi_*[\mathbf{X}, \mathbf{Y}] = [\phi_*\mathbf{X}, \phi_*\mathbf{Y}]. \tag{7.17}$$

No proof is required for this lemma since it follows straightforwardly from the definition (7.3) of the push-forward. However, the consequences of the lemma are far reaching. Indeed, it implies that the commutator of two left-invariant (respectively right-invariant) vector fields is still left-invariant (respectively right-invariant). So we have the following lemma.

Lemma 7.1.2. *The two sets* $\mathbb{G}_{L/R}$ *of left-invariant (respectively right-invariant) vector fields constitute two Lie subalgebras* $\mathbb{G}_{L/R} \subset \mathrm{Diff}_0(G)$*:*

$$\forall \mathbf{X}, \mathbf{Y} \in \mathbb{G}_{L/R}: \quad [\mathbf{X}, \mathbf{Y}] \in \mathbb{G}_{L/R}. \tag{7.18}$$

This being established we can now show that the left-invariant vector fields can be put into one-to-one correspondence with the elements of tangent space to the group manifold at the identity element, namely, with T_eG. From this correspondence it will follow that the Lie subalgebra $\mathbb{G}_L$ has dimension equal to the dimension n of the Lie group. The same correspondence can be established also for the right-invariant vector fields and the same conclusion about the dimension of the Lie algebra $\mathbb{G}_R$ can be deduced. The argument goes as follows.

Let $g : [0,1] \mapsto G$ be a path in G with initial point at the identity element $g(0) = e \in G$. In local coordinates on the group manifold the path g is described as follows:

$$t \in [0,1]; \quad G \ni g(t) = \mathfrak{g}(\alpha^1(t)\ldots\alpha^n(t)), \tag{7.19}$$

where α^i denote the group parameters and $\mathfrak{g}(\alpha)$ denotes the group element identified by the parameters α. Hence, we obtain

$$\frac{d}{dt}g(t) = \frac{d\alpha^i}{dt}\frac{\partial}{\partial\alpha^i}\mathfrak{g}(\alpha). \tag{7.20}$$

The set of derivatives

$$c^i = \left.\frac{d\alpha^i}{dt}\right|_{t=0} \tag{7.21}$$

constitutes the component of a tangent vector at the identity that we name $\mathbf{X}_e$:

$$T_eG \ni \mathbf{X}_e \equiv c^i\frac{\vec{\partial}}{\partial\alpha^i}. \tag{7.22}$$

Conversely, for each $\mathbf{X}_e \in T_eG$ we can construct a path $g(t)$ that admits $\mathbf{X}_e$ as tangent vector at the identity element. Given a tangent vector $\mathbf{t}$ at a point $p \in \mathcal{M}$ of a manifold, constructing a curve on $\mathcal{M}$ which goes through p and admits $\mathbf{t}$ as tangent vector in that point is a problem which admits solutions on any differentiable manifold, a fortiori on a Lie group manifold G. Let therefore $g(t)$ be such a path. With $\mathbf{X}_e$ we can associate a left-invariant vector field $\mathbf{X}_L$ defining the action of the latter on any smooth function f as follows:

$$\forall f \in \mathbb{C}^\infty(G) \quad \text{and} \quad \forall \rho \in G: \quad \mathbf{X}_L f(\rho) \equiv f(\rho g(t))\big|_{t=0}. \tag{7.23}$$

Applying the definition (7.3) the reader can immediately verify that $\mathbf{X}_L$ defined above is left-invariant. In a completely analogous way, with the same tangent vector $\mathbf{X}_e$ we can associate a right-invariant vector field $\mathbf{X}_R$, setting

$$\forall f \in \mathbb{C}^\infty(G) \quad \text{and} \quad \forall \rho \in G: \quad \mathbf{X}_R f(\rho) \equiv f(g(t)\rho)\big|_{t=0}. \tag{7.24}$$

In this way we have established that with each tangent vector at the identity we can associate both a left-invariant and a right-invariant vector field. On the other hand, since each vector field $\mathbf{X}$ is a section of the tangent bundle, namely, a map,

$$\forall p \in \mathcal{M}: \quad p \mapsto \mathbf{X}_p \in T_p\mathcal{M}, \tag{7.25}$$

it follows that each left-invariant or right-invariant vector fields singles out a tangent vector at the identity. The relevant point is that this double correspondence is an isomorphism of vector spaces. Indeed, we have the following lemma.

Lemma 7.1.3. *Let G be a Lie group and let $\mathbb{G}_{L/R}$ be the Lie algebra of left-invariant (respectively right-invariant) vector fields. The correspondence*

$$\forall \mathbf{X} \in \mathbb{G}_{L/R} \quad \pi : \mathbf{X} \mapsto \mathbf{X}_e \in T_e G \tag{7.26}$$

is an isomorphism of vector spaces.

Proof of Lemma 7.1.3. Let us first of all observe that the correspondence (7.26) is a linear map. Indeed, for $a, b \in \mathbb{R}$ and $X, Y \in \mathbb{G}_{L/R}$ we have

$$\pi(a\mathbf{X} + b\mathbf{Y}) = a\mathbf{X}_e + b\mathbf{Y}_e. \tag{7.27}$$

In order to show that π is an isomorphism we need to prove that π is both injective and surjective, or in other words, that

$$\ker \pi = 0; \quad \operatorname{Im} \pi = T_e G. \tag{7.28}$$

The second of the two conditions in (7.28) was already proved. Indeed, we have shown above that with each tangent vector $\mathbf{X}_e$ at the identity we can associate a left-invariant (right-invariant) vector field which reduces to that vector at $g = e$. It remains to be shown that $\ker \pi = 0$. This follows from a simple observation. If $\mathbf{X}$ is left-invariant, then its value at the group element $\gamma \in G$ is determined by its value at the identity by means of a left translation, namely,

$$\mathbf{X}_\gamma = L_{\gamma *} \mathbf{X}_e. \tag{7.29}$$

Therefore, if $\mathbf{X}_e = 0$ it follows that $\mathbf{X}_\gamma = 0$ for all group elements $\gamma \in G$ and hence $\mathbf{X} = 0$ as a vector field. Hence, there are no non-trivial vectors in the kernel of the map π and this concludes the proof of the lemma. □

We have therefore the following.

Corollary 7.1.1. *The dimensions of the two Lie algebras $\mathbb{G}_L$ and $\mathbb{G}_R$ are equal among themselves and equal to those of the abstract Lie algebra $\mathbb{G}$:*

$$\dim \mathbb{G}_L = \dim \mathbb{G}_R = \dim \mathbb{G} = \dim T_e G. \tag{7.30}$$

In this way we have shown the isomorphism $\mathbb{G}_L \sim \mathbb{G}_R \sim \mathbb{G}$ as vector spaces. The same isomorphism remains to be shown as Lie algebras.

Proof of Corollary 7.1.1. We can now complete the proof of Theorem 7.1.1. To this effect we argue in the following way. Utilizing the just established vector space isomorphism let us choose a basis for $T_e G$ which we denote as $\mathbf{t}_A$ ($A = 1, \dots, \dim G$). In this way we have

$$\forall \mathbf{t} \in T_e G : \quad \mathbf{t} = x^A \mathbf{t}_A, \quad x^A \in \mathbb{R}. \tag{7.31}$$

Relying on the constructions (7.23) and (7.24), with each element of the basis $\{\mathbf{t}_A\}$ we can associate the corresponding left (right)-invariant vector field:

$$\mathbf{T}_A^{(L)}(f)(\rho) = \frac{d}{dt} f(\rho g_A(t))\Big|_{t=0}, \tag{7.32}$$

$$\mathbf{T}_A^{(R)}(f)(\rho) = \frac{d}{dt} f(g_A(t)\rho)\Big|_{t=0}, \tag{7.33}$$

where $g_A(t)$ denotes a path in G passing through the identity and there admitting the vector $\mathbf{t}_A$ as tangent. The established vector space isomorphism guarantees that $\mathbf{T}_A^{(L)}$ and $\mathbf{T}_A^{(R)}$ constitute a basis respectively for $\mathbb{G}_L$ and $\mathbb{G}_R$. Hence, we can write

$$[\mathbf{T}_A^{(L)}, \mathbf{T}_B^{(L)}] = C_{AB}^{(L)C} \mathbf{T}_C^{(L)}, \tag{7.34}$$

$$[\mathbf{T}_A^{(R)}, \mathbf{T}_B^{(R)}] = C_{AB}^{(R)C} \mathbf{T}_C^{(R)}, \tag{7.35}$$

where $C_{AB}^{(L)C}$ and $C_{AB}^{(L)C}$ are constants that, a priori, might be completely different. The equations follow from the fact that we have established that both $\mathbb{G}_L$ and $\mathbb{G}_R$ are dimension n Lie algebras and the generators $\mathbf{T}_A^{(L/R)}$ constitute a basis.

Let us now calculate explicitly the commutators of the basis elements using their definitions. On any function $f : G \mapsto \mathbb{R}$, we find

$$\begin{aligned}[\mathbf{T}_A^{(L)}, \mathbf{T}_B^{(L)}](f)(\rho) &= \frac{d}{dt}\frac{d}{d\tau}[f(\rho g_B(t) g_A(\tau)) - f(\rho g_A(t) g_B(\tau))]\Big|_{t=\tau=0} \\ &= C_{AB}^{(L)C} \frac{d}{dt} f(\rho g_C(t))\Big|_{t=0},\end{aligned} \tag{7.36}$$

$$\begin{aligned}[\mathbf{T}_A^{(R)}, \mathbf{T}_B^{(R)}](f)(\rho) &= \frac{d}{dt}\frac{d}{d\tau}[f(g_A(t) g_B(\tau)\rho) - f(g_B(t) g_A(\tau)\rho)]\Big|_{t=\tau=0} \\ &= C_{AB}^{(R)C} \frac{d}{dt} f(\rho g_C(t))\Big|_{t=0}.\end{aligned} \tag{7.37}$$

The above equation (7.37) holds true at any point $\rho \in G$. Evaluating them at the identity $\rho = e$ and taking their sum we obtain

$$(C_{AB}^{(L)C} + C_{AB}^{(R)C}) \frac{d}{dt} f(\rho g_C(t))\Big|_{t=0} = 0, \tag{7.38}$$

which implies

$$C_{AB}^{(L)C} = -C_{AB}^{(R)C}. \tag{7.39}$$

This relation suffices to establish the Lie algebra isomorphism of $\mathbb{G}_L$ with $\mathbb{G}_R$. Indeed, they are isomorphic as vector spaces and their structure constants become identical under the very simple change of basis

$$\mathbf{T}_A^{(L)} \leftrightarrow -\mathbf{T}_A^{(R)}. \tag{7.40}$$

This concludes the proof of Theorem 7.1.1. Indeed, the last isomorphism advocated in that proposition amounts simply to a definition. □

Definition 7.1.3. The Lie algebra $\mathbb{G}_L$ of the left-invariant vector fields on the Lie group manifold G, isomorphic to that of the right-invariant ones $\mathbb{G}_R$, is named the Lie algebra $\mathbb{G}$ of the considered Lie group.

By explicit evaluation as in equation (7.37) we can also show that any left-invariant vector field commutes with any right-invariant one and vice versa. Indeed, we find

$$\begin{aligned}[\mathbf{T}_A^{(L)}, \mathbf{T}_B^{(R)}](f)(\rho) &= \frac{d}{dt}\frac{d}{d\tau}[f(g_B(t)\rho g_A(\tau)) - f(g_B(t)\rho g_A(\tau))]\Big|_{t=\tau=0} \\ &= 0.\end{aligned} \tag{7.41}$$

The interpretation of these relations is indeed very simple. The left-invariant vector fields happen to be the infinitesimal generators of the right translations, while the right-invariant ones generate the left translations. Hence, the vanishing of the above commutators just amounts to saying that the left-invariant ones are indeed invariant under left translations, while the right-invariant ones are insensitive to right translations.

7.2 Maurer–Cartan forms on Lie group manifolds

Let us now consider left-invariant (respectively right-invariant) 1-forms on the group manifold. They were defined in equation (7.15). Starting from the construction of the left (right)-invariant vector fields it is very easy to construct an independent set of n differential forms with the invariance property (7.15) that are in one-to-one correspondence with the generators of the Lie algebra $\mathbb{G}$. Let us consider the explicit form of $\mathbf{T}_A^{(L/R)}$ as first order differential operators:

$$\mathbf{T}_A^{(L/R)} = \overset{(l/r)}{\Sigma_A^i}(\alpha)\,\frac{\overrightarrow{\partial}}{\partial\alpha^i}. \tag{7.42}$$

According to the already introduced convention α^μ are the group parameters and the square matrix

$$\overset{(l/r)}{\Sigma_A^i}(\alpha) = \left.\underbrace{\begin{pmatrix} * & * & * & * \\ * & * & * & * \\ * & * & * & * \\ * & * & * & * \end{pmatrix}}_{A=1,\dots,n}\right\} \quad i = 1,\dots,n, \tag{7.43}$$

whose entries are functions of the parameters can be calculated starting from the constructive algorithm encoded in equation (7.33). In terms of the inverse of the above matrix denoted $\overset{(l/r)}{\Sigma}{}^A_\alpha(\alpha)$ such that

$$\overset{(l/r)}{\Sigma}{}^A_i(\alpha)\ \overset{(l/r)}{\Sigma}{}^i_B(\alpha) = \delta^A_B \tag{7.44}$$

we can define the following set of n differential 1-forms:

$$\sigma^A_{(L/R)} = \overset{(l/r)}{\Sigma}{}^A_i(\alpha)\, d\alpha^i, \tag{7.45}$$

which by construction satisfy the relations

$$\sigma^A_{(L/R)}\big(\mathbf{T}^{(L/R)}_B\big) = \delta^A_B. \tag{7.46}$$

From equation (7.46) it follows that all the forms $\sigma^A_{L/R}$ are left (respectively right)-invariant. Indeed, we have

$$\forall \gamma \in G: \quad L^*_\gamma \sigma^A_{(L)}\big(\mathbf{T}^{(L)}_B\big) \equiv \sigma^A_{(L)}\big(L_{\gamma *}\mathbf{T}^{(L)}_B\big) = \sigma^A_{(L)}\big(\mathbf{T}^{(L)}_B\big) = \delta^A_B, \tag{7.47}$$

which implies $L^*_\gamma \sigma^A_{(L)} = \sigma^A$ since both forms have the same values on a basis of sections of the tangent bundle as provided by the set of left-invariant vector fields $\mathbf{T}^{(L)}_B$. A completely identical proof holds obviously true for the right-invariant 1-forms $\sigma^A_{(R)}$ defined by the same construction.

We come therefore to the conclusion that on each Lie group manifold G we can construct both a left-invariant $\sigma_{(L)}$ and a right-invariant $\sigma_{(R)}$ Lie algebra valued 1-form defined as follows:

$$\mathbb{G} \ni \sigma_{(L)} = \sum_{A=1}^{n} \sigma^A_{(L)} \mathbf{T}_A, \tag{7.48}$$

$$\mathbb{G} \ni \sigma_{(R)} = \sum_{A=1}^{n} \sigma^A_{(R)} \mathbf{T}_A, \tag{7.49}$$

where, just as before, $\{\mathbf{T}_A\}$ denotes a basis of generators of the abstract Lie algebra $\mathbb{G}$.

One may wonder how the Lie algebra forms $\sigma_{(L/R)}$ could be constructed directly without going through the previous construction of the left (right)-invariant vector fields. The answer is very simple. Let $\mathfrak{g}(\alpha) \in G$ be a running element of the Lie group parameterized by the parameters α which constitute a coordinate patch for the corresponding group manifold and consider the 1-forms

$$\theta_L = \mathfrak{g}^{-1}\, d\mathfrak{g} = \mathfrak{g}^{-1}\partial_i \mathfrak{g}\, d\alpha^i; \quad \theta_R = \mathfrak{g}\, d\mathfrak{g}^{-1} = \mathfrak{g}\partial_i \mathfrak{g}^{-1}\, d\alpha^i. \tag{7.50}$$

It is immediate to check that such 1-forms are left- (respectively, right-)invariant. For instance, we have

$$L_\gamma^* \theta_L = (\gamma \mathfrak{g})^{-1} d(\gamma \mathfrak{g}) = \mathfrak{g}^{-1}\gamma^{-1}\gamma \, d\mathfrak{g} = \mathfrak{g}^{-1} \, d\mathfrak{g} = \theta_L. \tag{7.51}$$

What might not be immediately evident to the reader is why the left (right)-invariant 1-forms $\theta_{L/R}$ introduced in equation (7.50) should be Lie algebra valued. The answer is actually very simple. The relation between Lie algebras and Lie groups is provided by the exponential map: every element of a Lie group which lies in the branch connected to the identity can be represented as the exponential of a suitable Lie algebra element:

$$\forall \gamma \in \mathrm{G}_0 \subset G, \quad \exists X \in \mathbb{G} \setminus \gamma = \exp \mathbf{X}. \tag{7.52}$$

The Lie algebra element actually singles out an entire one-parameter subgroup:

$$\forall t \in \mathbb{R}, \quad G_0 \ni \gamma(t) = \exp[t\mathbf{X}]. \tag{7.53}$$

With an obvious calculation we obtain

$$\gamma^{-1}(t) d\gamma(t) = \mathbf{X}\, dt \in \mathbb{G}, \tag{7.54}$$

namely, the left-invariant 1-form associated with this parameter subgroup lies in the Lie algebra. This result extends to any group element, so that indeed the previously constructed Lie algebra valued left (right)-invariant 1-forms $\sigma_{L/R}$ and the θ-forms defined in equation (7.50) are just the very same objects:

$$\sigma_{L/R} = \theta_{L/R}. \tag{7.55}$$

All the above statements become much clearer when we consider classical groups whose elements are just matrices subject to some defining algebraic condition. Consider for instance the rotation group in N dimensions SO(N). All elements of this group are orthogonal N × N matrices:

$$\mathcal{O} \in \mathrm{SO(N)} \quad \Leftrightarrow \quad \mathcal{O}^T \mathcal{O} = \mathbf{1}_{\mathrm{N}\times\mathrm{N}}. \tag{7.56}$$

The elements of the orthogonal Lie algebra $\mathfrak{so}(\mathrm{N})$ are instead antisymmetric matrices:

$$\mathrm{A} \in \mathfrak{so}(\mathrm{N}) \quad \Leftrightarrow \quad \mathrm{A}^T + \mathrm{A} = \mathbf{0}_{\mathrm{N}\times\mathrm{N}}. \tag{7.57}$$

Calculating the transpose of the matrix $\Theta = \mathcal{O}^T \, d\mathcal{O}$ we immediately obtain

$$\Theta^T = \left(\mathcal{O}^T \, d\mathcal{O}\right)^T = d\mathcal{O}^T \mathcal{O} = -\mathcal{O}^T \, d\mathcal{O} = -\Theta \quad \Rightarrow \quad \Theta \in \mathfrak{so}(\mathrm{N}), \tag{7.58}$$

which proves that the left-invariant 1-form is indeed Lie algebra valued.

7.3 Maurer–Cartan equations

It is now of utmost interest to consider the following identity, which follows immediately from the definitions (7.50) and (7.55) of the left (right)-invariant 1-forms:

$$\mathfrak{F}[\sigma] \equiv d\sigma_{L/R} + \sigma_{L/R} \wedge \sigma_{L/R} = 0. \tag{7.59}$$

To prove the above statement it is sufficient to observe that $\mathfrak{g}^{-1}d\mathfrak{g} = -d\mathfrak{g}^{-1}\mathfrak{g}$.

The reason why we introduced a special name $\mathfrak{F}[\sigma]$ for the Lie algebra valued 2-form $d\sigma + \sigma \wedge \sigma$, which turns out to be zero in the case of left (right)-invariant 1-forms σ, is that precisely this combination will play a fundamental role in the theory of connections, representing their *curvature*. The equations in (7.59) are named the Maurer–Cartan equations of the considered Lie algebra $\mathbb{G}$ and they translate into the statement that left (right)-invariant 1-forms have *vanishing curvature*.

Since the forms σ are Lie algebra valued, they can be expanded along a basis of generators $\mathbf{T}_A$ for $\mathbb{G}$, as we already did in equation (7.49), and the same can be done for the curvature 2-form $\mathfrak{F}[\sigma]$, namely, we get

$$\begin{aligned} \mathfrak{F}[\sigma] &= \mathfrak{F}^A[\sigma]\mathbf{T}_A, \\ \mathfrak{F}^A[\sigma] &= d\sigma^A + \frac{1}{2}C_{BC}{}^{A}\sigma^B \wedge \sigma^C, \end{aligned} \tag{7.60}$$

where $C_{BC}{}^{A}$ are the structure constants of the Lie algebra,

$$[\mathbf{T}_B, \mathbf{T}_C] = C_{BC}{}^{A}\mathbf{T}_A, \tag{7.61}$$

and the Maurer–Cartan equations in (7.59) amount to the statement

$$\mathfrak{F}^A[\sigma_{L/R}] = 0. \tag{7.62}$$

7.4 Matrix Lie groups

All Lie groups can be viewed as groups whose elements are matrices. This occurs for most Lie groups in their very definition and it occurs for all Lie groups in their linear representations. For this reason we start recalling some properties of matrices that will help us very much to translate the above explained general notions into concrete cases.

7.4.1 Some properties of matrices

One can define a *matrix exponential* by means of a formal power series expansion: if A is a square matrix,

$$\exp(A) = \sum_{k=0}^{\infty} \frac{1}{k!} A^k. \tag{7.63}$$

For instance, consider the following 2×2 case:

$$m(\theta) = \theta\epsilon = \begin{pmatrix} 0 & \theta \\ -\theta & 0 \end{pmatrix} \quad \Rightarrow \quad \exp[m(\theta)] = \begin{pmatrix} \cos\theta & \sin\theta \\ -\sin\theta & \cos\theta \end{pmatrix}, \tag{7.64}$$

where θ is a real parameter. The above expression easily follows from (7.63) and from the fact that the Levi-Civita antisymmetric symbol ϵ_{ij}, seen as a matrix, obeys $\epsilon^2 = -\mathbf{1}$.

For matrix exponentials most properties of the usual exponential hold true, yet not all of them. For instance, if T, S are square matrices, we have

$$\exp(T)\exp(S) = \exp(T + S) \quad \Leftrightarrow \quad [T, S] = 0, \tag{7.65}$$

i. e., only if the two matrices commute. Otherwise, (7.65) generalizes to the so-called *Baker–Campbell–Hausdorff* formula:

$$\exp(T)\exp(S) = \exp\left(T + S + \frac{1}{2}[T, S] + \frac{1}{12}([T, [T, S]] + [[T, S], S]) + \cdots\right). \tag{7.66}$$

A property which is immediately verified is the following one:

$$\exp(U^{-1}TU) = U^{-1}\exp(T)U. \tag{7.67}$$

A very useful property of the determinants is the following:

$$\det(\exp(m)) = \exp(\operatorname{Tr} m), \tag{7.68}$$

or equivalently, by taking the logarithm (which for matrices is defined via a formal power series)

$$\det M = \exp(\operatorname{Tr}(\ln M)). \tag{7.69}$$

These relations can be easily proven for a diagonalizable matrix; indeed, the determinant and the trace are invariant under any change of basis. Assume that M has been diagonalized. If λ_i are its eigenvalues we get

$$\det M = \prod_i \lambda_i = \exp\left(\sum_i \ln \lambda_i\right) = \exp(\operatorname{Tr}(\ln M)). \tag{7.70}$$

The result can then be extended to generic matrices as it can be argued that every matrix can be approximated to any chosen accuracy by diagonalizable matrices.

Let us note that on the space of $n \times n$ matrices (with, say, complex entries) one can define a distance by

$$d(M,N) = \left(\sum_{i,j=1}^{n} |M_{ij} - N_{ij}|^2 \right)^{1/2}, \tag{7.71}$$

where M, M are two matrices. This is nothing else than considering the space of generic $n \times n$ matrices as the $\mathbb{C}^{n^2}$ space parameterized by the n^2 complex entries and endowing it with the usual topology.

7.4.2 Linear Lie groups

Consider a group of matrices such that:

(a) The elements of G within a neighborhood $\mathscr{S}$ of the identity matrix can be put into one-to-one correspondence with a neighborhood $\mathscr{B}$ of the origin in $\mathbb{R}^d$ (the correspondence is chosen so that the origin corresponds to the identity matrix). In other words, any matrix $M \in \mathscr{S}$ is parameterized by d *real* coordinates α^μ ($\mu = 1, \ldots d$): $M = M(\alpha)$.

(b) All the matrix elements $M_{ij}(\alpha)$ of a matrix $M(\alpha) \in \mathscr{S}$ are furthermore required to be *analytic* functions of the coordinates α, namely, they ought to admit an expression as a power series of the differences $\alpha^\mu - \alpha_0^\mu$, where $\alpha_0 \in \mathscr{B}$ are the coordinates of any other matrix in the set $\mathscr{S}$. Thus, all the derivatives $\partial M_{ij}/\partial \alpha^\mu, \partial^2 M_{ij}/\partial \alpha^\mu \partial \alpha^\nu, \ldots$ exist at all points within $\mathscr{B}$, and in particular at $\alpha = 0$.

It is not difficult to see that from the above conditions it follows that:

(i) The d matrices $\mathbf{T}_\mu$ defined by

$$\mathbf{T}_\mu = \left. \frac{\partial M}{\partial \alpha^\mu} \right|_{\alpha=0} \tag{7.72}$$

form the basis of a *real* vector space of dimension d. The matrices $\mathbf{T}_\mu$ provide the explicit form of the *infinitesimal generators* of the Lie algebra described above as vector fields on the group manifold.

(ii) If the product of $M(\alpha)$ and $M(\beta)$ both in $\mathscr{S}$ also belongs to $\mathscr{S}$, we have $M(\alpha)M(\beta) = M(\gamma)$ for some γ depending on α, β. The function $\gamma(\alpha, \beta)$ is analytic in both α and β. Also, the inverse matrix $[M(\alpha)]^{-1} = M(\eta)$, with $\eta(\alpha)$ analytic.

Item (ii) tells us that a group of matrices such that (a) and (b) hold satisfies the definition given above of a Lie group.

7.5 Bibliographical note

The main bibliographical sources for this chapter are the following books:

1. The first volume of [79].

2. The monograph [38].
3. The textbook [83].
4. The textbook [42].
5. The textbook [40].
6. The textbook [67].
7. The textbook [68].
8. The textbook [84].

8 Crystallographic groups and group extensions

> There are two forms of knowledge, one genuine and the other obscure; to the obscure one belong all the following objects: sight, hearing, smell, taste and touch. The other form is the genuine one and the objects of this are hidden.
>
> Sextus Empiricus

The present chapter focuses on a systematic treatment of the crystallographic groups: both point groups and space groups. This is the group theory utilized by chemists, chemical physicists, and mineralogists, yet it is here presented in a unified modern mathematical manner which overcomes the obsolete traditional notations and formats still currently employed in most of the chemistry/physics and crystallography literature. In particular, the derivation of the space groups is recast into the appropriate mathematical language of *group extensions and finite group cohomology*, while the classification of crystallographic lattice systems (Bravais lattices) is reduced to the problem of classifying the conjugacy classes of abstract n-dimensional point group embeddings into $\mathfrak{P}_\Lambda \hookrightarrow \mathrm{GL}(\mathrm{n}, \mathbb{Z})$. The method of Frobenius congruences is shown to be a convenient technique to calculate the explicit cohomology classes. While developing the setup of space-group extensions we briefly present also the concept of *universal classifying group* introduced in [71] and conceptually improved in [85]. Indeed, crystallographic groups play, after the pioneering work of [71], a relevant role in fluid dynamics of incompressible fluids and, in particular, in the construction of Beltrami flows, whose relation with *contact structures* [86, 87] and with *singular contact structures* [88, 89, 90, 91, 92, 93, 94, 95] is very fundamental and finds applications in diversified technological fields.

8.1 Lattices and crystallographic groups

In this section we consider the finite rotation groups from the point of view of crystallography, namely, as groups of automorphisms of certain lattices. To this effect we need first to introduce the very notion of lattice and then introduce the notion of crystallographic group.

8.1.1 Lattices

We begin by fixing our notations for space and momentum lattices that also define an n-torus T^n endowed with a flat metric structure, namely, with a symmetric positive definite inner product (see Definition 2.19).

https://doi.org/10.1515/9783111201535-008

Let us consider the standard $\mathbb{R}^n$ manifold[1] and introduce a basis of n linearly independent n-vectors that are not necessarily orthogonal to each other and of equal length:

$$\mathbf{w}_\mu \in \mathbb{R}^n, \quad \mu = 1, \dots n. \tag{8.1}$$

Any vector in $\mathbb{R}$ can be decomposed along such a basis and we have

$$\mathbf{r} = r^\mu \mathbf{w}_\mu. \tag{8.2}$$

The flat (constant) metric on $\mathbb{R}^n$ is defined by

$$g_{\mu\nu} = \langle \mathbf{w}_\mu, \mathbf{w}_\nu \rangle, \tag{8.3}$$

where $\langle\, , \rangle$ denotes the standard Euclidean scalar product. The space lattice Λ consistent with the metric (8.3) is the free Abelian group (with respect to sum) generated by the n basis vectors (8.1), namely,

$$\mathbf{q} \in \Lambda \subset \mathbb{R}^n \quad \Leftrightarrow \quad \mathbf{q} = q^\mu \mathbf{w}_\mu, \quad \text{where } q^\mu \in \mathbb{Z}. \tag{8.4}$$

The dual lattice $\Lambda^\star$ is defined by the property

$$\mathbf{p} \in \Lambda^\star \subset \mathbb{R}^n \quad \Leftrightarrow \quad \langle \mathbf{p}, \mathbf{q} \rangle \in \mathbb{Z}; \quad \forall \mathbf{q} \in \Lambda. \tag{8.5}$$

A basis for the dual lattice is provided by a set of n *dual vectors* $\mathbf{e}^\mu$ defined by the relations[2]

$$\langle \mathbf{w}_\mu, \mathbf{e}^\nu \rangle = \delta_\mu^\nu \tag{8.6}$$

so that

$$\forall \mathbf{p} \in \Lambda^\star \quad \mathbf{p} = p_\mu \mathbf{e}^\mu, \quad \text{where } p_\mu \in \mathbb{Z}. \tag{8.7}$$

8.1.2 The n-torus T^n

The n-torus is topologically defined as the product of n circles, namely,

$$\mathrm{T}^n \equiv \underbrace{\mathbb{S}^1 \times \dots \times \mathbb{S}^1}_{n\text{ times}} \equiv \underbrace{\frac{\mathbb{R}}{\mathbb{Z}} \times \dots \times \frac{\mathbb{R}}{\mathbb{Z}}}_{n\text{ times}}. \tag{8.8}$$

---

1 For the mathematically precise notion of manifold we refer the reader to Chapter 6.

2 In the sequel for the scalar product of two vectors we utilize also the equivalent shorter notation $\mathbf{a} \cdot \mathbf{b} = \langle \mathbf{a}, \mathbf{b} \rangle$.

Alternatively we can define the n-torus by modding $\mathbb{R}^n$ with respect to an n-dimensional lattice. In this case the n-torus comes out automatically equipped with a flat constant metric:

$$\mathrm{T}^n_g = \frac{\mathbb{R}^n}{\Lambda}. \tag{8.9}$$

According to (8.9) the flat Riemannian space[3] T^n_g is defined as the set of equivalence classes with respect to the following equivalence relation:

$$\mathbf{r}' \sim \mathbf{r} \quad \text{iff} \quad \mathbf{r}' - \mathbf{r} \in \Lambda. \tag{8.10}$$

The metric (8.3) defined on $\mathbb{R}^n$ is inherited by the quotient space and therefore it endows the topological torus (8.8) with a flat Riemannian structure. Seen from another point of view the space of flat metrics on T^3 is just the coset manifold $\mathrm{SL}(3,\mathbb{R})/\mathrm{O}(3)$ encoding all possible symmetric matrices, or alternatively all possible space lattices, each lattice being spanned by an arbitrary triplet of basis vectors (8.1).

8.1.3 Crystallographic groups and the Bravais lattices for $n = 3$ and $n = 2$

Every lattice Λ yields a metric g and every metric g singles out an isomorphic copy $\mathrm{SO}_g(3)$ of the continuous rotation group $\mathrm{SO}(n)$, which leaves it invariant:

$$M \in \mathrm{SO}_g(n) \quad \Leftrightarrow \quad M^T g M = g. \tag{8.11}$$

By definition $\mathrm{SO}_g(n)$ is the conjugate of the standard $\mathrm{SO}(n)$ in $\mathrm{GL}(n,\mathbb{R})$,

$$\mathrm{SO}_g(n) = \mathscr{S}\,\mathrm{SO}(n)\mathscr{S}^{-1}, \tag{8.12}$$

with respect to the matrix $\mathscr{S} \in \mathrm{GL}(n,\mathbb{R})$ which reduces the metric g to the Kronecker delta:

$$\mathscr{S}^T g \mathscr{S} = \mathbf{1}. \tag{8.13}$$

Notwithstanding this, a generic lattice Λ is not invariant with respect to any proper subgroup of the rotation group $\mathrm{G} \subset \mathrm{SO}_g(n) \equiv \mathrm{SO}(n)$. Indeed, by invariance of the lattice one understands the following condition:

$$\forall \gamma \in \mathrm{G} \quad \text{and} \quad \forall \mathbf{q} \in \Lambda: \quad \gamma \cdot \mathbf{q} \in \Lambda. \tag{8.14}$$

---

3 For the precise mathematical definition of Riemannian space we refer the reader to Chapter 15.

For $n = 3$, lattices that have a non-trivial symmetry group $G \subset SO(3)$ are those relevant to solid state physics and crystallography. There are 14 of them grouped in 7 classes that were already classified in the nineteenth century by Bravais. The symmetry group G of each of these Bravais lattices Λ is necessarily one of the finite subgroups of the three-dimensional rotation group O(3) that we discussed in Chapter 5. In the language universally adopted by chemistry and crystallography for each Bravais lattice Λ_B the corresponding invariance group G_B is named the *point group of the lattice*: $\mathfrak{P}_\Lambda$. In chemistry and crystallography by *point groups* one means all the discrete finite subgroups of O(3) that are obviously in correspondence with the ADE classification, yet, traditionally, they are given different names according to various notational systems among which the oldest and probably most frequently utilized is the *Schönflies*[4] *notation system.* Not all the point groups in this definition, which form an infinite set, are crystallographic in the three-dimensional sense, yet they can appear as rotation/reflection symmetries of molecules rather than crystals and for this reason they are considered. Schönflies classification is based on a description of the group by means of its explicit action on $\mathbb{R}^3$ via explicitly mentioned quantized rotations and reflections and not in terms of the group intrinsic structure, so that there are several isomorphisms and repetitions. One typical isomorphism, occurring several times, is that between the finite subgroups of the proper rotation group $\Gamma \subset SO(3)$ extended with some reflection $\Sigma \in O(3)/SO(3)$ and the binary extension $\Gamma_b \subset SU(2)$ which, abstractly, is a central extension of Γ. In order to clarify the relation between the rigorous mathematical treatment of finite subgroups of SO(3) provided by the ADE classification and the groups utilized by chemists and crystallographers it is convenient to focus on the translation vocabulary displayed in Table 8.1. Let us briefly discuss the definition of Schönflies groups as provided in chemical-crystallographic literature:

1. C_n, the cyclic group of order n, is generated by a rotation $A_n = \exp[\theta A_{\ell,m,n}]$ of an angle $\theta = 2\pi/n$ around some axis $\{\ell, m, n\}$ (see equation (5.14)). This group is obviously isomorphic to $\mathbb{Z}_n$.
2. S_{2n} is generated by the rotation A_{2n} multiplied by a reflection with respect to the plane orthogonal to the rotation axis, namely, it is generated by $T = A_{2n}\Sigma_h$, where Σ_h is the mentioned reflection. Clearly, since Σ_h commutes with A_{2n} the structure of this group is the same as the central extension of $\mathbb{Z}_n$, which is isomorphic to $\mathbb{Z}_{2n}$ and to $\Gamma_b[n, 1, 1]$ in the ADE classification.
3. C_{nh} is the group generated by a rotation A_n and Σ_h so, abstractly, it is isomorphic to $\mathbb{Z}_n \times \mathbb{Z}_2$ and hence to $\Gamma[n, 1, 1] \times \mathbb{Z}_2^q \subset O(3)$, where $\mathbb{Z}_2^q$ is the quotient group $O(3)/SO(3)$.
4. C_{nv} is the group generated by a rotation A_n and by a reflection Σ_v with respect to some plane that contains the rotation axis. Clearly it is isomorphic to the abstract group Dih_n and hence to $\Gamma[n, 2, 2]$ in the ADE classification system, yet in three dimensions it is a subgroup of O(3) and not of SO(3).

4 Arthur Moritz Schönflies (1853–1928) was a German mathematician working in crystallography.

Table 8.1: The points groups in $D = 3$ according with Schönflies nomenclature and their correspondence with the ADE classification system.

Schönflies notation	Abstract group	Order	ADE classif.	Presentation by generators and relations
C_n	$\mathbb{Z}_n$	n	$\Gamma[n,1,1]$	$(\mathrm{A} \mid \mathrm{A}^n = \mathbf{e})$
S_{2n}	$\mathbb{Z}_{2n}$	2n	$\Gamma_b[n,1,1]$	$(\mathscr{A},\mathscr{B},\mathscr{Z} \mid (\mathscr{A}\mathscr{B})^2 = \mathscr{A}^n = \mathscr{B}^2 = \mathscr{Z}\mathscr{Z}^2 = \mathbf{e})$
C_{nh}	$\mathbb{Z}_n \times \mathbb{Z}_2$	2n	$\Gamma[n,1,1] \times \mathbb{Z}_2$	$(\mathrm{A},\mathrm{Z} \mid \mathrm{A}^n = \mathrm{Z}^2 = \mathbf{e})$
C_{nv}	Dih_n	2n	$\Gamma[n,2,2]$	$(\mathrm{A},\mathrm{Z} \mid (\mathrm{AZ})^2 = \mathrm{A}^n = \mathrm{Z}^2 = \mathbf{e})$
D_n	Dih_n	2n	$\Gamma[n,2,2]$	$(\mathrm{A},\mathrm{B} \mid (\mathrm{AB})^2 = \mathrm{A}^k = \mathrm{B}^2 = \mathbf{e})$
D_{nd}	Dih_{2n}	4n	$\Gamma_b[n,2,2]$	$(\mathscr{A},\mathscr{B},\mathscr{Z} \mid (\mathscr{A}\mathscr{B})^2 = \mathscr{A}^n = \mathscr{B}^2 = \mathscr{Z}\mathscr{Z}^2 = \mathbf{e})$
D_{nh}	$\mathrm{Dih}_n \times \mathbb{Z}_2$	4n	$\Gamma[n,2,2] \times \mathbb{Z}_2$	$(\mathrm{A},\mathrm{B},\mathrm{Z} \mid (\mathrm{AB})^2 = \mathscr{A}^n = \mathscr{B}^2 = \mathbf{e}\mathrm{Z}^2 = \mathbf{e})$
T	$\mathrm{T}_{12} \sim \mathrm{A}_4$	12	$\Gamma[3,3,2]$	$(\mathrm{A},\mathrm{B} \mid (\mathrm{AB})^3 = \mathrm{A}^3 = \mathrm{B}^2 = \mathbf{e})$
T_d	$\mathrm{T}^{\mathrm{d}}_{24} \sim \mathrm{S}_4$	24	$\Gamma_b[3,3,2]$	$(\mathscr{A},\mathscr{B} \mid (\mathscr{A}\mathscr{B})^4 = \mathscr{A}^3 = \mathscr{B}^2 = \mathbf{e})$
O	$\mathrm{O}_{24} \sim \mathrm{S}_4$	24	$\Gamma[4,3,2]$	$(\mathrm{A},\mathrm{B} \mid (\mathrm{AB})^4 = \mathrm{A}^3 = \mathrm{B}^2 = \mathbf{e})$
O_d	$\mathrm{O}^{\mathrm{d}}_{48} \sim \mathrm{S}_4 \times \mathbb{Z}_2$	48	$\Gamma_b[4,3,2]$	$(\mathscr{A},\mathscr{B},\mathscr{Z} \mid (\mathscr{A}\mathscr{B})^4 = \mathscr{A}^3 = \mathscr{B}^2 = \mathscr{Z}\mathscr{Z}^2 = \mathbf{e})$
I	$\mathrm{I}_{60} \sim \mathrm{A}_5$	60	$\Gamma[5,3,2]$	$(\mathrm{A},\mathrm{B} \mid (\mathrm{AB})^5 = \mathrm{A}^3 = \mathrm{B}^2 = \mathbf{e})$
I_d	$\mathrm{I}^{\mathrm{d}}_{120} \sim \mathrm{A}_5 \times \mathbb{Z}_2$	120	$\Gamma_b[5,3,2]$	$(\mathscr{A},\mathscr{B},\mathscr{Z} \mid (\mathscr{A}\mathscr{B})^5 = \mathscr{A}^3 = \mathscr{B}^2 = \mathscr{Z}\mathscr{Z}^2 = \mathbf{e})$

5. D_n is the group generated by a rotation A_{n} around some axis $\{\ell, m, n\}$ and by a rotation of an angle π around a rotation axis perpendicular to rotation axis $\{\ell, m, n\}$
6. D_{nh} is the group S_{2n} described above, further extended with a reflection Σ_v with respect to a plane that contains the rotation axis of S_{2n}. Abstractly it is isomorphic to $\mathrm{Dih}_{2n} \sim \Gamma_b[n,2,2]$ and as a transformation group in three dimensions it is a subgroup of O(3) and not of SO(3).
7. D_{nd}, abstractly isomorphic to $\mathrm{Dih}_{\mathrm{n}} \times \mathbb{Z}_2 \sim \Gamma_b[n,2,2] \times \mathbb{Z}_2$, is obtained by adjoining to $\mathrm{D}_{\mathrm{n}} \sim \mathrm{Dih}_{\mathrm{n}}$ a reflection Σ_v with respect to a plane that contains the rotation axis of A_{n}. Hence, $D_{nd} \sim \Gamma[n,2,2] \times \mathbb{Z}_2 \subset \mathrm{O}(3)$.
8. T is the proper tetrahedral group $\mathrm{T}_{12} \equiv \Gamma[3,3,2] \subset \mathrm{SO}(3)$, which abstractly is isomorphic to the alternating group A_4.
9. T_d is the full tetrahedral group obtained by adjoining to the proper tetrahedral group a reflection and it corresponds to the central extension $\mathrm{T}^{\mathrm{d}}_{24} \sim \Gamma_b[3,3,2]$ which happens to be isomorphic to the symmetric group of four objects, $\mathrm{T}^{\mathrm{d}}_{24} \sim \mathrm{S}_4$.
10. O is the proper octahedral group $\mathrm{O}_{24} \equiv \Gamma[4,3,2] \subset \mathrm{SO}(3)$, which abstractly is isomorphic to the symmetric group on four objects, $\mathrm{O}_{24} \sim \mathrm{S}_4$.
11. O_d is the full octahedral group isomorphic obtained by adjoining to the proper octahedral group a reflection and it corresponds to the central extension $\mathrm{O}^{\mathrm{d}}_{48} \sim \Gamma_b[4,3,2]$.
12. I is the proper icosahedral group $\mathrm{I}_{60} \equiv \Gamma[5,3,2] \subset \mathrm{SO}(3)$, which abstractly is isomorphic to the alternating group on five objects $\mathrm{I}_{60} \sim \mathrm{A}_5$, namely, to the smallest simple group.
13. I_d is the full icosahedral group obtained by adjoining to the proper icosahedral group a reflection and it corresponds to the central extension $\mathrm{I}^{\mathrm{d}}_{120} \sim \Gamma_b[5,3,2]$.

As we already stressed above, not all of the Schönflies groups, notwithstanding the fact that they are named point groups in the chemical literature, are crystallographic either in two or three dimensions according to the following precise mathematical definition.

Definition 8.1.1. An abstract group Γ is named crystallographic in n dimensions if there exists an n-dimensional lattice Λ_n with basis vectors $\mathbf{w}_\mu$ such that:

1. There is an isomorphism

$$\omega : \Gamma \to H \subset \mathrm{O_g}(n), \tag{8.15}$$

   where $\mathrm{O_g}(n)$ is the conjugate of the n-dimensional rotation group respecting a metric g (see equation (8.13)).
2. The metric g is that defined by the basis vectors of the lattice Λ_n (see equation (8.3)).
3. All elements of H are $n \times n$ matrices with integer valued entries.

This is equivalent to the statement that Γ has an orthogonal action in $\mathbb{R}^n$ and preserves the lattice Λ_n.

Furthermore, in the opposite direction, by *proper point group* $\mathscr{P}_{\Lambda_n}$ of the lattice Λ_n in n dimensions we mean a subgroup $\mathscr{P}_{\Lambda_n} \subset \mathrm{SO_g}(n)$ which maps the lattice Λ_n into itself.

8.1.3.1 The seven Bravais classes and their associated point groups for $n = 3$

According to standard nomenclature, the seven classes of Bravais lattices in three dimensions are respectively named *triclinic, monoclinic, orthorhombic, tetragonal, rhombohedral, hexagonal,* and *cubic* (see Figure 8.1). Such classes are specified by giving the lengths of the basis vectors $\mathbf{w}_\mu$ and the three angles between them, in other words, by specifying the six components of the metric (8.3). The seven classes are illustrated in Table 8.1. As one verifies by inspecting such a table, it is evident that the proper point groups that appear in the seven lattice classes are either the cyclic groups $\mathbb{Z}_h$ with $h = 2, 3, 4$ or the dihedral groups Dih_k with $k = 3, 4, 6$ or the tetrahedral group T_{12} or the octahedral group O_{24}. Indeed, the $n = 3$ crystallographic proper point groups are, by definition, finite subgroups of the rotation group SO(3); hence, they must fall in the ADE classification. Yet not every finite rotation group is crystallographic. For instance, there is no lattice that is invariant under the icosahedral group and in general in a $n = 3$ point group there are no elements with orders different from two, three, four, or six.

8.1.3.2 The four Bravais classes in $n = 2$

The case of two-dimensional crystallographic lattice classes is even smaller. It is displayed in Figure 8.2. Once again such classes are specified by giving the lengths of the two basis vectors $\mathbf{w}_\mu$ and the angle between them, in other words, by specifying the three components of the metric (8.3). As we see, from the abstract viewpoint, the crystallographic groups in two dimensions are simply $\mathrm{C}_n \sim \mathbb{Z}_n, \mathrm{Dih}_n$ with $n = 1, 2, 3, 4, 6$, which

Crystal family	Lattice system	Point group (Schönflies notation)	14 Bravais lattices: Primitive (P)	Base-centered (S)	Body-centered (I)	Face-centered (F)
Triclinic (a)		C_i	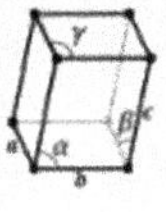 aP			
Monoclinic (m)		C_{2h}	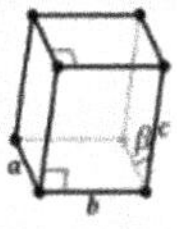 mP	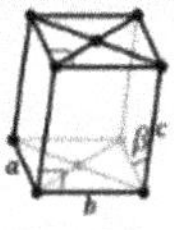 mS		
Orthorhombic (o)		D_{2h}	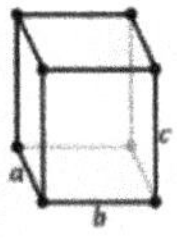 oP	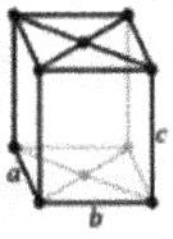 oS	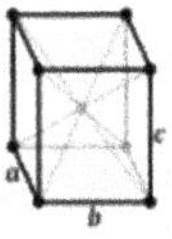 oI	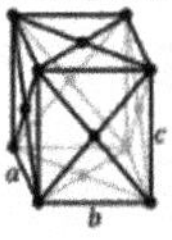 oF
Tetragonal (t)		D_{4h}	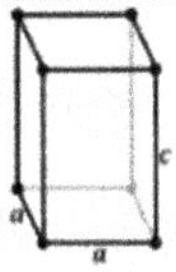 tP		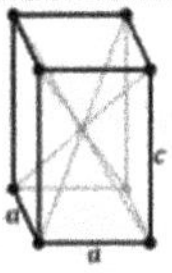 tI	
Hexagonal (h)	Rhombohedral	D_{3d}	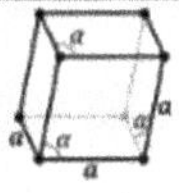 hR			
	Hexagonal	D_{6h}	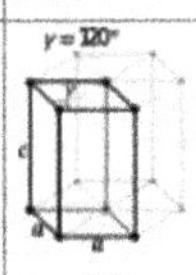 hP			
Cubic (c)		O_h	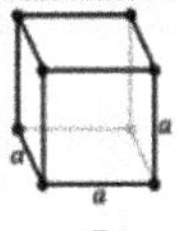 cP		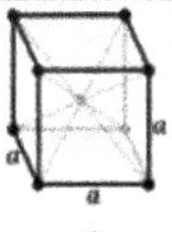 cI	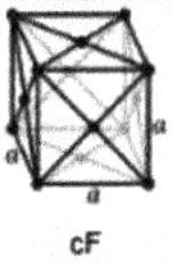 cF

Figure 8.1: The seven Bravais crystallographic lattice classes in three dimensions and the corresponding point groups in Schönflies notation.

makes nine groups taking into account the isomorphism $C_2 \sim \mathrm{Dih}_1 \sim \mathbb{Z}_2$. However, the action of the same abstract group on the two-dimensional lattice Λ can be different and we have a list of more cases. Furthermore, 9 is larger than 4, which means that the basis vectors $\mathbf{w}_\mu$ and the associated metric specify the lattice Λ but do not specify completely the available point groups. Indeed, each lattice Λ can be invariant under more than one

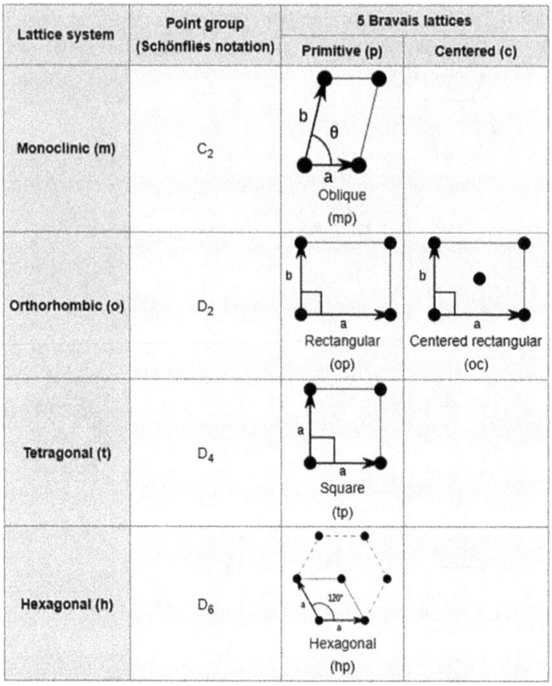
Lattice system
Point group (Schönflies notation)
5 Bravais lattices
Primitive (p)
Centered (c)
Monoclinic (m)
C_2
b
θ
a
Oblique
(mp)
Orthorhombic (o)
D_2
b
a
Rectangular
(op)
b
a
Centered rectangular
(oc)
Tetragonal (t)
D_4
a
a
Square
(tp)
Hexagonal (h)
D_6
120°
a
a
Hexagonal
(hp)

Figure 8.2: The four Bravais crystallographic lattice classes in two dimensions and the corresponding point groups in Schönflies notation.

point group. What is this relevant for? It is relevant for the group extensions that combine the considered point group and the lattice into a new infinite group $\mathfrak{G}$ containing the considered point group $\mathfrak{P}$ and an infinite normal subgroup $\mathfrak{T}$ isomorphic to the lattice. In order to illustrate this very interesting mathematical structure we focus on the case $n = 2$ and we begin by discussing the rigorous classification of two-dimensional point groups.

8.1.4 Rigorous mathematical classification of the point groups in two dimensions

As we see from Definition 8.1.1, every crystallographic group in n dimensions has a faithful representation in terms of $n \times n$ matrices with integer valued entries, or in other words, can be described as a finite subgroup $\Gamma \subset \mathrm{GL}(\mathrm{n}, \mathbb{Z})$. Providing a specific lattice basis corresponds to providing the explicit homomorphism:

$$h_c : \Gamma \longrightarrow \mathrm{GL}(\mathrm{n}, \mathbb{Z}). \tag{8.16}$$

The actual classification of crystallographic point groups is therefore the classification of the homomorphic images in equation (8.16) up to conjugation with respect to $\mathrm{GL}(\mathrm{n}, \mathbb{Z})$.

In two dimensions this rigorous method produces exactly 13 different subgroups of $\mathrm{GL}(2,\mathbb{Z})$:

(1) C_1. This is the identity group and it is represented inside $\mathrm{GL}(2,\mathbb{Z})$ by the identity matrix $\mathrm{Id}_2 = \left(\begin{smallmatrix}1 & 0\\ 0 & 1\end{smallmatrix}\right)$.

(2) C_2. This is the abstract $\mathbb{Z}_2$ group generated as a subgroup of $\mathrm{GL}(2,\mathbb{Z})$ by the matrix $\mathfrak{c}_2 = \left(\begin{smallmatrix}-1 & 0\\ 0 & -1\end{smallmatrix}\right)$ which squares to the identity $\mathfrak{c}_2^2 = \mathrm{Id}_2$ and corresponds to a rotation of an angle π.

(3) C_3. This is the abstract $\mathbb{Z}_3$ group generated inside $\mathrm{GL}(2,\mathbb{Z})$ by the matrix $\mathfrak{c}_3 = \left(\begin{smallmatrix}0 & -1\\ 1 & -1\end{smallmatrix}\right)$ which satisfies the relation $\mathfrak{c}_2^3 = \mathrm{Id}_2$ and corresponds to a rotation of an angle $2\pi/3$.

(4) C_4. This is the abstract $\mathbb{Z}_4$ group generated inside $\mathrm{GL}(2,\mathbb{Z})$ by the matrix $\mathfrak{c}_4 = \left(\begin{smallmatrix}0 & -1\\ 1 & 0\end{smallmatrix}\right)$ which satisfies the relation $\mathfrak{c}_4^4 = \mathrm{Id}_2$ and corresponds to a rotation of an angle $\pi/2$.

(5) C_6. This is the abstract $\mathbb{Z}_6$ group generated inside $\mathrm{GL}(2,\mathbb{Z})$ by the matrix $\mathfrak{c}_6 = \left(\begin{smallmatrix}1 & -1\\ 1 & 0\end{smallmatrix}\right)$ which satisfies the relation $\mathfrak{c}_6^6 = \mathrm{Id}_2$ and corresponds to a rotation of an angle $\pi/3$.

(6) Dih_1^p. This is the abstract $\mathrm{Dih}_1 \sim \mathbb{Z}_2$ group generated inside $\mathrm{GL}(2,\mathbb{Z})$ by the matrix $\mathfrak{d}_{1,p} = \left(\begin{smallmatrix}1 & 0\\ 0 & -1\end{smallmatrix}\right)$ which satisfies the relation $\mathfrak{d}_{1,p}^2 = \mathrm{Id}_2$ and represents a reflection in the y-direction.

(7) Dih_1^c. This is the abstract $\mathrm{Dih}_1 \sim \mathbb{Z}_2$ group generated inside $\mathrm{GL}(2,\mathbb{Z})$ by the matrix $\mathfrak{d}_{1,c} = \left(\begin{smallmatrix}0 & 1\\ 1 & 0\end{smallmatrix}\right)$ which satisfies the relation $\mathfrak{d}_{1,c}^2 = \mathrm{Id}_2$ and represents a reflection with respect to the $y = -x$ line.

(8) Dih_2^p. This is the abstract $\mathrm{Dih}_2 \sim \mathbb{Z}_2 \times \mathbb{Z}_2$ group generated inside $\mathrm{GL}(2,\mathbb{Z})$ by the two matrices $A_{2,p} = \left(\begin{smallmatrix}-1 & 0\\ 0 & -1\end{smallmatrix}\right)$ and $B_{2,p} = \left(\begin{smallmatrix}1 & 0\\ 0 & -1\end{smallmatrix}\right)$ satisfying the defining relations

$$A_{2,p}^2 = B_{2,p}^2 = (B_{2,p} \cdot A_{2,p})^2 = \mathrm{Id}_2.$$

(9) Dih_2^c. This is the abstract $\mathrm{Dih}_2 \sim \mathbb{Z}_2 \times \mathbb{Z}_2$ group generated inside $\mathrm{GL}(2,\mathbb{Z})$ by the two matrices $A_{2,c} = \left(\begin{smallmatrix}-1 & 0\\ 0 & -1\end{smallmatrix}\right)$ and $B_{2,c} = \left(\begin{smallmatrix}0 & 1\\ 1 & 0\end{smallmatrix}\right)$ satisfying the defining relations

$$A_{2,c}^2 = B_{2,c}^2 = (B_{2,c} \cdot A_{2,c})^2 = \mathrm{Id}_2.$$

(10) Dih_3^ℓ. This is the abstract Dih_3 group generated inside $\mathrm{GL}(2,\mathbb{Z})$ by the two matrices $A_{3,\ell} = \left(\begin{smallmatrix}0 & -1\\ 1 & -1\end{smallmatrix}\right)$ and $B_{3,\ell} = \left(\begin{smallmatrix}1 & 0\\ 1 & -1\end{smallmatrix}\right)$ satisfying the defining relations

$$A_{3,\ell}^3 = B_{3,\ell}^2 = (B_{3,\ell} \cdot A_{3,\ell})^2 = \mathrm{Id}_2.$$

(11) Dih_3^s. This is the abstract Dih_3 group generated inside $\mathrm{GL}(2,\mathbb{Z})$ by the two matrices $A_{3,s} = \left(\begin{smallmatrix}0 & -1\\ 1 & -1\end{smallmatrix}\right)$ and $B_{3,s} = \left(\begin{smallmatrix}1 & -1\\ 0 & -1\end{smallmatrix}\right)$ satisfying the defining relations

$$A_{3,s}^3 = B_{3,s}^2 = (B_{3,s} \cdot A_{3,s})^2 = \mathrm{Id}_2.$$

(12) Dih_4. This is the abstract Dih_4 group generated inside $\mathrm{GL}(2,\mathbb{Z})$ by the two matrices $A_4 = \left(\begin{smallmatrix}0 & -1\\ 1 & 0\end{smallmatrix}\right)$ and $B_4 = \left(\begin{smallmatrix}1 & 0\\ 0 & -1\end{smallmatrix}\right)$ satisfying the defining relations

$$A_4^4 = B_4^2 = (B_4 \cdot A_4)^2 = \mathrm{Id}_2. \tag{8.17}$$

(13) Dih_6. This is the abstract Dih_6 group generated inside $\mathrm{GL}(2,\mathbb{Z})$ by the two matrices $A_6 = \begin{pmatrix} 1 & -1 \\ 1 & 0 \end{pmatrix}$ and $B_6 = \begin{pmatrix} 1 & -1 \\ 0 & -1 \end{pmatrix}$ satisfying the defining relations

$$A_6^6 = B_6^2 = (B_6 \cdot A_6)^2 = \mathrm{Id}_2. \tag{8.18}$$

It is a simple exercise for the reader to check that the generators of the isomorphic but distinct groups like $\mathrm{Dih}_3{}^{\ell}$ and $\mathrm{Dih}_3{}^{s}$ cannot be both mapped one pair into the other pair by conjugation with any matrix $\begin{pmatrix} a & b \\ c & d \end{pmatrix} \in \mathrm{GL}(2,\mathbb{Z})$, namely, having integer entries.

8.2 The proper point groups: the octahedral group O_{24}

As mentioned above, given a lattice Λ in n dimensions, by proper point group of that lattice $\mathfrak{P}_\Lambda$ we mean the maximal finite subgroup of SO_g which leaves Λ invariant. Summarizing the observations of the previous sections and restricting our attention to $n = 3$, we conclude that:

(a) The point group $\mathfrak{P}$ must be a finite rotation group in $d = 3$; hence, it must belong to the list

$$\mathfrak{P} \in \{\mathbb{Z}_k, \mathrm{Dih}_k, \mathrm{T}_{12}, \mathrm{O}_{24}, \mathrm{I}_{60}\}. \tag{8.19}$$

(b) The order of any element $\gamma \in \mathfrak{P}$ belonging to the point group must be 2, 3, 4, or 6.

The intersection of these two conditions implies that

$$\mathfrak{P} \in \{\mathbb{Z}_{2,3,4,6}, \mathrm{Dih}_{3,4,6}, \mathrm{T}_{12}, \mathrm{O}_{24}\}. \tag{8.20}$$

Abstractly all the groups in the above list are subgroups of either one of the two maximal groups Dih_6 or O_{24} being, respectively, the point group of the hexagonal and cubic lattice.

In the next subsections, for the sake of illustration by means of a well-structured example, we restrict our attention to the largest possible proper point group, namely, that of the cubic lattice, which as we just said is O_{24}.

8.2.1 The cubic lattice and the octahedral point group

Let us now consider within the general frame presented above the cubic lattice. The cubic lattice is displayed in Figure 8.3.

The basis vectors of the cubic lattice Λ_{cubic} are

$$\mathbf{w}_1 = \{1,0,0\}; \quad \mathbf{w}_2 = \{0,1,0\}; \quad \mathbf{w}_3 = \{0,0,1\}, \tag{8.21}$$

which implies that the metric is just the Kronecker delta,

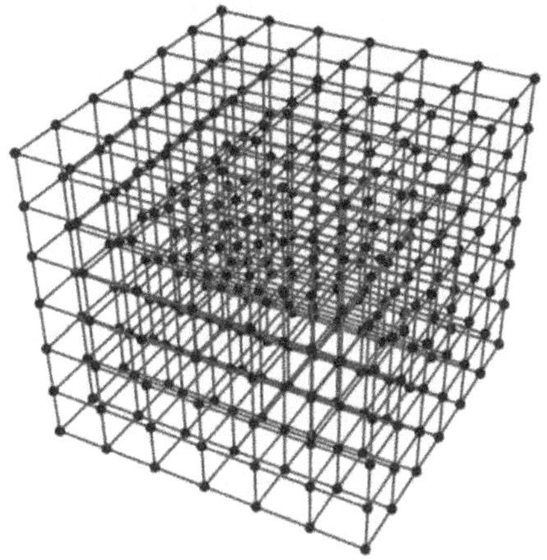

Figure 8.3: A view of the self-dual cubic lattice.

$$g_{\mu\nu} = \delta_{\mu\nu}, \tag{8.22}$$

and the basis vectors $\mathbf{e}^{\mu}$ of the dual lattice $\Lambda^{\star}_{\text{cubic}}$ coincide with those of the lattice Λ. Hence, the cubic lattice is self-dual:

$$\mathbf{w}_{\mu} = \mathbf{e}^{\mu} \quad \Rightarrow \quad \Lambda_{\text{cubic}} = \Lambda^{\star}_{\text{cubic}}. \tag{8.23}$$

The subgroup of the proper rotation group which maps the cubic lattice into itself is the octahedral group O_{24}, whose order is 24. The structure of O_{24} was described in Section 4.4.2. There, showing that it is a solvable group, we used the induction algorithm to deduce the main properties of its irreducible representations. In the next subsection we construct them directly.

8.2.2 Irreducible representations of the octahedral group

There are five conjugacy classes in O_{24}, and therefore according to theory there are five irreducible representations of the same group, which we name D_i, $i = 1, \ldots, 5$. We know from Section 4.4.2.8 that they have dimensions

$$\dim D_1 = 1; \quad \dim D_2 = 1; \quad \dim D_3 = 2; \quad \dim D_4 = 3; \quad \dim D_5 = 3. \tag{8.24}$$

Let us briefly describe them.

8.2.2.1 D_1: the identity representation

The identity representation which exists for all groups is that one where with each element of O we associate the number 1:

$$\forall\gamma \in \mathrm{O}_{24}: \quad D_1(\gamma) = 1. \tag{8.25}$$

Obviously the character of such a representation is

$$\chi_1 = \{1, 1, 1, 1, 1\}. \tag{8.26}$$

8.2.2.2 D_2: the quadratic Vandermonde representation

The representation D_2 is also one-dimensional. It is constructed as follows. Consider the following polynomial of order six in the coordinates of a point in $\mathbb{R}^3$ or T^3:

$$\mathfrak{V}(x, y, z) = (x^2 - y^2)(x^2 - z^2)(y^2 - z^2). \tag{8.27}$$

As one can explicitly check, under the transformations of the octahedral group listed in equation (4.105) the polynomial $\mathfrak{V}(x, y, z)$ is always mapped into itself modulo an overall sign. Keeping track of such a sign provides the form of the second one-dimensional representation, whose character is explicitly calculated to be the following one:

$$\chi_1 = \{1, 1, 1, -1, -1\}. \tag{8.28}$$

8.2.2.3 D_3: the two-dimensional representation

The representation D_3 is two-dimensional and it corresponds to a homomorphism,

$$D_3 : \mathrm{O}_{24} \to \mathrm{GL}(2, \mathbb{Z}), \tag{8.29}$$

which associates with each element of the octahedral group a 2×2 integer valued matrix. The homomorphism is completely specified by giving the two matrices representing the two generators:

$$D_3(\mathrm{A}) = \begin{pmatrix} 0 & 1 \\ -1 & -1 \end{pmatrix}; \quad D_3(\mathrm{B}) = \begin{pmatrix} 0 & 1 \\ 1 & 0 \end{pmatrix}. \tag{8.30}$$

The character vector of D_2 is easily calculated from the above information, and we have

$$\chi_3 = \{2, -1, 2, 0, 0\}. \tag{8.31}$$

8.2.2.4 D_4: the three-dimensional defining representation

The three-dimensional representation D_4 is simply the defining representation, where the generators A and B are given by the matrices in equation (4.106):

$$D_4(\mathrm{A}) = \mathrm{A}; \quad D_4(\mathrm{B}) = \mathrm{B}. \tag{8.32}$$

From this information the characters are immediately calculated, and we get

$$\chi_3 = \{3, 0, -1, -1, 1\}. \tag{8.33}$$

8.2.2.5 D_5: the three-dimensional unoriented representation

The three-dimensional representation D_5 is simply that where the generators A and B are given by the following matrices:

$$D_5(A) = \begin{pmatrix} 0 & 1 & 0 \\ 0 & 0 & 1 \\ 1 & 0 & 0 \end{pmatrix}; \quad D_5(B) = \begin{pmatrix} 0 & 1 & 0 \\ 1 & 0 & 0 \\ 0 & 0 & 1 \end{pmatrix}. \tag{8.34}$$

From this information the characters are immediately calculated, and we get

$$\chi_5 = \{3, 0, -1, 1, -1\}. \tag{8.35}$$

The table of characters is summarized in Table 8.2.

Table 8.2: Character table of the proper octahedral group.

	Class				
Irrep	$\{e, 1\}$	$\{C_3, 8\}$	$\{C_4^2, 3\}$	$\{C_2, 6\}$	$\{C_4, 6\}$
D_1, $\chi_1 =$	1	1	1	1	1
D_2, $\chi_2 =$	1	1	1	−1	−1
D_3, $\chi_3 =$	2	−1	2	0	0
D_4, $\chi_4 =$	3	0	−1	−1	1
D_5, $\chi_5 =$	3	0	−1	1	−1

8.3 The full tetrahedral group T_{12}^{d} and the octahedral group O_{24} are isomorphic

In this section we pursue a double goal. On the one hand, relying on an isomorphism, we show how the group theoretical lore exposed in the previous section for the point group of the cubic lattice $\mathfrak{P}_{\text{cubic}} = O_{24}$ can be utilized within the framework of a completely different interpretation. On the other hand, we take the opportunity to illustrate one classical application of group theory to *chemical–physical* problems, namely, the classification of the vibrational eigenmodes of molecules. Specifically, we consider the so-called XY_4 molecules, like *methane*, whose chemical formula is CH_4, that have an extended tetragonal symmetry (see Figure 8.4).

8.3.1 The vibrations of XY_4 molecules

The XY_4 molecules have four atoms of an element Y located at the vertices of a tetrahedron and one atom of the element X located at the center of the tetrahedron. The

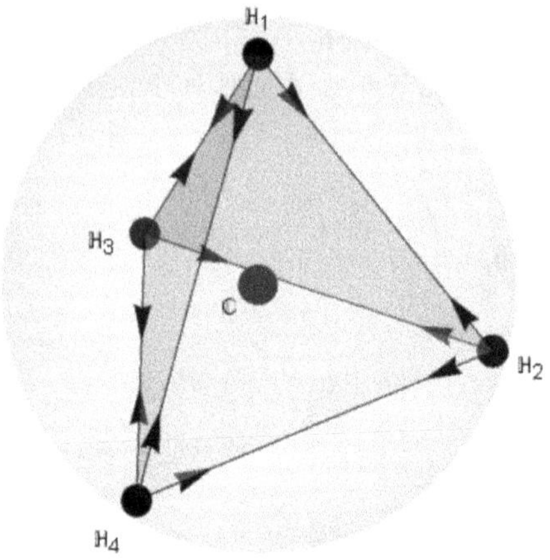

Figure 8.4: A molecule of methane is made up of four hydrogen atoms located at the vertices of a tetrahedron and one carbon atom located at the center of the tetrahedron. Each of the four hydrogens can move into three directions with respect to its rest position.

general problem of molecular vibrations can be formulated in the following way. We have five atoms in the considered molecule; let us name $\boldsymbol{q}_I = \{x_I, y_I, z_I\}$ their positions. The Hamiltonian of the corresponding dynamical system is of the form

$$H = \frac{1}{2}\sum_{I=0}^{4} m_I \left(\frac{d\boldsymbol{q}_I}{dt}\right)^2 + V(\boldsymbol{q}_I), \tag{8.36}$$

where $V(\boldsymbol{q}_I)$ is the potential describing interatomic forces. By hypothesis, in the case of XY_4 molecules, since the molecule exists, the positions $\boldsymbol{q}_I = \boldsymbol{p}_I$, where $\boldsymbol{p}_i$ $(i = 1, \dots, 4)$ are the vertices of the tetrahedron for the Y atoms, and $\boldsymbol{q}_0 = \boldsymbol{p}_0$, where $\boldsymbol{p}_0$ is the centroid of the same for the X atom, correspond to a stable minimum of the potential:

$$\partial_{\boldsymbol{q}_I} V(\boldsymbol{q}_I)|_{\boldsymbol{q}_I=\boldsymbol{p}_I} = 0, \quad I = 0, 1, \dots, 4. \tag{8.37}$$

Hence, we consider small fluctuations around this equilibrium position:

$$\boldsymbol{q}_I = \boldsymbol{p}_I + \boldsymbol{\xi}_I. \tag{8.38}$$

At second order in ξ_I we have the following Hamiltonian:

$$H = \frac{1}{2}\sum_{I=0}^{4} m_I \left(\frac{d\boldsymbol{\xi}_I}{dt}\right)^2 + \sum_{I,J=0}^{4}\sum_{\mu,\nu=1}^{3} A_{I,\mu,J,\nu}\xi_I^\mu \xi_J^\nu, \tag{8.39}$$

where

$$A_{I,\mu,J,\nu} \equiv \partial^2_{\boldsymbol{q}_I^\mu, \boldsymbol{q}_J^\nu} V|_{\boldsymbol{q}_I=\boldsymbol{p}_I}. \tag{8.40}$$

The problem is that of finding the eigenvalues and the eigenvectors of the 15×15 matrix $A_{I,\mu,J,\nu}$. Since the original Hamiltonian is invariant under the extended tetrahedral symmetry, this matrix must commute with elements of the group T_d and therefore it must be diagonal on each irreducible subspace in force of Schur's lemma. The eigenvalues are the vibrational frequencies that are physically observable. Mathematically we just have to decompose the 15-dimensional representation of the extended tetrahedral group into irreducible representations. In this section we just sketch such a procedure.

8.3.2 Structure of the full tetrahedral group

We begin with the proper tetrahedral group T_{12}, which is a subgroup of SO(3) and a symmetry of the tetrahedron. Consider the standard tetrahedron (see Figure 8.5) the coordinates of whose vertices are given by the following four 3-vectors:

$$\begin{aligned} \boldsymbol{V}_1 &= \{0, 0, 1\}, \\ \boldsymbol{V}_2 &= \left\{\sqrt{\frac{2}{3}}, \frac{\sqrt{2}}{3}, -\frac{1}{3}\right\}, \\ \boldsymbol{V}_3 &= \left\{-\sqrt{\frac{2}{3}}, \frac{\sqrt{2}}{3}, -\frac{1}{3}\right\}, \\ \boldsymbol{V}_4 &= \left\{0, -\frac{2\sqrt{2}}{3}, -\frac{1}{3}\right\}. \end{aligned} \tag{8.41}$$

Next we introduce the four rotations of 120 degrees around the four axes defined as the line that joins the center $\{0, 0, 0\}$ with each of the four vertices of the tetrahedron. These

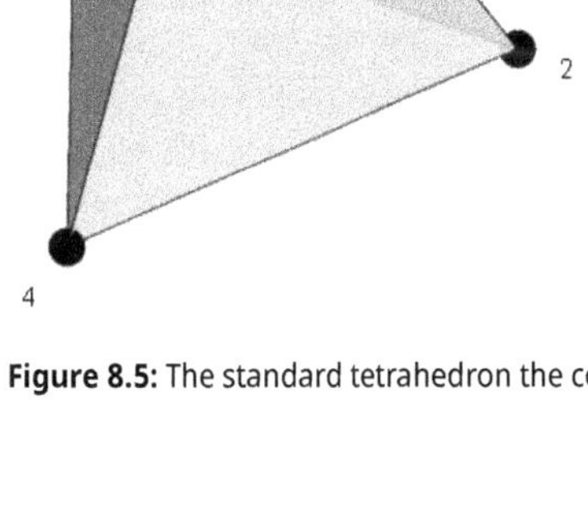

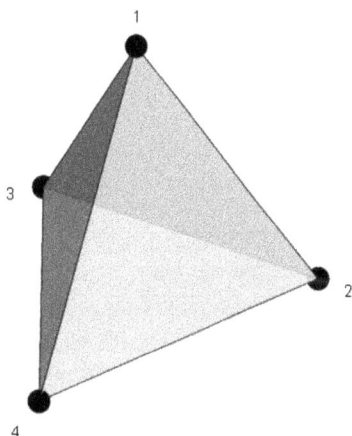

Figure 8.5: The standard tetrahedron the coordinates of whose vertices are displayed in equation (8.41).

finite elements of SO(3) are easily calculated recalling that the one-parameter group of rotations around any given axis $\mathbf{v} = \{v_1, v_2, v_3\}$ is given by the matrices

$$R_{\mathbf{v}}(\theta) \equiv \exp[\theta \mathbf{A}_{\mathbf{v}}], \quad \text{where } \mathbf{A}_{\mathbf{v}} = \left(\begin{array}{c|c|c} 0 & -v_3 & v_2 \\ \hline v_3 & 0 & -v_1 \\ \hline -v_2 & v_1 & 0 \end{array}\right). \tag{8.42}$$

Setting

$$R_i = R_{\mathbf{V}_i}\left(\frac{2\pi}{3}\right), \tag{8.43}$$

where $\mathbf{V}_i$ are the above defined vertices of the standard tetrahedron, we obtain the following four matrices:

$$\begin{aligned} R_1 &= \begin{pmatrix} -\frac{1}{2} & -\frac{\sqrt{3}}{2} & 0 \\ \frac{\sqrt{3}}{2} & -\frac{1}{2} & 0 \\ 0 & 0 & 1 \end{pmatrix}; \quad R_2 = \begin{pmatrix} \frac{1}{2} & \frac{\sqrt{3}}{2} & 0 \\ \frac{1}{2\sqrt{3}} & -\frac{1}{6} & -\frac{2\sqrt{2}}{3} \\ -\sqrt{\frac{2}{3}} & \frac{\sqrt{2}}{3} & -\frac{1}{3} \end{pmatrix}; \\ R_3 &= \begin{pmatrix} \frac{1}{2} & -\frac{1}{2\sqrt{3}} & \sqrt{\frac{2}{3}} \\ -\frac{\sqrt{3}}{2} & -\frac{1}{6} & \frac{\sqrt{2}}{3} \\ 0 & -\frac{2\sqrt{2}}{3} & -\frac{1}{3} \end{pmatrix}; \quad R_4 = \begin{pmatrix} -\frac{1}{2} & \frac{1}{2\sqrt{3}} & -\sqrt{\frac{2}{3}} \\ -\frac{1}{2\sqrt{3}} & \frac{5}{6} & \frac{\sqrt{2}}{3} \\ \sqrt{\frac{2}{3}} & \frac{\sqrt{2}}{3} & -\frac{1}{3} \end{pmatrix}. \end{aligned} \tag{8.44}$$

If we multiply these four matrices in all possible ways and we multiply their products among themselves, we generate a finite group with 12 elements that is named the proper tetrahedral group T_{12}. Abstractly we can show that it is isomorphic to a 12-element subgroup of the octahedral group extensively discussed in Section 8.2.1.

Let us now introduce also reflections, namely, elements of O(3) with determinant different from 1. In particular, let us add the following reflection:

$$S = \begin{pmatrix} -1 & 0 & 0 \\ 0 & 1 & 0 \\ 0 & 0 & 1 \end{pmatrix}, \tag{8.45}$$

which changes the sign of the first component of every vector. Looking at the explicit form of the vertices (8.41) we easily see that all of the five operations R_1, R_2, R_3, R_4, S are symmetry operations of the tetrahedron since they transform its vertices one into the other and they also do so with its faces and edges. Hence, also any product of these symmetry operations is equally a symmetry. The group generated by R_1, R_2, R_3, R_4, S has 24 elements. Organizing it into conjugacy classes we realize that it has the same structure as the octahedral group and actually it is isomorphic to $O_{24} \sim S_4$, whose structure was extensively discussed in Section 4.4.2. The explicit form of the 24 group elements identified by their action on the three Euclidean coordinates $\{x, y, z\}$ and organized into conjugation classes was displayed in equation (4.105); the multiplication table was provided in equation (4.107). Here we have a new representation of the same abstract group, where the same group elements, following the same multiplication table and therefore falling in the same conjugation classes, are identified by a different action on the three Euclidean coordinates which is displayed in Table 8.3.

Table 8.3: Action of the extended tetrahedral group T_d on $\mathbb{R}^3$.

Conjugacy class = 1

$1_1;\ \{x, y, z\}$

Conjugacy class = 2

$2_1;\ \{\frac{1}{6}(-3x - \sqrt{3}y + 2\sqrt{6}z), \frac{1}{6}(\sqrt{3}x + 5y + 2\sqrt{2}z), \frac{1}{3}(-\sqrt{6}x + \sqrt{2}y - z)\}$

$2_2;\ \{\frac{1}{6}(-3x + \sqrt{3}y - 2\sqrt{6}z), \frac{1}{6}(-\sqrt{3}x + 5y + 2\sqrt{2}z), \frac{1}{3}(\sqrt{6}x + \sqrt{2}y - z)\}$

$2_3;\ \{\frac{1}{2}(-x - \sqrt{3}y), \frac{1}{2}(\sqrt{3}x - y), z\}$

$2_4;\ \{\frac{1}{2}(-x + \sqrt{3}y), \frac{1}{2}(-\sqrt{3}x - y), z\}$

$2_5;\ \{\frac{1}{6}(3x - \sqrt{3}y + 2\sqrt{6}z), \frac{1}{6}(-3\sqrt{3}x - y + 2\sqrt{2}z), \frac{1}{3}(-2\sqrt{2}y - z)\}$

$2_6;\ \{\frac{1}{6}(3x + \sqrt{3}y - 2\sqrt{6}z), \frac{1}{6}(3\sqrt{3}x - y + 2\sqrt{2}z), \frac{1}{3}(-2\sqrt{2}y - z)\}$

$2_7;\ \{\frac{1}{2}(x - \sqrt{3}y), \frac{1}{6}(-\sqrt{3}x - y - 4\sqrt{2}z), \frac{1}{3}(\sqrt{6}x + \sqrt{2}y - z)\}$

$2_8;\ \{\frac{1}{2}(x + \sqrt{3}y), \frac{1}{6}(\sqrt{3}x - y - 4\sqrt{2}z), \frac{1}{3}(-\sqrt{6}x + \sqrt{2}y - z)\}$

Conjugacy class = 3

$3_1;\ \{-x, \frac{1}{3}(y - 2\sqrt{2}z), \frac{1}{3}(-2\sqrt{2}y - z)\}$

$3_2;\ \{-\frac{y+\sqrt{2}z}{\sqrt{3}}, \frac{1}{3}(-\sqrt{3}x - 2y + \sqrt{2}z), \frac{1}{3}(-\sqrt{6}x + \sqrt{2}y - z)\}$

$3_3;\ \{\frac{y+\sqrt{2}z}{\sqrt{3}}, \frac{1}{3}(\sqrt{3}x - 2y + \sqrt{2}z), \frac{1}{3}(\sqrt{6}x + \sqrt{2}y - z)\}$

Conjugacy class = 4

$4_1;\ \{\frac{1}{6}(-3x - \sqrt{3}y + 2\sqrt{6}z), \frac{1}{6}(3\sqrt{3}x - y + 2\sqrt{2}z), \frac{1}{3}(-2\sqrt{2}y - z)\}$

$4_2;\ \{\frac{1}{6}(-3x + \sqrt{3}y - 2\sqrt{6}z), \frac{1}{6}(-3\sqrt{3}x - y + 2\sqrt{2}z), \frac{1}{3}(-2\sqrt{2}y - z)\}$

$4_3;\ \{\frac{1}{2}(-x - \sqrt{3}y), \frac{1}{6}(\sqrt{3}x - y - 4\sqrt{2}z), \frac{1}{3}(-\sqrt{6}x + \sqrt{2}y - z)\}$

$4_4;\ \{\frac{1}{2}(-x + \sqrt{3}y), \frac{1}{6}(-\sqrt{3}x - y - 4\sqrt{2}z), \frac{1}{3}(\sqrt{6}x + \sqrt{2}y - z)\}$

$4_5;\ \{-\frac{y+\sqrt{2}z}{\sqrt{3}}, \frac{1}{3}(\sqrt{3}x - 2y + \sqrt{2}z), \frac{1}{3}(\sqrt{6}x + \sqrt{2}y - z)\}$

$4_6;\ \{\frac{y+\sqrt{2}z}{\sqrt{3}}, \frac{1}{3}(-\sqrt{3}x - 2y + \sqrt{2}z), \frac{1}{3}(-\sqrt{6}x + \sqrt{2}y - z)\}$

Conjugacy class = 5

$5_1;\ \{-x, y, z\}$

$5_2;\ \{\frac{1}{6}(3x - \sqrt{3}y + 2\sqrt{6}z), \frac{1}{6}(-\sqrt{3}x + 5y + 2\sqrt{2}z), \frac{1}{3}(\sqrt{6}x + \sqrt{2}y - z)\}$

$5_3;\ \{\frac{1}{6}(3x + \sqrt{3}y - 2\sqrt{6}z), \frac{1}{6}(\sqrt{3}x + 5y + 2\sqrt{2}z), \frac{1}{3}(-\sqrt{6}x + \sqrt{2}y - z)\}$

$5_4;\ \{\frac{1}{2}(x - \sqrt{3}y), \frac{1}{2}(-\sqrt{3}x - y), z\}$

$5_5;\ \{\frac{1}{2}(x + \sqrt{3}y), \frac{1}{2}(\sqrt{3}x - y), z\}$

$5_6;\ \{x, \frac{1}{3}(y - 2\sqrt{2}z), \frac{1}{3}(-2\sqrt{2}y - z)\}$

8.3.2.1 Generators of the extended tetrahedral group

Being isomorphic to the proper octahedral group, the extended tetrahedral group admits a presentation in terms of two generators that satisfy the same relations as given in equation (4.104), namely,

$$\mathscr{A}, \mathscr{B}: \quad \mathscr{A}^3 = \mathbf{e}; \quad \mathscr{B}^2 = \mathbf{e}; \quad (\mathscr{B}\mathscr{A})^4 = \mathbf{e}. \tag{8.46}$$

The explicit form of these generators is the following one:

$$\begin{aligned}
\mathscr{A} = 2_3 &= \begin{pmatrix} -\frac{1}{2} & -\frac{\sqrt{3}}{2} & 0 \\ \frac{\sqrt{3}}{2} & -\frac{1}{2} & 0 \\ 0 & 0 & 1 \end{pmatrix}, \\
\mathscr{B} = 5_3 &= \begin{pmatrix} \frac{1}{2} & \frac{1}{2\sqrt{3}} & -\sqrt{\frac{2}{3}} \\ \frac{1}{2\sqrt{3}} & \frac{5}{6} & \frac{\sqrt{2}}{3} \\ -\sqrt{\frac{2}{3}} & \frac{\sqrt{2}}{3} & -\frac{1}{3} \end{pmatrix}.
\end{aligned} \tag{8.47}$$

Relying on the established isomorphism we can utilize the irreducible representations of O_{24} discussed in Section 8.2.2 and the character table displayed in Table 8.2.

8.3.2.2 Creation of a basis of vectors for the displacements of the atoms

We begin by introducing a basis of vectors for the displacements of the four hydrogen atoms occupying the vertices of the tetrahedron. Instead of an orthonormal system, in each vertex of the tetrahedron we introduce three linear independent vectors aligned with the three edges of the tetrahedron departing from that vertex and pointing to the other three vertices. These 12 vectors are defined as follows:

$$\begin{aligned}
&\xi_{12} = \boldsymbol{V}_1 - \boldsymbol{V}_2; \quad \xi_{13} = \boldsymbol{V}_1 - \boldsymbol{V}_3; \quad \xi_{14} = \boldsymbol{V}_1 - \boldsymbol{V}_4; \\
&\xi_{21} = \boldsymbol{V}_2 - \boldsymbol{V}_1; \quad \xi_{23} = \boldsymbol{V}_2 - \boldsymbol{V}_3; \quad \xi_{24} = \boldsymbol{V}_2 - \boldsymbol{V}_4; \\
&\xi_{31} = \boldsymbol{V}_3 - \boldsymbol{V}_1; \quad \xi_{32} = \boldsymbol{V}_3 - \boldsymbol{V}_2; \quad \xi_{34} = \boldsymbol{V}_3 - \boldsymbol{V}_4; \\
&\xi_{41} = \boldsymbol{V}_4 - \boldsymbol{V}_1; \quad \xi_{42} = \boldsymbol{V}_4 - \boldsymbol{V}_2; \quad \xi_{43} = \boldsymbol{V}_4 - \boldsymbol{V}_3.
\end{aligned} \tag{8.48}$$

Let us organize them into a 12-array as follows:

$$\mathbb{V} = \{\xi_{12}, \xi_{13}, \xi_{14}, \xi_{23}, \xi_{24}, \xi_{21}, \xi_{31}, \xi_{32}, \xi_{34}, \xi_{41}, \xi_{42}, \xi_{43}\}. \tag{8.49}$$

Acting on these vectors the transformations of the extended tetrahedral group permute them in a way that can be easily reconstructed for all 24 elements of the group and in particular for the two generators $\mathscr{A}$ and $\mathscr{B}$, whose effect is displayed below:

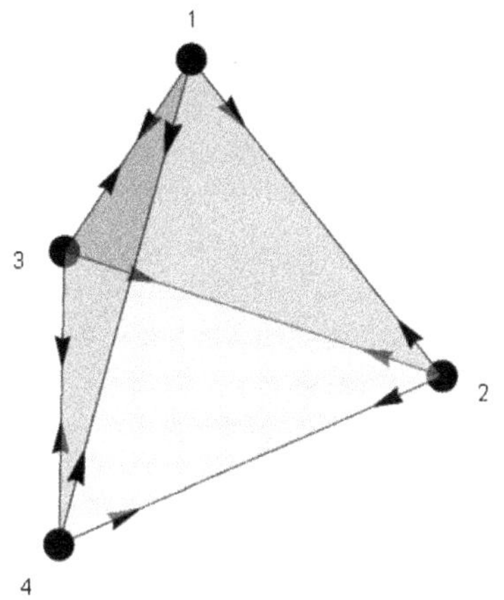

Figure 8.6: The 12 vectors given in equation (8.48) provide a complete basis for the displacements of the hydrogen atoms and are depicted here.

$$\mathscr{A}\begin{pmatrix}\xi_{12}\\ \xi_{13}\\ \xi_{14}\\ \xi_{23}\\ \xi_{24}\\ \xi_{21}\\ \xi_{31}\\ \xi_{32}\\ \xi_{34}\\ \xi_{41}\\ \xi_{42}\\ \xi_{43}\end{pmatrix}=\begin{pmatrix}\xi_{13}\\ \xi_{14}\\ \xi_{12}\\ \xi_{34}\\ \xi_{32}\\ \xi_{31}\\ \xi_{41}\\ \xi_{43}\\ \xi_{42}\\ \xi_{21}\\ \xi_{23}\\ \xi_{24}\end{pmatrix};\quad \mathscr{B}\begin{pmatrix}\xi_{12}\\ \xi_{13}\\ \xi_{14}\\ \xi_{23}\\ \xi_{24}\\ \xi_{21}\\ \xi_{31}\\ \xi_{32}\\ \xi_{34}\\ \xi_{41}\\ \xi_{42}\\ \xi_{43}\end{pmatrix}=\begin{pmatrix}\xi_{32}\\ \xi_{31}\\ \xi_{34}\\ \xi_{21}\\ \xi_{24}\\ \xi_{23}\\ \xi_{13}\\ \xi_{12}\\ \xi_{14}\\ \xi_{43}\\ \xi_{42}\\ \xi_{41}\end{pmatrix}. \tag{8.50}$$

The above result is encoded in two 12×12 integer valued matrices; altogether what we have done is to construct a 12-dimensional representation of the tetrahedral group that we name $\boldsymbol{D}_{12}$ (see Figure 8.6).

Using the standard formula (4.85) for multiplicities in the decomposition of reducible group representations into irreducible ones and the character table in Table 8.2, we immediately obtain

$$\boldsymbol{D}_{12} = \boldsymbol{D}_1 \oplus \boldsymbol{D}_2 \oplus 2\boldsymbol{D}_4 \oplus \boldsymbol{D}_5, \tag{8.51}$$

where we remind the reader that $\boldsymbol{D}_1$ is the singlet trivial representation, $\boldsymbol{D}_2$ is the unique two-dimensional representation, and $\boldsymbol{D}_4$ and $\boldsymbol{D}_5$ are the two inequivalent three-dimensional representations.

8.3.2.3 Chemical–physical interpretation

The chemical–physical implications of this decomposition are of utmost relevance. It means that the 12×12 matrix (8.40) can depend on at most five parameters corresponding to the five representations mentioned in equation (8.51). So far we have not yet considered the X atom in the centroid of the tetrahedron. We have discussed the motions of the Y atoms in the coordinate system centered at X. One of the two irreducible representations $\boldsymbol{D}_4$ clearly must correspond to the translation modes where all the Y atoms move coherently in the same direction. Interpreted in the reverse order these are the vibrations of the X atom with respect to the Y atoms, and this accounts for one of the five parameters. The other four parameters can be easily given a physical interpretation looking at the structure of the projection onto the irreducible representations. We consider a pair of them as an illustration.

The breathing mode

The identity representation $\boldsymbol{D}_1$ has a unique physical interpretation as breathing mode, namely, the overall contraction or enlargement of the molecule (contraction or expansion). To see this it suffices to consider the projection operator on the singlet representation. We obtain that the invariant space is provided by a 12-vector whose 12 entries are all equal. Physically interpreted this means that in this representation each of the four Y atoms (hydrogen in the case of methane) is displaced the same distance along the direction from the vertex where it sits when in equilibrium towards the centroid of the tetrahedron (the location of the carbon atom in the case of methane). Explicitly this means that the four displacement vectors are

$$\begin{aligned} \delta \boldsymbol{V}_1 &= \xi_{12} + \xi_{13} + \xi_{14}, \\ \delta \boldsymbol{V}_2 &= \xi_{23} + \xi_{24} + \xi_{21}, \\ \delta \boldsymbol{V}_3 &= \xi_{32} + \xi_{34} + \xi_{31}, \\ \delta \boldsymbol{V}_4 &= \xi_{41} + \xi_{42} + \xi_{42}. \end{aligned} \tag{8.52}$$

The effect on the molecule is represented in Figure 8.7.

The rotational modes

Next we provide the physical interpretation of the irreducible representation $\boldsymbol{D}_5$. Let us define the following four vectors (inside the 12-dimensional space spanned by linear combinations of the basis vectors (8.48)):

$$\begin{aligned} \boldsymbol{p}_1 &= \xi_{23} - \xi_{32} + \xi_{34} - \xi_{43} + \xi_{42} - \xi_{24}, \\ \boldsymbol{p}_2 &= \xi_{31} - \xi_{13} + \xi_{14} - \xi_{41} + \xi_{43} - \xi_{34}, \\ \boldsymbol{p}_3 &= \xi_{12} - \xi_{21} + \xi_{41} - \xi_{14} + \xi_{24} - \xi_{42}, \\ \boldsymbol{p}_4 &= \xi_{21} - \xi_{12} + \xi_{13} - \xi_{31} + \xi_{32} - \xi_{23}. \end{aligned} \tag{8.53}$$

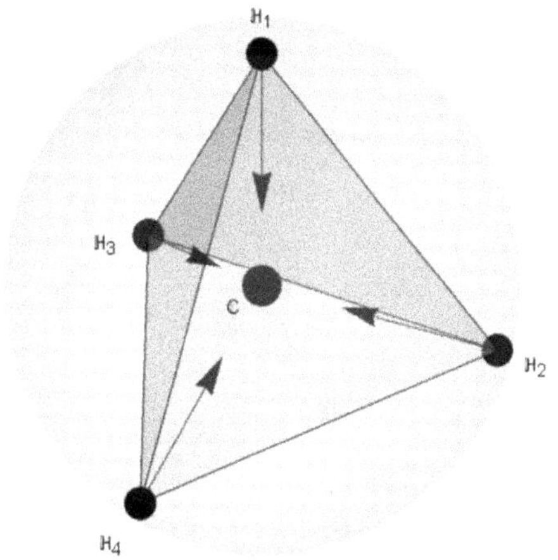

Figure 8.7: The one-dimensional singlet representation $\boldsymbol{D}_1$ corresponds to the breathing mode, where the molecule just expands or contracts in a conformal fashion, preserving all the angles. Geometrically one can inscribe the tetrahedron into a sphere $\mathbb{S}^2$, whose surface intersects all of the four vertices ξ_i. Increasing or decreasing the radius of such a sphere is what the breathing mode does.

By using equation (8.50) we easily check that they span an irreducible subspace and that they sum up to zero:

$$\sum_{i=1}^{4} \boldsymbol{p}_i = 0. \tag{8.54}$$

Calculation of the character of this three-dimensional representation shows that it corresponds to the irreducible $\boldsymbol{D}_5$. The physical interpretation of the four vectors (8.53) is visualized in Figure 8.8.

The displacements of the Y atom encoded in the vector $\boldsymbol{p}_i$ correspond to a rotation of the molecule around the axis passing through the centroid and the vertex i. As a result of such a rotation the face of the tetrahedron opposite to the vertex i rotates in the plane orthogonal to the mentioned axis.

The other modes

In a similar way one can study the geometry of the deformations associated with the remaining two irreducible representations, $\boldsymbol{D}_2$ and $\boldsymbol{D}_4$. They correspond to asymmetrical compressions and dilatations of the molecule that change its shape preserving however some of its features. We do not enter further into details being confident that what we have illustrated so far is already a quite convincing illustration of the foremost relevance of finite group theory in solid state and molecular physics. Observable data like vibrational normal modes are in one-to-one correspondence with the irreducible representations of the basic symmetry either of the crystal or of the molecule.

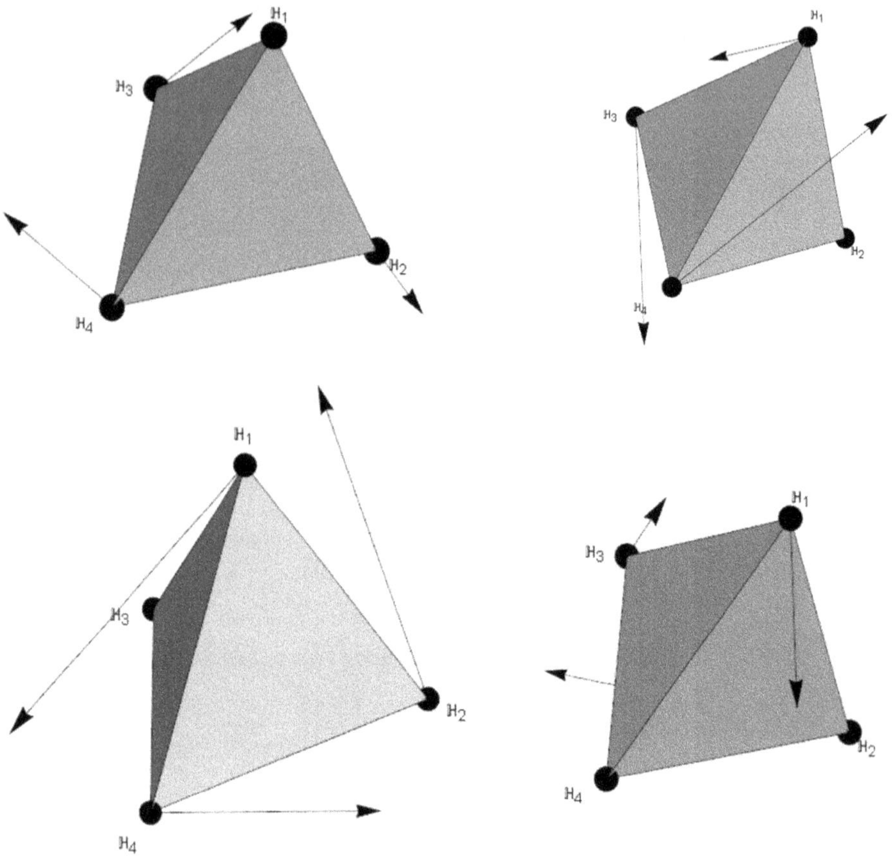

Figure 8.8: The three-dimensional representation $\boldsymbol{D}_5$ corresponds to the rotation modes. Each vector $\mathbf{p}_i$ ($i = 1, 2, 3, 4$) corresponds to the rotation of the entire molecule around the axis that joins the center with the i-th vertex as displayed in this figure. The four vectors are not independent, since their sum equals zero and this makes the representation three-dimensional.

8.4 Group extensions and space groups

We have arrived at the concept of *group extensions*, which once applied to lattices Λ and their associated point groups $\mathfrak{P}_\Lambda$ produce the construction directives for the so-called *space groups* of crystallography but which are much more general.

8.4.1 General theory of group extensions

We consider a group E with a normal subgroup N:

$$N \triangleleft E. \tag{8.55}$$

In the application of the mathematical machinery that we presently define to crystallography, E will be the space group $\mathfrak{S}$, while N will be the lattice Λ. The quotient group

$$G \simeq \frac{E}{N} \tag{8.56}$$

will be the point group $\mathfrak{P}_\Lambda$. In the general theory N is just required to be a normal subgroup that can be finite or infinite, and it is not required to be necessarily Abelian.

The mathematical construction one aims at is that of possible E groups given N and G.

The starting point is the following observation. If E is given, with each element g of E we can associate the following map:

$$\forall g \in E \quad \tau_g : N \to N, \quad \text{where } \forall n \in N \quad \tau_g(n) \equiv g \cdot n \cdot g^{-1} \in N. \tag{8.57}$$

Clearly, τ_g is an automorphism of the group N, since

$$\forall n_1, n_2 \in N : \quad \tau_g(n_1) \circ \tau_g(n_2) = \tau_g(n_1 \cdot n_2). \tag{8.58}$$

Furthermore, one has

$$\forall g, h \in E \quad \text{and} \quad \forall n \in N : \quad \tau_{g \cdot h}(n) = \tau_g \circ \tau_h(n). \tag{8.59}$$

Hence, if we denote by $\mathrm{Aut}(N)$ the automorphism group of N, we have a map

$$\phi : E \to \mathrm{Aut}(N) \tag{8.60}$$

given by τ, and this map is a homomorphism. For $m \in N$ the map $\tau_m \in \mathrm{Inn}(N) \subset \mathrm{Aut}(N)$ is an *inner automorphism* of N, while for $h \notin N$ the corresponding τ_h might belong either to the *inner* or to the *outer* automorphism group of N:

$$\mathrm{Out}(N) \equiv \frac{\mathrm{Aut}(N)}{\mathrm{Inn}(N)}, \tag{8.61}$$

which makes sense because for any group N the subgroup of inner automorphisms is a normal subgroup of $\mathrm{Aut}(N)$.

Strategic in our further arguments is the following.

Lemma 8.4.1. *Let $f : E \to H$ be a homomorphism and suppose that there exist two normal subgroups $N \lhd E$ and $W \lhd H$ such that $f(N) \subset W$. Then f induces a homomorphism between the two quotient groups:*

$$\hat{f} : \frac{E}{N} \to \frac{H}{W}. \tag{8.62}$$

Proof of Lemma 8.4.1. Let $\pi : H \to H/W$ be the projection. Let $g \in E$ and $n \in N$ and define

$$\hat{f}(g \cdot n) = \pi \circ f(g \cdot n) = \pi(f(g) \cdot f(n)) = \pi(f(g)). \tag{8.63}$$

The entire equivalence class $g \cdot N$ has a unique image, and this proves the lemma. □

8.4.1.1 Application of the lemma

If we consider the homomorphism in equation (8.60) we are in the conditions required by Lemma 8.4.1, since as we remarked $\phi(N) \subset \mathrm{Inn}(N)$. Hence there is a homomorphism:

$$\hat{\phi} : \frac{E}{N} \to \mathrm{Out}(N) \equiv \frac{\mathrm{Aut}(N)}{\mathrm{Inn}(N)}. \tag{8.64}$$

Theorem 8.4.1 (Twisted direct product). *Let $\lambda : G \to \mathrm{Aut}(N)$ be a homomorphism of a group G onto $\mathrm{Aut}(N)$, where N is another group. Then there is a group E with normal subgroup N and quotient group $E/N \sim G$, where $\sim$ means isomorphic. The associated map $\hat{\phi} : G \to \mathrm{Out}(N)$ is given by $\pi \circ \lambda$, where $\pi : \mathrm{Aut}(N) \to \mathrm{Out}(N)$ is the projection.*

Proof of Theorem 8.4.1. It suffices to define the product law on the Cartesian product $N \times G$ as follows:

$$\forall n, n' \in N \text{ and } \forall g, g' \in G : \quad (n, g) \bullet (n', g') = (n * \lambda(g)[n'], g \cdot g'), \tag{8.65}$$

where we have denoted by $*$ the product operation inside N, by $\cdot$ the product operation in G, and by $\bullet$ the new product operation in E. With such a definition of the internal binary operation it is straightforward to verify that $E = N \times G$ becomes a group, since all axioms are satisfied. □

8.4.1.2 An example with cyclic groups

We can provide an interesting example considering the automorphism group of the cyclic group $\mathbb{Z}_7$ which turns out to be $\mathbb{Z}_3$. The elements of $\mathbb{Z}_7$ are

$$\mathbb{Z}_7 = (0, 1, 2, 3, 4, 5, 6) + 7k, \quad k \in \mathbb{Z}. \tag{8.66}$$

Let $m \in \mathbb{Z}_7$ and consider the map

$$\xi : \mathbb{Z}_7 \to \mathbb{Z}_7; \quad \forall m \in \mathbb{Z}_7 : \xi(m) = 2 \times m + 7k', \quad k' \in \mathbb{Z}. \tag{8.67}$$

Considering the standard representatives of the equivalence classes we have

$$\begin{aligned} \xi &: (0, 1, 2, 3, 4, 5, 6) \to (0, 2, 4, 6, 1, 3, 5), \\ \xi^2 &: (0, 1, 2, 3, 4, 5, 6) \to (0, 4, 1, 5, 2, 6, 3), \\ \xi^3 &: (0, 1, 2, 3, 4, 5, 6) \to (0, 1, 2, 3, 4, 5, 6). \end{aligned} \tag{8.68}$$

Hence the operation on the elements of $\mathbb{Z}_7$ defined by the map ξ of equation (8.67) induces a homomorphism,

$$\lambda : \mathbb{Z}_3 \to \mathrm{Aut}(\mathbb{Z}_7), \tag{8.69}$$

explicitly defined as follows:

$$\begin{aligned} &\mathbb{Z}_3 = (0,1,2) + 3q; \quad q \in \mathbb{Z}, \\ &\forall x \in Z_3 : \quad \lambda_x \equiv \xi^x : \mathbb{Z}_7 \to \mathbb{Z}_7. \end{aligned} \tag{8.70}$$

Utilizing the previous Theorem 8.4.1 we can construct the group by writing the group law

$$(m \bmod 7, x \bmod 3) \bullet (n \bmod 7, y \bmod 3) \equiv (2^x n + m \bmod 7, x + y \bmod 3). \tag{8.71}$$

It is interesting to realize that although both $\mathbb{Z}_3$ and $\mathbb{Z}_7$ are Abelian, the twisted product E we have constructed is not Abelian. Furthermore, E can be generated by the following two elements,

$$X \equiv (0,1); \quad Y \equiv (1,0), \tag{8.72}$$

which satisfy the following relations:

$$X^3 = Y^7 = \mathbf{e} = (0,0); \quad X \bullet Y = Y^2 \bullet X = (1,2). \tag{8.73}$$

We will retrieve the above constructed group under the name of $\mathrm{G}_{21} \subset \mathrm{PSL}(2,7)$ as one of the two maximal subgroups of the simple group $\mathrm{PSL}(2,7)$ in Chapter 14.

8.5 Space groups

We come next to the issue of space groups associated with lattices that are also named *tessellation groups* and in two dimensions *wallpaper groups*. In physics/chemistry and in crystallography this is a very important issue and it leads to very much diversified applications. Yet it is a genuine and quite profound mathematical problem corresponding to an instance of the group extension constructions illustrated above, where the normal subgroup N is Abelian and it is provided by a lattice Λ in n dimensions (see Section 8.1.1 for the relevant definition). As discussed above, given the lattice Λ we can always determine the point group $\mathfrak{P}_\Lambda \subset \mathrm{SO}(\mathrm{n}, \mathbb{R})$ which maps the lattice into itself. Since the lattice Λ is an Abelian infinite discrete group isomorphic to $\mathbb{Z}^n$, the space group problem is that of constructing a group extension $\mathfrak{S}$ which admits Λ as a normal subgroup such that the quotient group $\mathfrak{S}/\Lambda$ is isomorphic to the point group $\mathfrak{P}_\Lambda$.

Utilizing the language of exact sequences we have

$$0 \underset{\mu_1}{\xrightarrow{\iota}} \Lambda \underset{\mu_2}{\xrightarrow{\iota}} \mathfrak{G} \underset{\mu_3}{\xrightarrow{\pi}} \mathfrak{P}_\Lambda \underset{\mu_4}{\xrightarrow{\pi}} \mathbf{1}. \tag{8.74}$$

The first map is the injection map of the neutral element of the Abelian lattice group Λ into the group it pertains to, the second map is the injection map of the abstract group $\Lambda \sim \mathbb{Z}^n$ as a normal subgroup in some group $\mathfrak{G}$, the third map is the projection onto the quotient $\mathfrak{P}_\Lambda \equiv \frac{\mathfrak{G}}{\Lambda}$, and the fourth map is the projection of the entire $\mathfrak{P}_\Lambda$ onto its neutral element **1**. The exactness property of the sequence

$$\ker(\mu_i) = \mathrm{Im}(\mu_{i+1}) \tag{8.75}$$

is evident from the description. Any time we succeed in realizing the middle term $\mathfrak{G}$ in such an exact sequence as that in equation (8.74) we say that $\mathfrak{G}$ is a *group extension* of the lattice group Λ by means of the point group $\mathfrak{P}_\Lambda$ which is supposed to have an automorphic action on Λ. The direct product is just one example of the realizations of such space group extensions, but it is not the only one as we argued in previous sections.

8.5.1 The basic ingredient of the construction: the lifting map

In order to construct the candidate elements of the group $\mathfrak{G}$ we need to work in a larger ambient group that should contain the automorphism group $\mathrm{Aut}(\Lambda)$ as a subgroup. If the lattice is n-dimensional, a convenient choice is provided by the general linear group in $(n+1)$ dimensions, namely, by the group $\mathrm{GL}(\mathrm{n}+1,\mathbb{R})$. We can explicitly inject the lattice group in this latter as follows:

$$\begin{aligned} L : \Lambda &\xrightarrow{L} \mathrm{GL}(\mathrm{n}+1,\mathbb{R}), \\ \forall \mathbf{m} \in \Lambda : L(\mathbf{m}) &= \left(\begin{array}{c|c} \mathbf{1}_{n\times n} & \mathbf{m} \\ \hline 0_{1\times n} & 1 \end{array} \right). \end{aligned} \tag{8.76}$$

The image in $\mathrm{GL}(\mathrm{n}+1,\mathbb{R})$ of the lattice group $L(\Lambda)$ is invariant with respect to conjugation with any element of the translation subgroup $T \subset \mathrm{GL}(\mathrm{n}+1,\mathbb{R})$ defined below:

$$\mathfrak{t} \in T \quad \Rightarrow \quad \mathfrak{t} = \left(\begin{array}{c|c} \mathbf{1}_{n\times n} & \mathbf{t} \\ \hline 0_{1\times n} & 1 \end{array} \right); \quad \mathbf{t} \in \mathbb{R}^n. \tag{8.77}$$

On the other hand, let us consider diagonal $(n+1) \times (n+1)$ matrices of the following type:

$$K_\gamma = \left(\begin{array}{c|c} \gamma_{n\times n} & \mathbf{0} \\ \hline 0_{1\times n} & 1 \end{array} \right); \quad \gamma_{n\times n} \in D[\mathfrak{P}_\Lambda, n], \tag{8.78}$$

where $D[\mathfrak{P}_\Lambda, n] \subset \mathrm{SO}_g(n)$ is the defining n-dimensional representation of the point group and $\mathrm{SO}_g(n)$ is the orthogonal group respecting the g metric defined by the lattice basis vectors (see equation (8.3)). By an immediate calculation we find

$$\forall \mathfrak{m} = \left(\begin{array}{c|c} 1 & \mathbf{m} \\ \hline 0_{1\times n} & 1 \end{array}\right) \in L(\Lambda): \quad K_\gamma \cdot \mathfrak{m} \cdot K_\gamma^{-1} = \left(\begin{array}{c|c} 1 & \gamma\mathbf{m} \\ \hline 0_{1\times n} & 1 \end{array}\right) \in L(\Lambda). \tag{8.79}$$

Hence, we can conclude that inside $\mathrm{GL}(n+1,\mathbb{R})$ the automorphism group $\mathrm{Aut}(\Lambda)$ is composed of all the matrices of the following form:

$$\mathrm{GL}(n+1,\mathbb{R}) \supset \mathrm{Aut}(\Lambda) = \left\{ \left(\begin{array}{c|c} \gamma & \mathbf{t} \\ \hline 0_{1\times n} & 1 \end{array}\right) \,\middle|\, \gamma \in D[\mathfrak{P}_\Lambda, n], \mathbf{t} \in \mathbb{R}^n \right\}. \tag{8.80}$$

This is the setup we need in order to discuss group extensions and group cohomology. The fundamental tool is the *lifting map*. Having determined in a concrete way the structure of the automorphism group of the lattice we can consider the following map:

$$\begin{aligned} L &: \mathfrak{P}_\Lambda \hookrightarrow \mathrm{Aut}(\Lambda), \\ \forall \gamma \in \mathfrak{P}_\Lambda \quad L(\gamma) &= \left(\begin{array}{c|c} \gamma & \mathbf{v}_\gamma \\ \hline 0_{1\times n} & 1 \end{array}\right); \quad \mathbf{v}_\gamma \in \mathbb{R}^n. \end{aligned} \tag{8.81}$$

The key point in the above equation is that the lifting map is not required to be a homomorphism of groups. One can define the following map:

$$\begin{aligned} f &: \mathfrak{P}_\Lambda \times \mathfrak{P}_\Lambda \to T \subset \mathrm{Aut}(\Lambda), \\ f(\gamma_1,\gamma_2) &= L(\gamma_1)L(\gamma_2)\,L(\gamma_1\cdot\gamma_2)^{-1}, \\ &\Downarrow \\ f(\gamma_1,\gamma_2) &= \left(\begin{array}{c|c} 1 & \overbrace{-\mathbf{v}_{\gamma_1\cdot\gamma_2} + \gamma_1\mathbf{v}_{\gamma_2} + \mathbf{v}_{\gamma_1}}^{\equiv \varphi^2(\gamma_1,\gamma_2)} \\ \hline 0_{1\times n} & 1 \end{array}\right) \end{aligned} \tag{8.82}$$

and verify that the lifting map is a group homomorphism if and only if $\forall \gamma_1,\gamma_2 \in \mathfrak{P}_\Lambda$, $f(\gamma_1,\gamma_2) = \mathrm{Id} \in T \subset \mathrm{Aut}(\Lambda)$, namely, if $\varphi^2(\gamma_1,\gamma_2) = 0$ for all pairs of elements of $\mathfrak{P}_\Lambda$.

8.5.2 The product law and the closure condition

Notwithstanding the fact that L is not necessarily a homomorphism, under a condition that we mention explicitly below we can construct a group $\mathfrak{S}$ by means of the following procedure. As a set we have

$$\mathfrak{S} = \mathfrak{P}_\Lambda \times \Lambda \tag{8.83}$$

and we uplift such a set to Aut(Λ) in the following way:

$$L : (\gamma, \mathbf{m}) \to L(\gamma, \mathbf{m}) \equiv L(\mathbf{m})L(\gamma) = \left(\begin{array}{c|c} \gamma & \mathbf{m} + \mathbf{v}_\gamma \\ \hline 0_{1\times n} & 1 \end{array}\right). \tag{8.84}$$

Utilizing the two identities

$$L(\gamma)L(\mathbf{m})L^{-1}(\gamma) = L(\gamma \cdot \mathbf{m}), \tag{8.85}$$

$$L(\gamma_1)L(\gamma_2) = f(\gamma_1, \gamma_2)L(\gamma_1 \cdot \gamma_2) \tag{8.86}$$

that follow from equations (8.79), (8.81), and (8.82), by means of a straightforward calculation we obtain

$$\begin{aligned} L(\gamma_1, \mathbf{m}_1)L(\gamma_2, \mathbf{m}_2) &= L(m_1 + \gamma_1 \cdot \mathbf{m}_2)f(\gamma_1, \gamma_2)L(\gamma_1 \cdot \gamma_2) \\ &= L(\mathbf{m}_1 + \gamma_2 \cdot \mathbf{m}_2 + \varphi^2(\gamma_1 + \gamma_2))L(\gamma_1 \cdot \gamma_2). \end{aligned} \tag{8.87}$$

From equation (8.87) we read off a new product law on the set $\mathfrak{S}$ defined in equation (8.83). We denote it with the symbol $\circ$:

$$(\gamma_1, \mathbf{m}_1) \circ (\gamma_2, \mathbf{m}_2) = (\gamma_1 \cdot \gamma_2, \mathbf{m}_1 + \gamma_2 \cdot \mathbf{m}_2 + \varphi^2(\gamma_1, \gamma_2)). \tag{8.88}$$

The question is whether with respect to this operation the set $\mathfrak{S}$ becomes a group, all the axioms being satisfied. With respect to this question we first note that if $\varphi^2(\gamma_1 + \gamma_2) = 0$, equation (8.88) is the direct product law and certainly in such a case S is a group. Secondly, we observe that in order to make sense even if $\varphi^2(\gamma_1 + \gamma_2) \neq 0$ it is necessary that it belongs to the lattice Λ. Hence, the first requirement on the choice of the vectors v_γ that we associate with each element of the group λ is that

$$\forall \gamma_1, \gamma_2 \in \mathfrak{P}_\Lambda : \quad \varphi^2(\gamma_1, \gamma_2) \in \Lambda. \tag{8.89}$$

Moreover, for associativity of the product and correct definition of the inverse of each element we find the following condition on $\varphi^2(\gamma_1 + \gamma_2)$ when we consider a third element $\gamma_3 \in \mathfrak{P}_\Lambda$:

$$\delta^3\varphi^2(\gamma_1, \gamma_2, \gamma_3) = \gamma_1 \cdot \varphi^2(\gamma_2, \gamma_3) - \varphi^2(\gamma_1 \cdot \gamma_2, \gamma_3) + \varphi^2(\gamma_1, \gamma_2 \cdot \gamma_3) - \varphi^2(\gamma_1, \gamma_2) = 0. \tag{8.90}$$

Equation (8.90) is identically satisfied by the map

$$\varphi^2(\gamma_1, \gamma_2) \equiv -\mathbf{v}_{\gamma_1 \cdot \gamma_2} + \gamma_1 \mathbf{v}_{\gamma_2} + \mathbf{v}_{\gamma_1} \tag{8.91}$$

that results from our uplifting construction. Hence, it remains only to find out how the vectors $\mathbf{v}_\gamma$ can be chosen in such a way as to fulfill the condition (8.89). We will do that by means of the so-called *Frobenius congruences*, yet, before addressing such a point, we pause to consider the meaning of equation (8.90). It is a closure condition inside an appropriate cohomology setup.

8.5.3 Finite group cohomology

Finite group cohomology can be defined in the general setup utilized at the beginning of the present section on group extensions, yet we illustrate it just in the case we are working on, namely, that of point groups and lattices. We recall the notions introduced in Section 6.9.3 and we begin by introducing the space $\mathscr{C}^n$ of cochains.

Definition 8.5.1. A cochain φ^n is a map

$$\varphi^n : \underbrace{\mathfrak{P}_\Lambda \times \mathfrak{P}_\Lambda \times \cdots \times \mathfrak{P}_\Lambda \times \mathfrak{P}_\Lambda}_{n\text{-times}} \to \Lambda. \tag{8.92}$$

We have a series of maps

$$\begin{aligned} \delta^{n+1} &: \mathscr{C}^n \to \mathscr{C}^{n+1}, \\ \delta^{n+1}\varphi^n(\gamma_1, \gamma_2, \ldots, \gamma_n, \gamma_{n+1}) &= \gamma_1 \cdot \varphi^n(\gamma_2, \ldots, \gamma_n, \gamma_{n+1}) \\ &\quad + \sum_{i=1}^{n} (-1)^i \varphi^n(\gamma_1, \ldots, \gamma_{i-1}, \gamma_i \cdot \gamma_{i+1}, \gamma_{+2}, \ldots \gamma_{n+1}) \\ &\quad + (-1)^{n+1}\varphi^n(\gamma_1, \ldots, \gamma_n). \end{aligned} \tag{8.93}$$

By meticulous straightforward calculation one can check that $\delta^{n+1} \circ \delta^n = 0$, so we realize the defining property of elliptic complexes:

$$\operatorname{Im} \delta^n \subset \ker \delta^{n+1}. \tag{8.94}$$

Definition 8.5.2. As in all other cohomological theories, $\varphi^n \in \ker \delta^{n+1}$ are named n-cocycles spanning the subspace $Z^n(\mathfrak{P}_\Lambda, \Lambda)$, while $\varphi^n \in \operatorname{Im} \delta^n$ are the n-coboundaries spanning the subspace $B^n(\mathfrak{P}_\Lambda, \Lambda)$ and the equivalence classes $Z^n(\mathfrak{P}_\Lambda, \Lambda) \bmod B^n(\mathfrak{P}_\Lambda, \Lambda)$ constitute the cohomology group $H^n(\mathfrak{P}_\Lambda, \Lambda)$.

From this point of view we see that the map $\varphi(\gamma_1, \gamma_2)$ of equation (8.91) is a 2-cocycle since it satisfies equation (8.90), yet at first sight it seems to be also a 2-coboundary. Indeed, we can use the general formula (8.93) and look at the expression of the most general 2-coboundary, which is the following:

$$\delta^2\varphi^1\gamma_1, \gamma_2 = \gamma_1 \cdot \varphi^1(\gamma_2) - \phi^1(\gamma_1 \cdot \gamma_2) + \phi^1(\gamma_1). \tag{8.95}$$

Setting $\varphi^1(\gamma) = \mathbf{v}_\gamma$ it seems that indeed $\varphi^2 = \delta^2\varphi^1$. The subtle point is that in order to be a coboundary $\mathbf{v}_\gamma$ should be elements of the lattice, $\mathbf{v}_\gamma \in \Lambda$. The catch of the whole construction and a fascinating aspect of space groups is that we try to determine a set of non-trivial vectors $\mathbf{v}_\gamma$ in the equivalence classes with respect to the equivalence relation

$$\forall \mathbf{v}, \mathbf{w} \in \mathbb{R}^n : \quad \mathbf{v} \sim \mathbf{w} \quad \text{iff} \quad \mathbf{v} - \mathbf{w} \in \Lambda \tag{8.96}$$

such that equation (8.89) is always satisfied.

8.5.4 Frobenius congruences

In order to derive the cohomology classes, a very convenient approach is provided by the time honored method of Frobenius congruences, which relies on the abstract presentation of the point group by means of generators and relations.

We consider the inclusion map of the group $\mathfrak{P}_\Lambda$ into $\mathrm{Aut}(\Lambda)$ provided by equation (8.81). The key observation is that in order to guarantee the quotient identification with the point group,

$$\frac{\mathfrak{S}}{\Lambda} \simeq \mathfrak{P}_\Lambda, \tag{8.97}$$

while performing the matrix product of two elements, in the translation block one has to take into account equivalence modulo lattice Λ, namely,

$$\left(\begin{array}{c|c} \gamma_1 & \mathfrak{c}_1 \\ \hline 0 & 1 \end{array}\right) \cdot \left(\begin{array}{c|c} \gamma_2 & \mathfrak{c}_2 \\ \hline 0 & 1 \end{array}\right) = \left(\begin{array}{c|c} \gamma_1 \cdot \gamma_2 & \gamma_1 \mathfrak{c}_2 + \mathfrak{c}_1 + \Lambda \\ \hline 0 & 1 \end{array}\right). \tag{8.98}$$

Utilizing the above notation, let the abstract point group be identified by some generators $J = \{J_1, \dots, J_q\}$ and some relations. For instance in the case of the finite subgroups of SO(3) we have $q = 2$ and the groups are singled out by the presentation in the second line of equation (5.50) where the triple of integers $\{k_1, k_2, k_3\}$ are those of the ADE classification:

$$\Gamma[k_1, k_2, k_3] = (\mathrm{A}, \mathrm{B} \mid (\mathrm{AB})^{k_1} = \mathrm{A}^{k_2} = \mathrm{B}^{k_3} = \mathbf{1}). \tag{8.99}$$

The case $k_1 = 4$, $k_2 = 3$, $k_3 = 2$ corresponds to the case of the octahedral point group O_{24} that we utilize here as an illustrative example. Focusing on that case, let moreover $\mathfrak{A}$, $\mathfrak{B}$ denote the explicit $n \times n$ matrix representations of the two abstract generators in the $n \times n$ dimensions in which the corresponding lattice Λ lives. In this case $n = 3$ and the lattice is the cubic one.

The next step consists of introducing translation deformations of the generators of the point group $\mathfrak{P}_\Lambda$ that cannot be eliminated by conjugation with elements of the normal subgroup $T \lhd \mathrm{Aut}(\Lambda)$ and satisfy the relations modulo the lattice normal subgroup. We write

$$\hat{\mathbf{A}} = \left(\begin{array}{c|c} \mathfrak{A} & \tau \\ \hline 0 & 1 \end{array}\right); \quad \hat{\mathbf{B}} = \left(\begin{array}{c|c} \mathfrak{B} & \sigma \\ \hline 0 & 1 \end{array}\right) \tag{8.100}$$

and we impose the following relations:

$$\begin{aligned} (\hat{\mathbf{A}}.\hat{\mathbf{B}})^{k_1} &= \left(\begin{array}{c|c} \mathbf{1} & \boldsymbol{m}_1 \\ \hline 0 & 1 \end{array}\right); \quad \hat{\mathbf{A}}^{k_2} = \left(\begin{array}{c|c} \mathbf{1} & \boldsymbol{m}_2 \\ \hline 0 & 1 \end{array}\right); \\ \hat{\mathbf{B}}^{k_3} &= \left(\begin{array}{c|c} \mathbf{1} & \boldsymbol{m}_3 \\ \hline 0 & 1 \end{array}\right); \quad \boldsymbol{m}_{1,2,3} \in \Lambda. \end{aligned} \tag{8.101}$$

In the next subsection we work out the example of O_{24} in detail.

8.5.4.1 Frobenius congruences for the octahedral group O_{24}

The explicit form of equation (8.100) in the considered case is the following one:

$$\hat{\mathbf{A}} = \left(\begin{array}{ccc|c} 0 & 1 & 0 & \tau_1 \\ 0 & 0 & 1 & \tau_2 \\ 1 & 0 & 0 & \tau_3 \\ \hline 0 & 0 & 0 & 1 \end{array}\right); \quad \hat{\mathbf{B}} = \left(\begin{array}{ccc|c} 0 & 0 & 1 & \sigma_1 \\ 0 & -1 & 0 & \sigma_2 \\ 1 & 0 & 0 & \sigma_3 \\ \hline 0 & 0 & 0 & 1 \end{array}\right). \tag{8.102}$$

Next we impose on the deformed generators the defining relations of O_{24}. By explicit calculation we find

$$\hat{\mathbf{A}}^3 = \left(\begin{array}{ccc|c} 1 & 0 & 0 & \tau_1+\tau_2+\tau_3 \\ 0 & 1 & 0 & \tau_1+\tau_2+\tau_3 \\ 0 & 0 & 1 & \tau_1+\tau_2+\tau_3 \\ \hline 0 & 0 & 0 & 1 \end{array}\right); \quad \hat{\mathbf{B}}^2 = \left(\begin{array}{ccc|c} 1 & 0 & 0 & \sigma_1+\sigma_3 \\ 0 & 1 & 0 & 0 \\ 0 & 0 & 1 & \sigma_1+\sigma_3 \\ \hline 0 & 0 & 0 & 1 \end{array}\right);$$

$$(\hat{\mathbf{B}}\hat{\mathbf{A}})^4 = \left(\begin{array}{ccc|c} 1 & 0 & 0 & 4\sigma_1+4\tau_3 \\ 0 & 1 & 0 & 0 \\ 0 & 0 & 1 & 0 \\ \hline 0 & 0 & 0 & 1 \end{array}\right)$$

so that we obtain the conditions

$$\tau_1+\tau_2+\tau_3 \in \mathbb{Z}; \quad \sigma_1+\sigma_3 \in \mathbb{Z}; \quad 4\sigma_1+4\tau_3 \in \mathbb{Z}, \tag{8.103}$$

which are the Frobenius congruences for the present case. Next we consider the effect of conjugation with the most general translation element of the group $T \lhd \mathrm{Aut}_{\Lambda_{\text{cubic}}}$. Just for convenience we parameterize the translation subgroup as

$$\mathrm{t} = \left(\begin{array}{ccc|c} 1 & 0 & 0 & a+c \\ 0 & 1 & 0 & b \\ 0 & 0 & 1 & a-c \\ \hline 0 & 0 & 0 & 1 \end{array}\right), \tag{8.104}$$

and we get

$$\mathbf{t}\hat{\mathbf{A}}\mathbf{t}^{-1} = \left(\begin{array}{ccc|c} 0 & 1 & 0 & a-b+c+\tau_1 \\ 0 & 0 & 1 & -a+b+c+\tau_2 \\ 1 & 0 & 0 & \tau_3 - 2c \\ \hline 0 & 0 & 0 & 1 \end{array}\right); \quad \mathbf{t}\hat{\mathbf{B}}\mathbf{t}^{-1} = \left(\begin{array}{ccc|c} 0 & 0 & 1 & 2c+\sigma_1 \\ 0 & -1 & 0 & 2b+\sigma_2 \\ 1 & 0 & 0 & \sigma_3 - 2c \\ \hline 0 & 0 & 0 & 1 \end{array}\right). \tag{8.105}$$

This shows that by using the parameters b, c we can always put $\sigma_1 = \sigma_2 = 0$, while using the parameter a we can put $\tau_1 = 0$ (this is obviously only one possible gauge choice, yet it is the most convenient) so that Frobenius congruences reduce to

$$\tau_2 + \tau_3 \in \mathbb{Z}; \quad \sigma_3 \in \mathbb{Z}; \quad 4\tau_3 \in \mathbb{Z}. \tag{8.106}$$

Equation (8.106) is of great momentum. It tells us that any non-trivial subgroup $\hat{\mathfrak{P}} \subset \mathrm{Aut}_{\Lambda_{\mathrm{cubic}}}$ which is isomorphic to the point group O_{24} but not conjugate to it contains point group elements extended with rational translations of the form $\mathfrak{c} = \{\frac{n_1}{4}, \frac{n_2}{4}, \frac{n_3}{4}\}$ with $n_i \in \mathbb{Z}$.

8.5.4.2 The example of the group GS$_{24}$

An example is provided by the group named GS$_{24}$ in [71, 85]. In the direct product realization of the point group $\mathfrak{P} = \mathrm{O}_{24}$ the generators $\mathfrak{A}$ and $\mathfrak{B}$ were specified in equations (4.104) and (4.106). In view of the Frobenius congruences let us set

$$\hat{\mathbf{A}} = \begin{pmatrix} 0 & 0 & 1 & 0 \\ 1 & 0 & 0 & \frac{1}{2} \\ 0 & 1 & 0 & \frac{1}{2} \\ 0 & 0 & 0 & 1 \end{pmatrix}; \quad \hat{\mathbf{B}} = \begin{pmatrix} 0 & 0 & 1 & \frac{3}{2} \\ 0 & -1 & 0 & \frac{1}{2} \\ 1 & 0 & 0 & \frac{1}{2} \\ 0 & 0 & 0 & 1 \end{pmatrix}. \tag{8.107}$$

By an immediate calculation we obtain

$$\begin{gathered} \hat{\mathbf{A}}^3 = \begin{pmatrix} 1 & 0 & 0 & 1 \\ 0 & 1 & 0 & 1 \\ 0 & 0 & 1 & 1 \\ 0 & 0 & 0 & 1 \end{pmatrix}; \quad \hat{\mathbf{B}}^2 = \begin{pmatrix} 1 & 0 & 0 & 2 \\ 0 & 1 & 0 & 0 \\ 0 & 0 & 1 & 2 \\ 0 & 0 & 0 & 1 \end{pmatrix}; \\ (\hat{\mathbf{B}} \cdot \hat{\mathbf{A}})^4 = \begin{pmatrix} 1 & 0 & 0 & 0 \\ 0 & 1 & 0 & 0 \\ 0 & 0 & 1 & 2 \\ 0 & 0 & 0 & 1 \end{pmatrix}. \end{gathered} \tag{8.108}$$

The above equation is interpreted by stating that

$$\hat{\mathbf{A}}^3 \in \Lambda \subset \mathfrak{S}_{\mathrm{GS}}; \quad \hat{\mathbf{B}}^2 \in \Lambda \subset \mathfrak{S}_{\mathrm{GS}}; \quad (\hat{\mathbf{B}} \cdot \hat{\mathbf{A}})^4 \in \Lambda \subset \mathfrak{S}_{\mathrm{GS}}, \tag{8.109}$$

where $\mathfrak{S}_{GS}$ is the space group in the exact sequence

$$0 \xrightarrow{\iota} \Lambda \xrightarrow{\iota} \mathfrak{S}_{\mathrm{GS}} \xrightarrow{\pi} \mathrm{GS}_{24} \xrightarrow{\pi} \mathbf{1} \tag{8.110}$$

and the lattice normal subgroup is realized within $\mathfrak{S}_{\mathrm{GS}}$ by all the matrices of the form

$$\mathfrak{S}_{\mathrm{GS}} \rhd \Lambda \ni \left(\begin{array}{ccc|c} 1 & 0 & 0 & n_1 \\ 0 & 1 & 0 & n_2 \\ 0 & 0 & 1 & n_3 \\ \hline 0 & 0 & 0 & 1 \end{array}\right); \quad n_i \in \mathbb{Z}. \tag{8.111}$$

The group GS_{24} is defined as the quotient group

$$\mathrm{GS}_{24} = \frac{\mathfrak{S}_{\mathrm{GS}}}{\Lambda} \sim \mathrm{O}_{24} \sim \mathrm{S}_4 \tag{8.112}$$

and $\mathfrak{S}_{\mathrm{GS}}$ is a group extension of the lattice group Λ by means of the abstract octahedral point group O_{24}, yet it is not a semi-direct product of the normal subgroup Λ with O_{24}. Indeed, the space group $\mathfrak{S}_{\mathrm{GS}}$ contains translations that do not belong to the cubic lattice.

8.5.4.3 The idea of the universal classifying group

In [71, 85], inspired by Frobenius congruences, a new idea was introduced that proved to be quite relevant in the hydrodynamical applications of crystallographic groups, yet it has a much more general significance. Instead of constructing explicitly various instances of space groups, in [71] the construction of a much larger discrete group associated with the pair $(\Lambda, \mathfrak{P}_\Lambda)$ of a lattice Λ and the largest group $\mathfrak{P}_\Lambda \subset \mathrm{SO}_g(n)$ that leaves such a lattice invariant was considered; the resulting group, named the *universal classifying group*, includes all the fractional translations allowed by Frobenius congruences and turns out to be the smallest group that includes all available space groups with the same data $(\Lambda, \mathfrak{P}_\Lambda)$, together with other subgroups that are not space groups in the crystallographic sense.

The universal classifying group for the cubic lattice: G_{1536}

The idea is best illustrated by the cubic case. Inspired by Frobenius congruences we just consider the subgroup of $\mathrm{Aut}(\Lambda_{\mathrm{cubic}})$ where translations are quantized in units of $\frac{1}{4}$ and then we take the quotient with respect to the lattice normal subgroup. Doing so, in each direction and modulo integers, there are just four translations, $0, \frac{1}{4}, \frac{1}{2}, \frac{3}{4}$, so that the translation subgroup reduces to $\mathbb{Z}_4 \otimes \mathbb{Z}_4 \otimes \mathbb{Z}_4$, which has a total of 64 elements. In this way we single out a discrete subgroup $\mathrm{G}_{1536} \subset \mathrm{Aut}(\Lambda_{\mathrm{cubic}})$ of order $24 \times 64 = 1536$, which is simply the semi-direct product of the point group O_{24} with $\mathbb{Z}_4 \otimes \mathbb{Z}_4 \otimes \mathbb{Z}_4$:

$$\mathfrak{G}_{\mathrm{cubic}} \supset \mathrm{G}_{1536} \simeq \mathrm{O}_{24} \ltimes (\mathbb{Z}_4 \otimes \mathbb{Z}_4 \otimes \mathbb{Z}_4). \tag{8.113}$$

The universal classifying group G_{1536} was exhaustively studied in [71, 85], and we refer the reader to such papers for further information.

8.5.5 Another example in the plane: the tetragonal lattice

As another simple example we consider in $n = 2$ dimensions the tetragonal lattice. With reference to Section 8.1.4 and to item 12 in the list of point groups we consider the dihedral group Dih_4 having the two generators A_4 and B_4 there displayed that satisfy the relations in equation (8.17). We can immediately write their translation deformation as

$$\hat{A}_4 = \begin{pmatrix} 0 & -1 & \alpha_1 \\ 1 & 0 & \alpha_2 \\ 0 & 0 & 1 \end{pmatrix}; \quad \hat{B}_4 = \begin{pmatrix} 1 & 0 & \beta_1 \\ 0 & -1 & \beta_2 \\ 0 & 0 & 1 \end{pmatrix}. \tag{8.114}$$

By means of conjugation with a generic translation we can put $\alpha_1 = \alpha_2 = 0$. Then the Frobenius congruence reduces to

$$2\beta_1 \in \mathbb{Z}; \quad \beta_1 - \beta_2 \in \mathbb{Z}. \tag{8.115}$$

Up to the lattice $\mathbb{Z} \times \mathbb{Z}$ there are only two solutions to the congruence equation (8.115), namely,

First solution	$\beta_1 = 0 + \mathbb{Z}$	$\beta_2 = 0 + \mathbb{Z}$
Second solution	$\beta_1 = \frac{1}{2} + \mathbb{Z}$	$\beta_2 = -\frac{1}{2} + \mathbb{Z}$

(8.116)

This simple calculation agrees with the results presented elsewhere on group cohomology stating that there are two distinct cohomology classes for the tetragonal lattice.

The famous 17 wallpaper groups of plane tessellations can be derived in such a simple way by writing the Frobenius congruences for each of the 13 point groups listed in Section 8.1.4. One cohomology class is always the trivial one corresponding to the direct product; when other classes exist, as in the present case, we have additional space groups.

8.6 Bibliographical note

Lattices and crystallographic groups are discussed in many textbooks not only in mathematics, but also in chemistry and crystallography. Two excellent mathematical books are the monographs by Humphrey [96, 23]. The explicit constructions concerning the octahedral group are taken from the already quoted paper [71] and from [85]; furthermore, the book [97] was very useful for the writing of the section on space groups of the present chapter.

9 Monodromy groups of differential equations

> ... the theory of differential equations is the most important discipline in modern mathematics.
>
> Sophus Lie, 1872

The present chapter is devoted to presenting a still different incarnation of discrete groups, namely, their appearance in the capacity of monodromy groups associated with linear differential equations in complex variables and with the analytic prolongation of their solutions around points of the complex plane which are singular points of the equation.

9.1 The role of differential equations in all branches of science

Differential equations are ubiquitous in all branches of science and are the typical way of formulating the basic laws of most theories. Indeed, the predictions of a theory are normally expressed by some functions $f_i(x_1, x_2, \dots)$ of one or more variables and the theory imposes a set of differential constraints that should be satisfied by those functions. The solutions of these equations obeying a complete set of boundary conditions are typically what encodes our knowledge of the mechanisms underlying certain processes.

The Newton laws of classical mechanics are provided by a set of differential equations of the second order that are named after Euler and Lagrange. It was Hamilton, the same who discovered quaternions, who substituted the second order Lagrange equations with a set of twice as many differential equations of the first order, inventing Hamiltonian mechanics. The very idea of *causality*, which is central to classical physics, is precisely rooted in the theory of differential equations: the boundary conditions are the *cause* of what will happen next and the way this cause acts is via *a rule* provided by a differential constraint.

The most clear formulation of classical determinism was provided by Laplace, who said:

> We may regard the present state of the universe as the effect of its past and the cause of its future. An intellect which at a certain moment would know all forces that set nature in motion, and all positions of all items of which nature is composed, if this intellect were also vast enough to submit these data to analysis, would embrace in a single formula the movements of the greatest bodies of the universe and those of the tiniest atom; for such an intellect nothing would be uncertain and the future just like the past would be present before its eyes.

As is well known, the advent of quantum mechanics undermined this kind of unshakeable faith in the relations between cause and effect, yet it did not overthrow the throne occupied by differential equations. Indeed, the basic law of quantum mechanics is once again a differential equation, namely, the Schrödinger equation for the wave function. Apart from that, differential equations pop up everywhere in engineering of all types,

https://doi.org/10.1515/9783111201535-009

in chemistry, in electronics, and in the social and economical sciences. Also in these environments the issue of the causal relation between the initial data and future developments is a basic one.

9.2 Groups in a new capacity

Independently from their realm of application, a different viewpoint emerges on differential equations when we look at them as differential constraints to be satisfied by *complex analytic functions*: the emphasis on the causal chain from boundary conditions to actual solutions is diminished in favor of the question about *singular points* of the possible solutions and *symmetry groups* that connect different solutions to one other.

As we already mentioned in the preface, Sophus Lie discovered the Lie groups while considering transformations that support the solution of differential equations. He had the highest respect for the latter and he wrote: *the theory of differential equations is the most important discipline in modern mathematics.*

The aspect of linear differential equations we will mostly discuss in the present chapter is the relation between singularities in the coefficients of the equation and singularities of its solutions.

The systematic classification of such singularities leads to an extremely rich theory that displays the following challenging features:

(a) Restricting the type of considered singularities to those named Fuchsian, one discovers, as solutions, the largest majority of the special functions of classical analysis that appear ubiquitously in all branches of pure and applied sciences.
(b) The so-called Fuchsian differential equations have unexpected deep relations with the theory of groups, both continuous and finite.
(c) The Fuchsian equations with three singular points, which can be moved around the Riemann sphere by the automorphism group $\mathrm{PSL}(2,\mathbb{C})$ of the latter, have profound relations with the Kleinian subgroups of $\mathrm{SU}(2)$ and with the algebraic geometry of tori and Riemann surfaces in general. The key item in this relation is the concept of monodromy group.

Hence the present chapter is devoted to enlightening the role of groups in such a capacity. As usual we focus on the simplest case in order to illustrate the basic ideas, namely, holomorphic functions of just one variable. Similar concepts apply to more complicated cases of differential equations living on other multi-dimensional complex varieties. Furthermore, utilizing an explicit example we show how a finite group, the tetrahedral T_{12} that we studied in Chapter 8 in its capacity as a crystallographic point group, reappears in a totally different incarnation in the context of differential equations related to an algebraic variety as the complex torus. Our aim is to illustrate the profound connection of the theory of linear differential equations with other branches of mathematics that connect analysis with group theory, complex algebraic geometry, and algebraic topology, in particular homology.

9.3 Ordinary differential equations

An ordinary differential equation of the N-th order is a relation of the form

$$F\left(z, u, \frac{du}{dz}, \dots, \frac{d^N u}{dz^N}\right) = 0. \tag{9.1}$$

Equation (9.1) is called a *linear differential equation* if F is a linear function of its arguments:

$$u, \frac{du}{dz}, \dots, \frac{d^N u}{dz^N}.$$

The most general form of such an equation is

$$q(z) + r(z)u + s(z)\frac{du}{dz} + \dots + t(z)\frac{d^N u}{dz^N} = 0 \quad (t(z) \neq 0). \tag{9.2}$$

9.4 Second order differential equations with singular points

For several reasons an outstanding subclass of homogeneous linear differential equations in one variable is that of the equations of the second order. Their general form is the following:

$$a(z)\frac{d^2 u}{dz^2} + b(z)\frac{du}{dz} + c(z)u(z) = 0; \quad a(z) \neq 0. \tag{9.3}$$

Let $u_1(z)$ and $u_2(z)$ be two solutions of equation (9.3). Since the equation is linear, also the linear combination $c_1 u_1 + c_2 u_2$ with $c_{1,2}$ two arbitrary constants is a solution of the same equation. If $u_{1,2}$ are linear independent, then any solution of (9.3) can be expressed as a linear combination thereof. The two constants reflect the need to fix boundary conditions in order to specify a unique solution; in the case of a second order equation such boundary conditions are necessarily two, for instance the value of the function and its first derivative at some reference point $z = z_0$. How do we establish the linear independence of two solutions? The answer to this question follows from the very definition of linear independence. Two solutions are linear independent *iff*

$$\alpha u_1(z) + \beta u_2(z) = 0 \quad \Rightarrow \quad \alpha = \beta = 0. \tag{9.4}$$

From equation (9.4), by derivation, we also have the following condition:

$$\alpha\frac{du_1(z)}{dz}+\beta\frac{du_2(z)}{dz}=0 \quad\Rightarrow\quad \alpha=\beta=0. \tag{9.5}$$

This means that the question of linear dependence of two solutions is formulated as the question whether a 2-vector $\mathbf{v} = \{\alpha,\beta\}$ might exist which is a null eigenvector of the matrix

$$\mathbf{W}(u_1,u_2)=\begin{pmatrix} u_1 & u_2 \\ u_1' & u_2' \end{pmatrix}.$$

Obviously no such null eigenvector exists *iff* the determinant of $\mathbf{W}$ does not vanish. This leads to the notion of *Wronskian* of two solutions,

$$W(u_1,u_2)\equiv \mathbf{W}(u_1,u_2)=u_1u_2'-u_2u_1', \tag{9.6}$$

and two solutions $u_{1,2}$ are linear independent and form a basis for the two-dimensional vector space of solutions *iff* their Wronskian does not vanish, $W(u_1,u_2)\neq 0$.

In the case $W(u_1,u_2)=0$, we have

$$\frac{u_2'}{u_2}=\frac{u_1'}{u_1} \quad\Rightarrow\quad \log(u_2)=\log(u_1)+\log(\text{const}) \quad\Rightarrow\quad u_2=\text{const}\times u_1.$$

Dividing equation (9.3) by the non-vanishing coefficient function $a(z)$ we can rewrite it in the following general form, which is traditional in the mathematical literature:

$$\frac{d^2u}{dz^2}+p(z)\frac{du}{dz}+q(z)u(z)=0. \tag{9.7}$$

We limit our considerations to the cases where the functions $p(z)$ and $q(z)$ are analytic in an open region of the complex plane $D\subset\mathbb{C}$, except perhaps at an enumerable number of points of D where these functions may have isolated singularities.

If $p(z)$ and $q(z)$ are analytic at a point $z_0\in D$, then z_0 is named an *ordinary point* of the differential equation. It can be proven (see for instance [81], p. 291 and following ones) that in a neighborhood of an ordinary point there always exists a solution $u(z)$ of the differential equation which is regular analytic,

$$u(z)=\sum_{n=0}^{\infty}c_n(z-z_0)^n, \tag{9.8}$$

and satisfies prescribed arbitrary boundary conditions. This means that in the neighborhood of a regular point there is a basis of solutions u_1, u_2 that are both holomorphic in that neighborhood. Since, by the Liouville theorem, the only holomorphic function over the entire Riemann surface is the constant, the functions $p(z)$, $q(z)$, if they are non-trivial, must have singularities somewhere. It is of utmost interest to study the solutions of the differential equation in the neighborhood of such singular points.

Definition 9.4.1. If z_0 is an isolated singularity of either $p(z)$ or $q(z)$, or both, z_0 is called a *singular point* of the equation.

In general, at least one of the solutions of equation (9.7) has a singularity at $z = z_0$. It turns out that the behavior of the solutions of equation (9.7) in the neighborhood of a singular point z_0 depends in a crucial way upon the nature of the singularities of $p(z)$ or $q(z)$ or both at that point.

Definition 9.4.2. If at $z = z_0$ the function $p(z) \simeq \frac{p_{-1}}{z-z_0} + \mathcal{O}((z-z_0))$ has a pole of order not greater than one and $q(z) \simeq \frac{q_{-2}}{(z-z_0)^2} + \frac{q_{-1}}{(z-z_0)} + \mathcal{O}((z-z_0))$ has there a pole not greater than 2, then z_0 is called a *regular singular point* of the equation. In all other cases z_0 is called an *irregular singular point*.

We will show that when z_0 is a regular singular point, the two fundamental solutions of equation (9.7) have either the form

$$\begin{aligned} u_1(z) &= (z-z_0)^{r_1} \sum_{n=0}^{\infty} c_n (z-z_0)^n, \\ u_2(z) &= (z-z_0)^{r_2} \sum_{n=0}^{\infty} d_n (z-z_0)^n, \end{aligned} \tag{9.9}$$

where $r_{1,2}$ are the two distinct roots of a quadratic polynomial $\mathscr{P}_I(r)$ constructed from the coefficients of the Laurent series representation of $p(z)$ and $q(z)$, or

$$\begin{aligned} u_1(z) &= (z-z_0)^{r_1} \sum_{n=0}^{\infty} c_n (z-z_0)^n, \\ u_2(z) &= (z-z_0)^{r_2} \sum_{n=0}^{\infty} d_n (z-z_0)^n + \text{const} \times \log(z-z_0) u_1(z). \end{aligned} \tag{9.10}$$

As we will see, the first case occurs when the two roots $r_{1,2}$ are distinct and do not differ by an integer number. The second case occurs when such a condition is not verified.

It can be shown that at irregular singular points at least one of the two solutions develops an essential singularity with an infinite number of negative terms in its Laurent expansion.

9.4.1 Solutions at regular singular points and the indicial equation

Let z_0 be a regular singular point and define

$$A(z) = (z-z_0)p(z); \quad B(z) = (z-z_0)^2 q(z). \tag{9.11}$$

By definition $A(z)$ and $B(z)$ are regular analytic in an open neighborhood D_0 of z_0 and there they admit a Taylor expansion:

$$A(z) = \sum_{n=0}^{\infty} a_n (z - z_0)^n; \quad B(z) = \sum_{n=0}^{\infty} b_n (z - z_0)^n;$$
$$a_n = \frac{1}{2\pi i} \oint_{z_0} \frac{A(z)}{(z - z_0)^{n+1}}\, dz; \quad b_n = \frac{1}{2\pi i} \oint_{z_0} \frac{B(z)}{(z - z_0)^{n+1}}\, dz, \tag{9.12}$$

where $\oint_{z_0}$ denotes the integral along a path contained in D_0 which encircles the singular point z_0.

We try to find a solution of the differential equation (9.7) in the form of a series:

$$u(z) = (z - z_0)^r \sum_{n=0}^{\infty} c_n (z - z_0)^n, \tag{9.13}$$

where r and c_n are constants to be determined. Inserting the ansatz (9.13) into the differential equation, after multiplication by $(z - z_0)^{2-r}$ we obtain

$$\begin{aligned} 0 &= \sum_{n=0}^{\infty} (n + r)(n + r - 1)(z - z_0)^n c_n \\ &+ \left\{ \sum_{m=0}^{\infty} a_m (z - z_0)^m \right\} \left\{ \sum_{\ell=0}^{\infty} c_\ell (\ell + r)(z - z_0)^\ell \right\} \\ &+ \left\{ \sum_{m=0}^{\infty} a_m (z - z_0)^m \right\} \left\{ \sum_{\ell=0}^{\infty} c_\ell (z - z_0)^\ell \right\}. \end{aligned} \tag{9.14}$$

Multiplying the power series in the first and in the second term of equation (9.14) one gets

$$\sum_{n=0}^{\infty} \left\{ (n + r)(n + r - 1)c_n + \sum_{m=0}^{n} [(m + r)a_{n-m} + b_{n-m}] \right\} (z - z_0)^n = 0, \tag{9.15}$$

so that we find the condition

$$0 = [(n + r)(n + r - 1) + (n + r)a_0 + b_0]c_n + \sum_{m=0}^{n-1} [(m + r)a_{n-m} + b_{n-m}]. \tag{9.16}$$

Introducing the functions

$$\begin{aligned} \lambda_0(x) &= x(x - 1) + xa_0 + b_0, \\ \lambda_i(x) &= \begin{cases} xa_i + b_i & i > 0, \\ 0 & i < 0, \end{cases} \end{aligned} \tag{9.17}$$

equation (9.16) can be rewritten as follows:

$$\forall n \geq 0: \quad \lambda_0(r + n)c_n = - \sum_{m=0}^{n-1} \lambda_{n-m}(r + m)c_m. \tag{9.18}$$

From equation (9.18) we reach the following conclusion. The case $n = 0$ implies

$$\lambda_0(r)c_0 = 0, \tag{9.19}$$

which requires $\lambda_0(r) = 0$ in order for the coefficient c_0 to remain arbitrary and non-vanishing. Indeed, by definition c_0 is just the first non-vanishing coefficient of the series, having already factored out $(z - z_0)^r$. This is the announced quadratic indicial equation

$$0 = \mathscr{P}(r) \equiv r(r-1) + ra_0 + b_0 = (r - r_1)(r - r_2). \tag{9.20}$$

Let us order the two roots in such a way that

$$\operatorname{Re} r_1 \geq \operatorname{Re} r_2. \tag{9.21}$$

With this convention it follows that $\lambda_0(r_1 + n)$ does not vanish for any integer $n > 0$. Hence, the recurrence relation (9.18) for the coefficients can always be solved for the next coefficient c_n once the coefficients $c_0, c_1, \ldots, c_{n-1}$ are known. That the resulting series is convergent in an open neighborhood of z_0 can be proven by showing that each of its terms is smaller than or equal to the corresponding term of a geometric series which is known to be convergent. We omit this proof, which can be found on p. 299 of [81]. In the case that $r_1 - r_2 \notin \mathbb{N}$ also $\lambda_0(r_2 + n)$ never vanishes for any positive integer $n > 0$ so that the same conclusion holds true also for the second root, and we obtain a second solution $u_2(z)$ as described in equation (9.9).

9.4.1.1 The second solution with a logarithmic singularity

In the case the two roots of the indicial equation differ by an integer, the solution of the recurrence relations breaks down at $n = N$, where $r_2 + N = r_1$, since there $\lambda_0(r_2 + N) = 0$. In this case, in order to find the second linear independent solution, we utilize the *method of variation of the constants* and we analyze the differential equation that has to be satisfied by the ratio of two solutions of equation (9.7). Setting

$$h(z) \equiv \frac{u_2(z)}{u_1(z)}, \tag{9.22}$$

we get

$$\frac{d}{dz}\log\left(\frac{dh}{dz}\right) = -p(z) - 2\frac{d}{dz}\log(u_1(z)). \tag{9.23}$$

By means of a first integration we find

$$\frac{d}{dz}\left(\frac{u_2}{u_1}\right) = \text{const} \times \frac{1}{u_1^2} \times \exp\left(-\int_{z_0}^{z} p(t)dt\right). \tag{9.24}$$

Since $p(z) = \frac{a_0}{z-z_0} + \sum_{m=0}^{\infty} a_m (z-z_0)^{m-1}$, we conclude that

$$-\int_{z_0}^{z} p(t)dt = -a_0 \log(z-z_0) - \sum_{n=1}^{\infty} \frac{a_n}{n}(z-z_0)^n. \tag{9.25}$$

Using the first of equation (9.10) for the solution u_1 we obtain

$$\frac{d}{dz}\left(\frac{u_2}{u_1}\right) = \text{const} \times (z-z_0)^{-a_0-2r_1} K(z), \tag{9.26}$$

$$K(z) = \frac{\exp(-\sum_{n=1}^{\infty} \frac{a_n}{n}(z-z_0)^n)}{\{\sum_{\ell=0}^{\infty} c_\ell (z-z_0)^n\}^2}. \tag{9.27}$$

The function $K(z)$ is certainly analytic in the neighborhood of z_0 and it can be developed in Taylor series:

$$K(z) = \sum_{m=0}^{\infty} k_m (z-z_0)^m. \tag{9.28}$$

Since $r_1 + r_2 = 1 - a_0$, as follows from equation (9.20) we get

$$2r_1 + a_0 = 1 + N, \tag{9.29}$$

which implies

$$\frac{d}{dz}\left(\frac{u_2}{u_1}\right) = \text{const} \times \left(\frac{d_0}{(z-z_0)^{N+1}} + \frac{d_1}{(z-z_0)^N} + \cdots + \frac{d_N}{z-z_0} + d_{N+1} + d_{N+2}(z-z_0) + \cdots\right). \tag{9.30}$$

By integration one obtains

$$\frac{u_2}{u_1} = \text{const} \times \log(z-z_0) + (z-z_0)^{-N} \sum_{\ell=0}^{\infty} f_\ell (z-z_0)^\ell, \tag{9.31}$$

where f_ℓ is a new set of coefficients. This is the proof of the second equation of (9.10).

9.4.2 Fuchsian equations with three regular singular points and the *P*-symbol

A *Fuchsian equation* is an equation that has only regular singular points. We consider the most general form of a Fuchsian equation where the number of singular points is just three. The choice of this number is not erratic. Three is the number of complex parameters of the Möbius group $\mathrm{PSL}(2,\mathbb{C})$ of conformal automorphisms of the compactified complex plane (or Riemann sphere) $\mathbb{C} \cup \{\infty\} = \mathbb{S}^2_{\text{Rie}}$ that has a transitive action on it. By means of a suitable conformal transformation any triplet of points $\{\xi_1, \xi_2, \xi_3\}$ of $\mathbb{S}^2_{\text{Rie}}$ can me mapped into the standard triplet $\{0, 1, \infty\}$. Indeed, if we introduce the 2×2 matrix

$$\begin{pmatrix} A & , & B \\ C & , & D \end{pmatrix} = \begin{pmatrix} \frac{\xi_2-\xi_3}{\sqrt{(\xi_1-\xi_3)(\xi_1-\xi_2)(\xi_3-\xi_2)}} & \frac{\xi_1(\xi_3-\xi_2)}{\sqrt{(\xi_1-\xi_3)(\xi_1-\xi_2)(\xi_3-\xi_2)}} \\ \frac{\xi_2-\xi_1}{\sqrt{(\xi_1-\xi_3)(\xi_1-\xi_2)(\xi_3-\xi_2)}} & \frac{\xi_3(\xi_1-\xi_2)}{\sqrt{(\xi_1-\xi_3)(\xi_1-\xi_2)(\xi_3-\xi_2)}} \end{pmatrix} \in \mathrm{SL}(2, \mathbb{C}), \tag{9.32}$$

which is unimodular since $AD - BC = 1$, the corresponding fractional linear map

$$\mathfrak{f}(z) \equiv \frac{Az + B}{Cz + D} = \frac{(\xi_3 - \xi_2)(z - \xi_1)}{(\xi_1 - \xi_2)(z - \xi_3)} \tag{9.33}$$

has the property that

$$\mathfrak{f}(\xi_1) = 0; \quad \mathfrak{f}(\xi_2) = 1; \quad \mathfrak{f}(\xi_3) = \infty. \tag{9.34}$$

In this way the parameters intrinsic to this type of equation are simply the characterization of the type of singularity occurring at each of the three singular points, their location being immaterial. As discussed in Section 9.4.1, a regular singular point is completely characterized by the two roots α, β of the corresponding indicial equation which is of the second degree. In this way one arrives at the provisional conclusion that any Fuchsian equation with three singular points should be characterized by a total of six complex parameters organized in a triplet of doublets $(\alpha, \beta)_i$ $(i = 1, 2, 3)$, namely, the two roots of the indicial equation for each of the three singularities. This leads to a very convenient notation due to Riemann and further elaborated by Papperitz that we anticipate. We write

$$P \begin{Bmatrix} \xi_1 & \xi_2 & \xi_3 & \\ \alpha_1 & \alpha_2 & \alpha_3 & z \\ \beta_1 & \beta_2 & \beta_3 & \end{Bmatrix} \tag{9.35}$$

to denote a solution of a differential equation of the second order in the variable z with three regular singular points located at $\xi_{1,2,3}$ having in each of them indices $(\alpha, \beta)_{1,2,3}$. Let us construct the form of such an equation that is named a *Papperitz–Riemann equation.* We start from the general form

$$u'' + p(z)u' + q(z)u = 0 \tag{9.36}$$

and assuming that all singular points ξ_i are finite we consider the condition on the coefficient functions $p(z)$, $q(z)$ for $\zeta = \infty$ to be a regular point,

$$p(z) \sim_{z\to\infty} \frac{2}{z}; \quad q(z) =_{z\to\infty} \mathcal{O}\left(\frac{1}{z^4}\right), \tag{9.37}$$

while in each point ξ_i the coefficient $p(z)$ must have a simple pole and $q(z)$ a double pole. The solution to this set of constraints is uniquely parameterized by six complex numbers C_i, D_i $(i = 1, \dots, 3)$, subject to one constraint:

$$\begin{aligned} p(z) &= \sum_{i=1}^{3} \frac{C_i}{z - \xi_i}, \\ q(z) &= \frac{1}{\prod_{i=1}^{3}(z - \xi_i)} \times \sum_{j=1}^{3} \frac{D_i}{z - \xi_i}, \\ 2 &= \sum_{i=1}^{3} C_i. \end{aligned} \tag{9.38}$$

Writing the indicial equation relative to each of the three singular points point we find

$$r^2 + (C_i - 1)r + \frac{D_i}{(\xi_i - \xi_k)(\xi_i - \xi_\ell)} = (r - \alpha_i)(r - \beta_i) = 0, \quad (i, k, \ell) \text{ in cyclic order,} \tag{9.39}$$

so that we have

$$\begin{aligned} C_i &= 1 - \alpha_i - \beta_i, \\ D_i &= \alpha_i \beta_i (\xi_i - \xi_k)(\xi_i - \xi_\ell); \quad i, k, \ell \text{ in cyclic order,} \\ 1 &= \sum_{i=1}^{3} (\alpha_i + \beta_i). \end{aligned} \tag{9.40}$$

The Papperitz–Riemann equation is therefore equation (9.36) with coefficient functions as in equation (9.38) upon substitution of equation (9.40).

Utilizing the P-symbol to denote a generic solution of the equation defined by the corresponding six exponents α_i, β_i and the three locations ξ_i of the singular points by means of a tedious yet straightforward calculation, one can verify the following fundamental symmetries of the Papperitz–Riemann equation:

(A) If r, s, t are three complex numbers satisfying the constraint $r + s + t = 0$, one has

$$(z - \xi_1)^r (z - \xi_2)^s (z - \xi_3)^t P \begin{Bmatrix} \xi_1 & \xi_2 & \xi_3 & \\ \alpha_1 & \alpha_2 & \alpha_3 & z \\ \beta_1 & \beta_2 & \beta_3 & \end{Bmatrix} = P \begin{Bmatrix} \xi_1 & \xi_2 & \xi_3 & \\ \alpha_1 + r & \alpha_2 + s & \alpha_3 + t & z \\ \beta_1 + r & \beta_2 + s & \beta_3 + t & \end{Bmatrix}. \tag{9.41}$$

This displays a translation symmetry of the indices, two of which can always be set to zero by a simple power rescaling of the searched for function $u(z)$.

(B) If $\mathfrak{g} = \begin{pmatrix} A & B \\ C & D \end{pmatrix} \in \mathrm{PSL}(2, \mathbb{C})$ is an arbitrary element of the Möbius group and we denote by $\mathfrak{g}(z) \equiv \frac{Az+B}{Cz+D}$ the corresponding conformal transformation of the complex plane, we have

$$P\begin{Bmatrix} \xi_1 & \xi_2 & \xi_3 & \\ \alpha_1 & \alpha_2 & \alpha_3 & z \\ \beta_1 & \beta_2 & \beta_3 & \end{Bmatrix} = P\begin{Bmatrix} \mathfrak{g}(\xi_1) & \mathfrak{g}(\xi_2) & \mathfrak{g}(\xi_3) & \\ \alpha_1 & \alpha_2 & \alpha_3 & \mathfrak{g}(z) \\ \beta_1 & \beta_2 & \beta_3 & \end{Bmatrix}. \tag{9.42}$$

Utilizing the above two transformations we can put the general Papperitz–Riemann equation into a standard form depending only on three parameters. Indeed, of the nine parameters $\xi_{1,2,3}$, $\alpha_{1,2,3}$, $\beta_{1,2,3}$, three can be disposed of by means of the conformal transformation (9.33) which brings the regular singular points to their standard position 0, 1, ∞ and another two can be disposed of by the arranging two indices being zero through equation (9.41). In this way we obtain

$$\begin{aligned} &P\begin{Bmatrix} \xi_1 & \xi_2 & \xi_3 & \\ \alpha_1 & \alpha_2 & \alpha_3 & z \\ \beta_1 & \beta_2 & \beta_3 & \end{Bmatrix} \\ &= \left(\frac{z-\xi_1}{z-\xi_3}\right)^{\alpha_1}\left(\frac{z-\xi_2}{z-\xi_3}\right)^{\alpha_2} P\begin{Bmatrix} 0 & 1 & \infty & \\ 0 & 0 & \alpha_3+\alpha_2+\alpha_1 & \mathfrak{f}(z) \\ \beta_1-\alpha_1 & \beta_2-\alpha_2 & \beta_3+\alpha_1+\alpha_2 & \end{Bmatrix}, \end{aligned} \tag{9.43}$$

where the transformation $\mathfrak{f}(z)$ is defined by equation (9.33). Then we introduce a renaming of the effective parameters, according to the definitions

$$a = \alpha_1+\alpha_2+\alpha_3; \quad b = \beta_3+\alpha_1+\alpha_2; \quad c = 1+\alpha_1-\beta_1, \tag{9.44}$$

and we conclude that the standard form of the Papperitz–Riemann equation, namely, of the Fuchsian equation with three regular singular points, is that encoded in the following P-symbol:

$$P\begin{Bmatrix} 0 & 1 & \infty & \\ 0 & 0 & a & z \\ 1-c & c-a-b & b & \end{Bmatrix}. \tag{9.45}$$

The corresponding differential equation is named the *hypergeometric equation.*

9.4.3 The hypergeometric equation and its solutions

As explained above, the hypergeometric equation is the standard form of a second order differential equation with three Fuchsian singular points encoded in the P-symbol of equation (9.45). It reads as follows:

$$z(1-z)\frac{d^2u}{dz^2} + [c-(a+b+1)z]\frac{du}{dz} - abu = 0. \tag{9.46}$$

Together with its generalizations to higher order it plays a very important role in all branches of mathematical physics, in complex and algebraic geometry, and also in engineering and other applied sciences. Its centrality is due to its deep relation with the Möbius group $\mathrm{PSL}(2,\mathbb{C})$ of projective transformations of the complex plane and to its discrete subgroups, related in turn to the algebraic topology of surfaces.

Following the general discussions of the previous sections we first consider the series solution of this equation, valid near the regular singular point at the origin. The roots of the indicial equation relative to the origin are 0 and $1-c$. Thus, there exists a solution that is analytic in the neighborhood of the origin which can be normalized to unity. We call this solution the *hypergeometric function* and we denote it $F(a,b;c;z)$. The singularities of $F(a,b;c;z)$ are located at the other two singular points of the equation, namely, at $z=1$ and at $z=\infty$. These are, in general, branch points, and in those cases we supplement the definition of $F(a,b;c;z)$ by taking the cut from $z=1$ to $z=\infty$ along the positive real axis.

Since $F(a,b;c;z)$ is analytic in the neighborhood of the origin, it may be represented there by a power series:

$$F(a,b;c;z)=\sum_{n=0}^{\infty}c_n z^n;\qquad c_0=1. \tag{9.47}$$

Inserting the ansatz (9.47) into equation (9.46), one obtains the recurrence relation

$$c_n=\frac{(a+n-1)(b+n-1)}{n(c+n-1)}c_{n-1}, \tag{9.48}$$

which is solved by

$$F(a,b;c;z)=\sum_{\ell=0}^{\infty}\frac{(a)_\ell(b)_\ell}{\ell!\,(c)_\ell}z^\ell, \tag{9.49}$$

$$(x)_\ell\equiv x(x+1)\dots(x+\ell-1)=\frac{\Gamma(x+\ell)}{\Gamma(\ell)};\qquad (x)_0=1. \tag{9.50}$$

The above formula is called the *hypergeometric series*. Clearly, its convergence radius is provided by the inequality $|z|<1$, since the nearest singularity to the origin is at $z=1$. The hypergeometric equation is invariant under the exchange $a\leftrightarrow b$, and this symmetry is obviously reflected in the solution, so that we have

$$F(a,b;c;z)=F(b,a;c;z). \tag{9.51}$$

Since the other root of the indicial equation is $1-c$, the second solution in the neighborhood of the origin is, provided that $1-c\notin\mathbb{Z}$, of the following type:

$$u_2=z^{1-c}g(z), \tag{9.52}$$

where $g(z)$ is a power series in z. Substituting (9.52) into (9.46) one finds that $g(z)$ satisfies the following equation:

$$z(1-z)\frac{d^2g}{dz^2}+\left[c-2+(a+b-2c+3)z\right]\frac{dg}{dz}-abg=0. \tag{9.53}$$

This is once again a hypergeometric equation of parameters:

$$a'=b-c+1;\quad b'=a-c+1;\quad c'=2-c. \tag{9.54}$$

Hence, in the neighborhood of $z=0$ a complete basis of solutions of the hypergeometric equation (9.46) is provided for $c\notin\mathbb{Z}$ by

$$\begin{aligned} u_1(z) &= F(a,b;c;z),\\ u_2(z) &= z^{1-c}F(b-c+1,a-c+1;2-c;z) \end{aligned} \tag{9.55}$$

In the case that $c\in\mathbb{Z}$, one element of the solution basis is the hypergeometric function $u_1(z)=F(a,b;c;z)$; the other needs to have a logarithmic singularity, and hence it must be of the form

$$u_2(z)=\log(z)F(a,b;c;z)+\sum_{n=0}^{\infty}c_nz^n. \tag{9.56}$$

9.4.3.1 The anharmonic group

Let us consider the following set of six 2×2 matrices:

$$\mathfrak{G}=\{\mathfrak{g}_1\dots,\mathfrak{g}_6\}, \tag{9.57}$$

where

$$\begin{aligned} \mathfrak{g}_1 &= \begin{pmatrix}1&0\\0&1\end{pmatrix}; & \mathfrak{g}_2 &= \begin{pmatrix}-1&1\\-1&0\end{pmatrix};\\ \mathfrak{g}_3 &= \begin{pmatrix}0&-1\\1&-1\end{pmatrix}; & \mathfrak{g}_4 &= \begin{pmatrix}-1&0\\-1&1\end{pmatrix};\\ \mathfrak{g}_5 &= \begin{pmatrix}0&1\\1&0\end{pmatrix}; & \mathfrak{g}_6 &= \begin{pmatrix}1&-1\\0&-1\end{pmatrix}. \end{aligned} \tag{9.58}$$

The set $\mathfrak{G}$ forms a group having the following multiplication table:

	$\mathfrak{g}_1$	$\mathfrak{g}_2$	$\mathfrak{g}_3$	$\mathfrak{g}_4$	$\mathfrak{g}_5$	$\mathfrak{g}_6$
$\mathfrak{g}_1$	$\mathfrak{g}_1$	$\mathfrak{g}_2$	$\mathfrak{g}_3$	$\mathfrak{g}_4$	$\mathfrak{g}_5$	$\mathfrak{g}_6$
$\mathfrak{g}_2$	$\mathfrak{g}_2$	$\mathfrak{g}_3$	$\mathfrak{g}_1$	$\mathfrak{g}_5$	$\mathfrak{g}_6$	$\mathfrak{g}_4$
$\mathfrak{g}_3$	$\mathfrak{g}_3$	$\mathfrak{g}_1$	$\mathfrak{g}_2$	$\mathfrak{g}_6$	$\mathfrak{g}_4$	$\mathfrak{g}_5$
$\mathfrak{g}_4$	$\mathfrak{g}_4$	$\mathfrak{g}_6$	$\mathfrak{g}_5$	$\mathfrak{g}_1$	$\mathfrak{g}_3$	$\mathfrak{g}_2$
$\mathfrak{g}_5$	$\mathfrak{g}_5$	$\mathfrak{g}_4$	$\mathfrak{g}_6$	$\mathfrak{g}_2$	$\mathfrak{g}_1$	$\mathfrak{g}_3$
$\mathfrak{g}_6$	$\mathfrak{g}_6$	$\mathfrak{g}_5$	$\mathfrak{g}_4$	$\mathfrak{g}_3$	$\mathfrak{g}_2$	$\mathfrak{g}_1$

(9.59)

This group is a finite subgroup of PSL(2, $\mathbb{C}$) and therefore of the conformal automorphism group of the complex plane, yet it is not a subgroup PSL(2, $\mathbb{Z}$). Indeed, as the reader can see, the determinants of the matrices in equation (9.58) are not all equal to 1 and they cannot be promoted to 1 by multiplication of the matrix with an integer matrix. This is instead possible within PSL(2, $\mathbb{C}$), since it suffices to multiply the matrix by an imaginary number. Yet the anharmonic group is related to PSL(2, $\mathbb{Z}$) in the following way. It is isomorphic to the quotient of PSL(2, $\mathbb{Z}$) with respect to one of his normal subgroups, the principal congruence subgroup Γ(2). Nevertheless, since the anharmonic group is part of the automorphism conformal group, we can consider the corresponding fractional linear transformations

$$\mathfrak{g}_i = \begin{pmatrix} A_i & B_i \\ C_i & D_i \end{pmatrix} \quad \Rightarrow \quad \mathfrak{g}_i(z) = \frac{A_i z + B_i}{C_i z + D_i}. \tag{9.60}$$

Doing so, we get a set of six transformations $\mathfrak{f}_i(z)$ with the property that the image through $\mathfrak{f}_i$ of each of the three singular points 0, 1, ∞ is again one of the same three, namely,

$$\forall \mathfrak{f}_i; \quad x \in \{0, 1, \infty\} \quad \Rightarrow \quad \mathfrak{f}_i(x) \in \{0, 1, \infty\}. \tag{9.61}$$

This means that with each element of the group $\mathfrak{G}$ we associate a permutation of the three *objects* $\{0, 1, \infty\}$ as follows:

$$w_i(z) = \mathfrak{g}_i(z) \quad \Rightarrow \quad \Pi \begin{pmatrix} 0 & 1 & \infty \\ \mathfrak{g}_i(0) & \mathfrak{g}_i(1) & \mathfrak{g}_i[\infty] \end{pmatrix}. \tag{9.62}$$

Explicitly we find the result displayed in Table 9.1. The arrows in Table 9.1 describe an isomorphism,

Table 9.1: Transformations of the anharmonic group.

$\mathfrak{g}_1(z) = z$	$\Rightarrow$	$\Pi\begin{pmatrix} 0 & 1 & \infty \\ 0 & 1 & \infty \end{pmatrix}$
$\mathfrak{g}_2(z) = \frac{z-1}{z}$	$\Rightarrow$	$\Pi\begin{pmatrix} 0 & 1 & \infty \\ \infty & 0 & 1 \end{pmatrix}$
$\mathfrak{g}_3(z) = \frac{1}{1-z}$	$\Rightarrow$	$\Pi\begin{pmatrix} 0 & 1 & \infty \\ 1 & \infty & 0 \end{pmatrix}$
$\mathfrak{g}_4(z) = \frac{z}{z-1}$	$\Rightarrow$	$\Pi\begin{pmatrix} 0 & 1 & \infty \\ 0 & \infty & 1 \end{pmatrix}$
$\mathfrak{g}_5(z) = \frac{1}{z}$	$\Rightarrow$	$\Pi\begin{pmatrix} 0 & 1 & \infty \\ \infty & 1 & 0 \end{pmatrix}$
$\mathfrak{g}_6(z) = 1 - z$	$\Rightarrow$	$\Pi\begin{pmatrix} 0 & 1 & \infty \\ 1 & 0 & \infty \end{pmatrix}$

$$\begin{aligned} \iota : \mathfrak{G} &\to S_3, \\ \forall \mathfrak{g}_1, \mathfrak{g}_\in \mathfrak{G} : \quad \iota(\mathfrak{g}_1 \cdot \mathfrak{g}_2) &= \iota(\mathfrak{g}_1) \cdot \iota(\mathfrak{g}_2), \\ \ker \iota &= \mathbf{e} \in \mathfrak{G}, \end{aligned} \tag{9.63}$$

between the group $\mathfrak{G}$, named the *anharmonic group*, and the permutation group of three objects that has the standard name of third symmetric group S_3.

Let us now reconsider equation (9.45) and apply it to the case of the transformations of the anharmonic group $\mathfrak{G}$; since after the transformation the singular points remain the same 0, 1, ∞, simply arranged in a different order, it follows that we still have a hypergeometric equation, just with permuted pairs of indices,

$$\begin{aligned} P\begin{Bmatrix} 0 & 1 & \infty & \\ 0 & 0 & a & z \\ 1-c & c-a-b & b & \end{Bmatrix} &= P\begin{Bmatrix} \Pi_i[0] & \Pi_i[1] & \Pi_i[\infty] & \\ 0 & 0 & a & \mathfrak{g}_i(z) \\ 1-c & c-a-b & b & \end{Bmatrix} \\ &= P\begin{Bmatrix} 0 & 1 & \infty & \\ \Pi_i^{-1}\begin{bmatrix} 0 \\ 1-c \end{bmatrix} & \Pi_i^{-1}\begin{bmatrix} 0 \\ c-a-b \end{bmatrix} & \Pi_i^{-1}\begin{bmatrix} a \\ b \end{bmatrix} & \mathfrak{g}_i(z) \end{Bmatrix}, \end{aligned} \tag{9.64}$$

having denoted by $\Pi_i[\dots]$ the permutation of the three objects 0, 1, ∞, induced by the i-th element of $\mathfrak{G}$. In this way one sees that the solutions of the hypergeometric equations can be expressed as hypergeometric functions of each of the six arguments

$$z, \frac{z-1}{z}, \frac{1}{1-z}, \frac{z}{z-1}, \frac{1}{z}, 1-z, \tag{9.65}$$

with new indices a_i, b_i, c_i that can be read off from equation (9.64). This observation implies that there are at least six different hypergeometric function solutions of the same hypergeometric equation. Yet this is not the end of the story. One additional set of transformations arises from the possibility of interchanging, within the Riemann symbol P, each α_i index with its corresponding index β_i. Indeed, α_i and β_i denote the two roots of the indicial equation at the i-th singular point and there is no intrinsic notion of what is the first and what is the second root. After one such transformation has been performed we can always reduce the new P-symbol to its standard hypergeometric form (9.45) by means of a transformation of type (9.41), if needed. This means that we have an extension of the anharmonic group by means of $2^3 = 8$ additional transformations of the type just mentioned. This seems to produce 48 different solutions of the hypergeometric equation provided in terms of hypergeometric functions of the six arguments in (9.65) with eight different sets of indices for each argument. Yet we should remember that any hypergeometric function $F(a, b; c, x) = F(b, a; c, x)$ is invariant under the exchange of a and b. This implies that only four of the above discussed eight transformations are effective, and this yields a total of 24 different solutions of the same hypergeometric equation in terms of hypergeometric functions. These solutions are named *Kummer's*

24 solutions. We can think of these solutions as the orbit of the original hypergeometric function $F(a, b; c; z)$ under the action of a group Γ of order 24 which is the semi-direct product of S_3 with $\mathbb{Z}_2 \times \mathbb{Z}_2$. One can show that Γ is isomorphic to the S_4 permutation group which is also the octahedral group O_{24}.

Obviously there must be relations among these 24 functions, since the space of solutions of the hypergeometric equation is in any case two-dimensional. These relations are extremely important because they provide the explicit token to perform the analytic continuation of the solutions from the original convergence disk $|z| < 1$ to the entire complex plane deprived of the singular points.

We mention a few of these relations whose coefficients can be derived by comparing the values of the two expressions of the same function at a pair of common points of their domain of definition. Three examples are given below:

$$
\begin{aligned}
F(a, b; c; z) &= (1 - z)^{c-a-b} F(c - a, c - b; c; z), \\
F(a, b; c; z) &= (1 - z)^{-a} F\left(a, c - b; c; \frac{z}{z - 1}\right), \\
F(a, b; c; z) &= (1 - z)^{-b} F\left(c - a, b; c; \frac{z}{z - 1}\right).
\end{aligned} \tag{9.66}
$$

Consider for instance the second of the identities in (9.66). That $(1-z)^{-a} F(a, c-b; c; \frac{z}{z-1})$ is a solution of the hypergeometric equation of parameters (a, b, c) is established by means of the action of the anharmonic group described above. Assuming $c \notin \mathbb{Z}$, a basis of solutions is provided by $u_1(z)$, $u_2(z)$ displayed in equation (9.55); hence, there must be two constants $c_{1,2}$ such that

$$
(1 - z)^{-a} F\left(a, c - b; c; \frac{z}{z - 1}\right) = c_1 u_1(z) + c_2 u_2(z). \tag{9.67}
$$

Analyzing the behavior of the function in the neighborhood of $z = 0$, we rapidly conclude that $c_1 = 1$, $c_2 = 0$. Another very important Kummer relation is the following one:

$$
\begin{aligned}
F(a, b; c; z) = {} & \frac{\Gamma(a + b - c)\Gamma(c)}{\Gamma(a)\Gamma(b)} (1 - z)^{c-a-b} F(c - a, c - b; c - a - b + 1; 1 - z) \\
& + \frac{\Gamma(c - a - b)\Gamma(c)}{\Gamma(c - a)\Gamma(c - b)} F(a, b; a + b - c + 1; 1 - z).
\end{aligned} \tag{9.68}
$$

The derivation works here in the opposite direction. By means of the anharmonic group transformations we arrive at the conclusion that the solutions of the hypergeometric equations of parameters a, b, and $a+b-c+1$ in the variable $\xi = 1-z$ are solutions of the hypergeometric equation of parameters a, b, and c in the variable z. Hence, we consider a basis of solutions of the equation with parameters a, b, and $a + b - c + 1$:

$$
\begin{aligned}
\mathbf{w}_1(1 - z) &= F(c - a, c - b; c - a - b + 1; 1 - z), \\
\mathbf{w}_2(1 - z) &= (1 - z)^{c-a-b} F(c - a, c - b; c - a - b + 1; 1 - z).
\end{aligned} \tag{9.69}
$$

Since $F(a, b; c; z)$ is a solution of the same equation, we must have

$$F(a, b; c; z) = c_1 \mathbf{w}_1(1 - z) + c_2 \mathbf{w}_2(1 - z) \tag{9.70}$$

for $c_{1,2}$ suitable constants. The values of $c_{1,2}$ are fixed to those displayed in equation (9.68) by comparing the behavior of the two expressions of the same function in the neighborhood of a common point of definition. In this case the determination of the two constants is more elaborate and we skip it.

9.5 An example of monodromy group: differential equations and topology

In Chapter 8 we studied crystallographic groups and in Section 8.3 we considered the tetrahedral group $T_{12} \sim A_4$ in its capacity of point group in three dimensions. Here we will meet the same abstract group in a totally different context as the quotient group of the so-called modular group of the torus $\Gamma = \mathrm{PSL}(2, \mathbb{Z})$ with respect to one of its normal subgroups $\Gamma(3)$ described in the next subsection. The normal subgroup $\Gamma(3)$, which is an example of a discrete, yet infinite group, will appear in the capacity of monodromy group of a certain differential equation. What monodromy means will become apparent in the discussion of the example.

9.5.1 The tetrahedral group in two dimensions and the torus

The connection with topology and differential equations occurs since the tetrahedral group T_{12} is isomorphic to the following quotient group:

$$T_{12} \sim \Gamma/\Gamma(3) = \mathrm{PSL}(2, \mathbb{Z}_3), \tag{9.71}$$

where Γ is the modular group, namely, the set of 2×2 integer valued, unimodular matrices (up to an overall sign)

$$\begin{aligned} &\gamma \in \Gamma \longrightarrow \gamma = \pm \begin{pmatrix} a & b \\ c & d \end{pmatrix}, \\ &a, b, c, d \in \mathbb{Z}, \quad ad - bc = 1, \\ &\Gamma = \mathrm{PSL}(2, \mathbb{Z}) = \frac{\mathrm{SL}(2, \mathbb{Z})}{\begin{pmatrix} \pm 1 & 0 \\ 0 & \pm 1 \end{pmatrix}}, \end{aligned} \tag{9.72}$$

while $\Gamma(3) \subset \Gamma$ is the *normal subgroup* of the modular group formed by those integer valued, unimodular matrices that are equivalent to the identity modulo 3

$$
\begin{aligned}
&\gamma \in \Gamma(3) \longrightarrow \gamma = \pm\begin{pmatrix} a & b \\ c & d \end{pmatrix}, \\
&a, b, c, d \in \mathbb{Z}, \quad ad - bc = 1, \\
&a = \pm 1 + 3k, \quad k \in \mathbb{Z}, \quad b = 3k, \quad k \in \mathbb{Z}, \\
&c = 3k, \quad k \in \mathbb{Z}, \quad d = \pm 1 + 3k, \quad k \in \mathbb{Z},
\end{aligned}
\tag{9.73}
$$

and $\mathrm{PSL}(2, \mathbb{Z}_3)$ is the modular group constructed on the field $\mathbb{Z}_3$ of integer numbers modulo 3.

Γ is named the *modular group* precisely because it is well known to be the complete symmetry group of the moduli space of the torus. A topological torus $\mathbb{S}^1 \otimes \mathbb{S}^1$ is described by two real coordinates ξ_1, ξ_2 defined modulo 2π, $\xi_i \approx \xi_i + 2\pi$. A complex torus $T_\tau = \frac{\mathbb{C}}{\Lambda_\tau}$ can be defined as the complex plane modded by the action of a lattice group Λ_τ (see Figure 9.1), namely, by identifying the complex numbers z through the equivalence relation

$$
z_1 \approx z_2 \quad \text{iff} \quad z_1 - z_2 = m + n\tau \in \Lambda_\tau; \quad m, n \in \mathbb{Z}. \tag{9.74}
$$

In this definition τ is a complex number with positive imaginary part ($\mathrm{Im}\,\tau > 0$) that parameterizes the possible complex structures of the topological torus and therefore provides a coordinatization of its moduli space. The way the complex structures are parameterized by τ can be made explicit by setting

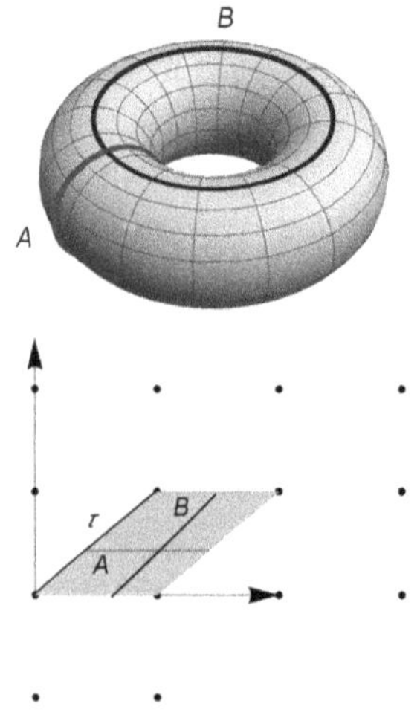

Figure 9.1: In the first image of this figure we display a topological torus $\mathbb{S}^1 \times \mathbb{S}^1$ represented as a donut-shaped surface immersed in three-dimensional space. The *A* and *B* cycles are the two generators of the first homology group. In the second image we illustrate the construction of the same topological torus as a complex variety. A complex torus of modulus τ is identified with $\mathbb{H}_+/\Lambda_\tau$, where Λ_τ is the lattice generated by 1 and a complex number τ with positive imaginary part. A representative of each equivalence class lies in the fundamental cell marked in dark color which is a parallelogram with identified opposite sides. The *A* and *B* cycles are visible as segments going from one side to its opposite.

$$z = \frac{1}{2\pi}(\xi_1 + \tau\xi_2). \tag{9.75}$$

If we deform τ by an infinitesimal transformation $\tau \longrightarrow \tau + \delta\tau$, there is no way of accounting for this deformation by means of a diffeomorphism $\xi_i = \xi_i'(\xi)$ contractible to the identity. There are, however, two global diffeomorphisms, corresponding to the two Dehn twists along the A and B cycles (see once again Figure 9.1):

$$\begin{aligned} D_a &: \begin{cases} \xi_1 \longrightarrow \xi_1 + \xi_2, \\ \xi_2 \longrightarrow \xi_2, \end{cases} \\ D_b &: \begin{cases} \xi_1 \longrightarrow \xi_1, \\ \xi_2 \longrightarrow \xi_1 + \xi_2, \end{cases} \end{aligned} \tag{9.76}$$

which can be reabsorbed by the following two transformations of the parameter τ:

$$\begin{aligned} D_a &: \tau \longrightarrow \tau + 1, \\ D_b &: \tau \longrightarrow \frac{\tau}{\tau + 1}. \end{aligned} \tag{9.77}$$

When we discussed the Riemann sphere, namely, the complex plane completed with the point at infinity $\mathbb{S}_{\text{Rie}} = \mathbb{C} \cup \{\infty\}$, we saw that it is invariant under the action of the projective group $\text{PSL}(2, \mathbb{C})$, which acts by means of fractional linear transformations. We also saw that the upper complex plane $\mathbb{H}_+ := \{z \in \mathbb{C} \mid \text{Im}\, z > 0\} \cup \{\infty\}$ is invariant under the action of the subgroup $\text{PSL}(2, \mathbb{R}) \subset \text{PSL}(2, \mathbb{C})$. Regarded as elements of $\text{PSL}(2, \mathbb{R})$ acting projectively on the complex upper plane,

$$\pm\begin{pmatrix} a & b \\ c & d \end{pmatrix} \in \text{PSL}(2, \mathbb{R}) : \tau \longrightarrow \frac{a\tau + b}{c\tau + d},$$

the two transformations in (9.77) correspond to the two matrices

$$\begin{aligned} D_a &\approx T = \begin{pmatrix} 1 & 1 \\ 0 & 1 \end{pmatrix}, \\ D_b &\approx TST = \begin{pmatrix} 1 & 0 \\ 1 & 1 \end{pmatrix}, \end{aligned}$$

which generate the modular group of equation (9.72). Hence, two tori whose moduli τ_1 and τ_2 are related by a modular transformation $\tau_1 = \gamma\tau_2 > (\gamma \in \Gamma)$ are *holomorphically equivalent* and have to be identified. This leads to the conclusion that the true moduli space of the torus is the quotient $\mathbb{H}_+/\Gamma$, $\mathbb{H}_+$ being the previously defined upper complex plane.

9.5.2 An algebraic representation of the torus by means of a cubic

Without considering the reasons why it is so, which would require entering algebraic geometry and the theory of Riemann surfaces, we kindly ask the reader to believe our word that a very interesting representation of the complex torus is algebraically provided by writing the following equation:[1]

$$0 = \mathcal{W}(x, y, z; \psi) \equiv \frac{1}{3}(x^3 + y^3 + z^3) - \psi xyz, \tag{9.78}$$

where $\mathcal{W}$ is a particular cubic homogeneous polynomial of x, y, z that we take to be the complex homogeneous coordinates of $\mathbb{P}^2$. This latter is the standard name of the two-dimensional complex projective space.[2] Hence, the one-dimensional complex torus is identified with the vanishing locus of the homogeneous polynomial $\mathcal{W}$ in $\mathbb{P}^2$. Indeed, the torus is just a particular instance of a compact Riemann surface and as such it must be described as the vanishing locus of some polynomial. In such a description, the complex coefficient ψ substitutes τ in the capacity of coordinate on the torus moduli space and it is of utmost interest to retrieve the relation between ψ and τ.

This can be done by means of the following argument.

First let us observe that in the description of the torus as the quotient of $\mathbb{H}_+$ with respect to the lattice Λ_τ, the two generators A and B of the homology group (the *cycles* displayed in Figure 9.1) are just given by $A = \{\xi_1 \in [0, 2\pi], \xi_2 = \text{const}\}$ and $B = \{\xi_1 = \text{const}, \xi_2 \in [0, 2\pi]\}$.

A closed holomorphic 1-form exists in this space and it is simply given by $\Omega^{(1)} = dz$. This form is cohomologically non-trivial because it admits no primitive. The coordinate z would be such a primitive, yet it is not a globally defined function since it is defined only up to an arbitrary lattice element. Hence, dz is not an exact differential form. It is very easy to construct the *periods* of $\Omega^{(1)}$, namely, the integrals of the latter along the two basic cycles:

$$\Pi_{a,b} = \int_{A,B} \Omega^{(1)}. \tag{9.79}$$

Utilizing equation (9.75), we immediately get

---

1 For a more detailed discussion of projective embedding of the torus via a cubic polynomial we refer the reader to another book by the present author in collaboration with his PhD student of that time [98].

2 By definition, the complex projective space $\mathbb{P}^n$ is the space of equivalences classes in $\mathbb{C}^{n+1}$ with respect to the equivalence relation $\{z_1, \dots, z_n\} \approx \{w_1, \dots, w_n\}$, *iff* $\{w_1, \dots, w_n\} = \lambda\{z_1, \dots, z_n\}$, where $\lambda \in \mathbb{C}^\star$, this latter being the group of all non-vanishing complex numbers.

$$\Pi_a = \int_A dz = \int_0^{2\pi} \frac{d\xi_1}{2\pi} = 1,$$
$$\Pi_b = \int_B dz = \int_0^{2\pi} \tau \frac{d\xi_2}{2\pi} = \tau, \tag{9.80}$$

so that we can identify the modulus parameter with the period ratio

$$\tau = \frac{\Pi_b}{\Pi_a} \tag{9.81}$$

and interpret the transformations in (9.77) as the effect on the ratio τ of the action of the diffeomorphisms in (9.76) on the homology basis (A, B). The diffeomorphisms in (9.77) generate the so-called *mapping class group*, namely, the quotient of the group of all diffeomorphisms with respect to the normal subgroup of those diffeomorphisms that can be continuously deformed into the identity map:

$$\mathscr{M}_{\text{class}} = \frac{\text{Diff}}{\text{Diff}_0}. \tag{9.82}$$

In this way *we recognize the modular group* Γ *as the homomorphic image on a canonical homology basis of 1-cycles of the mapping class group* (9.82). The connection between τ and ψ is introduced at this point relying on the fact that in the algebraic representation of the torus as the vanishing locus of the cubic polynomial (9.78), the periods of the holomorphic 1-form along the A, B cycles can be calculated as functions of the complex coefficient ψ by utilizing general methods of algebraic geometry. Actually, these methods reduce the calculation of the periods to the solution of an ordinary differential equation which, in this case, turns out to be of the second order and Fuchsian. Not too surprisingly, the differential equations obeyed by the periods of the holomorphic p-form in a p-dimensional complex variety are named *Picard–Fuchs equations*.

9.5.3 A differential equation enters the play and brings in a new group

In the case of the cubic torus, the afore mentioned Picard–Fuchs equation takes the following form, where Π denotes any period:

$$(1 - \psi^3)\frac{\partial^2 \Pi}{\partial \psi^2} - 3\psi^2 \frac{\partial \Pi}{\partial \psi} - \psi \Pi = 0. \tag{9.83}$$

Since the above equation is linear of the second order, it will admit a basis of solutions composed of two functions. It will be our care to adapt the basis in such a way that the basis functions represent the periods along a canonical basis of cycles A, B.

Let us note that this equation is Fuchsian with four regular singular points, namely,

$$\psi \in \{\xi, \xi^2, \xi^3, \infty\}; \quad \xi = \exp\left[\frac{2\pi}{3} i\right]. \tag{9.84}$$

Hence, it is not immediately of the hypergeometric type, yet it can be reduced to the hypergeometric case by a simple transformation. Performing the substitution

$$z = \psi^3; \quad \Pi(\psi) = \sqrt[3]{z} u(z) \tag{9.85}$$

in equation (9.83), we obtain

$$z(1-z)\frac{d^2u}{dz^2} + \frac{(6-15z)}{9}\frac{du}{dz} - \frac{1}{9}u = 0, \tag{9.86}$$

which is in the standard hypergeometric form with parameters $a = b = \frac{1}{3} c = \frac{2}{3}$. Since none of the parameters is integer, a basis of solutions in the neighborhood of $z = 0$ which also corresponds to $\psi = 0$ is provided by equation (9.55), namely, by

$$\begin{cases} u_1(z) = \frac{\Gamma^2(\frac{1}{3})}{\Gamma(\frac{2}{3})} F(\frac{1}{3}, \frac{1}{3}, \frac{2}{3}; z), \\ u_2(z) = \frac{\Gamma^2(\frac{2}{3})}{\Gamma(\frac{4}{3})} z^{\frac{1}{3}} F(\frac{2}{3}, \frac{2}{3}, \frac{4}{3}; z) \end{cases} \Rightarrow \begin{cases} \Pi_1(\psi) = \frac{\Gamma^2(\frac{1}{3})}{\Gamma(\frac{2}{3})} \psi F(\frac{1}{3}, \frac{1}{3}, \frac{2}{3}; \psi^3), \\ \Pi_2(\psi) = \frac{\Gamma^2(\frac{2}{3})}{\Gamma(\frac{4}{3})} \psi^2 F(\frac{2}{3}, \frac{2}{3}, \frac{4}{3}; \psi^3). \end{cases} \tag{9.87}$$

Other solutions defined in the neighborhood of the other singular points can then be obtained by Kummer's formulae discussed in the previous Section 9.4.3.1. We will do it presently. Before that, let us go back to the original equation (9.83) and discuss the concept of monodromy group.

9.5.3.1 The monodromy group

First of all, we note that in the neighborhood of all singular points $\psi = \xi^n$ for $n = 0, 1, 2$ the indicial equation of equation (9.83) is always very simple:

$$r^2 = 0. \tag{9.88}$$

The degeneracy of the root $r = 0$ implies that one solution $\Pi_1(\psi - \xi^n)$ is given in terms of a regular series, while the second is of the form

$$\Pi_2(\psi - \xi^n) = \log(\psi - \xi^n)\Pi_1(\psi - \xi^n) + \text{reg. series}.$$

We will see in a moment how to utilize Kummer's solutions of the hypergeometric equation in order to determine $\Pi_1(\psi - \xi^n)$. For our present argument the important thing is the presence of the logarithmic behavior of at least one of the solutions.

By definition, the monodromy group $\mathscr{M}on$ is the discrete group generated by the transvections γ_A around all singular points $A = (1, \xi, \xi^2, \infty)$. Given a basis

$$\{\Pi_0(\psi), \Pi_1(\psi)\}$$

of solutions of the differential equation (9.83) one considers the analytic continuation of each $\Pi_i(\psi)$ along a closed path γ_A encircling the singular point $\psi = \psi_A$. By $\gamma_A \Pi_i(\psi)$ we denote the result of this analytic continuation; it is still a solution of the same differential equation and hence it can be expressed as a linear combination of the solutions $\{\Pi_0(\psi), \Pi_1(\psi)\}$, namely,

$$\gamma_A \Pi_i(\psi) = (T_A)_i^{\ j} \Pi_j(\psi). \tag{9.89}$$

This defines a 2×2 matrix T_A representing the monodromy generator γ_A. Since the product $\gamma = \gamma_\infty \gamma_2 \gamma_1 \gamma_0$ is homotopically equivalent to a path that encircles no singular point, it follows that the monodromy generators satisfy the following basic relation:

$$T_\infty T_2 T_1 T_0 = \mathbf{1}. \tag{9.90}$$

In principle we should find the explicit form of all the T_A by considering the analytic continuation around all the singular points; however, taking into account the additional group $\Gamma_{\mathcal{W}}$ of duality transformations that preserve the algebraic equation defining the torus (9.78), our task is greatly simplified and reduces to calculating the monodromy transvection around a single singular point, say $\psi = \psi_0 = 1$. Let us see why.

By definition, the duality symmetries of $\mathcal{W}$ are those transformations of the modulus parameter ψ that can be reabsorbed by a unitary linear transformation of the projective coordinates $\mathbf{X} = \{x, y, z\}$ of the ambient space $\mathbb{P}^2$ in which the torus is algebraically immersed as the vanishing locus of $\mathcal{W}$. Indeed, if it happens that, for a certain transformation of the ψ-parameter

$$\psi' = f(\psi) \tag{9.91}$$

we can find a unitary 3×3 matrix $U(f)_\Lambda^{\ \Sigma}$ such that

$$\mathcal{W}(f(\psi); \mathbf{X}) = \mu(f) \mathcal{W}(\psi, U(f)\mathbf{X}), \tag{9.92}$$

where $\mu(f)$ is some complex number and $U(f)\mathbf{X}$ denotes the action of the matrix $U(f)$ on the vector $\mathbf{X}$, then the vanishing loci of $\mathcal{W}(f(\psi); \mathbf{X})$ and $\mathcal{W}(\psi; X)$ are just the same hypersurface in $\mathbb{P}^2$. This means that we must identify all the points of the ψ-complex plane that are related by transformations $f(\psi)$ with the above property. It is also clear that these transformations f form a group, namely, $\Gamma_{\mathcal{W}}$. Because of the same argument, any element $g \in \Gamma_{\mathcal{W}}$ must be a symmetry of the Picard–Fuchs equation (9.83); in particular, the image through g of a singular point must be another singular point. The property that singular points fill up $\Gamma_{\mathcal{W}}$ orbits is very handy in the construction of the monodromy generators. Indeed, if we construct a solution basis that has known and simple transformation properties under $\Gamma_{\mathcal{W}}$, then, once we have obtained the transvection generator T_0, we can retrieve all the others by a simple conjugation with $\Gamma_{\mathcal{W}}$.

One type of transformation includes the transformations

$$\psi' = f_j(\psi) = \xi^j \psi \quad (j = 0, 1, 2). \tag{9.93}$$

These transformations are compensated for by for instance the following $U(f_j)$ matrices:

$$\begin{aligned} U(f_0) &= \begin{pmatrix} 1 & 0 & 0 \\ 0 & 1 & 0 \\ 0 & 0 & 1 \end{pmatrix}, \\ U(f_1) &= \begin{pmatrix} \xi & 0 & 0 \\ 0 & 1 & 0 \\ 0 & 0 & 1 \end{pmatrix}, \\ U(f_2) &= \begin{pmatrix} \xi^2 & 0 & 0 \\ 0 & 1 & 0 \\ 0 & 0 & 1 \end{pmatrix}, \end{aligned} \tag{9.94}$$

as one can immediately check by looking at equation (9.78). These matrices satisfy equation (9.92) and are therefore good symmetries. Actually, there are many other choices of the matrices $U(f_j)$ that could compensate for the transformations (9.93) but they are all equivalent. We call A the generator of this duality $\mathbb{Z}_3$ group. Its action on ψ is given by the case $j = 1$ of equation (9.93). On the solutions of the Picard–Fuchs equation (9.83), which, as expected, are indeed invariant against this transformation, the operation A is represented by a suitable 2×2 matrix $\mathscr{A}$ possessing the property

$$\mathscr{A}^3 = \mathbf{1}. \tag{9.95}$$

Furthermore, under the action of this duality group, the singular points ψ_A of the Picard–Fuchs equation form two orbits: the first is composed of the third roots of the identity ψ_i $(i = 0, 1, 2)$, rotated one into the other by A; the second contains only ψ_∞, which is A-invariant. This fact explains why the indicial equation (9.88) is identical in all finite singular points. Relying on this, if we obtain the explicit form T_0 of the transvection around the singular point $\psi = 1$, we can write

$$T_j = \mathscr{A}^{-j} T_0 \mathscr{A}^j \quad (j = 1, 2), \tag{9.96}$$

which, combined with (9.90), also yields

$$T_\infty = (\mathscr{A} T_0)^{-1}. \tag{9.97}$$

The main strategy consists in choosing a solution basis where $\mathscr{A}$ is known a priori and then devising an analytic tool for the evaluation of T_0. However, this is not the end of the story since the duality group is enlarged by an additional generator and becomes non-Abelian. Consider the unitary 3×3 matrix

$$S = -\frac{i}{\sqrt{3}}\begin{pmatrix} 1 & 1 & 1 \\ 1 & \xi & \xi^2 \\ 1 & \xi^2 & \xi \end{pmatrix}. \tag{9.98}$$

A straightforward calculation reveals that

$$\mathscr{W}(\psi; \mathbf{SX}) = \mu_S(\psi)\mathscr{W}(\mathscr{S}\psi; \mathbf{X}), \tag{9.99}$$

where

$$\begin{aligned} \mu_S(\psi) &= \frac{i}{\sqrt{3}}(1-\psi), \\ \mathscr{S}\psi &= \frac{\psi+2}{\psi-1}. \end{aligned} \tag{9.100}$$

This shows that the transformation $\mathscr{S}$ defined by equation (9.100) is another generator of $\Gamma_{\mathscr{W}}$, in addition to the transformation $\mathscr{A}\psi = \xi\psi$. A direct check reveals that these generators satisfy

$$\mathscr{A}^3 = \mathscr{S}^2 = (\mathscr{A}\mathscr{S})^3 = \mathbb{1}. \tag{9.101}$$

Equation (9.101) is identical to equation (5.50) if we set $k_1 = 3$, $k_2 = 3$, $k_3 = 2$, $\mathscr{A} = \mathbf{A}$, and $\mathscr{S} = \mathbf{B}$. This shows that the duality group $\Gamma_{\mathscr{W}}$ of the cubic torus is isomorphic to the tetrahedral group $T_{12} \sim PSL(2, \mathbb{Z}_3)$.

The transvections around the singular points of the Picard–Fuchs equation must correspond to some global reparameterization of the surface $\mathscr{W}(X, \psi) = 0$. Hence, we infer that

$$\mathscr{M}on \subset \Gamma, \tag{9.102}$$

where Γ is the modular group of the torus, described in equation (9.72). Furthermore, reconsidering, in the case of the torus, equation (9.71), we can also conjecture that

$$\Gamma_{\mathscr{W}} = \frac{\Gamma}{\mathscr{M}on}. \tag{9.103}$$

This conjecture can indeed be verified in the case of the torus, where Γ is known a priori, by calculating $\mathscr{M}on$ and showing that $\mathscr{M}on \approx \Gamma(3)$.

9.5.3.2 Explicit calculation of the monodromy group of the cubic torus

We now focus on the second order Picard–Fuchs equation (9.83); its monodromy group can be derived as follows. As already pointed out, we recall that it is sufficient to compute the transvection T_0 around the singular point $\psi = 1$. In fact, applying to this particular case the general formula (9.95), the effect of a closed loop around $\psi = \xi$ and $\psi = \xi^2$ ($\xi =$

$e^{2\pi i/3}$) can be computed from the matrix T_0 by conjugation with $\mathscr{A}$, where $\mathscr{A}$ represents the operation $\psi \to \xi\psi$:

$$\begin{aligned} T_1 &= \mathscr{A}^{-1} T_0 \mathscr{A}, \\ T_2 &= \mathscr{A}^{-2} T_0 \mathscr{A}^2. \end{aligned} \tag{9.104}$$

Furthermore, a closed loop which encloses all the singular points, including ∞, is contractible, and therefore, according to equation (9.89), we can write

$$T_\infty T_2 T_1 T_0 = 1 \to T_\infty = (T_2 T_1 T_0)^{-1}. \tag{9.105}$$

To compute T_0 we start from the pair of independent solutions of the associated hypergeometric equation displayed in equation (9.87). These are defined in the neighborhood of $z \equiv \psi^3 = 0$. These two solutions can be analytically continued around $\psi^3 = 1$ by means of the anharmonic group that leads to a new hypergeometric equation of parameters $a = b = \frac{1}{3}$, $c = 1$. Since the new c is integer, as already observed, a pair of solutions is necessarily of the form

$$\begin{aligned} u_1(z) &= -\log(1-z) F\left(\frac{1}{3}, \frac{1}{3}, 1; 1-z\right) + B_1(1-z), \\ u_2(z) &= -\log(1-z) F\left(\frac{1}{3}, \frac{1}{3}, 1; 1-z\right) + B_2(1-z), \end{aligned} \tag{9.106}$$

where B_1 and B_2 are regular series around $\psi = 1 \Rightarrow z = 1$. Hence the transvection along a closed loop around $\psi = 1$ yields

$$\begin{pmatrix} u_1 \\ u_2 \end{pmatrix} \to \begin{pmatrix} u_1' \\ u_2' \end{pmatrix} = \begin{pmatrix} u_1 \\ u_2 \end{pmatrix} - 2\pi i F\left(\frac{1}{3}, \frac{1}{3}, 1; 1-z\right) \begin{pmatrix} 1 \\ 1 \end{pmatrix}. \tag{9.107}$$

The Kummer relation (9.68) among hypergeometric functions enables us to reexpress $F(\frac{1}{3}, \frac{1}{3}, 1; 1-\psi^3)$ in terms of the original basis (u_1, u_2) of solutions analytic in the regular point $\psi = 0$:

$$F\left(\frac{1}{3}, \frac{1}{3}, 1; 1-z\right) = \frac{\Gamma(\frac{1}{3})}{\Gamma^2(\frac{2}{3})} F\left(\frac{1}{3}, \frac{1}{3}, \frac{2}{3}; z\right) + \frac{\Gamma(-\frac{1}{3})}{\Gamma^2(\frac{1}{3})} F\left(\frac{2}{3}, \frac{2}{3}, \frac{4}{3}; z\right). \tag{9.108}$$

Therefore, using the relation $\Gamma(z)\Gamma(1-z) = \frac{\pi}{\sin \pi z}$ and $2\,\mathrm{Sin}(\frac{\Pi}{3}) = -\mathrm{tg}(\frac{\Pi}{3}) = -\sqrt{3}$, one obtains

$$\begin{pmatrix} u_1' \\ u_2' \end{pmatrix} = \begin{pmatrix} 1 + i\,\mathrm{tg}\,\frac{2\pi}{3} & -i\,\mathrm{tg}\,\frac{2\pi}{3} \\ i\,\mathrm{tg}\,\frac{2\pi}{3} & 1 - i\,\mathrm{tg}\,\frac{2\pi}{3} \end{pmatrix} \begin{pmatrix} u_1 \\ u_2 \end{pmatrix}, \tag{9.109}$$

i. e., the monodromy matrix around $\psi = 1$ is

$$T_0 = \begin{pmatrix} 1 - i\sqrt{3} & i\sqrt{3} \\ -i\sqrt{3} & 1 + i\sqrt{3} \end{pmatrix}. \tag{9.110}$$

To find T_1, T_2 we need to represent $\mathscr{A} : \psi \to a\psi$ on u_1, u_2. From (9.83) and (9.87) we see that under $\psi \to \xi\psi$ the differential operator is invariant while

$$\begin{pmatrix} u_1 \\ u_2 \end{pmatrix} \to \begin{pmatrix} 1 & 0 \\ 0 & \xi \end{pmatrix} \begin{pmatrix} u_1 \\ u_2 \end{pmatrix}. \tag{9.111}$$

Since we are interested in the projective representation of the monodromy group we may rescale our basis in such a way that $\det \mathscr{A} = 1$ (note that T_0 already satisfies $\det T_0 = 1$). Hence, we have

$$\mathscr{A} = \begin{pmatrix} \xi^{-1/2} & 0 \\ 0 & \xi^{1/2} \end{pmatrix} = \frac{1}{2} \begin{pmatrix} 1 - i\sqrt{3} & 0 \\ 0 & 1 - i\sqrt{3} \end{pmatrix},$$

and from (9.104) we find

$$T_1 = \begin{pmatrix} 1 - i\sqrt{3} & -\frac{3}{2} - \frac{i\sqrt{3}}{2} \\ -\frac{3}{2} + \frac{i\sqrt{3}}{2} & 1 + i\sqrt{3} \end{pmatrix}, \quad T_2 = \begin{pmatrix} 1 - i\sqrt{3} & \frac{3}{2} - \frac{i\sqrt{3}}{2} \\ \frac{3}{2} + \frac{i\sqrt{3}}{2} & 1 + i\sqrt{3} \end{pmatrix}. \tag{9.112}$$

In this way the monodromy group $\mathscr{M}on$ has been determined. In view of our previous discussion we know that on a well-adapted basis of homological cycles it should be a subgroup of $\mathrm{SL}(2, \mathbb{Z})$, so that we should be able to find a change of basis on the periods $\Pi_i(z) = \sqrt[3]{z}u_i(z)$ such that the entries of the generators T_0, T_1, T_2 are integer numbers. Actually, if our conjecture is correct we should find a basis where $\mathscr{M}on \approx \Gamma(3)$. The basis $(\mathscr{F}_1, \mathscr{F}_2)$ where this isomorphism becomes indeed manifest is obtained by means of the following linear transformation:

$$\begin{pmatrix} \mathscr{F}_1 \\ \mathscr{F}_2 \end{pmatrix} = M \begin{pmatrix} u_1 \\ u_2 \end{pmatrix} \stackrel{\text{def}}{=} \frac{1}{\pi} \begin{pmatrix} \frac{3i}{2} - \frac{3\sqrt{3}}{2} & -3i \\ \frac{3i}{2} - \frac{3}{2\sqrt{3}} & -\frac{3i}{2} + \frac{\sqrt{3}}{2} \end{pmatrix} \begin{pmatrix} u_1 \\ u_2 \end{pmatrix}.$$

The transformed $\mathscr{M}on$ generators $\hat{T}_i = MT_iM^{-1}$ take the following expressions:

$$\hat{T}_0 = \begin{pmatrix} 1 & 3 \\ 0 & 1 \end{pmatrix}; \quad \hat{T}_1 = \begin{pmatrix} -5 & 12 \\ -3 & 7 \end{pmatrix}; \quad \hat{T}_2 = \begin{pmatrix} -2 & 3 \\ -3 & 4 \end{pmatrix}; \tag{9.113}$$

$$\hat{T}_\infty \equiv (\hat{T}_2\hat{T}_1\hat{T}_0)^{-1} = \begin{pmatrix} 1 & 0 \\ -3 & 1 \end{pmatrix} \tag{9.114}$$

and are all, manifestly, elements of $\Gamma(3)$ according to the definition (9.73). Actually, by comparing with classical twentieth-century books we see that T_0, T_1, T_2, T_∞ are the classical four generators of $\Gamma(3)$ satisfying the relation (9.96). As already remarked above, the transformation $\mathscr{A} : \psi \to \xi\psi$ is obviously an invariance of the hypersurface equation $\mathscr{W}(X) = 0$, since it satisfies equation (9.92), and as a consequence it is an invariance of the differential operator (9.111). It is an element of the modular group but not of the monodromy group. In the basis $(\mathscr{F}_1, \mathscr{F}_2)$, the generator $\mathscr{A}$ takes the following form:

$$\hat{\mathscr{A}} = M\mathscr{A}M^{-1} = \begin{pmatrix} -1 & 3 \\ -1 & 2 \end{pmatrix}. \tag{9.115}$$

As one sees, $\hat{\mathscr{A}} \in \mathrm{PSL}(2,\mathbb{Z})$, but $\hat{\mathscr{A}} \notin \Gamma(3)$. This is precisely what we expected. Indeed, as stated in equation (9.71), the tetrahedral group T_{12}, isomorphic to the duality group $\Gamma_{\mathscr{W}}$, is also isomorphic to the quotient $\mathrm{PSL}(2,\mathbb{Z})/\Gamma(3)$; hence the two operations $\mathscr{A}$ and $\mathscr{S}$, in the basis $\mathscr{F}_1$, $\mathscr{F}_2$ of solutions of the Picard–Fuchs equation, must be represented by integer valued unimodular matrices that are not equivalent to the identity modulus 3. We were able to calculate the explicit form of $\mathscr{A}$. To find the form of $\mathscr{S}$ we should be able to reexpress $\mathscr{F}_i(\frac{\psi+2}{\psi-1})$ in terms of $\mathscr{F}_i(\psi)$; there are no known autotransformations of hypergeometric functions that do this for us and hence we should resort to explicit numerical calculations, but we do not feel it is necessary to do it for the purpose of our discussion. Indeed, by means of our present calculations we have already shown that the monodromy group of the cubic torus is $\mathscr{M}on = \Gamma(3)$, so that the conjectured relation (9.103) is indeed verified.

9.6 Conclusive remarks on this chapter

In the present chapter we have tried to illustrate how group theory enters the field of theory of linear differential equations in relation with monodromy groups. Actually, we have explicitly calculated the monodromy group of a specific differential equation of the hypergeometric type, showing that such a group turns out to be $\Gamma(3)$, i. e., one of the normal subgroups of the modular group $\Gamma = \mathrm{PSL}(2,\mathbb{Z})$. Although the derivation of this specific equation from algebraic geometry is beyond the scope of the present book, we mentioned without proof its origin in the description of a torus as the vanishing locus of a cubic polynomial in the projective space $\mathbb{P}^2$. This allowed us to show the relation between the torus homology and the solutions of the differential equation. The same techniques can be used for more complicated differential equations. We also presented the group theoretical basis for the existence of Kummer's 24 hypergeometric functions that are solutions of the same hypergeometric equations. Our goal was the illustration of the interplay between group theory and complex function analysis.

9.7 Bibliographical note

The sources for the material presented in this chapter are partly the already quoted unpublished lecture notes of the author for his innovative course *Mathematical Methods of Physics* and partly the book [98].

10 Structure of Lie algebras

> ... and when I had at last plucked it, the stalk was all frayed and the flower itself no longer seemed so fresh and beautiful. Moreover, owing to a coarseness and stiffness, it did not seem in place among the delicate blossoms of my nosegay. I threw it away feeling sorry to have vainly destroyed a flower that looked beautiful in its proper place. – But what energy and tenacity! With what determination it defended itself, and how dearly it sold its life! thought I, remembering the effort it had cost me to pluck the flower. –
>
> L. N. Tolstoy, Hadji Murat

10.1 Introduction

The present chapter addresses two questions: *What is the general structure of a Lie group*? And *can we classify Lie groups*? The first step in this direction consists in reducing the question from Lie groups to their Lie algebras. The second step consists in describing the general form of Lie algebras in terms of a semi-simple algebra (the Levi algebra) and a solvable radical. The third step consists in classifying simple Lie algebras and will be our concern in Chapter 11. In order to establish the first two steps on a firm basis we ought to introduce a few more fundamental mathematical concepts and constructions of a very general character. To these preliminaries we devote the next sections.

10.2 Linear algebra preliminaries

Let us consider a vector space V constructed over the field of complex numbers $\mathbb{C}$, whose dimension we denote by $\dim V = n$. We denote by $\mathrm{Hom}(V, V)$ the ring of all linear endomorphisms of V. In other words, an element $A \in \mathrm{Hom}(V, V)$ is a linear map:

$$\begin{aligned} &A : V \to V \\ &\qquad \forall \alpha, \beta \in \mathbb{C}, \forall \mathbf{v}, \mathbf{w} \in V : \quad A(\alpha\mathbf{v} + \beta\mathbf{w}) = \alpha A(\mathbf{v}) + \beta A(\mathbf{w}). \end{aligned} \tag{10.1}$$

As is well known, if $(\mathbf{e}_1, \ldots, \mathbf{e}_n)$ is a basis of V, in such a basis the endomorphism A is represented by the matrix A_{ij} determined by the condition

$$A(\mathbf{e}_j) = \mathbf{e}_i A^i{}_j. \tag{10.2}$$

Indeed, if $\{v^i\}$ are the components of the vector $\mathbf{v}$ in the basis $\{\mathbf{e}_i\}$, we have

$$A(v^j \mathbf{e}_j) = \mathbf{e}_i A^i{}_j v^j, \tag{10.3}$$

and hence the components of the vector $\mathbf{v}' \equiv A(\mathbf{v})$ are given by

$$v^{i\prime} = A^i{}_j v^j. \tag{10.4}$$

https://doi.org/10.1515/9783111201535-010

The association

$$A \mapsto A^i_{\ j} \tag{10.5}$$

is an isomorphism of $\mathrm{Hom}(V, V)$ onto the ring $M_n(\mathbb{C})$ of $n \times n$ matrices with complex coefficients. For simplicity, from now in this chapter, we no longer care about upper and lower indices of matrices, writing them all at the same level.

Definition 10.2.1. A matrix A_{ij} such that $A_{ij} = 0$ if $i > j$ is called *upper triangular*. A matrix such that $A_{ij} = 0$ if $i < j$ is called *lower triangular*. Finally, a matrix that is simultaneously upper and lower triangular is called *diagonal*.

We recall the concept of eigenvalue.

Definition 10.2.2. Let $A \in \mathrm{Hom}(V, V)$. A complex number $\lambda \in \mathbb{C}$ is called an *eigenvalue* of A if $\exists \mathbf{v} \in V$ such that

$$A\mathbf{v} = \lambda \mathbf{v}. \tag{10.6}$$

Definition 10.2.3. Let λ be an eigenvalue of the endomorphism $A \in \mathrm{Hom}(V, V)$. The set of vectors $\mathbf{v} \in V$ such that $A\mathbf{v} = \lambda\mathbf{v}$ is called the *eigenspace* $V_\lambda \subset V$ pertaining to the eigenvalue λ. It is obvious that it is a vector subspace.

As known from elementary courses in geometry and algebra, the possible eigenvalues of A are the roots of the secular equation

$$\det(\lambda \mathbf{1} - \mathscr{A}) = 0, \tag{10.7}$$

where $\mathbf{1}$ is the unit matrix and $\mathscr{A}$ is the matrix representing the endomorphism A in an arbitrary basis.

Definition 10.2.4. An endomorphism $N \in \mathrm{Hom}(V, V)$ is called nilpotent if there exists an integer $k \in \mathbb{N}$ such that

$$N^k = 0. \tag{10.8}$$

Lemma 10.2.1. *A nilpotent endomorphism has always the unique eigenvalue* $0 \in \mathbb{C}$.

Proof of Lemma 10.2.1. Let λ be an eigenvalue and let $\mathbf{v} \in V_\lambda$ be an eigenvector. We have

$$N^r \mathbf{v} = \lambda^r \mathbf{v}. \tag{10.9}$$

Choosing $r = k$ we obtain $\lambda^k = 0$, which necessarily implies $\lambda = 0$. □

Lemma 10.2.2. *Let* $N \in \mathrm{Hom}(V, V)$ *be a nilpotent endomorphism. In this case one can choose a basis* $\{\mathbf{e}_i\}$ *of* V *such that in this basis the matrix* N_{ij} *satisfies the condition* $N_{ij} = 0$ *for* $i \geq j$.

Proof of Lemma 10.2.2. Let $\mathbf{e}_1$ be a null eigenvector of N, namely, $N\mathbf{e}_1 = 0$, and let E_1 be the subspace of V generated by $\mathbf{e}_1$. From N we induce an endomorphism N_1 acting on the space V/E_1, namely, the vector space of equivalence classes of vectors in V modulo the relation

$$\mathbf{v} \sim \mathbf{w} \quad \Leftrightarrow \quad \mathbf{v} - \mathbf{w} = m\mathbf{e}_1, \quad m \in \mathbb{C}. \tag{10.10}$$

Also the new endomorphism $N_1 : V/E_1 \to V/E_1$ is nilpotent. If $\dim V/E_1 \neq 0$, then we can find another vector $\mathbf{e}_2 \in V$ such that $(\mathbf{e}_2 + E_1) \in V/E_1$ is an eigenvector of N_1. Continuing iteratively this process we obtain a basis $\mathbf{e}_1, \dots, \mathbf{e}_n$ of V such that

$$N\mathbf{e}_1 = 0; \quad N\mathbf{e}_p = 0 \bmod (\mathbf{e}_1, \dots, \mathbf{e}_{p-1}); \quad 2 \leq p \leq n, \tag{10.11}$$

where $(\mathbf{e}_1, \dots, \mathbf{e}_{p-1})$ denotes the subspace of V generated by the vectors $\mathbf{e}_1, \dots, \mathbf{e}_{p-1}$. In this basis the matrix representing N is triangular. Similarly, if N_{ij} is triangular with $N_{ij} = 0$ for $i \geq j$, then the corresponding endomorphism is nilpotent. □

Definition 10.2.5. Let $\mathscr{S} \subset \mathrm{Hom}(V, V)$ be a subset of the ring of endomorphisms and let $W \subset V$ be a vector subspace. The subspace W is named *invariant* with respect to $\mathscr{S}$ if $\forall S \in \mathscr{S}$ we have $SW \subset W$. The space V is named *irreducible* if it does not contain invariant subspaces.

Definition 10.2.6. A subset $\mathscr{S} \subset \mathrm{Hom}(V, V)$ is named *semi-simple* if every invariant subspace $W \subset V$ admits an orthogonal complement which is also invariant. In that case we can write

$$V = \bigoplus_{i=1}^{p} W_i, \tag{10.12}$$

where each subspace W_i is invariant.

A fundamental and central result in linear algebra which is essential for the further development of Lie algebra theory is the *Jordan decomposition theorem*, which we quote without proof.

Theorem 10.2.1. *Let $L \in \mathrm{Hom}(V, V)$ be an endomorphism of a finite-dimensional vector space V. Then the following unique Jordan decomposition exists:*

$$L = S_L + N_L, \tag{10.13}$$

where S_L is semi-simple and N_L is nilpotent. Furthermore, both S_L and N_L can be expressed as polynomials in L.

10.3 Types of Lie algebras and Levi decomposition

In the previous section we have discussed the notions of semi-simplicity and nilpotency for endomorphisms of vector spaces, namely, for matrices. Such notions can now be extended to entire Lie algebras. This is not surprising since Lie algebras admit linear representations where each of their elements is replaced by a matrix. In the present section we discuss *solvable*, *nilpotent*, and *semi-simple* Lie algebras. Solvable and nilpotent Lie algebras are those for which all linear representations are provided by triangular matrices. Semi-simple Lie algebras are those which do not admit any invariant subalgebra (or ideal) which is solvable. The main result in this section will be Levi's theorem, which is the counterpart for algebras of Jordan's decomposition theorem, Theorem 10.2.1, holding true for matrices.

Consider a Lie algebra $\mathbb{G}$ and define by

$$\mathscr{D}\mathbb{G} = [\mathbb{G}, \mathbb{G}] \tag{10.14}$$

the set of all elements $\mathbf{X} \in \mathbb{G}$ that can be written as the Lie bracket of two other elements $\mathbf{X} = [\mathbf{X}_1, \mathbf{X}_2]$. Clearly, $\mathscr{D}\mathbb{G}$ is an ideal in $\mathbb{G}$.

Definition 10.3.1. The sequence $\mathscr{D}^n\mathbb{G} = [\mathscr{D}^{n-1}\mathbb{G}, \mathscr{D}^{n-1}\mathbb{G}]$ of ideals

$$G \supset \mathscr{D}\mathbb{G} \supset \mathscr{D}^2\mathbb{G} \supset \cdots \supset \mathscr{D}^n\mathbb{G} \tag{10.15}$$

is called the *derivative series* of the Lie algebra $\mathbb{G}$.

10.3.1 Solvable Lie algebras

Definition 10.3.2. A Lie algebra $\mathbb{G}$ is called *solvable* if there exists an integer $n \in \mathbb{N}$ such that

$$\mathscr{D}^n\mathbb{G} = \{0\}. \tag{10.16}$$

Lemma 10.3.1. *A subalgebra $\mathbb{K} \subset \mathbb{G}$ of a solvable Lie algebra is also solvable.*

Proof of Lemma 10.3.1. Indeed, let G be solvable and let $\mathbb{K} \subset \mathbb{G}$ be a subalgebra. Clearly, $\mathscr{D}\mathbb{K} \subset \mathscr{D}\mathbb{G}$ and hence at every level n we have $\mathscr{D}^n\mathbb{K} \subset \mathscr{D}^n\mathbb{G}$, so the lemma follows. □

Definition 10.3.3. A Lie algebra $\mathbb{G}$ has the *chain property* if and only if for each ideal $\mathbb{H} \subset \mathbb{G}$ there exists an ideal $\mathbb{H}_1 \subset \mathbb{H}$ of the considered ideal which has codimension one in $\mathbb{H}$.

The above definition can be illustrated in the following way. Let $\mathbb{H}$ be the considered ideal in $\mathbb{G}$. If G has the chain property, then $\mathbb{H}$ can be written in the following way:

$$\mathbb{H} = \mathbb{H}_1 \oplus \lambda\mathbf{X}, \tag{10.17}$$

where $\mathbb{H}_1$ is a subspace of dimension

$$\dim \mathbb{H}_1 = \dim \mathbb{H} - 1 \tag{10.18}$$

and $\mathbf{X} \in \mathbb{H}$ ($\mathbf{X} \ni \mathbb{H}_1$) is an element that belongs to $\mathbb{H}$ but not to $\mathbb{H}_1$. Furthermore, we have

$$\forall \mathbf{Z} \in \mathbb{H}_1, \quad [\mathbf{Z}, \mathbf{X}] \in \mathbb{H}_1. \tag{10.19}$$

From this definition we obtain the following lemma.

Lemma 10.3.2. *A Lie algebra $\mathbb{G}$ is solvable if and only if it admits the chain property.*

Proof of Lemma 10.3.2. Let G be solvable and let us put $\dim \mathbb{G} = n$ and $\dim \mathscr{D}\mathbb{G} = m$. By the hypothesis of solvability we have $\mathscr{D}\mathbb{G} \neq \mathbb{G}$, so $n - m = p > 0$. Let us choose $p - 1$ linear independent elements $\{\mathbf{X}_1, \dots, \mathbf{X}_{p-1}\} \in \mathbb{G}$ such that $\mathbf{X}_i \notin \mathscr{D}\mathbb{G}$ and let us define the subspace

$$H_1 = \mathscr{D}\mathbb{G} + \lambda_1\mathbf{X}_1 + \cdots + \lambda_{p-1}\mathbf{X}_{p-1} \quad (\lambda_i \in \mathbb{C}). \tag{10.20}$$

By construction H_1 has codimension one and it is an ideal. This construction can be repeated for each ideal $\mathbb{H} \subset \mathbb{G}$ since it is solvable. Hence, $\mathbb{G}$ admits the chain property.

Conversely, let $\mathbb{G}$ be a Lie algebra admitting the chain property. Then we find a sequence of ideals

$$\mathbb{G} = \mathbb{G}^0 \supset \mathbb{G}^1 \supset \mathbb{G}^2 \supset \cdots \supset \mathbb{G}^n = \{0\} \tag{10.21}$$

such that $\mathbb{G}^r$ is an ideal in $\mathbb{G}^{r-1}$ of codimension one so that $\mathscr{D}^{r-1}\mathbb{G} \subset \mathbb{G}^r$. Hence, $\mathbb{G}$ is solvable. □

We can now state the most relevant property of solvable Lie algebras. This is encoded in the following Levi theorem and its corollary.

Theorem 10.3.1. *Let $\mathbb{G}$ be a solvable Lie algebra and let V be a finite-dimensional vector space over the field $\mathbb{F} = \mathbb{R}$, or $\mathbb{C}$, algebraically closed. Furthermore, let*

$$\pi : \mathbb{G} \to \mathrm{Hom}(V, V) \tag{10.22}$$

be a homomorphism of $\mathbb{G}$ on the algebra of linear endomorphisms of V. Then there exists a vector $\mathbf{v} \in V$ such that it is a simultaneous eigenvector for all elements $\pi(\mathbf{X})$ ($\forall \mathbf{X} \in \mathbb{G}$).

Proof of Theorem 10.3.1. The proof is constructed by induction. If $\dim \mathbb{G} = 1$, then there is just one endomorphism $\pi(g)$ and it necessarily admits an eigenvector. Suppose next that the theorem is true for each solvable algebra $\mathbb{K}$ of dimension

$$\dim \mathbb{K} < \dim \mathbb{G}. \tag{10.23}$$

Consider an ideal $\mathbb{H} \subset \mathbb{G}$ of codimension one:

$$\dim \mathbb{G} = \dim \mathbb{H} + 1. \tag{10.24}$$

Such an ideal exists because the Lie algebra is solvable and therefore admits the chain property. Write

$$\mathbb{G} = \mathbb{H} + \lambda \mathbf{X} \quad (\lambda \in \mathbb{F}), \tag{10.25}$$

where $\mathbf{X}$ is an element of $\mathbb{G}$ not contained in $\mathbb{H}$. By the induction hypothesis there exists a vector $\mathbf{e}_0 \in V$ such that

$$\forall \mathbf{H} \in \mathbb{H}: \quad \pi(\mathbf{H})\mathbf{e}_0 = \lambda(\mathbf{H})\mathbf{e}_0, \tag{10.26}$$

where $\lambda(\mathbf{H}) \in \mathbb{F}$ is an eigenvalue depending on the considered element $\mathbf{H}$. Define next the following vectors:

$$\mathbf{e}_p = [\pi(\mathbf{X})]^p \mathbf{e}_0, \quad p = 1, 2, \ldots. \tag{10.27}$$

The subspace $W \subset V$ spanned by the vectors $\mathbf{e}_p$ $(p \geq 0)$ is clearly invariant with respect to $\pi(\mathbf{X})$. We can also show that

$$\pi(\mathbf{H})\mathbf{e}_p = \lambda(\mathbf{H})\mathbf{e}_p \mod(\mathbf{e}_0, \ldots, \mathbf{e}_{p-1}) \quad (\forall \mathbf{H} \in \mathbb{H}). \tag{10.28}$$

Indeed, equation (10.28) is true for $p = 0$ and assuming it is true for p we get

$$\begin{aligned} \pi(\mathbf{H})\mathbf{e}_{p+1} &= \pi(\mathbf{H})\pi(\mathbf{X})\mathbf{e}_p = \pi([\mathbf{H}, X])\mathbf{e}_p + \pi(\mathbf{X})\pi(\mathbf{H})\mathbf{e}_p \\ &= \lambda([\mathbf{H}, \mathbf{X}])\mathbf{e}_p + \lambda(\mathbf{H})\mathbf{e}_{p+1} + \mathrm{mod}(\mathbf{e}_0, \ldots, \mathbf{e}_{p-1}). \end{aligned} \tag{10.29}$$

(Note that $[\mathbf{H}, \mathbf{X}] \in \mathbb{H}$.) Hence, we find

$$\pi(\mathbf{H})\mathbf{e}_{p+1} = \lambda(\mathbf{H})\mathbf{e}_{p+1} + \mathrm{mod}(\mathbf{e}_0, \ldots, \mathbf{e}_p). \tag{10.30}$$

It follows that the subspace W is invariant with respect to $\pi(\mathbb{G})$ and that

$$\mathrm{Tr}_W \, \pi(\mathbf{H}) = \lambda(\mathbf{H}) \dim W. \tag{10.31}$$

On the other hand, we have

$$\mathrm{Tr}_W(\pi([\mathbf{H}, \mathbf{X}])) = 0 \quad \Rightarrow \quad \lambda([\mathbf{H}, \mathbf{X}]) = 0. \tag{10.32}$$

Repeating the argument by induction, from the relation

$$\pi(\mathbf{H})\mathbf{e}_{p+1} = \pi([\mathbf{H}, \mathbf{X}])\mathbf{e}_p + \pi(\mathbf{X})\pi(\mathbf{H})\mathbf{e}_p \tag{10.33}$$

and the original definition of the eigenvalue $\lambda(\mathbf{H})$ in equation (10.26) we conclude that

$$\pi(\mathbf{H})\mathbf{e}_p = \lambda(\mathbf{H})\mathbf{e}_p \quad (p \geq 0). \tag{10.34}$$

This shows that $\forall \mathbf{H} \in \mathbb{H}$ we have $\pi(\mathbf{H}) = \lambda(\mathbf{H})\mathbf{1}$ on the vector subspace W. Choosing a vector $\mathbf{e}'_p \in W$ that is an eigenvector of $\pi(\mathbf{X})$ we find that it is a simultaneous eigenvector for all elements $\pi(\mathbf{X})$ ($\forall \mathbf{X} \in \mathbb{G}$). □

Corollary 10.3.1. *Let $\mathbb{G}$ be a solvable Lie algebra and let π be a linear representation of $\mathbb{G}$ on a finite-dimensional vector space V. Then there exists a basis $\{\mathbf{e}_1, \ldots, \mathbf{e}_n\}$ where every $\pi(\mathbf{X})$ ($\forall \mathbf{X} \in \mathbb{G}$) is a triangular matrix.*

Proof of Corollary 10.3.1. Let $\mathbf{e}_1$ be the simultaneous eigenvector of all $\pi(\mathbf{X})$. The representation π induces a new linear representation on the quotient vector space V/E_1, where $E_1 \equiv \lambda\mathbf{e}_1$. Hence applying Theorem 10.3.1 to this new representation we conclude that there is a common eigenvector $\mathbf{e}_2 + \lambda\mathbf{e}_1$ for all $\pi(\mathbf{X})$. Continuing in this way we obtain a basis such that

$$\pi(\mathbf{X})\mathbf{e}_i \equiv 0 \ \mathrm{mod}(\mathbf{e}_1, \mathbf{e}_2, \ldots, \mathbf{e}_i) \tag{10.35}$$

so that $\pi(\mathbf{X})$ is indeed upper triangular. □

We are now ready to introduce the concept of semi-simple Lie algebras and discuss their general properties.

10.3.2 Semi-simple Lie algebras

We introduce some more definitions.

Definition 10.3.4. Let $\mathbb{G}$ be a Lie algebra. An ideal $\mathbb{H} \subset \mathbb{G}$ is called *maximal* if there is no other ideal $\mathbb{H}' \subset \mathbb{G}$ such that $\mathbb{H}' \supset \mathbb{H}$ than $\mathbb{H}$ itself.

Definition 10.3.5. The maximal solvable ideal of a Lie algebra $\mathbb{G}$ is called the radical of $\mathbb{G}$ and it is denoted Rad $\mathbb{G}$.

Definition 10.3.6. A Lie algebra $\mathbb{G}$ is named semi-simple if and only if Rad $\mathbb{G} = 0$.

As an immediate consequence of the definition we have the following theorem.

Theorem 10.3.2. *A Lie algebra $\mathbb{G}$ is semi-simple if and only if it does not have any non-trivial Abelian ideal.*

Proof of Theorem 10.3.2. We have to show the equivalence of the following two propositions:

(a): $\mathbb{G}$ has a solvable ideal.
(b): $\mathbb{G}$ has an Abelian ideal.

(i) Let us show that b $\Rightarrow$ a. Let $\mathscr{I} \subset \mathbb{G}$ be the Abelian ideal. By definition we have $[\mathscr{I}, \mathscr{I}] = \mathscr{D}\mathscr{I} = 0$. Hence, $\mathscr{I}$ itself is a solvable ideal and this proves (a).
(ii) Let us now show that a $\Rightarrow$ b. To this effect let $\mathscr{I} \subset \mathbb{G}$ be the solvable ideal. By definition, since $\mathscr{I}$ is non-trivial, $\exists k \in \mathbb{N}$ such that $\mathscr{D}^{k-1}\mathscr{I} \neq 0$ and $\mathscr{D}^{k}\mathscr{I} = 0$. Then $\mathscr{D}^{k-1}\mathscr{I}$ is Abelian and being the derivative of an ideal is an ideal. Hence, (b) is true, and this concludes the proof of the theorem. □

10.3.3 Levi's decomposition of Lie algebras

We want to prove that any Lie algebra can be seen as the semi-direct product of its radical with a semi-simple Lie algebra. Such a decomposition is named the Levi decomposition, and this is what we want to illustrate in the present section. To this effect we need to introduce some preliminary notions. The first is the notion of *Lie algebra cohomology*. It is a further very relevant example of an algebraic construction that realizes the paradigm introduced in Section 6.9.3. The second is the equally important notion of semi-direct product.

10.3.3.1 Lie algebra cohomology

Let $\mathbb{G}$ be a Lie algebra and let $\rho : \mathbb{G} \to \mathrm{End}(V)$ be a representation of $\mathbb{G}$ on a complex, finite-dimensional vector space V. Let $V^s(\mathbb{G}, \rho)$ be the vector space of all antisymmetric linear maps

$$\theta : \underbrace{\mathbb{G} \times \mathbb{G} \times \cdots \times \mathbb{G}}_{s\text{ times}} \to V. \tag{10.36}$$

The spaces $V^s(\mathbb{G}, \rho)$ are the spaces of *s-cochains*. We can next define a coboundary operator d in the following way. Let $\theta \in V^s(\mathbb{G}, \rho)$ be an s-cochain. The value of the $(s+1)$-cochain $d\theta$ on any set of $s+1$ elements $X_1, \dots, X_{s+1}$ of the Lie algebra $\mathbb{G}$ is given by the following expression:

$$\begin{aligned} d\theta(\mathbf{X}_1, \mathbf{X}_2, \dots \mathbf{X}_{s+1}) &= \sum_{i=1}^{s+1} (-)^{i+1} \rho(\mathbf{X}_i)\theta(\mathbf{X}_1, \dots, \widehat{\mathbf{X}}_i, \dots, \mathbf{X}_{s+1}) \\ &\quad - \sum_{r=1}^{s+1} \sum_{q<r} (-1)^{r+q} \theta(\mathbf{X}_1, \dots, \widehat{\mathbf{X}}_q, \dots, \widehat{\mathbf{X}}_r, \dots, \mathbf{X}_{s+1}, [\mathbf{X}_q, \mathbf{X}_r]), \end{aligned} \tag{10.37}$$

where the hat on top of an X-element means that it is omitted. It is straightforward to verify that by applying a second time the coboundary operator d we obtain identically zero, namely,

$$d^2 = 0. \tag{10.38}$$

In particular, if we consider the 1-cochains $\theta^{[1]}$ that are maps $\mathbb{G} \to V$ from the Lie algebra to the vector space V, by applying the general definition (10.37) we obtain

$$\forall \mathbf{X}, \mathbf{Y} \in \mathbb{G} : \quad d\theta^{[1]}(\mathbf{X}, \mathbf{Y}) = \rho(\mathbf{X})\theta(\mathbf{Y}) - \rho(\mathbf{Y})\theta(\mathbf{X}) - \theta([\mathbf{X}, \mathbf{Y}]). \tag{10.39}$$

Given the coboundary operator d we have the usual definitions of an *elliptic complex*:
1. The space $C^{(n)}(\mathbb{G}, \rho)$ of n-cocycles is the vector space of all n-chains $\theta^{[n]}$ that are closed, $d\theta^{[n]} = 0$.
2. The space $B^{(n)}(\mathbb{G}, \rho)$ of n-coboundaries is the vector space of all n-chains $\theta^{[n]}$ that are exact, namely, that can be written as $\theta^{[n]} = d\phi^{[n-1]}$ for some $(n-1)$-chain $\phi^{[n-1]}$.
3. The n-th cohomology group $H^{[n]}(\mathbb{G}, \rho)$ of the Lie algebra $\mathbb{G}$ relative to the linear representation ρ is the quotient

$$H^{[n]}(\mathbb{G}, \rho) = \frac{C^{(n)}(\mathbb{G}, \rho)}{B^{(n)}(\mathbb{G}, \rho)}, \tag{10.40}$$

namely, it is the vector space whose equivalence classes are the n-cocycles modulo the n-coboundaries.

We have a useful general cohomological property of semi-simple Lie algebras that follows from the above definitions but whose proof we omit for brevity.

Theorem 10.3.3. *Let $\mathbb{G}$ be a semi-simple Lie algebra and let ρ be a linear representation of $\mathbb{G}$ on a finite-dimensional vector space V. Then the first two cohomology groups are trivial:*

$$H^{[1]}(\mathbb{G}, \rho) = H^{[2]}(\mathbb{G}, \rho) = 0. \tag{10.41}$$

10.3.3.2 Semi-direct product

We begin with two definitions.

Definition 10.3.7. Let $\mathbb{G}$ be a Lie algebra and let $\sigma : \mathbb{G} \to \mathbb{G}$ be an endomorphism of vector spaces. We say that σ is a *derivation* of the algebra if the following property holds true:

$$\forall \mathbf{X}, \mathbf{Y} \in \mathbb{G} : \quad \sigma([\mathbf{X}, \mathbf{Y}]) = [\sigma(\mathbf{X}), \mathbf{Y}] + [\mathbf{X}, \sigma(\mathbf{Y})]. \tag{10.42}$$

Definition 10.3.8. Let $\mathbb{Q}$ and $\mathbb{M}$ be two Lie algebras and let σ be a linear representation of $\mathbb{M}$ on $\mathbb{Q}$ such that $\forall \mathbf{Y} \in \mathbb{M}$ the map $\sigma(\mathbf{Y})$ is a derivation of $\mathbb{Q}$. Next, let $\mathbf{X}, \mathbf{X}'$ be elements of $\mathbb{Q}$ and let $\mathbf{Y}, \mathbf{Y}'$ be elements of $\mathbb{M}$. We define the Lie bracket of the ordered pair $(\mathbf{X}, \mathbf{Y})$ with the ordered pair $(\mathbf{X}', \mathbf{Y}')$ in the following way:

$$[(\mathbf{X},\mathbf{Y}),(\mathbf{X}',\mathbf{Y}')] = ([\mathbf{X},\mathbf{X}'] + \sigma(\mathbf{Y})\mathbf{X}' - \sigma(\mathbf{Y}')\mathbf{X}, [\mathbf{Y},\mathbf{Y}']). \tag{10.43}$$

With this definition of the Lie bracket, $\mathbb{Q} \times_\sigma \mathbb{M}$ becomes a Lie algebra, and it is called the semi-direct product of $\mathbb{Q}$ with $\mathbb{M}$ relative to the representation σ.

It is a straightforward exercise to check that the definition of the Lie bracket (10.43) is consistent and satisfies the Jacobi identity.

Let us now consider a Lie algebra $\mathbb{G}$ and let $\mathscr{Q} \subset \mathbb{G}$ be an ideal and $\mathbb{M} \subset \mathbb{G}$ a subalgebra such that, as vector spaces, we have the following orthogonal decomposition:

$$\mathbb{G} = \mathscr{Q} \oplus \mathbb{M} \quad \Rightarrow \quad \mathscr{Q} \cap \mathbb{M} = 0. \tag{10.44}$$

Obviously, $\mathbb{G}$ can be regarded as the semi-direct product of $\mathbb{Q}$ with $\mathbb{M}$. It suffices to use as derivation σ the internal derivation provided by the Lie bracket of $\mathbb{G}$:

$$\forall \mathbf{Y} \in \mathbb{M}, \forall \mathbf{X} \in \mathscr{Q}: \quad \sigma(\mathbf{Y})\mathbf{X} \equiv -[\mathbf{X},\mathbf{Y}]. \tag{10.45}$$

Definition 10.3.9. Let $\mathbb{G}$ be a Lie algebra. We say that $\mathbb{G}$ is *decomposed according to Levi* if there exists a subalgebra $\mathbb{L} \subset \mathbb{G}$ such that

$$\mathbb{G} = \mathbb{L} \times_{[,]} \operatorname{Rad}\mathbb{G}. \tag{10.46}$$

Obviously, since $\mathbb{G}/\operatorname{Rad}\mathbb{G}$ is semi-simple and $\mathbb{L} \sim \mathbb{G}/\operatorname{Rad}\mathbb{G}$, also $\mathbb{L}$ is semi-simple. It is called a *Levi subalgebra.*

Relying on this definition we can state the following fundamental theorem.

Theorem 10.3.4. *Let G be a Lie algebra and denote $\mathscr{Q} \equiv \operatorname{Rad}\mathbb{G}$. Every $\mathbb{G}$ admits Levi subalgebras. Furthermore, if $\mathbb{L} \subset \mathbb{G}$ is a Levi subalgebra of $\mathbb{G}$, it is also a Levi subalgebra of $\mathscr{D}\mathbb{G}$ and*

$$\mathscr{D}\mathbb{G} = [\mathscr{Q}, \mathbb{G}] \oplus \mathbb{L} \tag{10.47}$$

is a Levi decomposition of $\mathscr{D}\mathbb{G}$.

In order to prove Theorem 10.3.4 we need the following lemma.

Lemma 10.3.3. *Let $\mathbb{H}$ be a Lie algebra and let $\mathscr{Q}$ be its radical. If $\mathscr{A} \subset \mathbb{H}$ is an ideal such that $\mathbb{H}/\mathscr{A}$ is semi-simple, then $\mathscr{Q} \sqsubseteq \mathscr{A}$. Furthermore, if π is a homomorphism of $\mathbb{H}$ onto an algebra $\mathbb{H}'$, then $\pi(\mathscr{Q})$ is the radical of $\mathbb{H}'$.*

Proof of Lemma 10.3.3. Consider the natural map

$$\tau : \mathbb{H} \to \mathbb{H}/\mathscr{A} \tag{10.48}$$

that with each element $h \in \mathbb{H}$ associates the equivalence class $h + \mathscr{A}$. If $\mathscr{Q} \nsubseteq \mathscr{A}$, $\tau(\mathscr{Q})$ is a non-zero and solvable ideal of $\mathbb{H}/\mathscr{A}$. Indeed, under the homomorphism τ the ideal

$\mathcal{Q}$ flows into an ideal $\tau(\mathcal{Q})$. Furthermore, under the homomorphism we have $\mathcal{D}\tau(\mathcal{Q}) = \tau(\mathcal{D}\mathbb{G})$ so that if $\mathcal{Q}$ is solvable, the same is also true for $\tau(\mathcal{Q})$. The existence of a solvable ideal is in contradiction with the assumption that $\mathbb{H}/\mathcal{A}$ is semi-simple. Hence, $\mathcal{Q} \subseteq \mathcal{A}$, necessarily.

Let us arrive at the second part of the lemma. Let $\mathcal{Q}' = \mathrm{Rad}\,\mathbb{H}'$. The homomorphism π induces a homomorphism of $\mathbb{H}/\mathcal{Q}$ in $\mathbb{H}'/\pi(\mathcal{Q})$; hence, since $\mathbb{H}/\mathcal{Q}$ is semi-simple, also $\mathbb{H}/\pi(\mathcal{Q})$ is semi-simple. Therefore, relying on the previous result, $\mathcal{Q}' \subset \pi(\mathcal{Q})$. On the other hand, $\pi(\mathcal{Q})$ is a solvable ideal of $\mathbb{H}$. This implies $\pi(\mathcal{Q}) \subset \mathcal{Q}'$. We conclude $\pi(\mathcal{Q}) = \mathcal{Q}'$ and the lemma is proved. □

Let us now arrive at the proof of the main theorem, Theorem 10.3.4.

Proof of Theorem 10.3.4. The proof of Theorem 10.3.4 is by induction on the dimension of the radical $\dim \mathcal{Q}$. If $\dim \mathcal{Q} = 0$, then $\mathbb{G}$ is semi-simple and it is by itself a Levi subalgebra. Let us then assume that $\dim \mathcal{Q} \geq 1$ and that Levi subalgebras do exist for any Lie algebra $\mathbb{G}'$ such that $\dim \mathrm{Rad}\,\mathbb{G}' < \dim \mathrm{Rad}\,\mathbb{G}$. We consider two cases:

Case 1 The radical $\mathcal{Q}$ is non-Abelian, namely, $\mathcal{D}\mathcal{Q} \neq 0$.

As is well known, $\mathcal{D}\mathcal{Q}$ is by itself an ideal. Hence, we consider the Lie algebra $\mathbb{G}' \equiv \mathbb{G}/\mathcal{D}\mathcal{Q}$ and we let π be the natural map

$$\pi : \mathbb{G} \to \mathbb{G}' \equiv \mathbb{G}/\mathcal{D}\mathcal{Q}. \tag{10.49}$$

Relying on Lemma 10.3.3, we have $\mathcal{Q}' = \mathrm{Rad}\,\mathbb{G}' = \pi[\mathcal{Q}]$. Hence,

$$\mathcal{Q}' = \mathcal{Q}/\mathcal{D}\mathcal{Q} \quad \Rightarrow \quad \dim \mathcal{Q}' < \dim \mathcal{Q}. \tag{10.50}$$

By the induction hypothesis $\mathbb{G}'$ admits a Levi subalgebra $\mathcal{M}'$ and we have as vector spaces

$$\mathbb{G}' = \mathcal{Q}' \oplus \mathcal{M}'. \tag{10.51}$$

Define $\mathcal{M}_0 = \pi^{-1}(\mathcal{M}')$. We obtain

$$\mathbb{G} = \pi^{-1}(\mathbb{G}') = \mathcal{Q} \oplus \mathcal{M}_0. \tag{10.52}$$

Furthermore, it is true that $\mathcal{D}\mathcal{Q} = \mathcal{Q} \cap \mathcal{M}_0$ (indeed the common elements of $\mathcal{Q}$ and $\mathcal{M}_0$ must be contained in the kernel of π, namely, $\pi^{-1}(0)$). Hence, $\mathcal{D}\mathcal{Q}$ is a solvable ideal of $\mathcal{M}_0$ and since $\mathcal{M}_0/\mathcal{D}\mathcal{Q} \sim \mathcal{M}'$ = semi-simple algebra, we have $\mathcal{D}\mathcal{Q} \subset \mathrm{Rad}\,\mathcal{M}_0$. Yet by force of Lemma 10.3.3 we also have $\mathrm{Rad}\,\mathcal{M}_0 \subset \mathcal{D}\mathcal{Q}$, which implies $\mathrm{Rad}\,\mathcal{M}_0 = \mathcal{D}\mathcal{Q}$. Since $\mathcal{Q}$ is solvable by definition we have $\dim \mathcal{D}\mathcal{Q} < \dim \mathcal{Q}$. Then by the induction hypothesis we conclude that $\mathcal{M}_0$ admits a Levi decomposition,

$$\mathcal{M}_0 = \mathbb{L} \oplus \mathcal{D}\mathcal{Q}; \quad \mathbb{L} \cap \mathcal{D}\mathcal{Q} = 0, \tag{10.53}$$

from which we conclude

$$\mathbb{G} = \mathbb{L} \oplus \mathcal{Q}. \tag{10.54}$$

The theorem is proved in this case.

Case 2 The radical $\mathcal{Q}$ is Abelian, namely, $\mathcal{Q}\mathcal{Q} = 0$.

To prove the theorem in this case we have to use Theorem 10.3.3 stating that the second cohomology group of a semi-simple Lie algebra vanishes. Define $\mathbb{G}_1 = \mathbb{G}/\mathcal{Q}$ (so that $\mathbb{G}_1$ is semi-simple) and let π be the natural map of $\mathbb{G}$ onto $\mathbb{G}_1$. Let μ be any linear map of $\mathbb{G}_1$ into $\mathbb{G}$ such that $\pi \circ \mu = \text{id}$. For each $\mathbf{X}_1 \in \mathbb{G}_1$ define by $\rho(\mathbf{X}_1)$ the endomorphism $\text{ad}(\mathbf{X})|_{\mathcal{Q}}$, where $\mathbf{X}$ is such that $\pi(\mathbf{X}) = \mathbf{X}_1$. Since $\mathcal{Q}$ is Abelian, this is a well-posed definition. Indeed, $\mathbf{X}_1 = \mathbf{X} + \mathcal{Q}$ and $\forall q \in \mathcal{Q}$,

$$\rho(\mathbf{X}_1)q = [\mathbf{X} + \mathcal{Q}, q] = [\mathbf{X}, q]. \tag{10.55}$$

The map $\mathbf{X}_1 \mapsto \rho(\mathbf{X}_1)$ is a linear representation of the semi-simple Lie algebra $\mathbb{G}_1$ on the vector space $\mathcal{Q}$. Obviously, $\rho(\mathbf{X}_1) = \text{ad}\,\mu(\mathbf{X}_1)|_{\mathbb{G}}$. Next, define

$$\forall \mathbf{X}, \mathbf{Y} \in \mathbb{G}_1: \quad \theta(\mathbf{X}, \mathbf{Y}) \equiv [\mu(\mathbf{X}), \mu(\mathbf{Y})] - \mu([\mathbf{X}, \mathbf{Y}]). \tag{10.56}$$

Since π is a homomorphism and $\pi \circ \mu = \text{id}$, we have $\pi(\theta(\mathbf{X}, \mathbf{Y})) = 0 \Rightarrow \theta(\mathbf{X}, \mathbf{Y}) \in \mathcal{Q}$. This guarantees that $\theta \in V^2(\mathbb{G}_1, \rho)$ is a 2-cochain of the Lie algebra $\mathbb{G}_1$ relative to the representation ρ. By direct calculation, using the Jacobi identity we can immediately verify that θ is actually a 2-cycle, namely, $d\theta = 0$. Since the second cohomology group vanishes for semi-simple Lie algebras, $H^2(\mathbb{G}_1, \rho) = 0$, it follows that there exists a linear map

$$\nu : \mathbb{G}_1 \mapsto \mathcal{Q} \tag{10.57}$$

such that $d\nu = \theta$, namely,

$$[\mu(\mathbf{X}), \mu(\mathbf{Y})] - \mu([\mathbf{X}, \mathbf{Y}]) = [\mu(\mathbf{X}), \mu(\mathbf{Y})] - [\mu(\mathbf{X}), \nu(\mathbf{Y})] - [\mu(\mathbf{Y}), \nu(\mathbf{X})] - \nu([\mathbf{X}, \mathbf{Y}]). \tag{10.58}$$

Since $\nu(\mathbf{X}) \in \mathcal{Q}$, we have $[\nu(\mathbf{X}), \nu(\mathbf{Y})] = 0$. Hence, defining

$$\lambda(\mathbf{X}) = \mu(\mathbf{X}) - \nu(\mathbf{X}) \tag{10.59}$$

we see that $\lambda : \mathbb{G}_1 \mapsto \mathbb{G}_1$ is a homomorphism. It is also evident that $\pi \circ \lambda = \text{id}$. So we have found a map $\lambda : \mathbb{G}_1 \mapsto \mathbb{G}$ which is a homomorphism of algebras. It follows that $\lambda(\mathbb{G}_1) \subset \mathbb{G}$ is a subalgebra. Furthermore, by construction

$$\mathbb{G} = \mathcal{Q} \oplus \lambda(\mathbb{G}_1); \quad \mathcal{Q} \cap \lambda(\mathbb{G}_1) = 0. \tag{10.60}$$

Hence, $\lambda(\mathbb{G}_1)$ is a Levi subalgebra and we have completed the induction argument also in this case.

The theorem is proved. □

10.3.4 An illustrative example: the Galilei group

The invariance group of classical non-relativistic mechanics is the *Galilei group*, which consists of the following transformations on the space-time manifold whose points are labeled by the three space coordinates x^i and the time instant t:

$$\begin{pmatrix} x^i \\ t \end{pmatrix} \mapsto \begin{pmatrix} x^{i\prime} \\ t' \end{pmatrix}, \tag{10.61}$$

where

$$\begin{cases} x^{i\prime} = R^i{}_j x^j + v^i t + c^i, \\ t' = t + T \end{cases} \tag{10.62}$$

and

$$\begin{aligned} R^i{}_j &= \text{rotation matrix } RR^T = 1, \\ x^i &\mapsto x^i + c^i \quad \text{is a translation}, \\ x^i &\mapsto x^i + v^i t \quad \text{corresponds to a special Galilei transformation}, \\ t &\mapsto t + T \quad \text{corresponds to a time translation}. \end{aligned} \tag{10.63}$$

The total number of parameters is 10, just as for the relativistic Poincaré group. Let us write the corresponding Lie algebra. For the rotations we have the *angular momentum* generators

$$J_{ij} = x_i \partial_j - x_j \partial_i \rightarrow J_i = \epsilon_{ijk} x_j \partial_k \tag{10.64}$$

and for the space translations we have the *momentum generators*

$$P_i = \partial_i, \tag{10.65}$$

while the *Galilean boosts* are generated by

$$K_i = t\partial_i. \tag{10.66}$$

Finally, the *Hamiltonian* generates time translations:

$$H = \partial_t. \tag{10.67}$$

By explicit evaluation of the commutators we find that the Galilei–Lie algebra has the following structure:

$$\begin{aligned}
[J_i, J_j] &= \epsilon_{ijk} J_k; & [J_i, P_j] &= -\epsilon_{ijk} P_k; \\
[J_i, K_j] &= -\epsilon_{ijk} K_k; & [J_i, H] &= 0; \\
[P_i, H] &= 0; & [P_i, P_j] &= 0; \\
[K_i, H] &= -P_i; & [K_i, K_j] &= 0; \\
[P_i, K_j] &= 0.
\end{aligned} \tag{10.68}$$

We can ask the question whether the Galilei algebra $\mathbb{G}$ is *semi-simple*. The answer is no. Indeed, P_i ($i = 1, 2, 3$) generate an Abelian ideal; we can easily verify that $[P, X] \subset P$, $\forall \mathbf{X} \in \mathbb{G}$, so P is an ideal. Next we inquire whether $\mathbb{G}$ is *solvable*. The derivative algebra $D\mathbb{G}$ is made up of J_i, P_i, K_i. We easily verify, however, that $D^2\mathbb{G} = D\mathbb{G}$, so $\mathbb{G}$ is not solvable. On the other hand, if we consider the subalgebra $S^{(0)}$ generated by $\{P, K, H\}$, we see that

$$DS^{(0)} = S^{(1)} = \{P\}; \quad DS^{(1)} = \{0\}, \tag{10.69}$$

so that $S^{(0)}$ is solvable. The algebra generated by J_i is instead semi-simple. Hence, the Galilei algebra is, according to Levi's theorem, the direct product of a semi-simple algebra with a solvable one.

10.4 The adjoint representation and Cartan's criteria

Let us now introduce the concept of *adjoint representation* of a Lie algebra $\mathbb{G}$. Given such an algebra, with each element $\mathbf{X} \in \mathbb{G}$ we can associate a *linear endomorphism*,

$$\mathrm{ad}_{\mathbf{X}} : \mathbb{G} \to \mathbb{G}, \tag{10.70}$$

defined by

$$\forall \mathbf{Y} \in \mathbb{G} : \quad \mathrm{ad}_{\mathbf{X}}(\mathbf{Y}) \equiv [\mathbf{X}, \mathbf{Y}]. \tag{10.71}$$

If we choose a basis $\{T_A\}$ we immediately get

$$(\mathrm{ad}_{\mathbf{X}})_A{}^B = X^M f_{MA}{}^B, \tag{10.72}$$

where $f_{MA}{}^B$ are the Lie algebra structure constants, defined by

$$[T_A, T_B] = f_{AB}{}^C T_C. \tag{10.73}$$

Then we can introduce the bilinear symmetric Killing form of the Lie algebra,

$$\kappa : \mathbb{G} \otimes \mathbb{G} \to \mathbb{F}, \tag{10.74}$$

defined by

$$\forall \mathbf{X}, \mathbf{Y} \in \mathbb{G} : \quad \kappa(\mathbf{X}, \mathbf{Y}) = \operatorname{Tr}(\mathrm{ad}_{\mathbf{X}}\, \mathrm{ad}_{\mathbf{Y}}), \tag{10.75}$$

where $\mathbb{F}$ is the field over which the Lie algebra is constructed, namely, $\mathbb{F} = \mathbb{C}$ for complex Lie algebras and $\mathbb{F} = \mathbb{R}$ for real Lie algebras.

In a basis we obtain

$$\begin{aligned} \kappa(\mathbf{X}, \mathbf{Y}) &= (\mathrm{ad}_{\mathbf{X}})_A{}^B (\mathrm{ad}_{\mathbf{Y}})_B{}^A = \mathbf{X}^M \mathbf{Y}^N f_{MA}{}^B f_{NB}{}^A \\ &= \mathbf{X}^M \mathbf{Y}^N g_{MN}^{(\text{Killing})}, \end{aligned} \tag{10.76}$$

where the symmetric tensor $g_{MN}^{(\text{Killing})} = f_{MA}{}^B f_{NB}{}^A$ is called the Killing metric.

10.4.1 Cartan's criteria

Whether a Lie algebra is solvable or semi-simple is fully encoded in the properties of the Killing form, which therefore provides a very useful global tool to test the structure of the Lie algebra. That this is the case is established by two simple but very important theorems that go under the name of Cartan's criteria.

The first Cartan criterion establishes a test of solvability and is provided by the following theorem.

Theorem 10.4.1. *A Lie algebra $\mathbb{G}$ is solvable if and only if*

$$\forall \mathbf{X}, \mathbf{Y}, \mathbf{Z} \in \mathbb{G} \quad \kappa(\mathbf{X}, [\mathbf{Y}, \mathbf{Z}]) = 0. \tag{10.77}$$

For brevity we omit the proof of this theorem.

The second Cartan criterion, which uses the first in its own proof, is a test of semi-simplicity. It is given by the following theorem.

Theorem 10.4.2. *A Lie algebra $\mathbb{G}$ is semi-simple if and only if the Killing form $\kappa(,)$ on $\mathbb{G}$ is non-degenerate.*

Proof of Theorem 10.4.2. We recall that a bilinear form $\kappa(,)$ on a vector space $\mathbb{G}$ is degenerate if $\exists \mathbf{X} \in \mathbb{G}$ such that $\forall \mathbf{Y} \in \mathbb{G}$ we have $\kappa(\mathbf{X}, \mathbf{Y}) = 0$. In a basis $\mathbf{X}_i$ this implies that the determinant of the matrix $\kappa_{ij} = \kappa(\mathbf{X}_i, \mathbf{X}_j)$ vanishes, $\det \kappa_{ij} = 0$.

To prove the theorem we have to show that the following statements are both true:

(a) If $\mathbb{G}$ is semi-simple, then κ is non-degenerate.

(b) If κ is non-degenerate, then $\mathbb{G}$ is semi-simple.

Let us begin with case (a) and let us assume that κ is degenerate, namely, the set

$$B = \{\mathbf{X} : \kappa(\mathbf{X}, \mathbf{Y}) = 0, \forall \mathbf{Y} \in \mathbb{G}\} \tag{10.78}$$

contains non-trivial elements besides $\mathbf{0}$. We can immediately verify that B is an ideal of $\mathbb{G}$. Indeed, $\forall \mathbf{X} \in B$ and $\forall \mathbf{Z} \in \mathbb{G}$ we have $[\mathbf{X}, \mathbf{Y}] \in B$ since $\kappa([\mathbf{X}, \mathbf{Z}], \mathbf{Y}) = 0\ \forall \mathbf{Y} \in \mathbb{G}$. This follows from the properties of the Killing form that imply $\kappa([\mathbf{X}, \mathbf{Z}], \mathbf{Y}) = \kappa(\mathbf{X}, [\mathbf{X}, \mathbf{Y}]) = 0$. Next we can show that

$$\forall \mathbf{X}, \mathbf{X}' \in B : \quad \kappa(\mathbf{X}, \mathbf{X}') = \kappa_B(\mathbf{X}, \mathbf{X}'), \tag{10.79}$$

where $\kappa_B(,)$ denotes the restriction of the Killing form to ideal B. Indeed, given $\mathbf{Z} \in \mathbb{G}$ we have

$$\mathrm{ad}_{\mathbf{X}}\, \mathrm{ad}_{\mathbf{X}'}\, \mathbf{Z} = [\mathbf{X}, [\mathbf{X}', \mathbf{Z}]] \in B \quad \text{since } [\mathbf{X}', \mathbf{Z}] \in B. \tag{10.80}$$

This means that the image of the linear map $\mathrm{ad}_{\mathbf{X}}\, \mathrm{ad}_{\mathbf{X}'}$ is contained in the ideal B, which implies that the only contribution to the trace comes from its restriction to the subspace B. By our definition of the ideal B we have $\kappa(\mathbf{X}, \mathbf{X}') = 0$ for all $\mathbf{X}, \mathbf{X}' \in B$, which by the above argument implies also $\kappa_B(\mathbf{X}, \mathbf{X}') = 0$. Hence, the algebra $\mathbb{G}$ admits an ideal B whose Killing form is identically vanishing. By the first Cartan criterion, Theorem 10.4.1, it follows that the ideal B is solvable. Yet this contradicts the assumption that the Lie algebra $\mathbb{G}$ was semi-simple, so B necessarily contains only the zero element $\mathbf{0}$ and the Killing form is non-degenerate.

Let us turn to case (b). Assume that the Lie algebra $\mathbb{G}$ is not semi-simple and let us show that this implies that the Killing form is degenerate. If $\mathbb{G}$ is not semi-simple, there is a non-trivial solvable ideal $\mathscr{Q}$. By definition $\exists k \in \mathbb{N}$ such that

$$\mathscr{A} \equiv \mathscr{D}^k \mathscr{Q} \neq 0; \quad \mathscr{D}^{k+1} \mathscr{Q} = 0. \tag{10.81}$$

The subalgebra $\mathscr{A}$ is a non-trivial Abelian ideal. As the next step we show that

$$\forall \mathbf{X} \in \mathscr{A}, \forall \mathbf{Y} \in \mathbb{G} : \quad \kappa(\mathbf{X}, \mathbf{Y}) = \kappa_{\mathscr{A}}(\mathbf{X}, \mathbf{Y}). \tag{10.82}$$

Indeed, given $\mathbf{Z} \in \mathbb{G}$ we have $\mathrm{ad}_{\mathbf{X}} \circ \mathrm{ad}_{\mathbf{Y}}(\mathbf{Z}) = [\mathbf{X}, [\mathbf{Y}, \mathbf{Z}]] \in \mathscr{A}$, since $\mathscr{A}$ is an ideal. Hence, the image of $\mathrm{ad}_{\mathbf{X}} \circ \mathrm{ad}_{\mathbf{Y}}$ as $\mathbf{Z}$ varies in $\mathbb{G}$ takes values only in $\mathscr{A}$ and therefore its trace takes contributions only from $\mathscr{A}$. This suffices to prove that equation (10.82) is true. Next we observe that

$$\forall \mathbf{X} \in \mathscr{A}, \forall \mathbf{Y} \in \mathbb{G} \quad \text{we have } \kappa_{\mathscr{A}}(\mathbf{X}, \mathbf{Y}) = 0. \tag{10.83}$$

Indeed, given $\mathbf{X}' \in \mathscr{A}$ we have $\mathrm{ad}_{\mathbf{X}} \circ \mathrm{ad}_{\mathbf{Y}}(\mathbf{X}') = [\mathbf{X}, [\mathbf{Y}, \mathbf{X}']] = 0$, since both $\mathbf{X} \in \mathscr{A}$, $[\mathbf{Y}, \mathbf{X}'] \in \mathscr{A}$ and $\mathscr{A}$ is Abelian. Hence, there is no contribution to the trace. On the other hand, by force of equation (10.82) we conclude that $\kappa(\mathbf{X}, \mathbf{Y}) = 0$ for all $\mathbf{Y} \in \mathbb{G}$ and all

$\mathbf{X} \in \mathscr{A}$. This means that the Killing form is degenerate unless $\mathscr{A}$ is empty. So there cannot be any non-trivial solvable ideal and the algebra $\mathbb{G}$ has to be semi-simple. This concludes the proof of the theorem. □

10.5 Bibliographical note

The main bibliographical sources for this chapter are:
1. The textbook [39].
2. The textbook [83].
3. The textbook [38].
4. The textbook [80].
5. The textbook [76].
6. Volume 2 of [79].

11 Root systems and their classification

Tout ce qu'on invente est vrai, soi-en sure. La poésie est une chose aussi précise que la géométrie.

Gustave Flaubert, letter to Louise Colet

11.1 Cartan subalgebras

We consider a semi-simple Lie algebra $\mathbb{G}$ and we introduce the fundamental concept of Cartan subalgebra, which will be the primary instrument to set up the reduction of the Lie algebra to a canonical form and its identification in terms of a *root system.*

Definition 11.1.1. A Cartan subalgebra $\mathscr{H} \subset \mathbb{G}$ is a subalgebra that satisfies the following two defining properties:
(i) $\mathscr{H}$ is a *maximal Abelian subalgebra.*
(ii) $\forall H \in \mathscr{H}$ the map $\mathrm{ad}(H)$ is a *semi-simple endomorphism.*

First we prove that every semi-simple Lie algebra $\mathbb{G}$ has a Cartan subalgebra (frequently abbreviated as CSA). Then we show that if $\mathscr{H}_1$ and $\mathscr{H}_2$ are two CSAs, then they are isomorphic.

Let $H \in \mathbb{G}$ be an element of the semi-simple Lie algebra and let $\lambda_0, \lambda_1, \dots \lambda_r$ be the eigenvalues of $\mathrm{ad}(H)$. Define

$$g(H, \lambda_i) = \{X \in \mathbb{G} / \, \mathrm{ad}(H)X = \lambda_i X\} \tag{11.1}$$

as the subspace of $\mathbb{G}$ pertaining to the eigenvalue λ_i. We have

$$\mathbb{G} = \bigoplus_{i=0}^{r} g(H, \lambda_i). \tag{11.2}$$

Definition 11.1.2. An element $H_0 \in \mathbb{G}$ is named *regular* if

$$\dim g(H_0, 0) = \min_{X \in \mathbb{G}} (\dim g(X, 0)). \tag{11.3}$$

We have the following theorem.

Theorem 11.1.1. *If H_0 is a regular element, then $g(H_0, 0)$ is a Cartan subalgebra.*

Proof of Theorem 11.1.1. We have to show that:
(a) $g(H_0, 0)$ is a subalgebra.
(b) $g(H_0, 0)$ is a maximal Abelian subalgebra.
(c) If $H \in g(H_0, 0)$, then $\mathrm{ad}(H)$ is semi-simple as an endomorphism.

https://doi.org/10.1515/9783111201535-011

We begin by observing that

$$[g(Z,\lambda), g(Z,\mu)] \subset g(Z,\lambda+\mu), \tag{11.4}$$

which immediately follows from the Jacobi identities. This implies that $\mathscr{H} = g(H_0, 0)$ is a subalgebra. Next we prove that $\mathscr{H}$ is Abelian. To this effect let us denote by $0 = \lambda_0, \lambda_1, \lambda_2, \dots, \lambda_r$ the different eigenvalues of $\mathrm{ad}(H_0)$ and let us set

$$\mathbb{G}' = \bigoplus_{i=1}^{r} g(H, \lambda_i). \tag{11.5}$$

From equation (11.4) it follows that $[\mathscr{H}, \mathbb{G}'] \subset \mathbb{G}'$. Moreover, $\forall H \in \mathscr{H}$ let us denote by $\mathrm{ad}'(H)$ the restriction of $\mathrm{ad}(H)$ to the subspace $\mathbb{G}'$ and let us call $d(H) = \det[\mathrm{ad}'(H)]$ the determinant of such an endomorphism. By definition $d(H)$ is a polynomial function on the finite-dimensional vector space (algebra) $\mathscr{H}$; furthermore, by definition of the subspace $\mathbb{G}'$, the map $\mathrm{ad}'(H_0)$ has only non-vanishing eigenvalues, so that $d(H_0) \neq 0$. If a polynomial function vanishes on an open set, then it is identically zero. Since $d(H_0) \neq 0$, it follows that $d(H)$ is not identically zero and that its zeros are isolated. Calling S the set of elements of $\mathscr{H}$ for which $d(H) \neq 0$ we conclude that S is dense in $\mathscr{H}$. Let $H \in S \subset \mathscr{H}$. Since $\det[\mathrm{ad}'(h)] \neq 0$, it follows that all the null eigenvectors of $\mathrm{ad}(H)$, if any, are contained in $\mathscr{H}$. Hence, we have shown that

$$\forall H \in S: \quad g(H, 0) \subset \mathscr{H}. \tag{11.6}$$

Since the element H_0 is by hypothesis regular, we conclude that $g(H, 0) = \mathscr{H}$. Hence, it is proved that

$$\forall H \in S, \forall H_1 \in \mathscr{H}: \quad \mathrm{ad}(H)(H_1) = 0. \tag{11.7}$$

Hence, the restriction of $\mathrm{ad}(H)$ to the subalgebra $\mathscr{H}$ is nilpotent since it vanishes. Since S is dense in $\mathscr{H}$, by continuity it follows that

$$\forall H \in \mathscr{H}: \quad \mathrm{ad}_{\mathscr{H}}(H) = 0, \tag{11.8}$$

namely,

$$\forall H_1, H_2 \in \mathscr{H}: \quad [H_1, H_2] = 0. \tag{11.9}$$

This concludes the proof that $\mathscr{H} \equiv g(H_0, 0)$ is an Abelian subalgebra. By definition it is also maximal. Indeed, if there existed an element $X \notin g(H_0, 0)$ such that $[X, \mathscr{H}] = 0$ we would have a contradiction, since in particular $[X, H_0] = 0$, which implies $X \in g(H_0, 0)$.

Let us now show that if λ is a non-vanishing eigenvalue of $\mathrm{ad}(H_0)$, then every endomorphism $\mathrm{ad}(H)$ with $H \in \mathscr{H}$ maps the subspace $g(H_0, \lambda)$ into itself. Hence, denoting by $\mathrm{ad}_\lambda(H)$ the restriction of $\mathrm{ad}(H)$ to this subspace, $\mathrm{ad}_\lambda(H)$ is a *representation* of $\mathscr{H}$ on

$g(H_0,\lambda)$. Since $\mathrm{ad}_\lambda(H)$ is a family of commuting endomorphisms (solvable algebra, in particular), we can put all of them simultaneously in a triangular form, by choosing some appropriate basis $\mathbf{e}_1,\dots,\mathbf{e}_s$ of $g(H_0,\lambda)$. In this basis the semi-simple part of $\mathrm{ad}_\lambda(H)$ will be the diagonal part:

$$\mathrm{ad}_\lambda(H) = \begin{pmatrix} \alpha_1(H) & 0 & \cdots & \cdots & 0 \\ 0 & \alpha_2(H) & 0 & \cdots & 0 \\ \cdots & \cdots & \cdots & \cdots & \cdots \\ 0 & \cdots & 0 & \alpha_{s-1}(H) & 0 \\ 0 & \cdots & \cdots & 0 & \alpha_s(H) \end{pmatrix} + \text{nilpotent matrix.} \tag{11.10}$$

The diagonal elements $\alpha_i(H)$ are linear functions on $\mathscr{H}$ with the property that $\alpha_1(H_0) = \alpha_2(H_0) = \cdots = \alpha_s(H_0) = \lambda$. Let $\beta(H)$ be any linear function on $\mathscr{H}$ that takes the value $\beta(H_0) = \lambda$ at H_0. Let V_β be the subspace of $g(H_0,\lambda)$ spanned by those basis vectors $\mathbf{e}_i$ such that $\alpha_i(H) = \beta(H)$ ($\forall H \in \mathscr{H}$). By definition it follows that

$$\forall X \in V_\beta \quad \Rightarrow \quad \exists k \in \mathbb{N}/(\mathrm{ad}(H) - \beta(H)\mathbf{1})^k X = 0. \tag{11.11}$$

Indeed, once we have subtracted the diagonal part, what remains is nilpotent. In general we have

$$\mathbb{G} = \sum_i V_{\beta_i} \quad \text{for suitable } \beta_i. \tag{11.12}$$

Hence, if $\kappa(\,,\,)$ is the Killing form we can write

$$\forall H,H' \in \mathscr{H}: \quad \kappa(H,H') = \sum_i \beta_i(H)\beta_i(H')\dim V_{\beta_i}. \tag{11.13}$$

We decompose $\mathrm{ad}(H)$ à la Jordan,

$$\mathrm{ad}(H) = \underbrace{S(H)}_{\text{semi-simple}} + \underbrace{N(H)}_{\text{nilpotent}}, \tag{11.14}$$

and we recall that $S(H)$ is polynomial in $\mathrm{ad}(H)$. By construction the endomorphism $S(H)$ leaves each V_β subspace invariant and

$$S(H)X = \beta(H)X; \quad \forall X \in V_\beta. \tag{11.15}$$

Furthermore, since $[V_\alpha, V_\beta] \subset V_{\alpha+\beta}$, it follows that S is a derivation of the algebra. But for a semi-simple Lie algebra every derivation is internal; hence, $\exists Z \in \mathbb{G}$ such that $\mathrm{ad}(Z) = S(H)$. Since $S(H)$ commutes with all elements of $\mathscr{H}$, it follows that $Z \in \mathscr{H}$. In other words, $Z = H$ and this shows that $\mathrm{ad}(H)$ coincides with its semi-simple part. □

11.2 Root systems

Let $\mathscr{H} \subset \mathbb{G}$ be a Cartan subalgebra of the semi-simple Lie algebra $\mathbb{G}$. Consider an element $\alpha \in \mathscr{H}^\star$, namely, a linear functional

$$\begin{aligned} \alpha : \quad & \mathscr{H} \to \mathbb{C}, \\ \forall H_1, H_2 \in \mathscr{H}; \forall \lambda, \mu \in \mathbb{C} : \quad & \alpha(\lambda H_1 + \mu H_2) = \lambda\alpha(H_1) + \mu\alpha(H_2). \end{aligned} \tag{11.16}$$

Let us define the linear subspace $\mathbb{G}^\alpha \subset \mathbb{G}$:

$$\mathbb{G}^\alpha : \{X \in \mathbb{G} \setminus [H, X] = \alpha(H)X, \forall H \in \mathscr{H}\}. \tag{11.17}$$

If $\mathbb{G}^\alpha \neq \varnothing$ is not empty, then we say that $\alpha \in \mathscr{H}^\star$ is a *root* and $\mathbb{G}^\alpha$ is named the *corresponding subspace* of root α. Since, by definition, $\mathscr{H}$ is maximal Abelian, we have $\mathbb{G}^0 = \mathscr{H}$. On the other hand, from the Jacobi identity we immediately obtain

$$[\mathbb{G}^\alpha, \mathbb{G}^\beta] \subset \mathbb{G}^{\alpha+\beta} \quad \forall \alpha, \beta \in \mathscr{H}^\star. \tag{11.18}$$

Let us next denote by $\mathbf{\Phi}$ the set of all non-vanishing roots and by $\kappa(\,,\,)$ the Killing form. We have the following theorem.

Theorem 11.2.1. *The following statements are true:*

(i) *We have* $\mathbb{G} = \mathscr{H} \oplus \sum_{\alpha \in \mathbf{\Phi}} \mathbb{G}^\alpha$ *(direct sum).*

(ii) *We have* $\dim \mathbb{G}^\alpha = 1, \forall \alpha \in \mathbf{\Phi}$.

(iii) *Let* $\alpha, \beta \in \mathbf{\Phi}$ *be two roots such that* $\alpha + \beta \neq 0$. *Then the corresponding subspaces* $\mathbb{G}^\alpha$ *and* $\mathbb{G}^\beta$ *are mutually orthogonal with respect to the Killing form* $\kappa(\,,\,)$.

(iv) *The restriction of the Killing form* $\kappa(\,,\,)$ *to* $\mathscr{H} \otimes \mathscr{H}$ *is non-degenerate and for each root* $\forall \alpha \in \mathbf{\Phi}$ *there exists an element* $H_\alpha \in \mathscr{H}$ *of the Cartan subalgebra such that*

$$\kappa(H, H_\alpha) = \alpha(H) \quad \forall H \in \mathscr{H}. \tag{11.19}$$

(v) *If* $\alpha \in \mathbf{\Phi}$ *is a root, then also its negative is a root:* $-\alpha \in \mathbf{\Phi}$. *Furthermore, we have*

$$\begin{aligned} [\mathbb{G}^\alpha, \mathbb{G}^{-\alpha}] &= \mathit{const}\, H_\alpha, \\ \alpha(H_\alpha) &\neq 0. \end{aligned} \tag{11.20}$$

Proof of Theorem 11.2.1. We begin with point (i) in the above list and we show first that the sum is direct. If it were not, this would mean that there exists a linear relation

$$H^\star + \sum_i X_{\alpha_i} = 0, \tag{11.21}$$

where $H^\star \in \mathscr{H}$ and $X_{\alpha_i} \in \mathbb{G}^{\alpha_i}$. We can choose an element $H \in \mathscr{H}$ such that $\alpha_i(H) \neq 0$ for all the roots α_i. Indeed, the subset $N \subset \mathscr{H}$ on which all the roots α_i are different and non-vanishing is the complement of the union of a finite number of hyperplanes ($\alpha(H) = 0 \Leftrightarrow$ *hyperplane* $\ni H$). Hence, H with the required properties exists and, as a consequence, $H^\star$ and X_{α_i} belong to different eigenspaces of $\mathrm{ad}(H)$. As such they are linearly independent, which contradicts the assumption of equation (11.21). This shows that the sum of subspaces in statement (i) is direct. On the other hand, since $\mathrm{ad}_{\mathbb{G}}(\mathscr{H})$ is a set of semi-simple endomorphisms, it follows that $\mathbb{G}$ can be decomposed into eigenspaces and the relation advocated in statement (i) follows. Furthermore, if $\alpha(H_0) = 0$ for all roots $\alpha \in \mathbf{\Phi}$, then $H_0 = 0$. Indeed, by hypothesis we have $[H_0, X] = 0$, $\forall X \in \mathbb{G}$, and since the Lie algebra $\mathbb{G}$ is semi-simple this implies $H_0 = 0$.

Let us next prove statement (iii). To this effect we choose $X \in \mathbb{G}^\alpha$ and $Y \in \mathbb{G}^\beta$. With this choice the endomorphism $\mathrm{ad}(x).\,\mathrm{ad}(Y)$ maps the space $\mathbb{G}^\gamma$ into $\mathbb{G}^{\alpha+\beta+\gamma}$ and since $\alpha + \beta \neq 0$ we get

$$\mathbb{G}^\gamma \cap \mathbb{G}^{\alpha+\beta+\gamma} = 0. \tag{11.22}$$

Therefore, if we use a basis where every basis vector lies in some root subspace $\mathbb{G}^\gamma$ we immediately see that

$$\kappa(X, Y) \equiv \mathrm{Tr}\big(\mathrm{ad}(X).\,\mathrm{ad}(Y)\big) = 0, \tag{11.23}$$

which is what we wanted to show.

Next let us prove statement (iv). If $H_0 \in \mathscr{H}$ satisfies the condition

$$\kappa(H_0, H) = 0 \quad \forall H \in \mathscr{H}, \tag{11.24}$$

then as a consequence of statement (iii), which we have already proved, it follows that

$$\kappa(H_0, X) = 0 \quad \forall X \in \mathbb{G}. \tag{11.25}$$

This would imply that the Killing form $\kappa(,)$ is degenerate, in contradiction with Cartan's criterion for semi-simple Lie algebras. Hence, there are no vectors in $\mathscr{H}$ which are orthogonal to all vectors of $\mathscr{H}$, which proves the first part of statement (iv). Next choose a basis $\{H_i\}$ ($i = 1, \ldots \mathrm{rank}\,\mathbb{G}$) of the Cartan subalgebra[1] $\mathscr{H}$ and set

$$\kappa_{ij} = \kappa(H_i, H_j). \tag{11.26}$$

Writing $\forall H \in \mathscr{H}$, $H = h^i H_i$ we obtain $\alpha(H) = h^i \alpha_i$, where $\alpha_i \equiv \alpha(H_i)$. Statement (iv) advocates that we should be able to find an element $H_\alpha = \alpha^i H_i$ such that $\kappa(H, H_\alpha) = \alpha(H)$. In the chosen basis this means $H^i \alpha^j \kappa_{ij} = h^i \alpha_i$, which implies

$$\alpha^j \kappa_{ij} = \alpha_i. \tag{11.27}$$

1 We recall that the dimension of the CSA of a Lie algebra $\mathbb{G}$ is named the rank of $\mathbb{G}$.

Since κ_{ij} is a non-degenerate matrix we can always find its inverse and set

$$\alpha^j = (\kappa^{-1})^{ji}\alpha_i, \tag{11.28}$$

which concludes the proof of statement (iv).

Let us come to the proof of statement (v). Let us assume that $-\alpha \notin \mathbf{\Phi}$. This would imply that $\mathbb{G}^{-\alpha} = \mathbf{0}$. In this case an element $X \in \mathbb{G}^\alpha$ being orthogonal to all the other subspaces $\mathbb{G}^\beta$ would imply that

$$\kappa(X_\alpha, Y) = 0 \quad \forall Y \in \mathbb{G}. \tag{11.29}$$

In this case the Killing form would be degenerate, which is impossible for a semi-simple Lie algebra by the Cartan criterion. So $-\alpha \in \mathbf{\Phi}$. Let now H, X_α, and $X_{-\alpha}$ be arbitrary elements respectively in $\mathscr{H}$, $\mathbb{G}^\alpha$, and $\mathbb{G}^{-\alpha}$. Then by the properties of the Killing form we have

$$\begin{aligned}\kappa([X_\alpha, X_{-\alpha}], H) &= \kappa(X_{-\alpha}, [H, X_\alpha]H) \\ &= \kappa(X_{-\alpha}, X_\alpha)\alpha(H) \\ &= \kappa(X_{-\alpha}, X_\alpha)\kappa(H_\alpha, H),\end{aligned} \tag{11.30}$$

so we are forced to identify

$$[X_\alpha, X_{-\alpha}] = \kappa(X_{-\alpha}, X_\alpha)H_\alpha = \kappa(X_{-\alpha}, X_\alpha)\alpha^i H_i, \tag{11.31}$$

which concludes the proof of statement (v).

Finally let us prove statement (ii). Let us assume that $\dim \mathbb{G}^\alpha > 1$. In this case let us choose $X_\alpha \in \mathbb{G}^\alpha$ and $X_{-\alpha} \in \mathbb{G}^{-\alpha}$ such that

$$\kappa(X_{-\alpha}, X_\alpha) = 1. \tag{11.32}$$

If $\dim \mathbb{G}^\alpha > 1$ it follows that there exists $D_\alpha \in \mathbb{G}^\alpha$ such that

$$\kappa(D_\alpha, X_{-\alpha}) = 0. \tag{11.33}$$

Set $D_n = (\mathrm{ad}(X_\alpha))^n D_\alpha$ for $n = 0, 1, 2, \dots$. We have $D_n \in \mathbb{G}^{(n+1)\alpha}$ and hence

$$[H_\alpha, D_n] = \alpha(H)(n+1)D_n. \tag{11.34}$$

Furthermore, by induction we can show that

$$[X_{-\alpha}, D_n] = -n\frac{(n+1)}{2}\alpha(H_\alpha)D_{n-1}. \tag{11.35}$$

For $n = 0$ we have $[X_{-\alpha}, D_\alpha] = \kappa(D_\alpha, X_{-\alpha})H_\alpha = 0$. On the other hand, if equation (11.35) is true for n it follows that it is also true for $n+1$. Indeed,

$$\begin{aligned}[X_{-\alpha}, D_{n+1}] &= [X_{-\alpha}, [X_\alpha, D_n]] \\ &= -[X_\alpha, [D_n, X_{-\alpha}]] - [D_n, [X_{-\alpha}, X_\alpha]] \\ &= -n\frac{n+1}{2}\alpha(H_\alpha)D_n + (n+1)\alpha(H_\alpha)D_n \\ &= -(n+1)\frac{n+2}{2}\alpha(H_\alpha)D_n, \end{aligned} \tag{11.36}$$

which shows what we claimed. Therefore, if $D_0 = D_\alpha$ exists, also all the other D_n do exist and are non-vanishing. This implies that there are infinite roots $(n+1)\alpha$ and correspondingly infinite orthogonal subspaces $\mathbb{G}^{(n+1)\alpha}$; this is manifestly absurd since the dimension of the semi-simple Lie algebra $\mathbb{G}$ is finite. Hence, D_α cannot exist and the dimension of the subspace $\mathbb{G}^\alpha = 1$ as claimed.

This concludes the proof of the theorem. □

11.2.1 Final form of the semi-simple Lie algebra

Using the result provided by Theorem 11.2.1 we can now write a final general form of a semi-simple Lie algebra in terms of Cartan generators H_i and step operators E^α associated with the roots α. To this effect we normalize the Cartan subalgebra (CSA) generators in the following way:

$$\begin{aligned} \kappa(H_i, H_j) &= \delta_{ij} \quad \Rightarrow \quad H_\alpha = \alpha_i H_i, \\ \kappa(E^\alpha, E^{-\alpha}) &= 1, \\ \kappa(H_i, E^\alpha) &= 0. \end{aligned} \tag{11.37}$$

With this normalization the commutation relations of the *complex semi-simple Lie algebra* take the following general form:

$$\begin{aligned} [H_i, H_j] &= 0, \\ [H_i, E^\alpha] &= \alpha_i E^\alpha, \\ [E^\alpha, E^{-\alpha}] &= \alpha^i H_i, \\ [E^\alpha, E^\beta] &= \begin{cases} N(\alpha,\beta)E^{\alpha+\beta} & \text{if } \alpha+\beta \in \mathbf{\Phi}, \\ 0 & \text{if } \alpha+\beta \notin \mathbf{\Phi}, \end{cases} \end{aligned} \tag{11.38}$$

where $N(\alpha, \beta)$ is a coefficient that has to be determined using the Jacobi identities.

11.2.2 Properties of root systems

Let us now consider the properties of a root system associated with a semi-simple Lie algebra. We have the following theorem.

Theorem 11.2.2. *If $\alpha, \beta \in \Phi$ are two roots, then the following two statements are true:*
1. $2\frac{(\alpha,\beta)}{(\alpha,\alpha)} \in \mathbb{Z}$;
2. $\sigma_\alpha(\beta) \equiv \beta - 2\alpha\frac{(\alpha,\beta)}{(\alpha,\alpha)} \in \Phi$ *is also a root.*

The vector $\sigma_\alpha(\beta)$ defined above is called the reflection of β with respect to α and the second part of the thesis can be reformulated by saying that any root system Φ is invariant under reflection with respect to any of its elements.

Proof of Theorem 11.2.2. Let $\alpha, \beta \in \Phi$ be two roots and let us define the non-negative integer $j \in \mathbb{N}$ by means of the following conditions:

$$\begin{aligned} \gamma \equiv \beta + j\alpha \in \Phi, \\ \gamma + \alpha \notin \Phi. \end{aligned} \tag{11.39}$$

In other words, j is the maximal integer n for which $\beta + n\alpha$ is a root.

We know that $-\alpha$ is a root and hence we can conclude that

$$\begin{aligned} [E^{-\alpha}, E^{\gamma}] &= \hat{E}^{\gamma-\alpha}, \\ [E^{-\alpha}, \hat{E}^{\gamma-\alpha}] &= \hat{E}^{\gamma-2\alpha}, \\ &\vdots \end{aligned} \tag{11.40}$$

where $\hat{E}^{\gamma-n\alpha}$ denotes some element in the one-dimensional subspace pertaining to the root $\gamma - n\alpha$. Since the number of roots is necessarily finite, it follows that there exists some positive integer $g \in \mathbb{N}$ such that

$$[E^{-\alpha}, \hat{E}^{\gamma-g\alpha}] = \hat{E}^{\gamma-(g+1)\alpha} = 0. \tag{11.41}$$

In general, due to the one-dimensionality of each root space we can set

$$[E^{\alpha}, \hat{E}^{\gamma-(n+1)\alpha}] = \mu_{n+1}\hat{E}^{\gamma-n\alpha}, \tag{11.42}$$

where μ_{n+1} is some normalization factor. From the Jacobi identities we immediately obtain a recursion relation satisfied by these normalization factors. Indeed,

$$\begin{aligned} [E^{\alpha}[E^{-\alpha}, \hat{E}^{\gamma-n\alpha}]] &= -[E^{-\alpha}[\hat{E}^{\gamma-n\alpha}, E^{\alpha}]] - [E^{\gamma-n\alpha}[E^{\alpha}, E^{-\alpha}]] \\ &= \mu_n\hat{E}^{\gamma-n\alpha} + \alpha^i[H_i, \hat{E}^{\gamma-n\alpha}] \\ &= (\mu_n + (\gamma,\alpha) - n(\alpha,\alpha))\hat{E}^{\gamma-n\alpha}, \end{aligned} \tag{11.43}$$

which implies the recursion relation

$$\mu_{n+1} = \mu_n + (\gamma,\alpha) - n(\alpha,\alpha). \tag{11.44}$$

Since by hypothesis $\gamma + \alpha$ is not a root, we have

$$[E^\alpha, E^\gamma] = \mu_0 E^{\gamma+\alpha} = 0, \quad \text{namely, } \mu_0 = 0. \tag{11.45}$$

This allows to solve the recursion relation explicitly, yielding

$$\mu_n = n(\alpha,\gamma) - \frac{n(n-1)}{2}(\alpha,\alpha). \tag{11.46}$$

Since at the other end of the chain we have assumed that $\gamma - (g+1)\alpha$ is not a root, we conclude that $\mu_{g+1} = 0$ and hence

$$(g+1)\left\{(\alpha,\gamma) - \frac{g}{2}(\alpha,\alpha)\right\} = 0. \tag{11.47}$$

This implies that $2\frac{(\gamma,\alpha)}{(\alpha,\alpha)} = g \in \mathbb{N}$. Hence, for each pair of roots $\alpha\beta$ there exists a non-negative integer $j \geq 0$ such that $\gamma = \beta + j\alpha$ is a root and

$$(\alpha,\beta) = (\alpha,\gamma) - j(\alpha,\alpha) = \left(\frac{g}{2} - j\right)(\alpha,\alpha), \tag{11.48}$$

namely,

$$2\frac{(\alpha,\beta)}{(\alpha,\alpha)} = g - 2j \in \mathbb{Z} \quad \text{(positive or negative).} \tag{11.49}$$

This concludes the first part of our proof. Let us now consider the string of roots that we have constructed to make the above argument:

$$\begin{aligned} \beta_0 &= \gamma = \beta + j\alpha, \\ \beta_1 &= \gamma - \alpha = \beta + (j-1)\alpha, \\ \beta_2 &= \gamma - 2\alpha = \beta + (j-2)\alpha, \\ &\vdots \end{aligned} \tag{11.50}$$

$$\beta_g = \gamma - g\alpha = \beta + (j-g)\alpha. \tag{11.51}$$

Since $2\frac{(\gamma,\alpha)}{(\alpha,\alpha)} = g$, it is evident by means of the replacement

$$\beta \mapsto \beta - 2\alpha\frac{(\beta,\alpha)}{(\alpha,\alpha)} \tag{11.52}$$

that the string (11.51) is simply reflected into itself: $\beta_g \mapsto \beta_0, \beta_{g-1} \mapsto \beta_1, \ldots$. So we proved not only that if β and α are roots, then the reflection of β with respect to α is a root, but also that the entire string of α through β is invariant under such reflection. □

Collecting together the properties of the roots that we have so far we can axiomatize the notion of root system in the following way.

Definition 11.2.1. Let $\mathbb{E}$ be a Euclidean space of dimension ℓ. A subset $\mathbf{\Phi} \subset \mathbb{E}$ is called a *root system* if:
1. $\mathbf{\Phi}$ is finite, spans $\mathbb{E}$, and does not contain **0**.
2. If $\alpha \in \mathbf{\Phi}$ the only multiples of α contained in $\mathbf{\Phi}$ are $\pm\alpha$.
3. $\forall \alpha, \beta \in \mathbf{\Phi}$ we have $2\frac{(\alpha,\beta)}{(\alpha,\alpha)} \in \mathbb{Z}$.
4. $\forall \alpha, \beta \in \mathbf{\Phi}$ we have $\sigma_\alpha(\beta) \equiv \beta - 2\alpha\frac{(\alpha,\beta)}{(\alpha,\alpha)} \in \mathbf{\Phi}$.

11.2.2.1 Angles between the roots

It is convenient to introduce the following notation of a *hook product*:

$$\langle \beta, \alpha \rangle \equiv 2\frac{(\beta, \alpha)}{(\alpha, \alpha)}. \tag{11.53}$$

From Theorem 11.2.2 we have learned that $\langle \beta, \alpha \rangle \in \mathbb{Z}$, but at the same time also $\langle \alpha, \beta \rangle \in \mathbb{Z}$. Hence, we conclude that

$$\langle \beta, \alpha \rangle \langle \alpha, \beta \rangle = 4\cos^2\theta_{\alpha\beta} \in \mathbb{Z}, \tag{11.54}$$

where $\theta_{\alpha\beta}$ is the angle between the two roots.

This implies that the angles between the roots are quantized and the available cases are listed in the following table:

$\langle \alpha, \beta \rangle$	$\langle \beta, \alpha \rangle$	θ	$\frac{\|\beta\|^2}{\|\alpha\|^2}$
0	0	$\frac{\pi}{2}$	undetermined
1	1	$\frac{\pi}{3}$	1
−1	−1	$\frac{2\pi}{3}$	1
1	2	$\frac{\pi}{4}$	2
−1	−2	$\frac{3\pi}{4}$	2
1	3	$\frac{\pi}{6}$	3
−1	−3	$\frac{5\pi}{6}$	3

(11.55)

As one sees, also the ratio between the squared lengths of two roots that are not orthogonal to each other are quantized; we have three possibilities: 1, 2, and 3. As we will see in the sequel, these ratios are very relevant and distinguish the Lie algebras in two categories: the *simply laced* and the *non-simply laced* algebras.

A very useful lemma which is just a consequence of the above arguing is the following one.

Lemma 11.2.1. *Let $\alpha, \beta \in \mathbf{\Phi}$ be two non-proportional roots. If $(\alpha, \beta) > 0$, then $\alpha - \beta \in \mathbf{\Phi}$ is a root. If $(\alpha, \beta) < 0$, then $\alpha + \beta \in \mathbf{\Phi}$ is a root.*

Proof of Lemma 11.2.1. Looking at Table (11.55) we see that if $(\alpha, \beta) > 0$, then one or the other of the hook products $\langle \alpha, \beta \rangle$ or $\langle \beta, \alpha \rangle$ is equal to 1. In the first case $\alpha - \beta = \sigma_\beta(\alpha)$ and then it is a root. In the second case $\beta - \alpha = \sigma_\alpha(\beta)$ is a root but also its negative, namely, $\alpha - \beta$, is a root. To prove the lemma in the case $(\alpha, \beta) < 0$ it suffices to apply the same argument to $\beta' = -\beta$. □

11.3 Simple roots, the Weyl group, and the Cartan matrix

The next steps in our discussion are all rather simple constructions of Euclidean geometry whose consequences are however quite far reaching. We begin with the notion of simple roots.

Definition 11.3.1. Given a root system $\mathbf{\Phi} \subset \mathbb{E}^\ell$ in a Euclidean space of dimension ℓ, a set Δ of exactly ℓ roots is called a simple root basis if:

1. Δ is a basis for the entire $\mathbb{E}^\ell$.
2. Every root $\alpha \in \mathbf{\Phi}$ can be written as a linear combination of the elements α_i whose coefficients are either all positive or all negative integers:

$$\alpha = \sum_{i=1}^{\ell} k^i \alpha_i; \quad k^i \in \begin{cases} \mathbb{Z}_+, \\ \text{or } \mathbb{Z}_-. \end{cases} \tag{11.56}$$

The vectors α_i contained in Δ are called the simple roots of $\mathbf{\Phi}$.

What might look surprising at first sight is the following.

Theorem 11.3.1. *Every root system $\mathbf{\Phi}$ admits a basis Δ of simple roots.*

Proof of Theorem 11.3.1. The proof is a constructive one since it indicates a precise algorithm to construct the simple roots.

Let us consider any vector $\gamma \in \mathbb{E}^\ell$ which does not lie in any of the hyperplanes Π_α orthogonal to the roots $\alpha \in \mathbf{\Phi}$. Since the roots are in a finite number, such vectors certainly exist. The hyperplane Π_γ separates the set of roots in two subsets:

$$\mathbf{\Phi} = \mathbf{\Phi}_+ \cup \mathbf{\Phi}_-. \tag{11.57}$$

The set of *positive roots* $\mathbf{\Phi}_+$ is comprised of all those roots α for which $(\gamma, \alpha) > 0$. The set of *negative roots* $\mathbf{\Phi}_-$ is comprised of all those for which the reverse is true. A positive root $\alpha \in \mathbf{\Phi}_+$ is named *decomposable* if it can be written as the sum of two other positive roots: $\alpha = \beta_1 + \beta_2$ with $\beta_{1,2} \in \mathbf{\Phi}_+$. Otherwise it is named *indecomposable*.

The searched for simple root basis is provided by the set of indecomposable positive roots that we call Δ_γ. To demonstrate such a statement we proceed through the following steps.

Step 1: We begin by proving that every element $\alpha \in \Phi_+$ is a linear combination with coefficients in $\mathbb{Z}_+$ of the roots in Δ_γ. We proceed by *reductio ad absurdum*. Suppose that there are roots $\alpha \in \Phi_+$ which do not admit such a decomposition. Choose among them that one for which (γ, α) is minimal. Since by hypothesis $\alpha \notin \Delta_\gamma$, it follows that there exists $\beta_{1,2} \in \Phi_+$ such that $\alpha = \beta_1 + \beta_2$ and hence $(\gamma, \alpha) = (\gamma, \beta_1) + (\gamma, \beta_2)$. But each of the two contributions in the above sum is positive and we are at a contradiction: if $\beta_{1,2}$ both admit a decomposition in terms of the simple roots with positive integral coefficients, then also α does, which contradicts the initial hypothesis. If on the other hand one of the two does not, then for that one, say β_1, the scalar product with γ is smaller than the same scalar product for α. This contradicts the hypothesis that for α it was minimal. Hence, exceptions do not exist and the statement is true.

Step 2: If $\alpha, \beta \in \Delta_\gamma$, then necessarily $(\alpha, \beta) \leq 0$. Otherwise, by force of Lemma 11.2.1 $\alpha - \beta \in \Phi$. In that case either $\alpha - \beta$ or $\beta - \alpha$ are positive roots of Φ_+. In the first case $\alpha = \beta + (\alpha - \beta)$ is shown to be decomposable. In the second case it is $\beta = (\beta - \alpha) + \alpha$ that is shown to be decomposable. In either case the hypothesis is contradicted. Hence, $(\alpha, \beta) \leq 0$ is true.

Step 3: The vectors in Δ_γ are linearly independent. Let us prove this by *reductio ad absurdum*. Suppose that there is a vanishing linear combination of the vectors in Δ_γ, namely, $0 = \sum_i \alpha_i$. Separating the coefficients into two subsets, $p_i > 0$ (the positive ones) and $n_i < 0$ (the negative ones), the same condition can be written as $\epsilon = \eta$, where $\epsilon = \sum_i p_i \alpha_i$ and $\eta = -\sum_i n_i \alpha_j$. Because of what we proved in Step 2 we have $(\epsilon, \epsilon) = \sum_{ij} p_i p_j (\alpha_i, \alpha_j) \leq 0$. Hence, $\epsilon = 0$ and $p_i = 0$. Similarly for η. Therefore, the conjectured linear combination does not exist and all the vectors in Δ_γ are linearly independent.

Step 4: Combining Step 3 and Step 1 we arrive at the conclusion that Δ_γ spans Φ and then also $\mathbb{E}$. Furthermore, for each root the decomposition involves either non-negative or non-positive integers. Hence, Δ_γ is indeed a simple root basis and the theorem is proved. □

From now on we can associate with every complex simple Lie algebra its root system Φ and with every root system its simple root basis Δ. Furthermore, each root system singles out a well-defined finite group, named the Weyl group, that is obtained by combining the reflections with respect to all the roots.

Definition 11.3.2. Let Φ be a root system in dimension ℓ. The Weyl group of Φ, denoted $\mathscr{W}(\Phi)$, is the finite group generated by the reflections σ_α, $\forall \alpha \in \Phi$.

Since for any two vectors $\mathbf{v}, \mathbf{w} \in \mathbb{E}$ we have

$$(\sigma_\alpha(\mathbf{v}), \sigma_\alpha(\mathbf{w})) = (\mathbf{v}, \mathbf{w}), \tag{11.58}$$

it follows that the Weyl group, which is finite, is always a finite subgroup of the rotation group in ℓ dimensions:

$$\mathscr{W}(\boldsymbol{\Phi}) \subset \mathrm{SO}(\ell). \tag{11.59}$$

11.4 Classification of the irreducible root systems

Having established that all possible irreducible root systems $\boldsymbol{\Phi}$ are uniquely determined (up to isomorphisms) by the Cartan matrix,

$$C_{ij} = \langle \alpha_i, \alpha_j \rangle \equiv 2\frac{(\alpha_i, \alpha_j)}{(\alpha_j, \alpha_j)}, \tag{11.60}$$

we can classify *all the complex simple Lie algebras* by classifying all possible Cartan matrices.

11.4.1 Dynkin diagrams

Each Cartan matrix can be given a graphical representation in the following way. With each simple root α_i we associate a circle ○ as in Figure 11.1 and then we link the i-th circle with the j-th circle by means of a line which is *simple*, *double*, or *triple*, depending on whether

$$\langle \alpha_i, \alpha_j \rangle \langle \alpha_j, \alpha_i \rangle = 4\cos^2\theta_{ij} = \begin{cases} 1, \\ 2, \\ 3, \end{cases} \tag{11.61}$$

having denoted by θ_{ij} the angle between the two simple roots α_i and α_j. The corresponding graph is named a *Coxeter graph*.

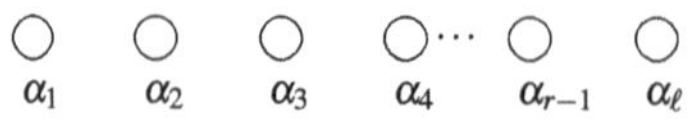

Figure 11.1: The simple roots α_i are represented by circles.

If we consider the simplest case of two-dimensional Cartan matrices we have the four possible Coxeter graphs depicted in Figure 11.2. Given a Coxeter graph, if it is *sim-*

$a_1 \times a_1$ ○ ○

a_2 ○—○

$b_2 \sim c_2$ ○═○

g_2 ○≣○

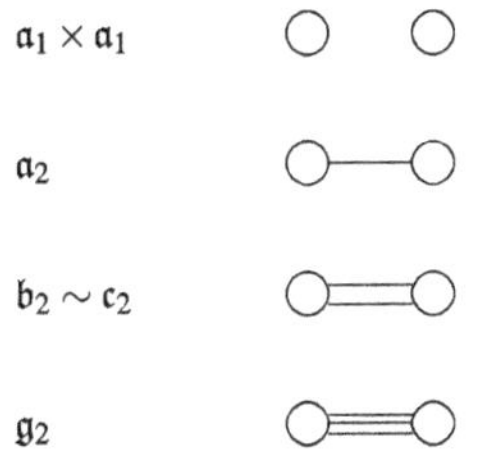

Figure 11.2: The four possible Coxeter graphs with two vertices.

ply laced, namely, if there are only simple lines, then all the simple roots appearing in such a graph have the same *length* and the corresponding Cartan matrix is completely identified. On the other hand, if the Coxeter graph involves double or triple lines, then in order to identify the corresponding Cartan matrix, we need to specify which of the two roots sitting at the endpoints of each multiple line is the *long* root and which is the *short* one. This can be done by associating an arrow with each multiple line. By convention we decide that this *arrow points in the direction of the short root*. A Coxeter graph equipped with the necessary arrows is called a *Dynkin diagram*. Applying this convention to the case of the Coxeter graphs of Figure 11.2 we obtain the result displayed in Figure 11.3. The one-to-one correspondence between the Dynkin diagram and the associated Cartan matrix is illustrated by considering in some detail the case b_2 of Figure 11.3. By definition of the Cartan matrix we have

$$2\frac{(a_1, a_2)}{(a_2, a_2)} = 2\frac{|a_1|}{|a_2|} \cos\theta = -2,$$
$$2\frac{(a_2, a_1)}{(a_1, a_1)} = 2\frac{|a_2|}{|a_1|} \cos\theta = -1,$$

(11.62)

$a_1 \times a_1$ ○ ○   = $\begin{pmatrix} 2 & 0 \\ 0 & 2 \end{pmatrix}$

a_2 ○—○   = $\begin{pmatrix} 2 & -1 \\ -1 & 2 \end{pmatrix}$

b_2 ○⇒○   = $\begin{pmatrix} 2 & -2 \\ -1 & 2 \end{pmatrix}$

c_2 ○⇐○   = $\begin{pmatrix} 2 & -1 \\ -2 & 2 \end{pmatrix}$

g_2 ○⇛○   = $\begin{pmatrix} 2 & -3 \\ -1 & 2 \end{pmatrix}$

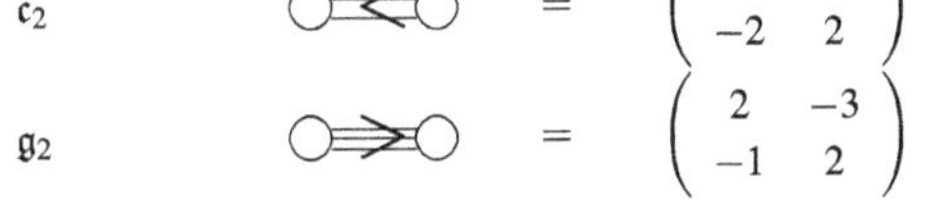

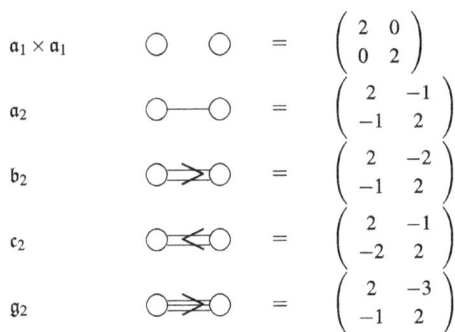

Figure 11.3: The distinct Cartan matrices in two dimensions (and therefore the simple algebras in rank 2) correspond to the Dynkin diagrams displayed above. We have distinguished a b_2 and a c_2 matrix since they are the limiting case for $\ell = 2$ of two series of Cartan matrices, the b_ℓ and the c_ℓ series, which for $\ell > 2$ are truly different. However, b_2 is the transpose of c_2, so they correspond to isomorphic algebras obtained from one another by renaming the two simple roots $a_1 \leftrightarrow a_2$.

so we conclude

$$|\alpha_1|^2 = 2|\alpha_2|^2, \tag{11.63}$$

which shows that α_1 is a long root, while α_2 is a short one. Hence, the arrow in the Dynkin diagram pointing towards the short root α_2 tells us that the matrix element C_{12} is −2 while the matrix element C_{21} is −1. The opposite happens in the case of $\mathfrak{c}_2$.

11.4.2 The classification theorem

Having clarified the notation of Dynkin diagrams, the basic classification theorem of *complex simple Lie algebras* is the following.

Theorem 11.4.1. *If* Φ *is an irreducible system of roots of rank* ℓ*, then its Dynkin diagram is either one of those shown in Figure* 11.4 *or for special values of* ℓ *it is one of those shown in Figure* 11.5. *There are no other irreducible root systems besides these ones.*

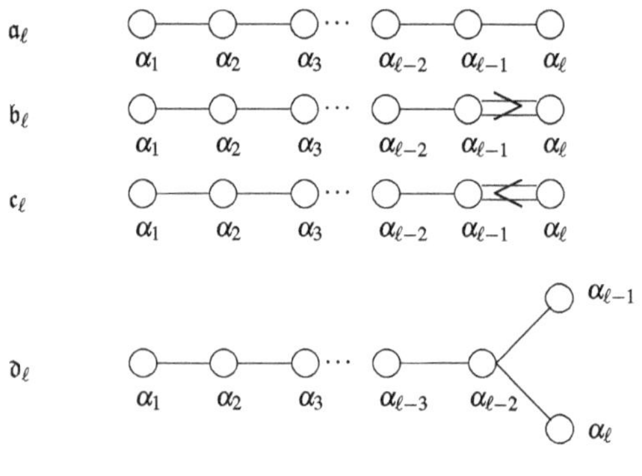

Figure 11.4: The Dynkin diagrams of the four infinite families of classical simple algebras.

Proof of Theorem 11.4.1. Let us consider a Euclidean space E, and in E let us consider a set of vectors

$$\mathscr{U} = \{\epsilon_1, \epsilon_2, \ldots, \epsilon_\ell\} \tag{11.64}$$

that satisfy the following three conditions:

$$\begin{aligned} (\epsilon_i, \epsilon_i) &= 1, \\ (\epsilon_i, \epsilon_j) &\leq 0, \quad i \neq j, \\ 4(\epsilon_i, \epsilon_j)^2 &= 0, 1, 2, 3, \quad i \neq j. \end{aligned} \tag{11.65}$$

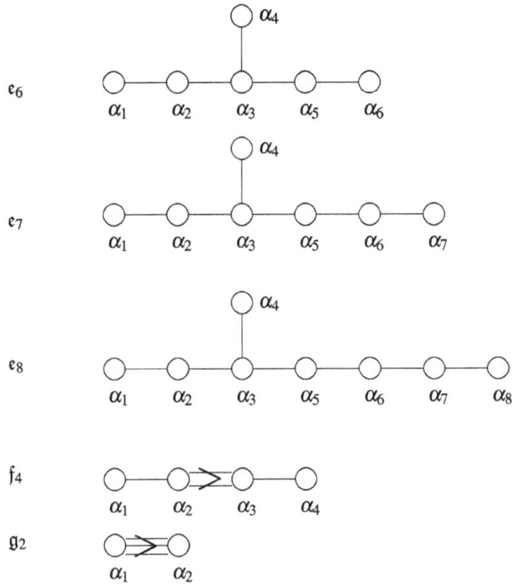

Figure 11.5: The Dynkin diagrams of the five exceptional algebras.

Such a system of vectors is called *admissible*. It is clear that each admissible system of vectors singles out a Coxeter graph Γ. Indeed, the vectors ϵ_i correspond to the simple roots α_i divided by their norm:

$$\epsilon_i = \frac{\alpha_i}{\sqrt{|\alpha_i|^2}}. \tag{11.66}$$

Our task is that of classifying all connected Coxeter graphs.

We proceed through a series of steps.

Step 1: We note that by deleting a subset of vectors ϵ_i in an admissible system those that are left still form an admissible system whose Coxeter graph is obtained from the original one by deleting the corresponding vertices and all the lines that end in these vertices.

Step 2: The number of pairs of vertices that are connected by at least one line is strictly less than the number of vectors ϵ_i, that is, strictly less than ℓ. Indeed, let us set $\epsilon = \sum_{i=1}^{\ell} \epsilon_i$ and observe what follows. Since all ϵ_i are independent, we have $\epsilon \neq 0$. Hence,

$$0 < (\epsilon, \epsilon) = \ell + 2\sum_{i<j} (\epsilon_i, \epsilon_j). \tag{11.67}$$

If the i-th vertex is joined to the j-th vertex, we have $4(\epsilon_i, \epsilon_j)^2 = 1, 2, 3$. Hence, we can conclude that in this case

$$2(\epsilon_i, \epsilon_j) \leq -1. \tag{11.68}$$

On the other hand, if the i-th vertex is not joined to the j-th vertex, we have $2(\epsilon_i, \epsilon_j) = 0$. Denoting by N_J the number of pairs of vertices joined by at least one line, we conclude that

$$0 < (\epsilon, \epsilon) < \ell - N_J \quad \Rightarrow \quad N_J \leq \ell - 1, \tag{11.69}$$

which is what we have asserted.

Step 3: The Coxeter graph Γ cannot contain any loop. Indeed, if a loop existed this would constitute a subgraph Γ' for which the number of pairs joined by a line, N_J, would be larger than the number of vertices, and this we have shown is impossible.

Step 4: The number of lines that end up in any vertex is at most three. Indeed, let $\epsilon \in \mathscr{U}$ and let us denote by $\eta_1, \eta_2, \ldots, \eta_k$ the vectors connected to ϵ by some link. In other words, we have $(\epsilon, \eta_i) < 0$ $(\forall \eta_i)$. Since there are no loops in the graph, it follows that no η_i can be connected to any other η_j, that is, $(\eta_i, \eta_j) = 0\ \forall i \neq j$. Since $\mathscr{U}$ is a set of linearly independent vectors, there must exist a unit vector η_0 in the vector span of $\epsilon, \eta_1, \ldots, \eta_k$, which is orthogonal to $\eta_1, \ldots, \eta_k$. Obviously the projection of such a vector η_0 on ϵ is non-vanishing, that is, $(\epsilon, \eta_0) \neq 0$. The set $\eta_0, \eta_1, \ldots, \eta_k$ makes an orthogonal basis for the linear span of the vectors $\epsilon, \eta_1, \ldots, \eta_k$, and we can write

$$\begin{aligned} \epsilon &= \sum_{i=0}^{k} (\epsilon, \eta_i)\eta_i, \\ 1 = (\epsilon, \epsilon) &= \sum_{i=0}^{k} (\epsilon, \eta_i)^2. \end{aligned} \tag{11.70}$$

This reasoning implies that $\sum_{i=1}^{k} (\epsilon, \eta_i)^2 < 1$ and hence

$$4 \sum_{i=1}^{k} (\epsilon, \eta_i)^2 < 4. \tag{11.71}$$

On the other hand, $4(\epsilon, \eta_i)^2$ is precisely the number of lines that link η_i to ϵ, so equation (11.71) is precisely the statement we wanted to prove in *Step 4*.

Step 5: The only connected Coxeter graph that contains a triple line is the $\mathfrak{g}_2$ graph of Figure 11.2. This immediately follows from *Step 4*.

Step 6: Let $\{\epsilon_1, \ldots, \epsilon_k\} \subset \mathscr{U}$ be a subset of vectors corresponding to a simple line as in Figure 11.6. Then the subset $\mathscr{U}' \equiv \{\mathscr{U} - \{\epsilon_1, \ldots, \epsilon_k\}\} \cup \{\epsilon\}$, where $\epsilon \equiv \sum_{i=1}^{k} \epsilon_i$ is still an admissible system. Graphically the operation of making the transition from the admissible system $\mathscr{U}$ to the admissible system $\mathscr{U}'$ corresponds to collapsing the entire simple line to a single vertex. That this statement is true can be proved in the following way. That the vectors composing $\mathscr{U}'$ are linearly independent is obvious. By hypothesis of a simple chain we have

$$2(\epsilon_i, \epsilon_{i+1}) = -1, \quad 1 \le i \le k-1, \tag{11.72}$$

so

$$(\epsilon, \epsilon) = k + 2\sum_{i<j}(\epsilon_i, \epsilon_j) = k - (k-1) = 1, \tag{11.73}$$

and hence ϵ is a unit vector. Furthermore, each $\eta \in \mathscr{U} - \{\epsilon_1, \dots, \epsilon_k\}$ can be joined to at most one of the vectors $\epsilon_1, \dots, \epsilon_k$; otherwise, we would generate a loop. Hence, we have either $(\eta, \epsilon) = 0$ or $(\eta, \epsilon_i) \neq 0$ for some value of i. In any case we conclude $4(\eta, \epsilon_i)^2 = 0, 1, 2, 3$, which is what makes $\mathscr{U}'$ an admissible system.

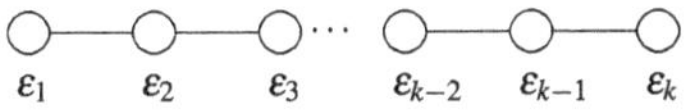

Figure 11.6: A simple line Coxeter graph.

Step 7: A Coxeter graph cannot contain subgraphs of the form displayed in Figure 11.7. Indeed, in all these three cases, by using the property shown in *Step 6* and collapsing a simple chain we obtain a graph that contains a vertex where four lines converge. This was shown to be forbidden in *Step 4*.

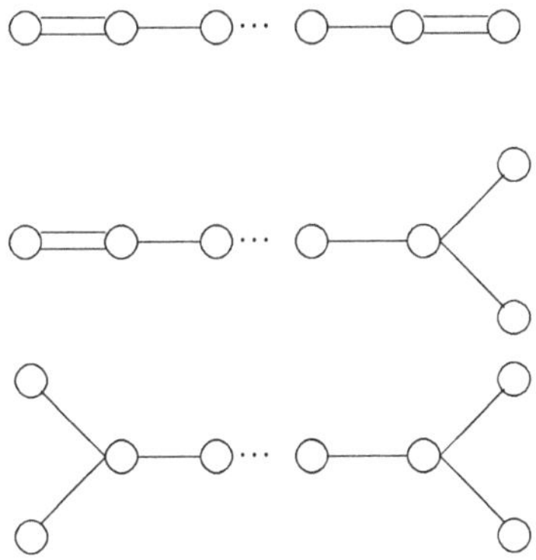

Figure 11.7: Prohibited subgraphs.

Step 8: Relying on the properties we have so far proven we are left with four types of possible Coxeter graphs, namely, (i) the simple chains of length ℓ corresponding to the $\mathfrak{a}_\ell$ Dynkin diagrams of Figure 11.4, (ii) the $\mathfrak{g}_2$ graph of Figure 11.2, (iii) the graphs of Figure 11.8 with a double line, and finally (iv) the graphs of Figure 11.9 with a node.

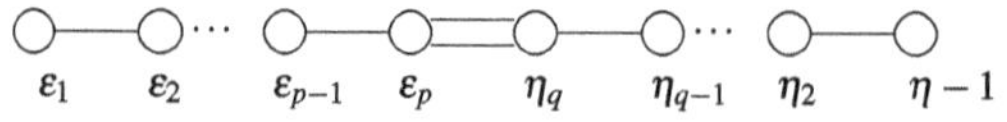

Figure 11.8: Coxeter graph with a double link that is preceded by a simple chain of length p and followed by a simple chain of length q.

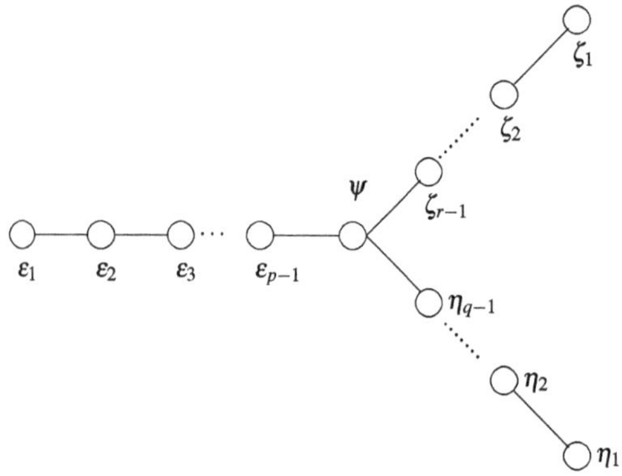

Figure 11.9: Coxeter graph with a node. The unit vector in the node is called ψ, while the unit vectors along the three simple lines departing from the node are respectively called $\epsilon_1, \dots, \epsilon_{p-1}, \eta_1, \dots, \eta_{q-1}, \zeta_1, \dots, \zeta_{r-1}$. The graph is characterized by the three integer numbers p, q, r, which are the lengths of the three simple lines departing from the node.

Step 9: If we consider the graphs of the type shown in Figure 11.8, there are only two solutions, namely,

$$\begin{aligned} p = 2; \quad q = 2 \quad &\Rightarrow \quad \mathfrak{f}_4 \ \text{Dynkin diagram}, \\ p = \ell \in \mathbb{N}; \quad q = 1 \quad &\Rightarrow \quad \mathfrak{b}_\ell \ \text{or} \ \mathfrak{c}_\ell \ \text{Dynkin diagrams}. \end{aligned} \tag{11.74}$$

Indeed, let us set $\epsilon = \sum_{i=1}^{p} i\epsilon_i$ and $\eta = \sum_{i=1}^{q} i\eta_i$. By the hypothesis of simple chains we have $2(\epsilon_i, \epsilon_{i+1}) = -1$, $2(\eta_i, \eta_{i+1}) = -1$ and all the other pairs of vectors are mutually orthogonal. In this way we obtain

$$\begin{aligned} (\epsilon, \epsilon) &= \sum_{i=1}^{p} i^2 - \sum_{i=1}^{p-1} i(i-1) = p\frac{p-1}{2}, \\ (\eta, \eta) &= \sum_{i=1}^{q} i^2 - \sum_{i=1}^{q-1} i(i-1) = q\frac{q-1}{2}, \end{aligned} \tag{11.75}$$

and since by the hypothesis of double line we have $4(\epsilon_p, \eta_q)^2 = 2$ it follows that

$$(\epsilon, \eta)^2 = p^2 q^2 (\epsilon_p, \eta_q)^2 = \frac{1}{2} p^2 q^2. \tag{11.76}$$

On the other hand, from the triangular Schwarz inequality of Euclidean geometry we have

$$\begin{aligned} (\epsilon, \eta) &< (\epsilon, \epsilon)(\eta, \eta), \\ &\Downarrow \\ (p-1)(q-1) &< 2, \end{aligned} \tag{11.77}$$

which for positive integers p, q admits only the two solutions advocated in equation (11.74). The first solution leads to the Dynkin diagram of the exceptional Lie algebra $\mathfrak{f}_4$, while the second solution leads to the two infinite series of classical Lie algebras $\mathfrak{b}_\ell$ and $\mathfrak{c}_\ell$.

Step 10: Let us finally consider the Coxeter graphs of the type shown in Figure 11.9. We claim that the only possible solutions are:

$$(p, q, r) = \begin{cases} (\ell, 1, 1) & \Rightarrow \quad \mathfrak{a}_\ell \quad \text{Dynkin diagrams}, \quad \ell \in \mathbb{N}, \\ (\ell - 2, 2, 2) & \Rightarrow \quad \mathfrak{d}_\ell \quad \text{Dynkin diagrams}, \quad 4 \le \ell \in \mathbb{N}, \\ (3, 3, 2) & \Rightarrow \quad \mathfrak{e}_6 \quad \text{Dynkin diagram}, \\ (4, 3, 2) & \Rightarrow \quad \mathfrak{e}_7 \quad \text{Dynkin diagram}, \\ (5, 3, 2) & \Rightarrow \quad \mathfrak{e}_8 \quad \text{Dynkin diagram}. \end{cases} \tag{11.78}$$

To prove this statement we follow a strategy similar to that used in the proof of *Step 9*. We define the following three vectors:

$$\epsilon = \sum_{i=1}^{p-1} i\epsilon_i; \quad \eta = \sum_{i=1}^{q-1} i\eta_i; \quad \sum_{i=1}^{r-1} i\zeta_i. \tag{11.79}$$

Clearly, ϵ, η, and ζ are mutually orthogonal, and ψ, the vector in the node, is not in the subspace generated by ϵ, η, and ζ. Hence, if in the linear span of $\{\psi, \epsilon, \eta, \zeta\}$ we construct a vector γ that is orthogonal to $\{\epsilon, \eta, \zeta\}$, we obtain $(\gamma, \psi) \neq 0$. Normalizing this vector to 1 we can write

$$\psi = (\psi, \gamma)\gamma + \frac{(\psi, \epsilon)}{\sqrt{(\epsilon, \epsilon)}}\epsilon + \frac{(\psi, \eta)}{\sqrt{(\eta, \eta)}}\eta + \frac{(\psi, \zeta)}{\sqrt{(\zeta, \zeta)}}\zeta, \tag{11.80}$$

and we obtain

$$(\psi, \psi) = 1 = (\psi, \gamma)^2 + \frac{(\psi, \epsilon)^2}{(\epsilon, \epsilon)} + \frac{(\psi, \eta)^2}{(\eta, \eta)} + \frac{(\psi, \zeta)^2}{(\zeta, \zeta)}, \tag{11.81}$$

which implies the inequality

$$1 > \frac{(\psi,\epsilon)^2}{(\epsilon,\epsilon)} + \frac{(\psi,\eta)^2}{(\eta,\eta)} + \frac{(\psi,\zeta)^2}{(\zeta,\zeta)}. \tag{11.82}$$

By the definition of the Coxeter graph in Figure 11.9 we have

$$\begin{aligned} (\psi,\epsilon) = (p-1)(\epsilon_{p-1},\psi) \quad \Rightarrow \quad (\psi,\epsilon)^2 &= \frac{(p-1)^2}{4}, \\ (\epsilon,\epsilon) &= \frac{p(p-1)}{2}, \end{aligned} \tag{11.83}$$

and similarly for the scalar products associated with the other chains. Inserting these results into (11.82) we obtain the Diophantine inequality

$$\frac{1}{p} + \frac{1}{q} + \frac{1}{r} > 1, \tag{11.84}$$

whose independent solutions are those displayed in equation (11.78). To this effect it is sufficient to note that (11.84) has an obvious permutational symmetry in the three numbers p, q, and r. To avoid double counting of solutions we break this symmetry by setting $p \geq q \geq r$ and then we see that the only possibilities are those listed in equation (11.78). □

Having concluded the proof of the classification theorem, we can look back and compare the just obtained results with those summarized in Section 5.2.4. The anticipated correspondence between finite rotation subgroups and simply laced Lie algebras should now be clear.

11.5 Identification of the classical Lie algebras

In the previous sections we have classified the allowed Dynkin diagrams and hence the allowed simple root systems. We have not shown that all of them do indeed exist. This is what we do in the present section by explicit construction. Furthermore, we identify the classical or exceptional complex Lie algebra that corresponds to each of the constructed root systems.

11.5.1 The $\mathfrak{a}_\ell$ root system and the corresponding Lie algebra

The Dynkin diagram is recalled in Figure 11.10. We want to perform the explicit construction of a root system that admits a basis corresponding to such a diagram. To this effect let us consider the $(\ell+1)$-dimensional Euclidean space $\mathbb{R}^{\ell+1}$ and let us denote by $\epsilon_1, \dots, \epsilon_{\ell+1}$ the unit vectors along the $\ell+1$ axes:

$\mathfrak{a}_\ell$

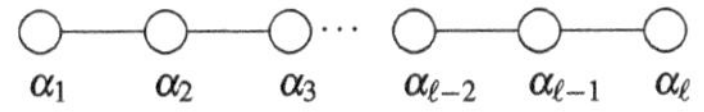

Figure 11.10: The Dynkin diagram of $\mathfrak{a}_\ell$ type.

$$\epsilon_1 = \begin{pmatrix} 1 \\ 0 \\ \cdots \\ \cdots \\ 0 \end{pmatrix}, \quad \epsilon_2 = \begin{pmatrix} 0 \\ 1 \\ 0 \\ \cdots \\ 0 \end{pmatrix} \quad \cdots \quad \epsilon_{\ell+1} = \begin{pmatrix} 0 \\ 0 \\ \cdots \\ \cdots \\ 1 \end{pmatrix}. \tag{11.85}$$

Define the vector $v = \epsilon_1 + \epsilon_2 + \cdots + \epsilon_{\ell+1}$:

$$v = \begin{pmatrix} 1 \\ 1 \\ 1 \\ \cdots \\ 1 \end{pmatrix}. \tag{11.86}$$

Define $\mathbb{I} \subset \mathbb{R}^{\ell+1}$ to be the $(\ell+1)$-dimensional cubic lattice immersed in $\mathbb{R}^{\ell+1}$:

$$\mathbb{I} = \{x \in \mathbb{R}^{\ell+1} / x = n^i \epsilon_i,\ n^i \in \mathbb{Z}\}. \tag{11.87}$$

In the cubic lattice $\mathbb{I}$ consider the sublattice

$$\mathbb{I}' = \mathbb{I} \cap E, \tag{11.88}$$

where E is the hyperplane of vectors orthogonal to the vector v:

$$E = \{y \in \mathbb{R}^{\ell+1} / (v, y) = 0\}. \tag{11.89}$$

Finally, in the sublattice $\mathbb{I}'$ consider the finite set of vectors whose norm is $\sqrt{2}$:

$$\mathbf{\Phi} = \{\alpha \in \mathbb{I}' / (\alpha, \alpha) = 2\}. \tag{11.90}$$

Theorem 11.5.1. *The above defined set* $\mathbf{\Phi}$ *is a root system, and it corresponds to the* $\mathfrak{a}_\ell$ *Dynkin diagram.*

Proof of Theorem 11.5.1. To prove this proposition let us first summarize the properties of $\mathbf{\Phi}$. We have

$$\alpha \in \mathbf{\Phi} \Rightarrow \begin{cases} (1) & \alpha = n^i \epsilon_i, & & n^i \in \mathbb{Z}, \\ (2) & (\alpha, v) = 0 & \Leftrightarrow & \sum_{i=1}^{\ell+1} n^i = 0, \\ (3) & (\alpha, \alpha) = 2 & \Leftrightarrow & \sum_{i=1}^{\ell+1} (n^i)^2 = 2. \end{cases} \tag{11.91}$$

These Diophantine equations have the following solutions:

$$\alpha = \epsilon_i - \epsilon_j \quad (i \neq j). \tag{11.92}$$

The number of such solutions is equal to twice the number of pairs (ij) in $(\ell + 1)$-dimensional space:

$$\#\alpha = 2\frac{1}{2}(\ell+1)(\ell+1-1) = \ell(\ell+1) = (\ell+1)^2 - 1 - \ell. \tag{11.93}$$

We verify that this finite set of vectors is a root system. First we check that for all pairs $\alpha, \beta \in \mathbf{\Phi}$ their hook product is an integer. Indeed, we have

$$\begin{aligned}\langle\alpha,\beta\rangle &= 2\frac{(\alpha,\beta)}{(\beta,\beta)} = (\alpha,\beta)\\ &= (\epsilon_i - \epsilon_j, \epsilon_k - \epsilon_\ell) = \delta_{ik} - \delta_{jk} - \delta_{i\ell} + \delta_{j\ell} \in \mathbb{Z}.\end{aligned} \tag{11.94}$$

Second we check that the reflection of any candidate root $\beta \in \mathbf{\Phi}$ with respect to any other candidate root $\alpha \in \mathbf{\Phi}$ belongs to the same set $\mathbf{\Phi}$:

$$\begin{aligned}\sigma_\alpha(\beta) &= \beta - (\alpha,\beta)\\ &= \epsilon_k - \epsilon_\ell - (\delta_{ik} - \delta_{jk} - \delta_{i\ell} + \delta_{j\ell})(\epsilon_i - \epsilon_j).\end{aligned} \tag{11.95}$$

If (k,ℓ) are both different from (i,j), then $\sigma_\alpha(\beta) = \beta \in \mathbf{\Phi}$, so the statement is true. If $k = i$, then necessarily $k \neq j$ and $i \neq \ell$, so

$$\sigma_\alpha(\beta) = \epsilon_k - \epsilon_\ell - (\delta_{ik} + \delta_{jl})(\epsilon_i - \epsilon_j). \tag{11.96}$$

If $j \neq \ell$, then

$$\sigma_\alpha(\beta) = \epsilon_k - \epsilon_\ell - (1)(\epsilon_i - \epsilon_j) = \epsilon_j - \epsilon_\ell \in \mathbf{\Phi}. \tag{11.97}$$

If $j = \ell$, then

$$\sigma_\alpha(\beta) = \epsilon_k - \epsilon_\ell - 2(\epsilon_k - \epsilon_\ell) = \epsilon_\ell - \epsilon_k \in \mathbf{\Phi}, \tag{11.98}$$

which exhausts all possible cases.

Hence, in $\mathbb{R}^{\ell+1}$ we have constructed a root system of $\ell(\ell+1)$ roots.

Consider the roots

$$\alpha_i = \epsilon_i - \epsilon_{i+1} \quad (i = 1, \dots, \ell). \tag{11.99}$$

These roots are clearly linearly independent and given a root $\alpha \in \mathbf{\Phi}$ it can be expressed as a linear combination of α_i. We subdivide the set of roots into a positive and a negative set according to the following rule:

$$
\begin{aligned}
&\mathbf{\Phi} = \mathbf{\Phi}_+ \cup \mathbf{\Phi}_-, \\
&\alpha \in \mathbf{\Phi}_+ : \qquad \{\alpha = \epsilon_i - \epsilon_j,\ i < j\}, \\
&\alpha \in \mathbf{\Phi}_- : \qquad \{\alpha = \epsilon_j - \epsilon_i,\ i < j\}.
\end{aligned}
\tag{11.100}
$$

Clearly, positive roots can be written as

$$\alpha = \epsilon_i - \epsilon_j = \alpha_i + \alpha_{i+1} + \cdots + \alpha_{j-1} \tag{11.101}$$

and consequently have integer positive components in the $\{\alpha, \dots, \alpha_\ell\}$ basis. Negative roots have negative integer components. It follows that $\{\alpha, \dots, \alpha_\ell\}$ form a basis of simple roots.

Let us compute the Cartan matrix:

$$
\begin{aligned}
(\alpha_i, \alpha_j) &= (\epsilon_i - \epsilon_j, \epsilon_j - \epsilon_{j+1}) \quad (i < j) \\
&= \delta_{ij} - \delta_{i+1,j} - \delta_{i,j+1} + \delta_{i+1,j+1}.
\end{aligned}
\tag{11.102}
$$

If	$i = j$	$(\alpha_i, \alpha_i) = 2$
If	$j = i + 1$	$(\alpha_i, \alpha_{i+1}) = -1$
If	$j = i - 1$	$(\alpha_i, \alpha_{i-1}) = -1$

Hence we have precisely the Dynkin diagram of Figure 11.10.

This concludes the proof of our theorem. □

Theorem 11.5.2. *The root system $\mathfrak{a}_\ell$ corresponds to the complex Lie algebra $\mathfrak{sl}(\ell + 1, \mathbb{C})$ of traceless matrices in $\ell + 1$ dimensions.*

Proof of Theorem 11.5.2. Note that the dimension of the $\mathfrak{sl}(\ell + 1, \mathbb{C})$ Lie algebra is

$$\dim \mathfrak{sl}(\ell + 1, \mathbb{C}) = (\ell + 1)^2 - 1, \tag{11.103}$$

since on the $(\ell+1)\times(\ell+1)$ matrix A we just impose one scalar condition, namely, $\operatorname{Tr} A = 0$. This agrees with the number of roots in the system $\mathbf{\Phi}$,

$$\operatorname{card} \mathbf{\Phi} = \ell(\ell + 1) = (\ell + 1)^2 - \ell - 1, \tag{11.104}$$

if the rank of $\mathfrak{sl}(\ell + 1, \mathbb{C})$ is precisely ℓ:

$$\operatorname{card} \mathbf{\Phi} = \dim \mathbb{G} - \dim \mathscr{H}, \tag{11.105}$$

$\mathscr{H}$ being the Cartan subalgebra.

This is indeed the case. Let e_{ij} denote the $(\ell+1)\times(\ell+1)$ matrix whose entries are all zero except for the ij-th entry, which is 1,

$$e_{ij} = \begin{pmatrix} 0 & 0 & \cdots & \cdots & \cdots & 0 & \\ 0 & 0 & \cdots & \cdots & \cdots & 0 & \\ \cdots & \cdots & 0 & 0 & 0 & \cdots & \\ \cdots & \cdots & 0 & 1 & 0 & \cdots & i\text{-th} \\ \cdots & \cdots & 0 & 0 & 0 & \cdots & \\ 0 & 0 & \cdots & \cdots & \cdots & 0 & \\ & & & j\text{-th} & & & \end{pmatrix}, \tag{11.106}$$

and define

$$\begin{aligned} H_i &= e_{ii} - e_{\ell+1,\ell+1} \quad (i = 1,\dots,\ell), \\ \mathscr{E}_{ij} &= e_{ij} \quad (i \neq j). \end{aligned} \tag{11.107}$$

Since

$$e_{ij}\cdot e_{km} = \delta_{jk}e_{im}, \tag{11.108}$$

we have

$$\begin{aligned} [H_i, H_j] &= 0, \\ [H_i, \mathscr{E}_{rs}] &= \delta_{ir}e_{is} - \delta_{si}e_{ri} - \delta_{\ell+1,r}e_{\ell+1,s} + \delta_{s,\ell+1}e_{r,\ell+1} \\ &= (\delta_{ir} - \delta_{si} - \delta_{\ell+1,r} + \delta_{s,\ell+1})\mathscr{E}_{rs}. \end{aligned} \tag{11.109}$$

Now observe that a basis for the space E of vectors orthogonal to $v = (1,1,\dots,1)$ is provided by

$$u_i = \epsilon_i - \epsilon_{\ell+1} \quad (i = 1,\dots,\ell). \tag{11.110}$$

Indeed, this is a system of ℓ linearly independent vectors in an ℓ-dimensional space. Hence, we can identify

$$(\delta_{ir} - \delta_{si} - \delta_{\ell+1,r} + \delta_{s,\ell+1}) = (\epsilon_r - \epsilon_s, u_i); \quad (r,s = 1,\dots,\ell+1) \quad (i = 1,\dots,\ell). \tag{11.111}$$

This implies that with every Cartan subalgebra element $h = \omega^r H_r$ we can associate the vector $\omega \equiv \omega^r u_r \in E$ and with every root $\epsilon_r - \epsilon_s$ we can associate the linear functional

$$[\epsilon_r - \epsilon_s](\omega) = (\epsilon_r - \epsilon_s, \omega). \tag{11.112}$$

With such identifications, $\mathfrak{sl}(\ell+1,\mathbb{C})$ is cast into the canonical Weyl form of equation (11.38) and our theorem is proved. □

11.5.2 The $\mathfrak{d}_\ell$ root system and the corresponding Lie algebra

The Dynkin diagram is recalled in Figure 11.11. We want to perform the explicit construction of a root system that admits a basis corresponding to such a diagram. To this effect let us consider $E = \mathbb{R}^\ell$ and the set

$$\Phi \equiv \{\alpha \in \mathbb{I}_\ell \mid (\alpha, \alpha) = 2\}, \tag{11.113}$$

namely, the set of all length $\sqrt{2}$ vectors in the cubic lattice:

$$\alpha = n^i \epsilon_i; \quad \sum_{i=1}^{\ell} (n^i)^2 = 2. \tag{11.114}$$

The solution of the above constraint is given by all the vectors of the following type:

$$\alpha = \pm(\epsilon_i \pm \epsilon_j), \quad i \neq j. \tag{11.115}$$

These vectors are easily counted. We have

$$\operatorname{card} \Phi = 2\ell(\ell - 1). \tag{11.116}$$

In a completely similar way to the previous case one verifies that Φ is a root system. The simple roots can be chosen as follows:

$$\begin{aligned}
\alpha_1 &= \epsilon_1 - \epsilon_2, \\
\alpha_2 &= \epsilon_2 - \epsilon_3, \\
&\vdots \\
\alpha_{\ell-1} &= \epsilon_{\ell-1} - \epsilon_\ell, \\
\alpha_\ell &= \epsilon_{\ell-1} + \epsilon_\ell,
\end{aligned} \tag{11.117}$$

which clearly corresponds to the chosen Dynkin diagram in Figure 11.11.

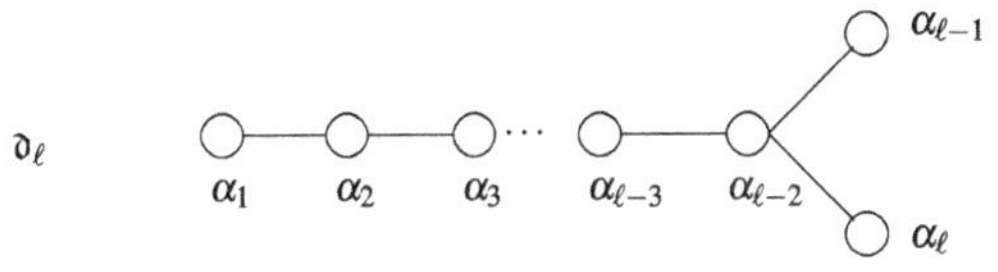

Figure 11.11: The Dynkin diagram of $\mathfrak{d}_\ell$ type.

Next we identify the corresponding Lie algebra with $\mathfrak{so}(2\ell, \mathbb{C})$. The procedure is the following. Consider in 2ℓ dimensions the following symmetric matrix:

$$\mathfrak{Q} \equiv \left(\begin{array}{c|c} \mathbf{0}_{\ell\times\ell} & \mathbf{1}_{\ell\times\ell} \\ \hline \mathbf{1}_{\ell\times\ell} & \mathbf{0}_{\ell\times\ell} \end{array} \right). \tag{11.118}$$

The Lie algebra $\mathfrak{so}(2\ell, \mathbb{C})$ can be identified as the set of $2\ell \times 2\ell$ matrices satisfying the $\mathfrak{Q}$-antisymmetricity constraint:

$$\mathfrak{so}(2\ell, \mathbb{C}) = \left\{ A = \left(\begin{array}{c|c} a_{11} & a_{12} \\ \hline a_{21} & a_{21} \end{array} \right) \middle| A^T \mathfrak{Q} + \mathfrak{Q} A = 0 \right\}. \tag{11.119}$$

The constraint is easily solved by setting

$$a_{22} = -a_{11}^T; \quad a_{21} = -a_{21}^T; \quad a_{12} = -a_{12}^T. \tag{11.120}$$

We leave to the reader the easy task of explicitly verifying that a Cartan–Weyl basis spanning the space of all matrices of type (11.119) is the following one:

$$\begin{aligned} H_i &= e_{ii} - e_{\ell+i,\ell+i}, \\ \mathcal{E}^{\epsilon_i - \epsilon_j} &= e_{ji} - e_{i+\ell,j+\ell}, \\ \mathcal{E}^{\epsilon_i + \epsilon_j} &= e_{i+\ell,j} - e_{j+\ell,i}, \\ \mathcal{E}^{-\epsilon_i - \epsilon_j} &= e_{j,i+\ell} - e_{i,j+\ell}. \end{aligned} \tag{11.121}$$

The diagonal matrices H_i are the Cartan generators. The commutator of each of the step operators $\mathcal{E}^\alpha$ with each of the Cartan operators yields the appropriate answer according to equation (11.38). This concludes the proof that the root systems of type $\mathfrak{d}_\ell$ are associated with the Lie algebras $\mathfrak{so}(2\ell, \mathbb{C})$ of the orthogonal groups in even dimensions.

11.5.3 The $\mathfrak{b}_\ell$ root system and the corresponding Lie algebra

The Dynkin diagram is recalled in Figure 11.12. We want to perform the explicit construction of a root system that admits a basis corresponding to such a diagram. To this effect let us consider $E = \mathbb{R}^\ell$. Within this Euclidean space the candidate root system is defined by the following set of vectors:

$$\mathbf{\Phi} \equiv \underbrace{\{\pm\epsilon_i \pm \epsilon_j\}}_{2\ell(\ell-1)} \cup \underbrace{\{\pm\epsilon_i\}}_{2\ell}, \tag{11.122}$$

where in the underbrace we have indicated the number of vectors for each of the two types. Hence, we have

$\mathfrak{b}_\ell$

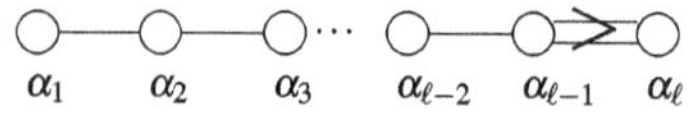

Figure 11.12: The Dynkin diagram of $\mathfrak{b}_\ell$ type.

$$\operatorname{card}\mathbf{\Phi} = 2\ell^2. \tag{11.123}$$

In a completely similar way to the previous cases one can verify that $\mathbf{\Phi}$ is a root system. The simple roots can be chosen as follows:

$$\begin{aligned}
\alpha_1 &= \epsilon_1 - \epsilon_2,\\
\alpha_2 &= \epsilon_2 - \epsilon_3,\\
&\vdots\\
\alpha_{\ell-1} &= \epsilon_{\ell-1} - \epsilon_\ell,\\
\alpha_\ell &= \epsilon_\ell,
\end{aligned} \tag{11.124}$$

which clearly corresponds to the chosen Dynkin diagram in Figure 11.12.

Next we identify the corresponding Lie algebra with $\mathfrak{so}(2\ell+1,\mathbb{C})$. The procedure is similar to that used in the previous case. Consider in $2\ell+1$ dimensions the following symmetric matrix:

$$\mathfrak{Q} \equiv \left(\begin{array}{c|c|c} 1 & \mathbf{0}_{1\times\ell} & \mathbf{0}_{1\times\ell}\\ \hline \mathbf{0}_{\ell\times1} & \mathbf{0}_{\ell\times\ell} & \mathbf{1}_{\ell\times\ell}\\ \hline \mathbf{0}_{\ell\times1} & \mathbf{1}_{\ell\times\ell} & \mathbf{0}_{\ell\times\ell} \end{array}\right). \tag{11.125}$$

The Lie algebra $\mathfrak{so}(2\ell+1,\mathbb{C})$ can be identified as the set of $(2\ell+1)\times(2\ell+1)$ matrices satisfying the $\mathfrak{Q}$-antisymmetricity constraint:

$$A^T\mathfrak{Q} + \mathfrak{Q}A = 0, \tag{11.126}$$

whose solution in terms of independent subblocks is

$$A = \begin{pmatrix} 0 & v_1 & v_2\\ -v_2^T & a & b\\ -v_1^T & -b^T & -a^T \end{pmatrix}. \tag{11.127}$$

We leave to the reader the easy task of explicitly verifying that a Cartan–Weyl basis spanning the space of all matrices of type (11.127) is the following one:

$$\begin{aligned}
H_i &= e_{i+1,i+1} - e_{\ell+i+1,\ell+i+1},\\
\mathcal{E}^{\epsilon_i-\epsilon_j} &= e_{j+1,i+1} - e_{i+\ell+1,j+\ell+1},\\
\mathcal{E}^{\epsilon_i+\epsilon_j} &= e_{i+\ell+1,j+1} - e_{j+\ell+1,i+1},\\
\mathcal{E}^{-\epsilon_i-\epsilon_j} &= e_{j+1,i+\ell+1} - e_{i+1,j+\ell+1},\\
\mathcal{E}^{\epsilon_i} &= e_{1,i+1} - e_{i+\ell+1,1},\\
\mathcal{E}^{-\epsilon_i} &= e_{i+1,1} - e_{1,i+\ell+1}.
\end{aligned} \tag{11.128}$$

The diagonal matrices H_i are the Cartan generators. The commutator of each of the step operators $\mathcal{E}^{\alpha}$ with each of the Cartan operators yields the appropriate answer according to equation (11.38). This concludes the proof that the root systems of type $\mathfrak{b}_\ell$ are associated with the Lie algebras $\mathfrak{so}(2\ell+1,\mathbb{C})$ of the orthogonal groups in odd dimensions.

11.5.4 The $\mathfrak{c}_\ell$ root system and the corresponding Lie algebra

The Dynkin diagram is recalled in Figure 11.13. We want to perform the explicit construction of a root system that admits a basis corresponding to such a diagram. To this effect let us consider $E = \mathbb{R}^\ell$. Within this Euclidean space the candidate root system is defined by the following set of vectors:

$$\Phi \equiv \underbrace{\{\pm\epsilon_i \pm \epsilon_j\}}_{2\ell(\ell-1)} \cup \underbrace{\{\pm 2\epsilon_i\}}_{2\ell}, \tag{11.129}$$

where in the underbrace we have indicated the number of vectors for each of the two types. Hence, we have

$$\operatorname{card}\Phi = 2\ell^2. \tag{11.130}$$

In a completely similar way to the previous cases one can verify that Φ is a root system. The simple roots can be chosen as follows:

$$\begin{aligned} \alpha_1 &= \epsilon_1 - \epsilon_2, \\ \alpha_2 &= \epsilon_2 - \epsilon_3, \\ &\vdots \\ \alpha_{\ell-1} &= \epsilon_{\ell-1} - \epsilon_\ell, \\ \alpha_\ell &= 2\epsilon_\ell, \end{aligned} \tag{11.131}$$

which clearly corresponds to the chosen Dynkin diagram in Figure 11.13.

$\mathfrak{c}_\ell$

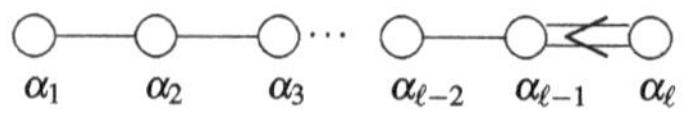

Figure 11.13: The Dynkin diagram of $\mathfrak{c}_\ell$ type.

Next we identify the corresponding Lie algebra with $\mathfrak{sp}(2\ell,\mathbb{C})$. The procedure is similar to that used in the previous case. Consider in 2ℓ dimensions the following antisymmetric matrix:

$$\mathfrak{Q} \equiv \left(\begin{array}{c|c} \mathbf{0}_{\ell\times\ell} & \mathbf{1}_{\ell\times\ell} \\ \hline -\mathbf{1}_{\ell\times\ell} & \mathbf{0}_{\ell\times\ell} \end{array}\right). \tag{11.132}$$

The Lie algebra $\mathfrak{sp}(2\ell,\mathbb{C})$ can be identified as the set of $2\ell \times 2\ell$ matrices satisfying the $\mathfrak{Q}$-antisymmetricity constraint:

$$A^T\mathfrak{Q} + \mathfrak{Q}A = 0, \tag{11.133}$$

whose solution in terms of independent subblocks is

$$\begin{aligned} A &= \left(\begin{array}{c|c} a_{11} & a_{12} \\ \hline a_{21} & a_{21} \end{array}\right), \\ a_{22} &= -a_{11}^T; \quad a_{21} = a_{21}^T; \quad a_{12} = a_{12}^T. \end{aligned} \tag{11.134}$$

We leave to the reader the easy task of explicitly verifying that a Cartan–Weyl basis spanning the space of all matrices of type (11.127) is the following one:

$$\begin{aligned} H_i &= e_{i,i} - e_{\ell+i,\ell+i}, \\ \mathcal{E}^{\epsilon_i-\epsilon_j} &= e_{i,j} - e_{j+\ell,i+\ell} \quad (i<j), \\ \mathcal{E}^{-\epsilon_i+\epsilon_j} &= e_{j,i} - e_{i+\ell,j+\ell} \quad (i<j), \\ \mathcal{E}^{\epsilon_i+\epsilon_j} &= e_{i,j+\ell} + e_{j,i+\ell}, \\ \mathcal{E}^{-\epsilon_i-\epsilon_j} &= e_{i+\ell,j} + e_{j+\ell,i}, \\ \mathcal{E}^{2\epsilon_i} &= e_{i,i+\ell}, \\ \mathcal{E}^{-2\epsilon_i} &= e_{i+\ell,i}. \end{aligned} \tag{11.135}$$

The diagonal matrices H_i are the Cartan generators. The commutator of each of the step operators $\mathcal{E}^\alpha$ with each of the Cartan operators yields the appropriate answer according to equation (11.38). This concludes the proof that the root systems of type $\mathfrak{c}_\ell$ are associated with the Lie algebras $\mathfrak{sp}(2\ell,\mathbb{C})$ of the symplectic groups in even dimensions.

11.5.5 The exceptional Lie algebras

We should now construct, as Cartan did, the explicit form of the exceptional Lie algebras. We postpone this to Chapter 13. Each exceptional Lie algebra will be studied in relation with a particular homogeneous manifold of interest in physics.

11.6 Bibliographical note

The main bibliographical sources for the present chapter are:
- Unpublished lecture notes by the author, dating back to the time when he usually gave the Lie algebra course at SISSA (beginning of the 1990s).
- The textbook [96].

- The textbook [39].
- The textbook [40].
- The monograph [38].

Furthermore, let us mention that a detailed historical account of how Lie algebras were discovered by Lie, independently classified by Killing (who also discovered exceptional Lie algebras), and finally systematized by Cartan, stressing also the key role played in such developments by Klein and Engels, is presented in the book by the present author [28].

12 Lie algebra representation theory

My work always tried to unite the truth with the beautiful, but when I had to choose one or the other, I usually chose the beautiful . . .

Hermann Weyl

12.1 Linear representations of a Lie algebra

The general concept of linear representations of groups was introduced in Section 4.3. It equally applies to discrete groups as to Lie groups. The reader is invited to revisit Definitions 4.3.1, 4.3.2, 4.3.3, and 4.3.4, which apply without any change to continuous groups.

What is new in the case of Lie groups is that linear group representations can be constructed first as Lie algebra linear representations, using next the exponential map (7.63), described in Section 7.4, in order to obtain the matrix representation of finite group elements from that of Lie algebra elements.

In full analogy to the above quoted definitions that apply to groups we have the following ones.

Definition 12.1.1. Let $\mathbb{G}$ be a finite-dimensional Lie Algebra and let V be a vector space of dimension n. Any homomorphism:

$$D : \mathbb{G} \to \mathrm{Hom}(V, V) \tag{12.1}$$

such that

$$\forall A, B \in \mathbb{G} : \quad D([A, B]) = [D(A), D(B)] \tag{12.2}$$

is called a linear representation of dimension n of the Lie algebra $\mathbb{G}$. In the left hand side of equation (12.2), the symbol $[\ ,\ ]$ denotes the abstract Lie bracket, while in the right hand side of the same equation it denotes the commutator of two linear maps applied in sequence.

If the vectors $\{\mathbf{e}_i\}$ form a basis of the vector space V, then each Lie algebra element $\gamma \in \mathbb{G}$ is mapped into an $n \times n$ matrix $D_{ij}(\gamma)$ such that

$$D(\gamma).\mathbf{e}_i = \mathbf{e}_j D_{ji}(\gamma), \tag{12.3}$$

and we see that the homomorphism D can also be rephrased as the following one:[1]

1 Let us recall that by $\mathfrak{gl}(n, \mathbb{F})$ one refers to the Lie algebra of arbitrary $n \times n$ matrices.

https://doi.org/10.1515/9783111201535-012

$$D : G \to \mathfrak{gl}(n, \mathbb{F}), \quad \mathbb{F} = \begin{cases} \mathbb{R}, \\ \mathbb{C}, \end{cases} \tag{12.4}$$

where the field $\mathbb{F}$ is that of the real or complex numbers, depending on whether V is a real or complex vector space. Correspondingly we say that D is a *real* or *complex representation.*

Definition 12.1.2. Let $D : G \to \mathrm{Hom}(V, V)$ be a linear representation of a Lie algebra $\mathbb{G}$. A vector subspace $W \subset V$ is said to be *invariant* iff

$$\forall \gamma \in \mathbb{G}, \forall \mathbf{w} \in W : \quad D(\gamma).\mathbf{w} \in W. \tag{12.5}$$

Definition 12.1.3. A linear representation $D : G \to \mathrm{Hom}(V, V)$ of a Lie algebra $\mathbb{G}$ is named *irreducible* iff the only invariant subspaces of V are $\mathbf{0}$ and V itself.

In other words, a representation is irreducible if it does not admit any proper invariant subspace.

Definition 12.1.4. A linear representation $D : \mathbb{G} \to \mathrm{Hom}(V, V)$ that admits at least one proper invariant subspace $W \subset V$ is called *reducible.* A reducible representation $D : G \to \mathrm{Hom}(V, V)$ is called *fully reducible* iff the orthogonal complement $W^\perp$ of any invariant subspace W is also invariant.

12.1.1 Weights of a representation and the weight lattice

The classification of complex Lie algebras in terms of Cartan subalgebras and root systems provides very powerful weaponry for the explicit construction of linear representations of Lie algebras. In this context the contributions of Hermann Weyl have been extremely important and the role of the Weyl group is quite essential.

Given the root system $\mathbf{\Phi}$ of rank ℓ and its basis of simple roots α_i $(i = 1, \ldots, \ell)$ we can define the root lattice

$$\Lambda_{\mathrm{root}} \subset \mathbb{R}^\ell / \mathbf{v} \in \Lambda_{\mathrm{root}} \quad \Leftrightarrow \quad \mathbf{v} = n^i \alpha_i, \quad n^i \in \mathbb{Z}. \tag{12.6}$$

As explained in Section 8.1.1, given any lattice Λ we can define its dual $\Lambda^\star$ by means of the definition (8.5). Concretely, given a basis of Λ, one introduces a dual basis as in equation (8.6) and takes the linear combinations of these dual basis vectors with integer valued coefficients. In view of this general construction a naturally arising question concerns the interpretation of the dual of the root lattice. This latter is named the *weight lattice* and it is of utmost relevance for *representation theory*:

$$\Lambda_{\mathrm{weight}} \equiv \Lambda^\star_{\mathrm{root}}. \tag{12.7}$$

Explicitly one defines the basis of simple weights dual to the basis of simple roots through the condition

$$\langle \lambda^i, \alpha_j \rangle \equiv 2\frac{(\lambda^i, \alpha_j)}{(\alpha_j, \alpha_j)} = \delta^i_j \tag{12.8}$$

and sets

$$\Lambda_{\text{weight}} \subset \mathbb{R}^\ell / \mathbf{w} \in \Lambda_{\text{weight}} \quad \Leftrightarrow \quad \mathbf{w} = n_i \lambda^i, \quad n_i \in \mathbb{Z}. \tag{12.9}$$

As we presently show, linear representations Γ of the Lie algebra $\mathbb{G}$ associated with the considered root system $\mathbf{\Phi}$ are essentially encoded in certain precisely defined subsets $\Pi_\Gamma \subset \Lambda_{\text{weight}}$ called the set of *weights of the representation.*

12.1.1.1 The maximally split and the maximal compact real sections of a simple Lie algebra $\mathbb{G}(\mathbb{C})$

So far, in our discussion of Lie algebras $\mathbb{G}$ we have assumed that, as vector spaces, they are defined over the complex numbers $\mathbb{C}$. An important topic that we will address later is that of real sections $\mathbb{G}_r \subset \mathbb{G}(\mathbb{C})$. Given the simple Lie algebra generators in the canonical Cartan–Weyl form, $T_A = \{H_i, E^\alpha, E^{-\alpha}\}$, the question is which restrictions on the imaginary and real parts of the coefficients c^A of Lie algebra elements $c^A T_A$ can be introduced that are consistent with the Lie bracket and produce a real Lie algebra $\mathbb{G}_r$. Furthermore, one would like to know how many such real sections do exist up to isomorphism.

Here we just introduce two real sections that are simply and universally defined for all simple Lie algebras:

(a) *The maximally split real section* $\mathbb{G}_{\max}$. This is defined by assuming that the allowed coefficients c^A are all real. In any linear representation of $\mathbb{G}_{\max}$ the matrices representing

$$T_A \equiv \{H_i, E^\alpha, E^{-\alpha}\} \tag{12.10}$$

are all *real.* From the representations of $\mathbb{G}_{\max}$, by taking linear combinations of the generators with complex coefficients one obtains all the linear representations of the complex Lie algebra $\mathbb{G}(\mathbb{C})$.

(b) *The maximally compact real section* $\mathbb{G}_c$. This real section, whose exponentiation produces a compact Lie group, is obtained by allowing linear combinations with real coefficients of the set of generators:

$$\mathfrak{T}_A \equiv \{i H_i, i(E^\alpha + E^{-\alpha}), (E^\alpha - E^{-\alpha})\}. \tag{12.11}$$

In all linear representations of $\mathbb{G}_c$ the matrices representing the generators $\mathfrak{T}_A$ are *anti-Hermitian.*

One easily obtains the Hermitian matrix representation of the generators $\mathfrak{T}_A$ from the real representation of the generators T_A, and vice versa. It also follows that the matrices representing $E^{-\alpha}$ are the transpose of those representing E^{α}.

A very useful instrument in the explicit construction of matrix representations that has also important consequences for later developments is provided by the notion of Borel subalgebra. Starting from the Cartan–Weyl basis, if one considers the subset of generators

$$\mathrm{Bor}[\mathbb{G}] = \mathrm{span}\{H_i, E^{\alpha}\}; \quad \alpha > 0, \tag{12.12}$$

we see that it corresponds to a *solvable subalgebra of* $\mathbb{G}$. Hence, according to Corollary 10.3.1 every representation of $\mathrm{Bor}[\mathbb{G}]$ can be put into an upper triangular form. This gives a powerful construction criterion for the fundamental representation. We just construct an upper triangular representation of the $\mathrm{Bor}[\mathbb{G}]$ subalgebra and then we promote it to a representation of the full Lie algebra $\mathbb{G}$ by setting

$$E^{-\alpha} = (E^{\alpha})^T. \tag{12.13}$$

Furthermore, in view of the above discussion, the representations of the real sections $\mathbb{G}_{\max}$ and $\mathbb{G}_{\mathbf{c}}$ can be considered together and on that we rely in the following.

Let us assume that

$$\Gamma : \mathbb{G}_c \Longrightarrow \mathrm{Hom}(V, V) \tag{12.14}$$

is a linear representation of the compact section which, by the above argument, can be turned into a representation of the maximally split real section. From the linearity we conclude that

$$\Gamma(H_i) = -i\Gamma(iH_i) \quad (i = 1, \ldots, \ell) \tag{12.15}$$

are a set of ℓ *commuting Hermitian matrices*. Therefore, we can choose a basis in the vector space V where all $\Gamma(H_i)$ are diagonal matrices. This is equivalent to the statement that the vector space V of dimension n can be decomposed into a direct sum of subspaces:

$$V = \bigoplus_{\lambda\in\Pi_\Gamma} V_\lambda \quad (\dim V_\lambda \geq 1), \tag{12.16}$$

where $\Pi_\Gamma \subset \mathscr{H}^\star$ is a subset of linear functionals on the Cartan subalgebra $\mathscr{H}$ that is characteristic of the considered representation Γ,

$$\lambda : \mathscr{H} \to \mathbb{C}, \tag{12.17}$$

and $\forall \mathbf{v} \in V_\lambda$ we have

$$\Gamma(h)\mathbf{v} = \lambda(h)\mathbf{v}; \quad \forall h \in \mathscr{H}. \tag{12.18}$$

The set Π_Γ is necessarily a finite set if Γ is a finite-dimensional representation. It is the *set of weights of the representation* as we already anticipated above. The dimension of V_λ is called the weight multiplicity and it is denoted $m(\lambda)$:

$$m(\lambda) = \dim V_\lambda. \tag{12.19}$$

It follows that

$$n = \dim V = \sum_{\lambda \in \Pi_\Gamma} m(\lambda). \tag{12.20}$$

Since the Cartan subalgebra is endowed with a natural scalar product induced by the Killing metric $\kappa(,)$ restricted to $\mathscr{H}$, the elements of $\mathscr{H}^\star$ can be represented by r-dimensional vectors that belong to the same Euclidean space which includes the roots. Let $\{H_i\}$ be the basis of $\mathscr{H}$ in which $\kappa(H_i, H_j) = \delta_{ij}$ and let us define the components of the λ vector by

$$\lambda_i = \lambda(H_i). \tag{12.21}$$

Furthermore, let us utilize the bracket notation for the vectors belonging to V_λ:

$$\begin{aligned} &|\lambda, p\rangle; \quad p = 1, \dots, m(\lambda), \\ \langle \mu, r | \lambda, p\rangle &= \delta_{\lambda\mu}\delta_{rp}, \\ \Gamma(h)|\lambda, p\rangle &= \lambda(h)|\lambda, p\rangle. \end{aligned} \tag{12.22}$$

Then we have the following.

Lemma 12.1.1. *Let $\lambda \in \Pi_\Gamma$ be a weight and let $\alpha \in \boldsymbol{\Phi}$ be a root. Then $\alpha + \lambda \in \Pi_\Gamma$ is another weight if there exists a vector $|\lambda\rangle \in V_\lambda$ such that*

$$\Gamma(E^\alpha)|\lambda\rangle \neq 0. \tag{12.23}$$

Proof of Lemma 12.1.1. To prove the statement it suffices to set

$$|\lambda + \alpha\rangle = \Gamma(E^\alpha)|\lambda\rangle \tag{12.24}$$

and verify that

$$\begin{aligned} \Gamma(h)|\lambda + \alpha\rangle &= [\Gamma(h), \Gamma(E^\alpha)]|\lambda\rangle + \Gamma(E^\alpha)\Gamma(h)|\lambda\rangle \\ &= (\alpha(h) + \lambda(h))|\lambda + \alpha\rangle, \end{aligned} \tag{12.25}$$

which shows that $(\alpha + \lambda)$ is a weight. ◻

12.1.1.2 The subalgebra $A_1 \subset G$ associated with a positive root and the properties of the weights

With each positive root $\alpha \in \Phi_+$ of a root system we can associate a subalgebra $\mathbb{G} \supset A_1 \sim \mathfrak{su}(2)$ which plays an important role in determining the properties of weights of a representation. Given the canonical Cartan–Weyl form of the simple Lie algebra (11.38), let us define

$$\mathscr{H}_\alpha = \frac{2}{(\alpha,\alpha)}\alpha^i H_i; \quad \mathscr{E}_\alpha^\pm = \sqrt{\frac{2}{(\alpha,\alpha)}}E^{\pm\alpha}. \tag{12.26}$$

We obtain the commutation relations

$$\begin{aligned}[\mathscr{H}_\alpha, \mathscr{H}_\alpha] &= 0,\\ [\mathscr{H}_\alpha, \mathscr{E}_\alpha^\pm] &= \pm 2\mathscr{E}_\alpha^\pm,\\ [\mathscr{E}_\alpha^+, \mathscr{E}_\alpha^-] &= \mathscr{H}_\alpha.\end{aligned} \tag{12.27}$$

This algebra is isomorphic to the angular momentum algebra

$$\begin{aligned}[J_3, J_3] &= 0,\\ [J_3, J^\pm] &= \pm J^\pm,\\ [J^+, J^-] &= 2J_3\end{aligned} \tag{12.28}$$

upon the identifications

$$\mathscr{H}_\alpha = 2J_3; \quad \mathscr{E}_\alpha^\pm = J^\pm. \tag{12.29}$$

Since any irreducible $\mathbb{G}$ representation necessarily decomposes into a sum of irreducible representations of each $A_1(\alpha)$ subalgebra and in each irreducible representation of A_1 the eigenvalues of J_3 are necessarily either integer or half-integer,[2] it follows that for any root α the eigenvalues of H_α must be integer. This provides the relation between the weights of the representation defined above and the weight lattice introduced in equations (12.9) and (12.8). Indeed, from the outlined reasoning we conclude

$$\begin{aligned} H_\alpha|\lambda, i\rangle &= \lambda(H_\alpha)|\lambda, i\rangle,\\ \lambda(H_\alpha) = 2\frac{(\lambda,\alpha)}{(\alpha,\alpha)} &\equiv \langle\lambda,\alpha\rangle \in \mathbb{Z},\\ &\Downarrow\\ \lambda &\in \Lambda_{\text{weight}}.\end{aligned} \tag{12.30}$$

2 Here we use the properties of angular momenta that are known to any student of physics, mathematics, or engineering from the course of elementary quantum mechanics. A representation of spin j contains $(2j+1)$ states with third component $m = -j, -j+1, \dots, j-1, j$, where j is integer or half-integer.

So the weights of a representation Π_Γ constitute a finite subset of vectors lying in the weight lattice. We have the following very simple but very important theorem.

Theorem 12.1.1. *Given any finite-dimensional representation Γ of a simple Lie algebra $\mathbb{G}$, the set of its weights Π_Γ is invariant with respect to the Weyl group defined in Definition 11.3.2 (Section 11.3). Explicitly we have*

$$\begin{aligned} &\forall \alpha \in \mathbf{\Phi}: \quad \lambda \in \Pi_\Gamma \quad \Rightarrow \quad \sigma_\alpha(\lambda) \in \Pi_\Gamma, \\ &m(\sigma_\alpha(\lambda)) = m(\lambda). \end{aligned} \tag{12.31}$$

Proof of Theorem 12.1.1. Let us note that

$$\sigma_\alpha \lambda(h) = \lambda(h) - \langle \lambda, \alpha \rangle \alpha(h) = \lambda(h) - \lambda(H_\alpha)\alpha(h) \tag{12.32}$$

and let us define

$$\beta = \begin{cases} \alpha & \text{if } \lambda(H_\alpha) \geq 0, \\ -\alpha & \text{if } \lambda(H_\alpha) < 0. \end{cases} \tag{12.33}$$

With such a notation we always have $\lambda(H_\beta) \geq 0$ and $\forall h \in \mathscr{H}_{CSA}$, $\sigma_\alpha\lambda(h) = \sigma_\beta\lambda(h)$. The number $\lambda(H_\beta) = R \in \mathbb{Z}_+$ is by construction a non-negative integer. If $R = 0$ the theorem is proved. Hence, we consider the case $R > 0$.

We have

$$\sigma_\alpha \lambda(h) = \lambda(h) - R\beta(h). \tag{12.34}$$

The linear functional $\sigma_\alpha\lambda(h)$ is a weight of the representation with multiplicity $m(\sigma_\beta\lambda) = m(\lambda)$ if

$$[\Gamma(E^{-\beta})]^R |\lambda, p\rangle \neq 0; \quad p = 1, \dots, m(\lambda). \tag{12.35}$$

Indeed, if A, B are two linear operators, we have

$$[A, B] = xB \quad \Rightarrow \quad [A, B^k] = kxB; \quad \forall k \in \mathbb{Z}_+, \tag{12.36}$$

so that

$$\Gamma(h)[\Gamma(E^{-\beta})]^R |\lambda, p\rangle = (\lambda(h) - R\beta(h))[\Gamma(E^{-\beta})]^R |\lambda, p\rangle \tag{12.37}$$

and

$$[\Gamma(E^{-\beta})]^R |\lambda, p\rangle \equiv |\lambda - R\beta, p\rangle \in V_{\lambda - R\beta} \tag{12.38}$$

unless $[\Gamma(E^{-\beta})]^R|\lambda, p\rangle = 0$. Let us now consider the representation of the $A_1(\beta)$ Lie algebra induced by Γ, where $|\lambda, p\rangle$ is an eigenvector of J_3 with eigenvalue $\frac{1}{2}R$, since

$$J_3|\lambda, p\rangle = \frac{1}{2}\Gamma(H_\beta)|\lambda, p\rangle = \frac{1}{2}R|\lambda, p\rangle. \tag{12.39}$$

By general properties of the angular momentum representations, in the same representation there must also be the eigenstate of angular momentum $J_3 = -\frac{1}{2}R$, and this is precisely the state $[\Gamma(E^{-\beta})]^R|\lambda, p\rangle$. Hence,

$$[\Gamma(E^{-\beta})]^R|\lambda, p\rangle \neq 0, \tag{12.40}$$

and the theorem is proved. □

Next let us generalize the concept of α-string through β and consider the weights of the form $\lambda + k\alpha$, where λ is a weight, α is a root, and k is an integer. We have another theorem.

Theorem 12.1.2. *Let α be a non-vanishing root and let $\lambda \in \Pi_\Gamma$ be a weight of a representation Γ of a semi-simple Lie algebra $\mathbb{G}$. Then there exist two integers p, q such that $\lambda + k\alpha \in \Pi_\Gamma$ is a weight of the representation for all the integers $-p \leq k \leq q$. Furthermore, we have*

$$p - q = \langle\lambda, \alpha\rangle \equiv 2\frac{(\lambda, \alpha)}{(\alpha, \alpha)}. \tag{12.41}$$

Proof of Theorem 12.1.2. Using the above definition (12.33) for β we conclude that for $k > 0$, $\lambda + k\beta$ is a weight if and only if

$$0 \neq (\Gamma(E^\beta))^k|\lambda\rangle \tag{12.42}$$

for some $|\lambda\rangle \in V_\lambda$. On the other hand, this happens only if $\lambda(H_\beta)+k\beta(H_\beta)$ is an eigenvalue of H_β in the restriction of the given representation to the subalgebra $A_1(\beta)$. Similarly, in the case $k < 0$ one argues for $(\Gamma(E^{-\beta}))^{-k}|\lambda\rangle$. Therefore, let us suppose that the largest irreducible representation of $A_1(\beta)$ contained in Γ is of spin j and of dimension $2j + 1$. In this case the possible eigenvalues of $\frac{1}{2}H_\beta$ are extended from j to $-j$ with an integer jump from one to the next one. Hence, k takes all values in the interval determined by

$$\begin{aligned} \lambda(H_\beta) + q &= j, \\ \lambda(H_\beta) - p &= -j \end{aligned} \tag{12.43}$$

and we have

$$p - q = \lambda(H_\beta) = \langle\lambda, \beta\rangle, \tag{12.44}$$

as we wanted to show. □

As we can see from the above results, all properties of linear representations of higher-dimensional algebras follow from the properties of the representations of the angular momentum algebra (12.28) that every student of physics learns in elementary courses on quantum mechanics.

12.1.1.3 Irreducible representations and maximal weights

One important notion that applies to the weights of a representation is their partial ordering that relies on the following.

Definition 12.1.5. Given two weights λ, μ of an irreducible representation Γ of a simple Lie algebra $\mathbb{G}$ we say that λ is larger than μ if and only if their difference is a sum of positive roots:

$$\lambda \succ \mu \quad \Leftrightarrow \quad \lambda - \mu = \sum_{\alpha_k \in \Phi_+} \alpha_k. \tag{12.45}$$

The main theorem of representation theory (whose proof we omit) is the following.

Theorem 12.1.3. *If the irreducible representation Γ of a simple Lie algebra $\mathbb{G}$ is finite-dimensional, then the set of weights Π_Γ admits a maximal weight Λ such that any other weight $\lambda \in \Pi_\Gamma$ is $\lambda \prec \Lambda$. Consequently, we have*

$$\Gamma(E^\alpha)|\Lambda\rangle = 0. \tag{12.46}$$

Furthermore, we have the following.

Corollary 12.1.1. *The highest weights of all existing irreducible representations belong to the Weyl chamber $\mathfrak{W}$, which is the convex hull defined by the following conditions:*

$$\mathbf{v} \in \mathfrak{W} \quad \Leftrightarrow \quad 2\frac{(\mathbf{v}, \alpha_i)}{(\alpha_i, \alpha_i)} > 0, \quad i = 1, \dots, \ell, \tag{12.47}$$

where α_i are the simple roots.

Theorem 12.1.3 and its corollary are much more than an existence statement. They provide an actual construction algorithm for the weights of the representation. Starting from any dominant weight Λ in the Weyl chamber and imposing the condition (12.46) we can begin the construction of the representation by acting on $|\Lambda\rangle$ with the step-down operators $\Gamma(E^{-\alpha})$, obtaining in this way the states $|\Lambda - k\alpha\rangle$ of the α-strings through Λ. The last value of k is determined, for each string, by $\langle\Lambda, \alpha\rangle$. Next, by acting once again with the operators $\Gamma(E^{-\alpha})$ one derives all the states and the weights of the representation.

12.1.1.4 Weyl's character formula

Weyl brought Lie group theory to perfection by providing the analog of characters of finite group theory.

In the case of finite groups, characters are associated with the conjugacy classes of group elements. In the case of Lie groups, every group element $\gamma \in \mathrm{G}$ is conjugate to some element $h \in \mathrm{T} \subset \mathrm{G}$, where, by definition, the maximal torus $\mathrm{T} \equiv \exp[\mathcal{H}]$ is the exponential of the Cartan subalgebra. This statement can be easily understood if we consider any linear representation of G. Given a group element γ we can always rotate it to a basis where it is diagonal; in that basis it lies in T. Then the character of a representation can be defined as the following linear functional on the Cartan subalgebra:

$$\forall h \in \mathcal{H}: \quad \chi_\Gamma(h) \equiv \mathrm{Tr}_\Gamma(\exp[h]) = \sum_{\lambda \in \Pi_\Gamma} m(\lambda) \exp[\lambda(h)], \tag{12.48}$$

where the sum is extended to all the weights of the representation with their multiplicities.

Weyl demonstrated that the character can be written by means of a universal formula whose only input is the maximal weight of the representation. His result is the following:

$$\chi_\Gamma = \frac{\sum_{w \in W} \epsilon(w) \exp[\Lambda_\Gamma + \rho]}{\exp[\rho] \prod_{\alpha \in \Phi_+} (1 - \exp[-\alpha])}, \tag{12.49}$$

where:
1. $\rho \equiv \frac{1}{2}(\sum_{\alpha \in \Phi_+} \alpha)$ is half of the sum of positive roots.
2. W denotes the Weyl group and $\epsilon(w)$ denotes the determinant of the action of the Weyl group element w on the Cartan subalgebra $\mathcal{H} \subset \mathbb{G}$.
3. Λ_Γ is the highest weight of the irreducible representation.

Weyl's formula (12.49) is to be interpreted in the linear functional sense. Weights and roots are all linear functionals on the Cartan subalgebra $\mathcal{H}$ and the character defined by (12.49) is also a functional on $\mathcal{H}$.

From equation (12.49) considering the limiting character $h \to 0$ Weyl obtained the formula that gives the dimension of the irreducible representation with highest weight Λ_Γ:

$$\dim \Gamma = \frac{\prod_{\alpha > 0} (\alpha, \Lambda_\Gamma + \rho)}{\prod_{\alpha > 0} (\alpha, \rho)}. \tag{12.50}$$

Although Weyl's formulae are not always friendly for explicit calculations since one has to cancel vanishing factors between the numerator and the denominator, they have an outstanding conceptual relevance. They explicitly show that complete information about any linear representation is encoded in its highest weight vector and what are the possible dominant weights is completely fixed by the weight lattice, ultimately by the Cartan matrix.

Next we turn to illustrative examples.

12.2 Discussion of tensor products and examples

In this section we illustrate the concepts introduced in the previous ones by means of some examples that will be worked out in detail. For pedagogical effectiveness we will mainly consider Lie algebras of rank $r = 2$ where the roots and the weights are easily visualized in a Euclidean plane so that they can be intuitively grasped.

One of our goals is to compare the universal Weyl–Dynkin approach to representations with the natural approach in terms of tensors with fixed symmetry patterns.

The underlying question is a very important and conceptual one. The following digression tries to illustrate the issue at stake.

12.2.1 Tensor products and irreps

Every simple Lie algebra $\mathbb{G}$ possesses a faithful representation of smallest dimension that we can dub the *fundamental representation* Γ^{fun}, and a simple way of constructing higher-dimensional representations is that of taking tensor products of that representation. Assume that the fundamental representation is of dimension n:

$$\Gamma^{\text{fun}} : \mathbb{G} \to \text{Hom}(V_n, V_n). \tag{12.51}$$

If we take the tensor product

$$V_{m|n} \equiv \underbrace{V_n \otimes V_n \otimes \cdots \otimes V_n}_{m \text{ times}}, \tag{12.52}$$

equation (12.51) canonically induces a new representation of dimension n^m:

$$\Gamma_{m|n} : \mathbb{G} \to \text{Hom}(V_{m|n}, V_{m|n}). \tag{12.53}$$

If $\mathbf{e}_i$ ($i =, 1, \dots, n$) are a set of basis vectors of V_n, then a natural basis for $V_{m|n}$ is provided by

$$|i_1 \dots i_m\rangle \equiv \mathbf{e}_{i_1} \otimes \cdots \otimes \mathbf{e}_{i_m} \tag{12.54}$$

and the matrices of the induced representation $\Gamma_{m|n}$ are constructed as follows:

$$\forall \mathfrak{g} \in \mathbb{G} \quad \langle j_1 \dots j_m | \Gamma_{m|n}(\mathfrak{g}) | i_1 \dots i_m\rangle = \sum_{p=1}^{m} \delta_{j_1,i_1} \delta_{j_2,i_2} \dots \delta_{j_{p-1},i_{p-2}} \Gamma^{\text{fun}}_{j_p,i_p}(\mathfrak{g}) \delta_{j_{p+1},i_{p+2}} \dots \delta_{j_m,i_m}. \tag{12.55}$$

Typically the representation $\Gamma_{m|n}$ is reducible since one can find several invariant subspaces, one way of finding invariant subspaces being the construction of irreducible tensors with respect to permutations.

A generic element of the vector space $V_{m|n}$ is of the form

$$V_{m|n} \ni \mathbf{w} = w^{i_1 \dots i_m} |i_1 \dots i_m\rangle. \tag{12.56}$$

Since the definition of equation (12.55) is invariant under any permutation of the m indices, we can impose that the coefficients $w^{i_1 \dots i_m}$ transform into an irreducible representation of the symmetric group $\mathscr{S}_m$ operating on the set of indices: the corresponding subspace of $V_{m|n}$ will be an invariant subspace under the Lie algebra $\mathbb{G}$. The simplest example is provided by the singlet representation of the symmetric group, namely, by the full symmetrization of the indices. Let us assume that the coefficients $w^{i_1 \dots i_m}$ have the following property:

$$\forall P \in \mathscr{S}_m : \quad w^{P(i_1) \dots P(i_m)} = w^{i_1 \dots i_m}. \tag{12.57}$$

The subspace $SV_{m|n} \subset V_{m|n}$ made up of those vectors whose components have property (12.57) is an invariant subspace. What is the dimension of such a subspace? Answering this question amounts to determining how many different m-tuples of n numbers are there up to permutations. We easily find

$$\dim SV_{m|n} = \frac{n \cdot (n+1) \cdot (n+2) \dots (n+m)}{m!}. \tag{12.58}$$

Another simple example corresponds to the alternating representation of $\mathscr{S}_m$ or, if you prefer such a wording, to complete antisymmetrization. Suppose that the coefficients $w^{i_1 \dots i_m}$ have the following property:

$$\forall P \in \mathscr{S}_m : \quad w^{P(i_1) \dots P(i_m)} = (-)^{\delta_P} w^{i_1 \dots i_m}, \tag{12.59}$$

where δ_P denotes the parity of the permutation. The subspace $AV_{m|n} \subset V_{m|n}$ made up of those vectors whose components have property (12.59) is an invariant subspace. What is the dimension of such a subspace? Answering this question amounts to determining how many different m-tuples of n numbers there are, without repetitions. We easily find

$$\dim AV_{m|n} = \frac{n \cdot (n-1) \cdot (n-2) \dots (n-m)}{m!}. \tag{12.60}$$

In Section 4.2.4 we discussed the conjugacy classes of the symmetric group $\mathscr{S}_m$ and we showed that they are in one-to-one correspondence with the partitions of m into integers, which can be represented by Young tableaux (see equation (4.13)). On the other hand, we know from the general principles of finite group theory that the irreducible representations of a finite group G are in one-to-one correspondence with its conjugacy classes. It follows that the irreducible representations of the symmetric group S_m are in one-to-one correspondence with the Young tableaux one can construct out of m boxes and that to each Young tableau corresponds an irreducible tensor spanning an invariant subspace under the full considered Lie algebra $\mathbb{G}$.

Let us consider a Young tableau like

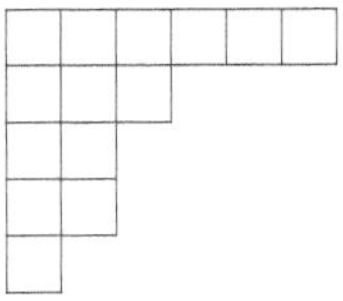

(12.61)

and let us fill its m boxes with the numbers $1, 2, \ldots, m$ in any preferred order. For instance, we can arrange the 14 symbols at our disposal in the example of equation (12.61) as follows:

1	2	3	4	5	6
7	8	9			
10	11				
12	13				
14					

(12.62)

Once the tableau has been fixed, we consider two types of permutations. *Horizontal permutations p* are permutations which interchange only symbols in the same row. *Vertical permutations q* interchange only symbols in the same column. Then we construct the following index operators:

$$P = \sum_{p} p \quad \text{symmetrizer,} \tag{12.63}$$

$$Q = \sum_{q} \delta_q q \quad \text{antisymmetrizer,} \tag{12.64}$$

where the sum is extended respectively to all the horizontal and to all the vertical permutations encoded in the tableau. Finally we define the Young operator

$$Y = QP \tag{12.65}$$

that can be applied to any tensor with m indices once these latter have been divided into blocks corresponding to the rows of the tableau. The resulting tensor forms an irreducible representation of the symmetric group and certainly belongs to an invariant subspace in the tensor product of m fundamental representations Γ^{fun} of any Lie algebra $\mathbb{G}$. In particular, the symmetric tensor introduced in equation (12.57) corresponds to the following Young tableau composed of a single row

1	2	·	·	·	m

(12.66)

while the antisymmetric tensor discussed in equation (12.59) corresponds to the Young tableau composed of a single column:

1
2
·
·
m

(12.67)

The question is whether these irreducible tensors always correspond to irreducible representations of $\mathbb{G}$ and whether all irreps of $\mathbb{G}$ can be obtained in this way.

The answer to these questions is not universal and it depends on the type of considered Lie algebras. Let us first discuss the case of classical Lie algebras.

12.2.1.1 Classical Lie algebras

($\mathfrak{a}_\ell$) For the algebras $\mathfrak{a}_\ell$ that correspond to the matrix algebras $\mathfrak{sl}(\ell+1,\mathbb{C})$ the answer to both questions is positive. There is only one fundamental representation, namely, the defining one in $\ell+1$ dimensions, and the set of all possible irreducible representations is in one-to-one correspondence with the set of irreducible m-tensors obtained by means of the Young tableau discussed above. Every *highest weight* module or *irrep* can be uniquely identified with a corresponding irreducible tensor associated with a Young tableau.

($\mathfrak{c}_\ell$) For the algebras $\mathfrak{c}_\ell$ that correspond to the matrix algebras $\mathfrak{sp}(2\ell,\mathbb{C})$ the answer is more elaborated. There is only one fundamental representation, which is the defining one in 2ℓ dimensions, but the candidate irreducible tensors obtained acting with a Young operator on the multiple tensor product of that fundamental irrep are not guaranteed to be fully irreducible with respect to $\mathfrak{sp}(2\ell,\mathbb{C})$. The reason is that differently from the previous case there exists a symplectic invariant antisymmetric tensor $\mathbb{C}_{\alpha\beta}$ by means of which any antisymmetric pair of indices of a multiple tensor can be contracted giving rise to an invariant symplectic trace. To obtain an irreducible tensor corresponding to a truly irreducible representation of the Lie algebra $\mathfrak{sp}(2\ell,\mathbb{C})$ one has to enforce the additional condition that all the symplectic traces vanish.

($\mathfrak{b}_\ell,\mathfrak{d}_\ell$) The algebras $\mathfrak{b}_\ell$ correspond to the orthogonal Lie algebras in odd dimensions $\mathfrak{so}(2\ell+1,\mathbb{C})$, while the algebras D_ℓ correspond to the same in even dimensions $\mathfrak{so}(2\ell,\mathbb{C})$. In both cases we do not have a single fundamental representation; rather, we have two of them. One is the defining representation Γ^{vec}, which we call the vector representation, and the other is the spinor representation Γ^{spin}, which is a peculiarity of the orthogonal Lie algebras. For $\mathfrak{so}(n,\mathbb{C})$ with $n \leq 6$ the dimension of the spinor representation is smaller than or equal to the dimension of the vector representation, yet for $n > 6$ the dimension of the spinor representation becomes larger than n and it grows exponentially with it. While the vector representation is typically contained in the square of the spinor one, there is no way to construct it generically as an irreducible symmetric or antisymmetric tensor in spinors. Hence, the general truth is that for orthogonal groups there are two types of representations:

1. the bosonic ones, which are obtained, as in the symplectic case, applying Young tableau operators to tensor products of the vector representation and setting to zero all symmetric traces taken with the symmetric invariant matrix $\eta_{\alpha\beta}$,

2. the spinor ones, which are obtained taking the tensor product of any bosonic irreducible one with the spinor representation and setting to zero the gamma-traces that can be constructed with the gamma-matrices.

In conclusion, for classical algebras the irreducible highest weight representations can be identified with irreducible tensors or tensor-spinors, but one has to be careful with standard traces, symplectic traces, or gamma-traces.

12.2.1.2 Exceptional Lie algebras

Since the exceptional Lie algebras have no classical definition in terms of matrix algebras, the irreducible representations cannot be identified with irreducible tensors starting from a fundamental representation. The matrix representations have to be constructed *ad hoc* case by case.

12.2.1.3 The value of the highest weight module construction

From the point of view of the highest weight module construction, all representations are on the same footing and there is no essential distinction between classical and exceptional Lie algebras. This shows that although more abstract, this approach is the deepest and most algorithmic one and it encodes the essence of the underlying mathematical structure. In Weyl's spirit, it provides that symbolic system which is the real mathematical truth behind the screen.[3]

Let us now turn to some examples.

12.2.2 The Lie algebra $\mathfrak{a}_2$, its Weyl group, and examples of its representations

In order to illustrate the general ideas discussed in previous sections, we begin with an analysis of the rank $r = 2$ Lie algebra $\mathfrak{sl}(3, \mathbb{R})$ and its fundamental representation. This Lie algebra is the maximally split real section of the $\mathfrak{a}_2$ Lie algebra, encoded in the Dynkin diagram of Figure 12.1.

As discussed in previous sections, from the representations of the maximally split real section we easily obtain those of the compact real section, which, in this case, is $\mathfrak{su}(3)$.

The root system has rank 2 and it is composed of the six vectors displayed below and pictured in Figure 12.1:

3 To appreciate the spirit of this last sentence the reader is referred to Weyl's lecture *On the mathematical way of thinking* fully analyzed and discussed in the philosophical-historical book [28] by the present author.

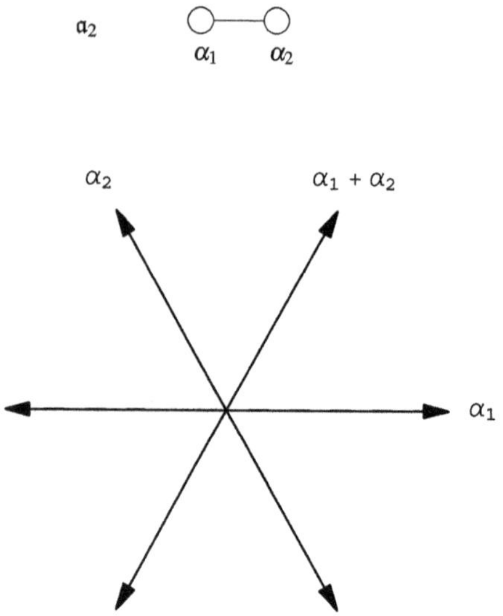

Figure 12.1: The $\mathfrak{a}_2$ Dynkin diagram and root system.

$$\Phi_{\mathfrak{a}_2} = \begin{cases} \alpha_1 = (\sqrt{2}, 0), \\ \alpha_2 = (-\frac{1}{\sqrt{2}}, \sqrt{\frac{3}{2}}), \\ \alpha_1 + \alpha_2 = (\frac{1}{\sqrt{2}}, \sqrt{\frac{3}{2}}), \\ -\alpha_1 = (-\sqrt{2}, 0), \\ -\alpha_2 = (\frac{1}{\sqrt{2}}, -\sqrt{\frac{3}{2}}), \\ -\alpha_1 - \alpha_2 = (-\frac{1}{\sqrt{2}}, -\sqrt{\frac{3}{2}}). \end{cases} \tag{12.68}$$

The simple roots are α_1 and α_2.

A complete set of generators for the Lie algebra $\mathfrak{sl}(3, \mathbb{R})$ is provided by the following 3×3 matrices:

$$\begin{aligned} H_1 &= \begin{pmatrix} \frac{1}{\sqrt{2}} & 0 & 0 \\ 0 & -\frac{1}{\sqrt{2}} & 0 \\ 0 & 0 & 0 \end{pmatrix}; \quad H_2 = \begin{pmatrix} \frac{1}{\sqrt{6}} & 0 & 0 \\ 0 & \frac{1}{\sqrt{6}} & 0 \\ 0 & 0 & -\sqrt{\frac{2}{3}} \end{pmatrix}; \\ E^{\alpha_1} &= \begin{pmatrix} 0 & 1 & 0 \\ 0 & 0 & 0 \\ 0 & 0 & 0 \end{pmatrix}; \quad E^{\alpha_2} = \begin{pmatrix} 0 & 0 & 0 \\ 0 & 0 & 1 \\ 0 & 0 & 0 \end{pmatrix}; \\ E^{\alpha_1+\alpha_2} &= \begin{pmatrix} 0 & 0 & 1 \\ 0 & 0 & 0 \\ 0 & 0 & 0 \end{pmatrix}, \\ E^{-\alpha_1} &= (E^{\alpha_1})^T; \quad E^{-\alpha_2} = (E^{\alpha_2})^T; \quad E^{-\alpha_1-\alpha_2} = (E^{\alpha_1+\alpha_2})^T, \end{aligned} \tag{12.69}$$

where $H_{1,2}$ are the two Cartan generators and E^{α} are the step operators associated with the corresponding roots.

As a useful exercise, let us calculate the eight generators of the compact subalgebra $\mathfrak{su}(3)$. We have

$$
\begin{aligned}
h_1 &= iH_1 = \begin{pmatrix} \frac{i}{\sqrt{2}} & 0 & 0 \\ 0 & -\frac{i}{\sqrt{2}} & 0 \\ 0 & 0 & 0 \end{pmatrix}, \\
h_2 &= iH_2 = \begin{pmatrix} \frac{i}{\sqrt{6}} & 0 & 0 \\ 0 & \frac{i}{\sqrt{6}} & 0 \\ 0 & 0 & -i\sqrt{\frac{2}{3}} \end{pmatrix},
\end{aligned}
\tag{12.70}
$$

$$
\begin{aligned}
r_1 &= E^{\alpha_1} - E^{-\alpha_1} = \begin{pmatrix} 0 & 1 & 0 \\ -1 & 0 & 0 \\ 0 & 0 & 0 \end{pmatrix}, \\
r_2 &= E^{\alpha_2} - E^{-\alpha_2} = \begin{pmatrix} 0 & 0 & 0 \\ 0 & 0 & 1 \\ 0 & -1 & 0 \end{pmatrix}, \\
r_3 &= E^{\alpha_1+\alpha_2} - E^{-\alpha_1-\alpha_2} = \begin{pmatrix} 0 & 0 & 1 \\ 0 & 0 & 0 \\ -1 & 0 & 0 \end{pmatrix},
\end{aligned}
\tag{12.71}
$$

and

$$
\begin{aligned}
\ell_1 &= i(E^{\alpha_1} + E^{-\alpha_1}) = \begin{pmatrix} 0 & i & 0 \\ i & 0 & 0 \\ 0 & 0 & 0 \end{pmatrix}, \\
\ell_2 &= i(E^{\alpha_2} + E^{-\alpha_2}) = \begin{pmatrix} 0 & 0 & 0 \\ 0 & 0 & i \\ 0 & i & 0 \end{pmatrix}, \\
\ell_3 &= i(E^{\alpha_1+\alpha_2} + E^{-\alpha_1-\alpha_2}) = \begin{pmatrix} 0 & 0 & i \\ 0 & 0 & 0 \\ i & 0 & 0 \end{pmatrix}.
\end{aligned}
\tag{12.72}
$$

Any linear combination with real coefficients of $\{h_1, h_2, r_1, r_2, r_3, \ell_1, \ell_2, \ell_3\}$ is an anti-Hermitian matrix and an element of the compact Lie algebra $\mathfrak{su}(3)$ in its fundamental representation.[4]

4 In the physical literature of the 1960s and 1970s the eight generators $\{h_1, h_2, r_1, r_2, r_3, \ell_1, \ell_2, \ell_3\}$ were named the Gell-Mann lambda-matrices.

The Borel–Lie algebra is composed of the operators

$$\mathrm{Bor}[\mathfrak{sl}(3,\mathbb{R})] = \mathrm{span}\{H_1, H_2, E^{\alpha_1}, E^{\alpha_2}, E^{\alpha_1+\alpha_2}\}, \tag{12.73}$$

and it is clearly represented by upper triangular matrices.

Let us now remark that the two real sections $\mathfrak{sl}(3,\mathbb{R})$ and $\mathfrak{su}(3)$ share a common maximal compact subalgebra,

$$\mathfrak{so}(3) \equiv \mathbb{H} \subset \mathbb{G} \equiv \mathfrak{sl}(3,\mathbb{R}), \tag{12.74}$$

defined by the following generators:

$$\begin{aligned}\mathbb{H} &= \mathrm{span}\{J_1, J_2, J_3\}\\ &\equiv \left\{\frac{1}{\sqrt{2}}(E^{\alpha_1} - E^{-\alpha_1}), \frac{1}{\sqrt{2}}(E^{\alpha_2} - E^{-\alpha_2}), \frac{1}{\sqrt{2}}(E^{\alpha_1+\alpha_2} - E^{-\alpha_1-\alpha_2})\right\}.\end{aligned} \tag{12.75}$$

This is a general phenomenon. The maximally split real section $\mathbb{G}_{\max}$ always contains a maximal compact subalgebra:

$$\mathbb{H}_{\text{compact}} = \mathrm{span}(E^{\alpha} - E^{-\alpha}); \quad \alpha > 0. \tag{12.76}$$

The orthogonal splitting

$$\mathbb{G}_{\max} = \mathbb{H}_{\text{compact}} \oplus \mathbb{K}, \tag{12.77}$$

where

$$\mathbb{K} = \mathrm{span}\{H_i, (E^{\alpha} - E^{-\alpha})\}, \tag{12.78}$$

plays an important role in our subsequent discussion of symmetric spaces and in other geometrical issues.

12.2.2.1 The Weyl group of $\mathfrak{a}_2$

Let us now consider a generic element of the Cartan subalgebra, namely, a diagonal matrix of the form

$$\mathrm{CSA} \ni \mathscr{C}(\{\lambda_1, \lambda_2\}) = \begin{pmatrix} \lambda_1 & 0 & 0 \\ 0 & \lambda_2 & 0 \\ 0 & 0 & -\lambda_1 - \lambda_2 \end{pmatrix}. \tag{12.79}$$

By definition the Weyl group maps the Cartan subalgebra into itself, so that we have

$$\forall w \in \mathscr{W} \subset \mathrm{SO}(3): \quad w^T \mathscr{C}(\{\lambda_1, \lambda_2\}) w = \mathscr{C}(w\{\lambda_1, \lambda_2\}) \in \mathrm{CSA}, \tag{12.80}$$

where $\mathscr{C}(w\{\lambda_1, \lambda_2\})$ denotes the diagonal matrix of type (12.79) with eigenvalues λ_1', λ_2', $-\lambda_1' - \lambda_2'$ obtained from the action of the Weyl group on the original ones. In the case of the Lie algebras A_n the Weyl group is the symmetric group $\mathscr{S}_{n+1}$ and its action on the eigenvalues $\lambda_1, \lambda_2, \ldots, \lambda_n, \lambda_{n+1} = -\sum_{i=1}^{n} \lambda_i$ is just that of permutations on these $n+1$ eigenvalues. For $\mathfrak{a}_2$ we have $\mathscr{S}_3$, whose order is six. The six group elements can be enumerated in the following way:

$$
\begin{aligned}
w_1 &= \begin{pmatrix} 1 & 0 & 0 \\ 0 & 1 & 0 \\ 0 & 0 & 1 \end{pmatrix}; \quad (\lambda_1, \lambda_2, \lambda_3) \mapsto (\lambda_1, \lambda_2, \lambda_3), \\
w_2 &= \begin{pmatrix} 0 & 1 & 0 \\ 1 & 0 & 0 \\ 0 & 0 & 1 \end{pmatrix}; \quad (\lambda_1, \lambda_2, \lambda_3) \mapsto (\lambda_2, \lambda_1, \lambda_3), \\
w_3 &= \begin{pmatrix} 0 & 0 & 1 \\ 0 & 1 & 0 \\ 1 & 0 & 0 \end{pmatrix}; \quad (\lambda_1, \lambda_2, \lambda_3) \mapsto (\lambda_3, \lambda_2, \lambda_1),
\end{aligned}
\tag{12.81}
$$

$$
\begin{aligned}
w_4 &= \begin{pmatrix} 1 & 0 & 0 \\ 0 & 0 & 1 \\ 0 & 1 & 0 \end{pmatrix}; \quad (\lambda_1, \lambda_2, \lambda_3) \mapsto (\lambda_1, \lambda_3, \lambda_2), \\
w_5 &= \begin{pmatrix} 0 & 0 & 1 \\ 1 & 0 & 0 \\ 0 & 1 & 0 \end{pmatrix}; \quad (\lambda_1, \lambda_2, \lambda_3) \mapsto (\lambda_2, \lambda_3, \lambda_1), \\
w_6 &= \begin{pmatrix} 0 & 1 & 0 \\ 0 & 0 & 1 \\ 1 & 0 & 0 \end{pmatrix}; \quad (\lambda_1, \lambda_2, \lambda_3) \mapsto (\lambda_3, \lambda_1, \lambda_2).
\end{aligned}
\tag{12.82}
$$

12.2.2.2 The weight lattice and the weights of some irreps

Next we discuss some of the irreducible representations of $\mathfrak{a}_2$ in relation with their weights.

The weight lattice of $\mathfrak{a}_2$ is defined as

$$\Lambda_{\text{weight}}[\mathfrak{a}_2] \ni n_1 \lambda^1 + n_2 \lambda^2; \quad n_{1,2} \in \mathbb{Z}, \tag{12.83}$$

where the two fundamental weights have the following explicit analytic form:

$$\lambda^1 = \left\{ \frac{1}{\sqrt{2}}, \frac{1}{\sqrt{6}} \right\}, \tag{12.84}$$

$$\lambda^2 = \left\{ 0, \sqrt{\frac{2}{3}} \right\}. \tag{12.85}$$

They are shown in Figure 12.2. Given the fundamental weights, one immediately determines the Weyl chamber, which is displayed in Figure 12.3.

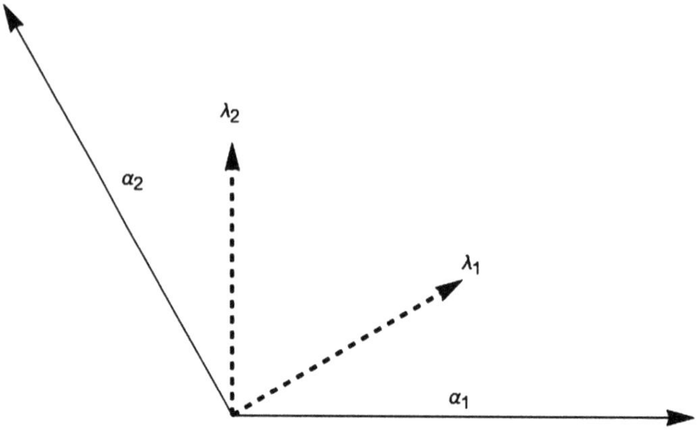

Figure 12.2: The fundamental weights of the $\mathfrak{a}_2$ Lie algebra, compared to the simple roots.

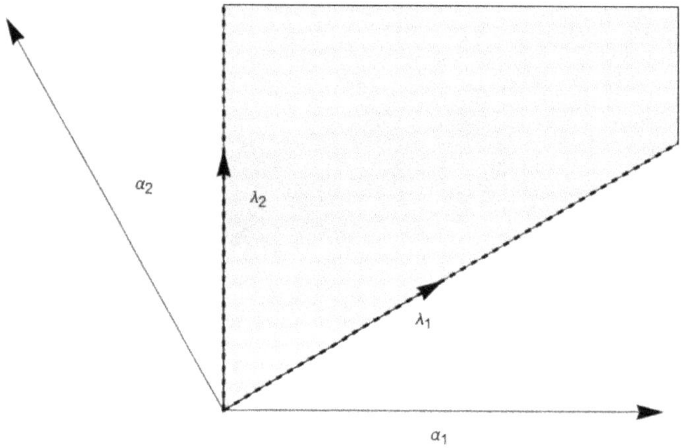

Figure 12.3: The Weyl chamber of the $\mathfrak{a}_2$ Lie algebra is, in the above picture, the infinite shaded region.

12.2.2.3 The weights of the fundamental representation

The fundamental representation of the $\mathfrak{sl}(3, \mathbb{R})$ Lie algebra or equivalently of the $\mathfrak{su}(3)$ algebra is three-dimensional, and we have already constructed the matrix realization of its generators in equation (12.69). Let us work out its weights. These are easily seen from the explicit form of the Cartan matrices displayed in equation (12.69). Indeed, setting

$$\begin{aligned}\Pi^{\text{triplet}} &= \{\Lambda^1, \Lambda^2, \Lambda^3\},\\ \Lambda^1 &= \lambda^1 && \Rightarrow \quad (\mathbf{1}, \mathbf{0}),\\ \Lambda^2 &= -\lambda^1 + \lambda^2 && \Rightarrow \quad (-\mathbf{1}, \mathbf{1}),\\ \Lambda^3 &= -\lambda^2 && \Rightarrow \quad (\mathbf{0}, -\mathbf{1}),\end{aligned} \tag{12.86}$$

where the boldfaced components on the right are those with respect to the fundamental weight basis (the Dynkin labels), we have

$$\begin{aligned} |\Lambda^1\rangle &= \{\mathbf{1}, \mathbf{0}, \mathbf{0}\}; \quad h^i H_i|\Lambda^1\rangle = h^i \Lambda_i^1 |\Lambda^1\rangle, \\ |\Lambda^2\rangle &= \{\mathbf{0}, \mathbf{1}, \mathbf{0}\}; \quad h^i H_i|\Lambda^2\rangle = h^i \Lambda_i^2 |\Lambda^2\rangle, \\ |\Lambda^3\rangle &= \{\mathbf{0}, \mathbf{0}, \mathbf{1}\}; \quad h^i H_i|\Lambda^3\rangle = h^i \Lambda_i^3 |\Lambda^3\rangle, \end{aligned} \tag{12.87}$$

as the reader can easily verify looking at equation (12.69).

The three weights of the fundamental representation are displayed in Figure 12.4. As mentioned in the figure caption, the highest weight of this representation is clearly singled out: it is

$$\lambda_{\text{max}} = \Lambda_1 = \lambda^1, \tag{12.88}$$

and from this simple information, using the structure of the root system we might construct all the matrices of the representation, which are those displayed in equation (12.69).

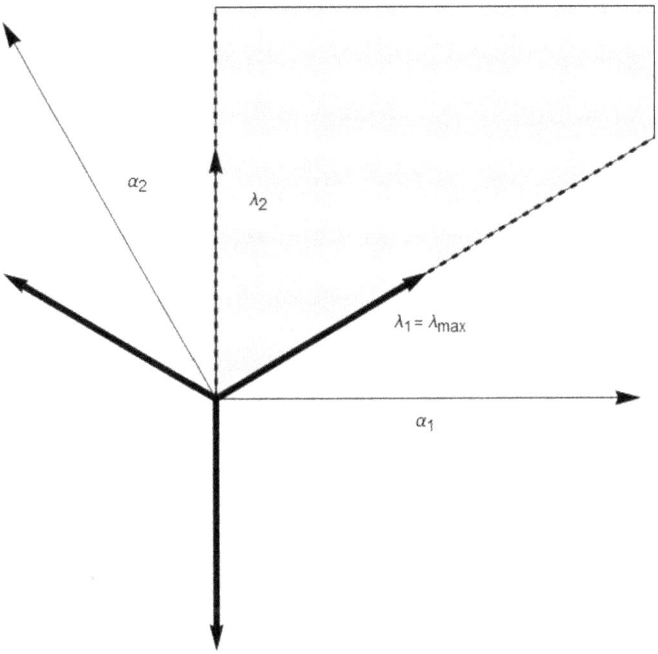

Figure 12.4: The three weights of the fundamental representation of the $\mathfrak{a}_2$ Lie algebra correspond to the three thicker arrows. The highest weight is clearly λ_1, the only positive one lying in the Weyl chamber, actually on its boundary.

12.2.2.4 The weights of the adjoint representation: the octet

The next representation that we consider is the adjoint representation, namely, the octet. This representation is already encoded in the canonical Cartan–Weyl structure of the Lie algebra and it is eight-dimensional. It is interesting to reexamine it from the point of view of the weights that we can immediately predict. The eight weights are indeed the weight $\{0,0\}$ with multiplicity $m(\{0,0\}) = 2$ corresponding to the two Cartan generators and the six roots with multiplicity $m = 1$. These latter can be expressed in terms of the fundamental weights as follows. Setting

$$\Pi^{\text{octet}} = \{\Lambda^1, \Lambda^2, \Lambda^3, \Lambda^4, \Lambda^5, \Lambda^6, \Lambda^7\}, \tag{12.89}$$

we have

$$\begin{aligned}
\Lambda^1 &= \alpha_1 + \alpha_2 = \lambda^1 + \lambda^2 & &\Rightarrow & \Lambda^1 &= (\mathbf{1}, \mathbf{1}),\\
\Lambda^2 &= \alpha_1 = 2\lambda^1 - \lambda^2 & &\Rightarrow & \Lambda^2 &= (\mathbf{2}, -\mathbf{1}),\\
\Lambda^3 &= \alpha_2 = -\lambda^1 + 2\lambda^2 & &\Rightarrow & \Lambda^3 &= (-\mathbf{1}, \mathbf{2}),\\
\Lambda^4 &= \{0,0\} = \{0,0\} & &\Rightarrow & \Lambda^4 &= (\mathbf{0}, \mathbf{0}),\\
\Lambda^5 &= -\alpha_2 = \lambda^1 - 2\lambda^2 & &\Rightarrow & \Lambda^5 &= (\mathbf{1}, -\mathbf{2}),\\
\Lambda^6 &= -\alpha_1 = -2\lambda^1 + \lambda^2 & &\Rightarrow & \Lambda^6 &= (-\mathbf{2}, \mathbf{1}),\\
\Lambda^7 &= -\alpha_1 - \alpha_2 = -\lambda^1 - \lambda^2 & &\Rightarrow & \Lambda^7 &= (-\mathbf{1}, -\mathbf{1}),
\end{aligned} \tag{12.90}$$

where the boldfaced components are the integer ones in the basis of the fundamental weights (the Dynkin labels of the weight). Recalling that the height of a weight is the sum of such components we see that the weights are correctly weakly ordered, since

$$\Lambda^1 \succ \Lambda^2 \succeq \Lambda^3 \succ \Lambda^4 \succ \Lambda^5 \succeq \Lambda^6 \succ \Lambda^7. \tag{12.91}$$

The positive components of the highest weight are named the *Dynkin labels of the representation*. In this case the highest weight is provided by

$$\lambda_{\max} = \alpha_1 + \alpha_2 = \lambda^1 + \lambda^2 \tag{12.92}$$

and the Dynkin labels of the adjoint (octet) representation are $(1,1)$. The weights of the octet representation are displayed in Figure 12.5.

12.2.2.5 The weights of a 15-dimensional representation

As a final example we consider the weights of the representation whose maximal weight is the following:

$$\lambda_{\max} = 2\lambda^1 + \lambda^2; \quad \Rightarrow \quad \lambda_{\max} = (\mathbf{2}, \mathbf{1}). \tag{12.93}$$

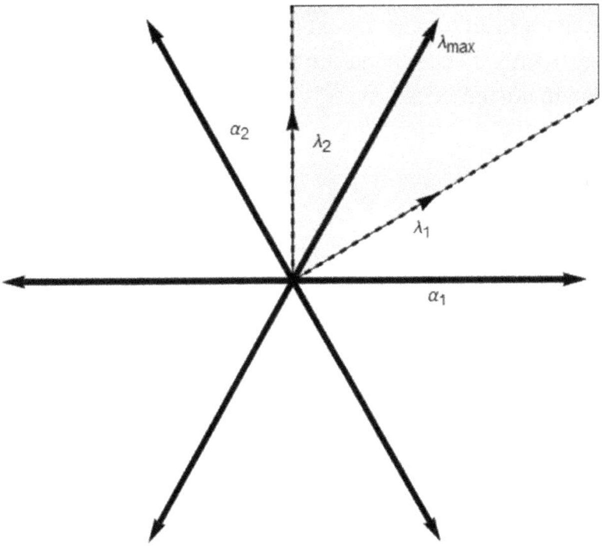

Figure 12.5: The six non-vanishing weights of the adjoint representation of the $\mathfrak{a}_2$ Lie algebra correspond to the six thicker arrows. The highest weight is clearly $\lambda^1 + \lambda^2$, the only positive one lying in the Weyl chamber.

Utilizing the Weyl group symmetry of the weight set and the α through λ string technique we can easily work out the full set of weights together with their multiplicities. The result is

$$\Pi_\Gamma = \{\Lambda_1, \Lambda_2, \Lambda_3, \Lambda_4, \Lambda_5, \Lambda_6, \Lambda_7, \Lambda_8, \Lambda_9, \Lambda_{10}, \Lambda_{11}, \Lambda_{12}\}, \tag{12.94}$$

where the list of ordered weights with their multiplicity is displayed below:

(12.95)

Name				Orth. comp.		Dynk. lab	mult.
Λ_1	$=$	$2\lambda^1 + \lambda^2$	$=$	$\{\sqrt{2}, 2\sqrt{\frac{2}{3}}\}$	$=$	$(\mathbf{2}, \mathbf{1})$	1
Λ_2	$=$	$3\lambda^1 - \lambda^2$	$=$	$\{\frac{3}{\sqrt{2}}, \frac{1}{\sqrt{6}}\}$	$=$	$(\mathbf{3}, \mathbf{-1})$	1
Λ_3	$=$	$2\lambda^2$	$=$	$\{0, 2\sqrt{\frac{2}{3}}\}$	$=$	$(\mathbf{0}, \mathbf{2})$	1
Λ_4	$=$	λ^1	$=$	$\{\frac{1}{\sqrt{2}}, \frac{1}{\sqrt{6}}\}$	$=$	$(\mathbf{1}, \mathbf{0})$	2
Λ_5	$=$	$-2\lambda^1 + 3\lambda^2$	$=$	$\{-\sqrt{2}, 2\sqrt{\frac{2}{3}}\}$	$=$	$(\mathbf{-2}, \mathbf{3})$	1
Λ_6	$=$	$2\lambda^1 - 2\lambda^2$	$=$	$\{\sqrt{2}, -\sqrt{\frac{2}{3}}\}$	$=$	$(\mathbf{2}, \mathbf{-2})$	1
Λ_7	$=$	$-\lambda^1 + \lambda^2$	$=$	$\{-\frac{1}{\sqrt{2}}, \frac{1}{\sqrt{6}}\}$	$=$	$(\mathbf{-1}, \mathbf{1})$	2
Λ_8	$=$	$-\lambda^2$	$=$	$\{0, -\sqrt{\frac{2}{3}}\}$	$=$	$(\mathbf{0}, \mathbf{-1})$	2
Λ_9	$=$	$-3\lambda^1 + 2\lambda^2$	$=$	$\{-\frac{3}{\sqrt{2}}, \frac{1}{\sqrt{6}}\}$	$=$	$(\mathbf{-3}, \mathbf{2})$	1
Λ_{10}	$=$	$-3\lambda^2 + \lambda^1$	$=$	$\{\frac{1}{\sqrt{2}}, -\frac{5}{\sqrt{6}}\}$	$=$	$(\mathbf{1}, \mathbf{-3})$	1
Λ_{11}	$=$	$-2\lambda^1$	$=$	$\{-\sqrt{2}, -\sqrt{\frac{2}{3}}\}$	$=$	$(\mathbf{-2}, \mathbf{0})$	1
Λ_{12}	$=$	$-\lambda^1 - 2\lambda^2$	$=$	$\{-\frac{1}{\sqrt{2}}, -\frac{5}{\sqrt{6}}\}$	$=$	$(\mathbf{-1}, \mathbf{-2})$	1

The weight set is displayed in Figure 12.6. Counting the multiplicity 2 of the three shortest weights it appears that this representation has dimension 15. The question is how we can understand this representation in the tensorial way.

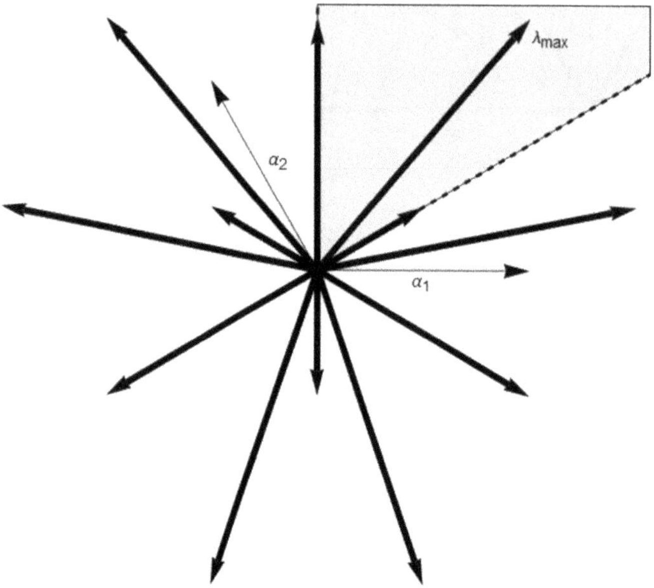

Figure 12.6: The 12 weights of the irreducible representation of the $\mathfrak{a}_2$ Lie algebra corresponding to the maximal weight $(\mathbf{2},\mathbf{1})$ are given by the 12 thicker arrows. The highest weight is clearly $2\lambda^1 + \lambda^2$. It is the largest of the three lying in the Weyl chamber.

The answer is the following. Those displayed in equation (12.95) and in Figure 12.6 are the weights of the representation corresponding to the following irreducible tensor:

$$\mathbf{15} \equiv \begin{array}{|c|c|c|}\hline & & \\ \hline \\ \cline{1-1}\end{array} \tag{12.96}$$

For pedagogical reasons we dwell on the details of these identifications. First let us introduce four 3-vectors transforming in the fundamental representation of $\mathfrak{sl}(3,\mathbb{C})$:

$$\mathbf{v} = \begin{pmatrix} v_1 \\ v_2 \\ v_3 \end{pmatrix}; \quad \mathbf{w} = \begin{pmatrix} w_1 \\ w_2 \\ w_3 \end{pmatrix}; \quad \mathbf{z} = \begin{pmatrix} z_1 \\ z_2 \\ z_3 \end{pmatrix}; \quad \mathbf{t} = \begin{pmatrix} t_1 \\ t_2 \\ t_3 \end{pmatrix}. \tag{12.97}$$

Next we construct the irreducible tensor in the following way:

$$\begin{array}{|c|c|c|}\hline \mathbf{v} & \mathbf{w} & \mathbf{z} \\ \hline \mathbf{t} \\ \cline{1-1}\end{array} = \begin{array}{|c|c|c|}\hline \mathbf{v} & \mathbf{w} & \mathbf{z} \\ \hline \end{array} \otimes \begin{array}{|c|}\hline \mathbf{t} \\ \hline \end{array} - \begin{array}{|c|c|c|c|}\hline \mathbf{v} & \mathbf{w} & \mathbf{z} & \mathbf{t} \\ \hline \end{array} \tag{12.98}$$

where the Young tableau is, as explained above, shorthand for the application to the product $v_i w_j z_k t_k$ of the row symmetrization and column antisymmetrization. The possible components of the tensor are obtained by replacing in each of the boxes one of the three components of each vector **v**, **w**, **z**, and **t**. In this way the tensor components are quartic polynomials in the components $v_i w_j z_k t_k$. These polynomials can be expressed in a basis of 15 independent ones that correspond to a basis of the irreducible representation. A possible basis is provided by the polynomials corresponding to the tensor components displayed in Table 12.1. The simplest way to verify that this is indeed the proper correspondence relies on finite group transformations. Given a Cartan–Lie algebra element

Table 12.1: Components of the irreducible tensor corresponding to the weights of the 15-dimensional representation defined by the highest weight $(\mathbf{2}, \mathbf{1})$.

Weight	tensor comp.	weight mult.
Λ_1	$\begin{array}{lll}1&1&1\\2\end{array}$	1
Λ_2	$\begin{array}{lll}1&1&1\\3\end{array}$	1
Λ_3	$\begin{array}{lll}1&1&2\\2\end{array}$	1
Λ_4	$\begin{array}{lll}1&1&2\\3\end{array}$, $\begin{array}{lll}1&1&3\\2\end{array}$	2
Λ_5	$\begin{array}{lll}1&2&2\\2\end{array}$	1
Λ_6	$\begin{array}{lll}1&1&3\\3\end{array}$	1
Λ_7	$\begin{array}{lll}1&2&2\\3\end{array}$, $\begin{array}{lll}1&2&3\\2\end{array}$	2
Λ_8	$\begin{array}{lll}1&2&3\\3\end{array}$, $\begin{array}{lll}1&3&3\\2\end{array}$	2
Λ_9	$\begin{array}{lll}2&2&2\\3\end{array}$	1
Λ_{10}	$\begin{array}{lll}1&3&3\\3\end{array}$	1
Λ_{11}	$\begin{array}{lll}2&2&3\\3\end{array}$	1
Λ_{12}	$\begin{array}{lll}2&3&3\\3\end{array}$	1

$$\mathfrak{h} = h_1 H_1 + h_2 H_2; \quad \mathbf{h} \equiv \{h_1, h_2\}, \tag{12.99}$$

the corresponding finite group element has the form

$$\mathfrak{T} = \exp[\mathfrak{h}] = \begin{pmatrix} \frac{h_1}{\sqrt{2}} + \frac{h_2}{\sqrt{6}} & 0 & 0 \\ 0 & \frac{h_2}{\sqrt{6}} - \frac{h_1}{\sqrt{2}} & 0 \\ 0 & 0 & -\sqrt{\frac{2}{3}} h_2 \end{pmatrix}, \tag{12.100}$$

and under this group transformation we have

$$\mathbf{v}' = \mathfrak{T}\mathbf{v}; \quad \mathbf{w}' = \mathfrak{T}\mathbf{w}; \quad \mathbf{z}' = \mathfrak{T}\mathbf{z}; \quad \mathbf{t}' = \mathfrak{T}\mathbf{t}. \tag{12.101}$$

Replacing the unprimed vectors with the primed ones in the expression for the tensor

$$\begin{array}{|c|c|c|} \hline a & b & c \\ \hline d \\ \cline{1-1} \end{array} \tag{12.102}$$

we find that each of its components mentioned in Table 12.1 transforms by multiplication with an exponential factor:

$$\exp[\Lambda \cdot \mathbf{h}], \tag{12.103}$$

where Λ is the corresponding weight vector also mentioned in Table 12.1. This proves the correspondence between weights and tensor components.

12.2.3 The Lie algebra $\mathfrak{sp}(4, \mathbb{R}) \simeq \mathfrak{so}(2,3)$, its fundamental representation, and its Weyl group

The next example that we consider is that of the Lie algebras $\mathfrak{c}_2$ and $\mathfrak{b}_2$ that are accidentally isomorphic. This isomorphism is very convenient for our pedagogical purposes since it allows us to discuss in a simple way an instance of spin representation.

When we restrict our attention to maximally split real sections, the isomorphism between the complex $\mathfrak{c}_2$ and $\mathfrak{b}_2$ Lie algebras translates into the isomorphism between the following two real Lie algebras:

$$\mathfrak{so}(2,3) \simeq \mathfrak{sp}(4, \mathbb{R}). \tag{12.104}$$

On the contrary, if we focus on maximally compact real sections, the implied isomorphism between compact algebras is the following one:

$$\mathfrak{so}(5) \simeq \mathfrak{usp}(4). \tag{12.105}$$

The Lie algebra $\mathfrak{usp}(4)$ is by definition composed of all those 4×4 matrices that are simultaneously anti-Hermitian and symplectic, in the sense that they preserve an anti-symmetric real matrix which squares to $-\mathbf{1}$.

Considering the $\mathfrak{b}_2$ and $\mathfrak{c}_2$ formulations at the same time we discover that the fundamental representation of $\mathfrak{sp}(4, \mathbb{C})$ is the spinor representation of $\mathfrak{so}(5, \mathbb{C})$. In this way we can easily work out all the weights of the spinor representations to be compared with those of the vector representation. The structure of the spinor weights is rather general for all orthogonal Lie algebras. Diagrams are displayed in Figure 12.7. Because of the isomorphism we can rely on either formulation in terms of 4×4 symplectic matrices or 5×5 pseudo-orthogonal matrices to obtain a fundamental representation of the Lie algebra. The first will prove to be the spinor representation of the second, as we anticipated.

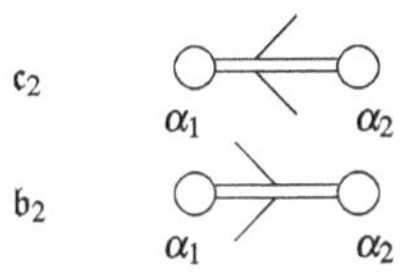

Figure 12.7: The Dynkin diagram of the $\mathfrak{c}_2 \sim \mathfrak{b}_2$ Lie algebra. The exchange of the α_1 with the α_2 root maps one Dynkin diagram into the other and this is sufficient to prove the isomorphism.

In the symplectic $\mathfrak{sp}(4)$ interpretation, the $\mathfrak{c}_2$ root system can be realized by the following eight two-dimensional vectors:

$$\mathbf{\Phi}_{\mathfrak{c}_2} = \{\pm\epsilon^i \pm \epsilon^j, \pm 2\epsilon^i\}, \tag{12.106}$$

where ϵ^i $(i = 1, 2)$ denotes a basis of orthonormal unit vectors. In the pseudo-orthogonal $\mathfrak{so}(2, 3)$ interpretation of the same algebra, the $\mathfrak{b}_2$ root system is instead realized by the following eight vectors:

$$\mathbf{\Phi}_{\mathfrak{b}_2} = \{\pm\epsilon^i \pm \epsilon^j, \pm\epsilon^i\}. \tag{12.107}$$

The two root systems are displayed in Figure 12.8. Because of the isomorphism between the two Lie algebras, we can use only one of the two root systems and consider the other as belonging to the weight lattice of the first. We choose to utilize the $\mathfrak{c}_2$ realization and we introduce the simple root basis

$$\begin{aligned} \alpha_1 &= \{1, -1\}, \\ \alpha_2 &= \{0, 2\}, \end{aligned} \tag{12.108}$$

leading to the Cartan matrix

$$\mathbb{C} = \begin{pmatrix} 2 & -1 \\ -2 & 2 \end{pmatrix}. \tag{12.109}$$

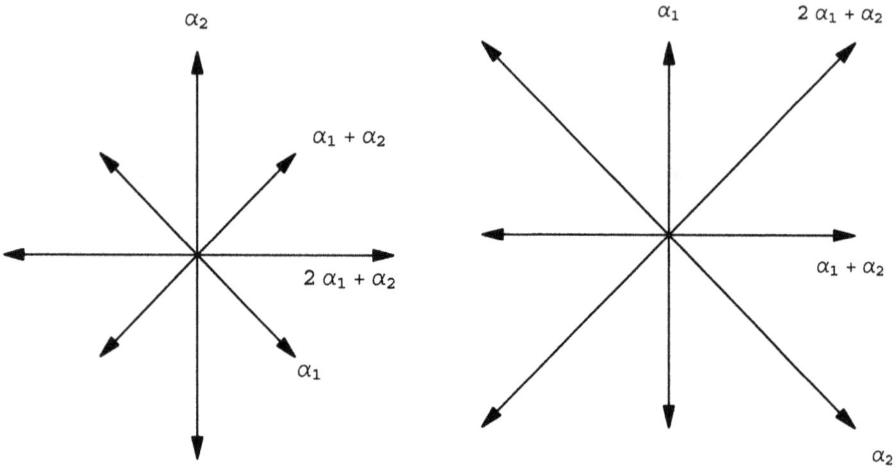

Figure 12.8: The $\mathfrak{c}_2$ and $\mathfrak{b}_2$ root systems. They are related by the exchange of the long and the short roots.

The fundamental weights satisfying the necessary relation

$$2\frac{(\lambda^i, \alpha_j)}{(\alpha_j, \alpha_j)} = \delta^i_j \tag{12.110}$$

are easily determined; one finds

$$\begin{aligned} \lambda^1 &= \{1, 0\}, \\ \lambda^2 &= \{1, 1\}. \end{aligned} \tag{12.111}$$

The complete set of positive roots is written below in the simple root and in the fundamental weight basis:

$$\begin{aligned} \alpha_1 &= 2\lambda_1 - \lambda_2, \\ \alpha_2 &= -2\lambda_1 + 2\lambda_2, \\ \alpha_1 + \alpha_2 &= \lambda_2, \\ 2\alpha_1 + \alpha_2 &= 2\lambda_1. \end{aligned} \tag{12.112}$$

From equation (12.112) we also read off the Dynkin labels of the adjoint representation that are those of the highest root, namely, $(2, 0)$. The simple roots and the simple weights are displayed in Figure 12.9.

12.2.3.1 The Weyl group $\mathfrak{sp}(4, \mathbb{R})$

Abstractly the Weyl group $\mathrm{Weyl}(\mathfrak{c}_2)$ of the Lie algebra $\mathfrak{sp}(4, \mathbb{R})$ is given by $(\mathbb{Z}_2 \times \mathbb{Z}_2) \ltimes S_2$, which is just the dihedral group Dih_4. The eight elements $w_i \in \mathrm{Weyl}(\mathfrak{c}_2)$ of the Weyl group can be described by their action on the two Cartan fields h_1, h_2:

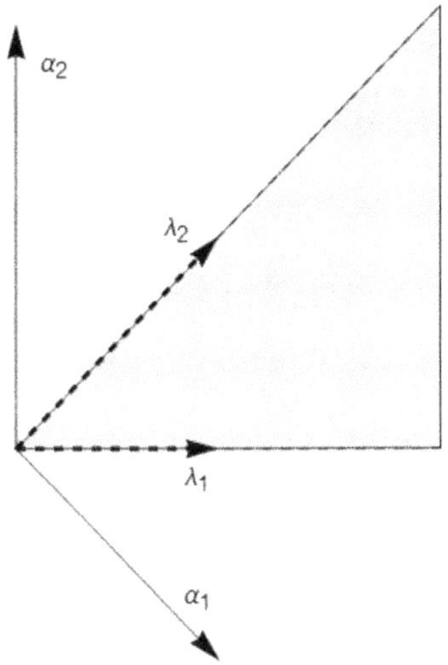

Figure 12.9: The simple roots and the simple weights of the $\mathfrak{c}_2$ Lie algebra. The shaded region is the Weyl chamber.

$$\begin{aligned}
w_1 &: (h_1, h_2) \to (h_1, h_2), \\
w_2 &: (h_1, h_2) \to (-h_1, -h_2), \\
w_3 &: (h_1, h_2) \to (-h_1, h_2), \\
w_4 &: (h_1, h_2) \to (h_1, -h_2), \\
w_5 &: (h_1, h_2) \to (h_2, h_1), \\
w_6 &: (h_1, h_2) \to (h_2, -h_1), \\
w_7 &: (h_1, h_2) \to (-h_2, h_1), \\
w_8 &: (h_1, h_2) \to (-h_2, -h_1).
\end{aligned} \tag{12.113}$$

We leave to the reader the explicit verification that the reflections along the four positive roots generate the order eight group explicitly displayed in equation (12.113).

12.2.3.2 Construction of the $\mathfrak{sp}(4, \mathbb{R})$ Lie algebra

The most compact way of presenting our basis is the following. According to our general strategy let us begin with the Borel–Lie algebra $\mathrm{Bor}[\mathfrak{sp}(4, \mathbb{R})]$. Abstractly the most general element of this algebra is given by

$$\mathscr{T} = h_1 \mathscr{H}_1 + h_2 \mathscr{H}_2 + e_1 E^{\alpha_1} + e_2 E^{\alpha_2} + e_3 E^{\alpha_1+\alpha_2} + e_4 E^{2\alpha_1+\alpha_2}. \tag{12.114}$$

If we write the explicit form of $\mathscr{T}$ as a 4×4 upper triangular symplectic matrix

$$\mathcal{T}_{\text{sym}} = \begin{pmatrix} h_1 & e_1 & e_3 & -\sqrt{2}e_4 \\ 0 & h_2 & \sqrt{2}e_2 & e_3 \\ 0 & 0 & -h_2 & -e_1 \\ 0 & 0 & 0 & -h_1 \end{pmatrix} \in \mathfrak{sp}(4,\mathbb{R}) \tag{12.115}$$

which satisfies the condition

$$\mathcal{T}_{\text{sym}}^T \begin{pmatrix} \mathbf{0}_2 & \sigma_1 \\ -\sigma_1 & \mathbf{0}_2 \end{pmatrix} + \begin{pmatrix} \mathbf{0}_2 & \sigma_1 \\ -\sigma_1 & \mathbf{0}_2 \end{pmatrix} \mathcal{T}_{\text{sym}} = 0; \quad \sigma_1 = \begin{pmatrix} 0 & 1 \\ 1 & 0 \end{pmatrix}, \tag{12.116}$$

all the generators of the solvable algebra are defined in the four-dimensional symplectic representation. Moreover, also the generators associated with negative roots are defined by transposition:

$$\forall \alpha \in \mathbf{\Phi}; \quad E^{-\alpha} = \left[E^{\alpha}\right]^T. \tag{12.117}$$

By writing the same Lie algebra element (12.114) as a 5×5 matrix

$$\mathcal{T}_{so} = \begin{pmatrix} h_1+h_2 & -\sqrt{2}e_2 & -\sqrt{2}e_3 & -\sqrt{2}e_4 & 0 \\ 0 & h_1-h_2 & -\sqrt{2}e_1 & 0 & \sqrt{2}e_4 \\ 0 & 0 & 0 & \sqrt{2}e_1 & \sqrt{2}e_3 \\ 0 & 0 & 0 & h_2-h_1 & \sqrt{2}e_2 \\ 0 & 0 & 0 & 0 & -h_1-h_2 \end{pmatrix} \in \mathfrak{so}(2,3) \tag{12.118}$$

which satisfies the condition

$$\mathcal{T}_{so}^T \left(\begin{array}{cc|c|cc} 0 & 0 & 0 & 0 & 1 \\ 0 & 0 & 0 & 1 & 0 \\ \hline 0 & 0 & 1 & 0 & 0 \\ \hline 0 & 1 & 0 & 0 & 0 \\ 1 & 0 & 0 & 0 & 0 \end{array}\right) + \left(\begin{array}{cc|c|cc} 0 & 0 & 0 & 0 & 1 \\ 0 & 0 & 0 & 1 & 0 \\ \hline 0 & 0 & 1 & 0 & 0 \\ \hline 0 & 1 & 0 & 0 & 0 \\ 1 & 0 & 0 & 0 & 0 \end{array}\right) \mathcal{T}_{so} \equiv \mathcal{T}_{so}^T \eta + \eta \mathcal{T}_{so} = 0, \tag{12.119}$$

we define the same generators also in the five-dimensional pseudo-orthogonal representation. The choice of the invariant metric displayed in equation (12.119) is that which guarantees the upper triangular structure of the solvable Lie algebra generators.[5] This allows to define the generators associated with negative roots in the same way as in equation (12.117) and the five-dimensional representation is fully constructed.

12.2.3.3 Weights of the vector, spinor, and adjoint representations

Relying on the results of the previous section we have an explicit realization of three linear representations of the same complex Lie algebra: a four-dimensional one, a five-

5 The eigenvalues of the matrix η are $(-1,-1,1,1,1,)$. This shows that by means of a change of basis the matrices fulfilling equation (12.119) span the $\mathfrak{so}(2,3)$ Lie algebra, according to its conventional definition.

dimensional one, and the 10-dimensional adjoint representation. It is interesting to construct the weights of these three representations and compare them.

12.2.3.4 Weights of the four-dimensional spinor representation

The weights of the fundamental representation of $\mathfrak{sp}(4, \mathbb{C})$ which corresponds to the spinor representation of $\mathfrak{so}(3, 2, \mathbb{C}) \sim \mathfrak{so}(5, \mathbb{C})$ are easily read off from the explicit form of the matrix representation (12.115). They are listed below:

Dynk. lab	Orth. comp.	mult.
$\{1, 0\}$	$\{1, 0\}$	1
$\{-1, 1\}$	$\{0, 1\}$	1
$\{1, -1\}$	$\{0, -1\}$	1
$\{-1, 0\}$	$\{-1, 0\}$	1

(12.120)

where the first column provides their components in the fundamental weight basis, while the second column gives their expression in an orthonormal basis. The weights are graphically shown in Figure 12.10.

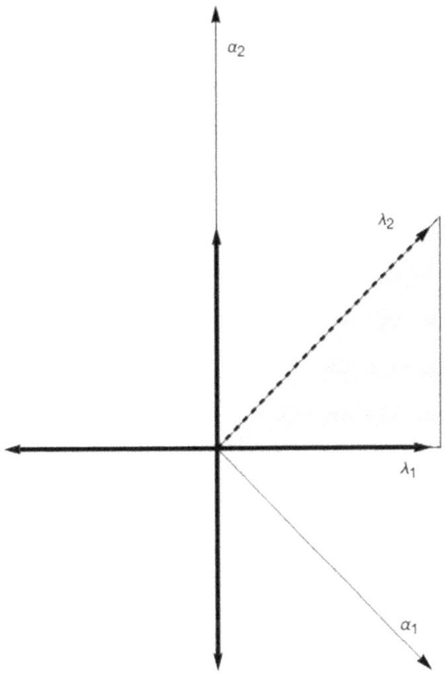

Figure 12.10: The weights of the spinor representation of $\mathfrak{so}(5)$, which is the fundamental representation of $\mathfrak{sp}(4)$. The shaded region is the Weyl chamber. One sees that the fundamental weight λ^1 is just the highest weight of the spinor representation.

12.2.3.5 Weights of the five-dimensional vector representation

The weights of the fundamental vector representation of $\mathfrak{so}(5, \mathbb{C})$ are easily read off from the explicit form of the matrix representation (12.118). They are listed below:

Dynk. lab	Orth. comp.	mult.
$\{0, 1\}$	$\{1, 1\}$	1
$\{2, -1\}$	$\{1, -1\}$	1
$\{0, 0\}$	$\{0, 0\}$	1
$\{-2, 1\}$	$\{-1, 1\}$	1
$\{0, -1\}$	$\{-1, -1\}$	1

(12.121)

where the first column provides their components in the fundamental weight basis, while the second column gives their expression in an orthonormal basis. The weights are graphically shown in Figure 12.11.

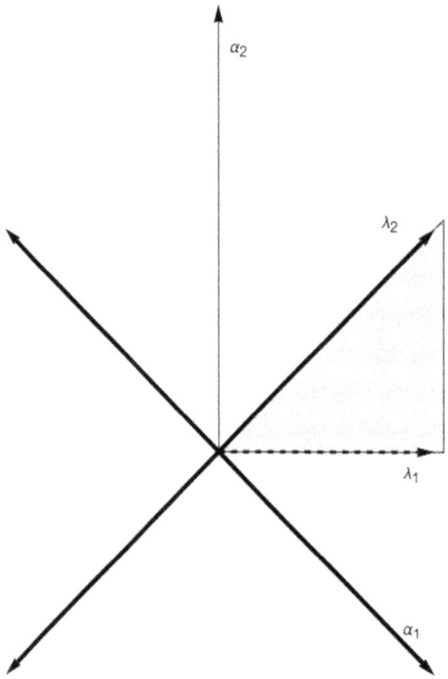

Figure 12.11: The weights of the vector defining the representation of $\mathfrak{so}(5)$. The shaded region is the Weyl chamber. One sees that the fundamental weight λ^2 is just the highest weight of the vector.

12.2.3.6 Weights of the 10-dimensional adjoint representation

The weights of the adjoint representation of $\mathfrak{so}(5,\mathbb{C}) \sim \mathfrak{sp}(4,\mathbb{C})$ are easily read off from the transcription of the roots in terms of fundamental weights (12.112). They are listed below:

Dynk. lab	Orth. comp.	mult.
$\{2,0\}$	$\{2,0\}$	1
$\{0,1\}$	$\{1,1\}$	1
$\{2,-1\}$	$\{1,-1\}$	1
$\{-2,2\}$	$\{0,2\}$	1
$\{0,0\}$	$\{0,0\}$	2
$\{2,-2\}$	$\{0,-2\}$	1
$\{-2,1\}$	$\{-1,1\}$	1
$\{0,-1\}$	$\{-1,-1\}$	1
$\{-2,0\}$	$\{-2,0\}$	1

(12.122)

where the first column provides their components in the fundamental weight basis, while the second column gives their expression in an orthonormal basis. The weights are graphically shown in Figure 12.12.

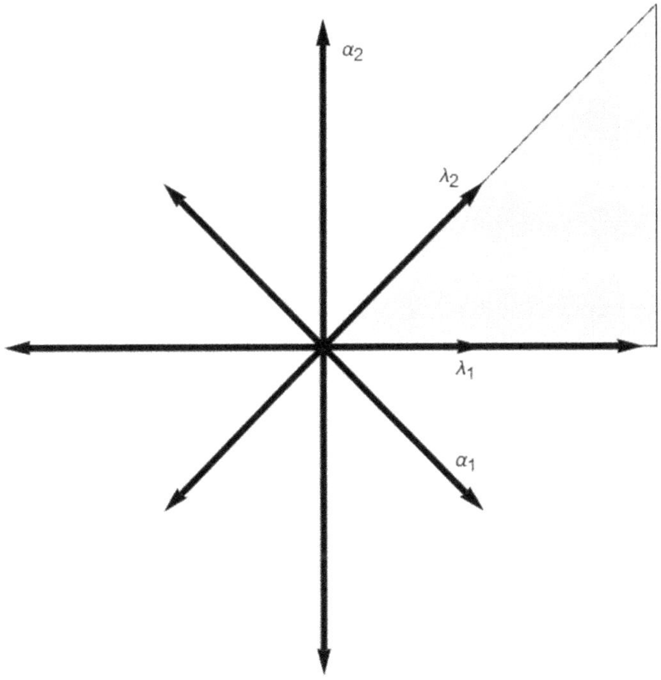

Figure 12.12: The weights of the adjoint representation of $\mathfrak{so}(5)$. The shaded region is the Weyl chamber. One sees that the weight $2\lambda^1 = 2\alpha_1 + \alpha_2$ is the highest weight of the adjoint representation.

12.2.3.7 Spinor interpretation of the four-dimensional representation

We have repeatedly claimed that the four-dimensional representation is the spinor representation with respect to the $\mathfrak{so}(5,\mathbb{C})$ Lie algebra. Which argument do we have to support such a claim? The answer is that to prove this it suffices to construct a well-adapted set of gamma-matrices.

Let us recall equations (12.119) and (12.116) and let us set

$$\eta = \left(\begin{array}{cc|c|cc} 0 & 0 & 0 & 0 & 1 \\ 0 & 0 & 0 & 1 & 0 \\ \hline 0 & 0 & 1 & 0 & 0 \\ \hline 0 & 1 & 0 & 0 & 0 \\ 1 & 0 & 0 & 0 & 0 \end{array}\right) \quad ; \quad \mathbf{C} = \begin{pmatrix} \mathbf{0}_2 & \sigma_1 \\ -\sigma_1 & \mathbf{0}_2 \end{pmatrix}. \tag{12.123}$$

The $\mathfrak{so}(5,\mathbb{C}) \sim \mathfrak{so}(2,3,\mathbb{C})$ algebra is spanned by the 4×4 matrices $\mathscr{O}$ that satisfy the equation

$$\mathscr{O}^T \eta + \eta \mathscr{O} = 0, \tag{12.124}$$

while the isomorphic $\mathfrak{sp}(4,\mathbb{C})$ algebra is spanned by the 5×5 matrices $\mathscr{S}$ that satisfy the equation

$$\mathscr{S}^T \mathbf{C} + \mathbf{C}\mathscr{S} = 0. \tag{12.125}$$

In order to discuss the spinor representation we have to introduce a set of gamma-matrices Γ_a which satisfy the appropriate Clifford algebra:

$$\{\Gamma_a, \Gamma_b\} = 2\eta_{ab}. \tag{12.126}$$

Furthermore, we remark that the gamma-matrices should be antisymmetric with respect to the symplectic matrix $\mathbf{C}$,

$$\Gamma_a^T \mathbf{C} - \mathbf{C}\Gamma_a = 0, \tag{12.127}$$

and $\mathbf{C}$-traceless:

$$\mathrm{Tr}(\Gamma_a \mathbf{C}) = 0. \tag{12.128}$$

The reasoning is the following: the space of general 4×4 matrices is 16-dimensional and it corresponds to the tensor product of two fundamental representations of $\mathfrak{sp}(4)$. The symmetric part (with respect to the matrix $\mathbf{C}$) spans the adjoint representation and has dimension 10. The antisymmetric part, which has dimension six, further splits into a $\mathbf{C}$-trace which is by definition a singlet and into a five-dimensional irreducible representation that must necessarily be the fundamental representation of $\mathfrak{so}(5)$. The gamma-matrices are just the projection operators onto this five-dimensional representation. In other words, if $\mathfrak{s}_{1,2}$ are two spinors, we obtain a vector by setting

$$\mathbf{v}^a = \mathfrak{s}_1^T \mathbf{C}\Gamma^a \mathfrak{s}_2. \tag{12.129}$$

We are supposed to solve the three constraints (12.126), (12.127), and (12.128). An ansatz for the solution is inspired by the knowledge of the weights. Since the directions $(1, 2, 3, 4, 5)$ in vector spaces are in one-to-one correspondence with the weights of the five-dimensional representation and the directions $(1, 2, 3, 4)$ in spinor space are in one-to-one correspondence with the weights of the four-dimensional representation, it suffices to check which spinor weights add up to which vector weight and we immediately know which entries of each gamma-matrix are different from zero. In this way we find the following solution:

$$\begin{aligned}
\Gamma_1 &= \begin{pmatrix} 0 & 0 & 1 & 0 \\ 0 & 0 & 0 & -1 \\ 0 & 0 & 0 & 0 \\ 0 & 0 & 0 & 0 \end{pmatrix}; \quad \Gamma_2 = \begin{pmatrix} 0 & -1 & 0 & 0 \\ 0 & 0 & 0 & 0 \\ 0 & 0 & 0 & -1 \\ 0 & 0 & 0 & 0 \end{pmatrix}; \\
\Gamma_3 &= \begin{pmatrix} 1 & 0 & 0 & 0 \\ 0 & -1 & 0 & 0 \\ 0 & 0 & -1 & 0 \\ 0 & 0 & 0 & 1 \end{pmatrix}; \quad \Gamma_4 = \begin{pmatrix} 0 & 0 & 0 & 0 \\ -2 & 0 & 0 & 0 \\ 0 & 0 & 0 & 0 \\ 0 & 0 & -2 & 0 \end{pmatrix}; \\
\Gamma_5 &= \begin{pmatrix} 0 & 0 & 0 & 0 \\ 0 & 0 & 0 & 0 \\ 2 & 0 & 0 & 0 \\ 0 & -2 & 0 & 0 \end{pmatrix}.
\end{aligned} \tag{12.130}$$

Introducing the generators of the $\mathfrak{so}(5)$ Lie algebra in the spinor representation,

$$\mathfrak{t}_{ab} = -\frac{1}{4}[\Gamma_a, \Gamma_b], \tag{12.131}$$

we verify that all of them satisfy

$$\mathfrak{t}_{ab}^T C + C\mathfrak{t}_{ab} = 0, \tag{12.132}$$

namely, they are elements of the $\mathfrak{sp}(4)$ Lie algebra and constitute a basis for it. This is as much as is needed to conclude that the fundamental of $\mathfrak{sp}(4)$ is just the spinor representation of $\mathfrak{so}(5)$.

12.2.3.8 Maximal compact subgroup of the maximally split real section

In view of our previous remarks about the maximal compact subalgebra and its orthogonal complement, let us see what is their structure in the present case.

The orthonormal generators of the $\mathbb{K}$ subspace are as follows:

$$
\begin{aligned}
\mathrm{K}_1 &= \mathscr{H}_1, \\
\mathrm{K}_2 &= \mathscr{H}_2, \\
\mathrm{K}_3 &= \frac{1}{\sqrt{2}}(E^{\alpha_1} + (E^{\alpha_1})^T), \\
\mathrm{K}_4 &= \frac{1}{\sqrt{2}}(E^{\alpha_2} + (E^{\alpha_2})^T), \\
\mathrm{K}_5 &= \frac{1}{\sqrt{2}}(E^{\alpha_1+\alpha_2} + (E^{\alpha_1+\alpha_2})^T), \\
\mathrm{K}_6 &= \frac{1}{\sqrt{2}}(E^{\alpha_1+2\alpha_2} + (E^{\alpha_1+2\alpha_2})^T),
\end{aligned}
\tag{12.133}
$$

and those of the maximal compact subalgebra $\mathbb{H} = \mathfrak{u}(2)$ are as follows:

$$
\begin{aligned}
\mathrm{J}_1 &= \frac{1}{\sqrt{2}}(E^{\alpha_1} - (E^{\alpha_1})^T), \\
\mathrm{J}_2 &= \frac{1}{\sqrt{2}}(E^{\alpha_2} - (E^{\alpha_2})^T), \\
\mathrm{J}_3 &= \frac{1}{\sqrt{2}}(E^{\alpha_1+\alpha_2} - (E^{\alpha_1+\alpha_2})^T), \\
\mathrm{J}_4 &= \frac{1}{\sqrt{2}}(E^{\alpha_1+2\alpha_2} - (E^{\alpha_1+2\alpha_2})^T).
\end{aligned}
\tag{12.134}
$$

Those above span a Lie algebra $\mathfrak{u}(2) = \mathfrak{u}(1) \oplus \mathfrak{su}(2)$.

12.3 Conclusions for this chapter

As the examples provided in the present chapter and those about exceptional Lie algebras to be discussed in the next one should clearly illustrate, although deterministic and implicitly defined by the Dynkin diagram, the actual construction of large Lie algebras is far from being a trivial matter and involves a series of strategies and long calculations that are best done by means of computer codes. One deals with large matrices that are difficult to display on paper and the best approach is to save the constructions in electronic libraries that can be utilized in subsequent calculations. It is not surprising that it took such a giant of mathematics as Élie Cartan to explicitly construct the fundamental representations of the exceptional Lie algebras, especially at a time when computers were not available.

From another point of view, the existing mathematical literature usually presents the construction of Lie algebra representations in a very compact format that is not of too friendly use to physicists, differential geometers, and other scientists concerned with their application to the problems inherent to their field of activity. As we stressed, it is not only a question of convenience but also a conceptual one. In the architecture of Lie algebras and their representations there are deep and significant aspects that are easily lost if you are not looking at them in the proper way, motivated by those questions that are posed, for instance by the various special geometries implied by supersymmetry or by complex system analysis. The explicit construction of the exceptional and non-exceptional Lie algebras in the light of supergravity is one of the motivations that pushed the present author to write the more advanced book [72] and its historical twin [28]. The aim was to present the *mathematics of symmetry* in a conceptually unified yet practical and computational way, at many stages different from the conventional approaches of most textbooks [96, 38, 83, 39]. The general transversal view of the present book suggests that the advanced constructions of special geometries presented in [72], although initially motivated by supersymmetry, might have in the future a wider spectrum of diversified applications also in fields different from theoretical physics.

12.4 Bibliographical note

In addition to the already quoted textbooks [96, 38, 83, 39], the most relevant bibliographical source for the first part of this chapter is the textbook by Cornwell [40]. The examples analyzed in full-fledged fashion in later sections are partially based on private notes of the author and partially have their basis in a research paper written by him in collaboration with A. Sorin [99].

13 Exceptional Lie algebras

> There are more things in heaven and earth, Horatio
> Than are dreamt of in your philosophy
>
> William Shakespeare, Hamlet, scene 5, 166–167

It was Killing who, through his own classification of the root systems, first discovered the possible existence of the exceptional Lie algebras, yet their concrete existence was proved only later by Cartan, who was able to construct the fundamental representation of all of them.

In this chapter we review the explicit construction of the fundamental representation of exceptional Lie algebras. The latter are very important in supergravity and superstring theory, yet in view of the more general scope of the present book, which aims at providing the reader with all the fundamental principles necessary to address disparate applications of group theory and symmetric spaces, we do not enter all the details that are relevant to those theoretical physicists interested in the special homogeneous geometries associated with supergravity. Such details can be found in the specialized book [72]. Here our approach is more general and we just sketch the methods to construct the fundamental representations.

13.1 The exceptional Lie algebra $\mathfrak{g}_2$

In this section we study the smallest of the five exceptional algebras, namely, $\mathfrak{g}_2$, and we explicitly exhibit its fundamental representation, which is seven-dimensional.

Our presentation is aimed not only at showing that $\mathfrak{g}_2$ exists, but also at enlightening some features of its structure that will turn out to be general within a certain algebraic scheme that encompasses an entire set of classical and exceptional Lie algebras relevant for the *special geometries* implied by supergravity and superstrings.

Before the advent of supergravity, exceptional Lie algebras were viewed by physicists as some mathematical extravagance good only for a Dickensian *Old Curiosity Shop*. Supergravity quite surprisingly showed that all exceptional Lie algebras have a distinct and essential role to play in the connected web of gravitational theories that one obtains through dimensionality reduction and coupling of matter multiplets in diverse dimensions. Furthermore, there is an inner algebraic structure of the exceptional algebras, shared with other classical algebras, that appears to be specially prepared to fit the geometrical yields of supersymmetry. This provides a new structural viewpoint motivated by physics that, in Weyl's spirit, encodes a deep truth, at the same time physical and mathematical, the distinction being somewhat irrelevant. For these aspects we refer the reader to the published book [72]. Here we just focus on the construction issue with two motivations; the first is to derive concrete results utilizable in several contexts, the second, more educational, is to illustrate in full detail all the steps needed to arrive at such results. Hence, let us turn to the specific topic of the present section.

https://doi.org/10.1515/9783111201535-013

The complex Lie algebra $\mathfrak{g}_2(\mathbb{C})$ has rank 2 and it is defined by the 2×2 Cartan matrix encoded in the following Dynkin diagram:

$\mathfrak{g}_2$  $= \begin{pmatrix} 2 & -1 \\ -3 & 2 \end{pmatrix}.$

The $\mathfrak{g}_2$ root system $\boldsymbol{\Phi}$ consists of the following six positive roots plus their negatives:

$$\begin{aligned} \alpha_1 &= \alpha_1 = (1,0); \quad \alpha_2 = \alpha_2 = \frac{\sqrt{3}}{2}(-\sqrt{3},1); \\ \alpha_3 &= \alpha_1 + \alpha_2 = \frac{1}{2}(-1,\sqrt{3}); \quad \alpha_4 = 2\alpha_1 + \alpha_2 = \frac{1}{2}(1,\sqrt{3}); \\ \alpha_5 &= 3\alpha_1 + \alpha_2 = \frac{\sqrt{3}}{2}(\sqrt{3},1); \quad \alpha_6 = 3\alpha_1 + 2\alpha_2 = (0,\sqrt{3}). \end{aligned} \tag{13.1}$$

The two fundamental weights are easily derived and have the following form:

$$\lambda^1 = \{1,\sqrt{3}\}; \quad \lambda^2 = \left\{0, \frac{2}{\sqrt{3}}\right\}. \tag{13.2}$$

Simple roots, fundamental weights, and the Weyl chamber are displayed in Figure 13.1. Figure 13.2 instead displays the entire root system. The fundamental representation of the Lie algebra is identified as the one which admits the fundamental weight λ^1 as highest weight. Using the Weyl group symmetry and the α through λ string technique one derives all the weights of the seven-dimensional fundamental representation:

Name		Dynk. lab		Orth. comp.	mult.
Λ_1	=	$\{1,0\}$	$\Rightarrow$	$\{1,\sqrt{3}\}$	1
Λ_2	=	$\{-1,1\}$	$\Rightarrow$	$\{-1,-\frac{1}{\sqrt{3}}\}$	1
Λ_3	=	$\{2,-1\}$	$\Rightarrow$	$\{2,\frac{4}{\sqrt{3}}\}$	1
Λ_4	=	$\{0,0\}$	$\Rightarrow$	$\{0,0\}$	1
Λ_5	=	$\{-2,1\}$	$\Rightarrow$	$\{-2,-\frac{4}{\sqrt{3}}\}$	1
Λ_6	=	$\{1,-1\}$	$\Rightarrow$	$\{1,\frac{1}{\sqrt{3}}\}$	1
Λ_7	=	$\{-1,0\}$	$\Rightarrow$	$\{-1,-\sqrt{3}\}$	1

(13.3)

The six non-vanishing weights are displayed in Figure 13.3. Given this information we are ready to derive the fundamental representation of the algebra. According to our general strategy we are supposed to construct 7×7 upper triangular matrices spanning the Borel subalgebra of the maximally split real section $g_{2(2)}$ of $\mathfrak{g}_2(\mathbb{C})$:

$$\mathrm{Bor}[\mathfrak{g}_2] = \mathrm{span}\{H_1, H_2, E^{\alpha_1}, E^{\alpha_2}, \ldots, E^{\alpha_6}\}. \tag{13.4}$$

As for all maximally split algebras, the Cartan generators H_i and the step operators E^{α} associated with each positive root α can be chosen completely real in all representations. In the fundamental seven-dimensional representation the explicit form of the

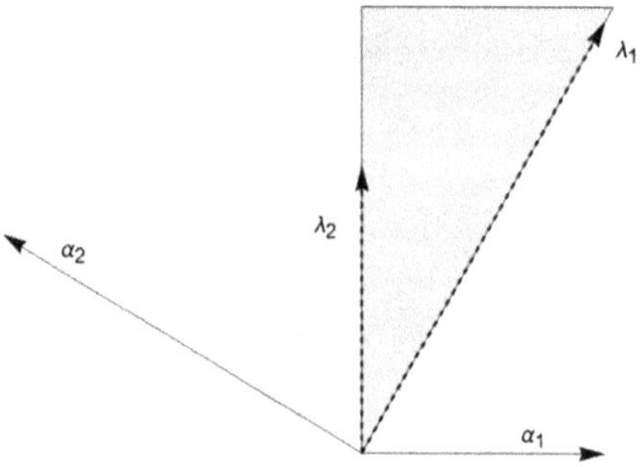

Figure 13.1: The simple roots and the fundamental weights of the $\mathfrak{g}_2$ Lie algebra. The shaded region is the Weyl chamber.

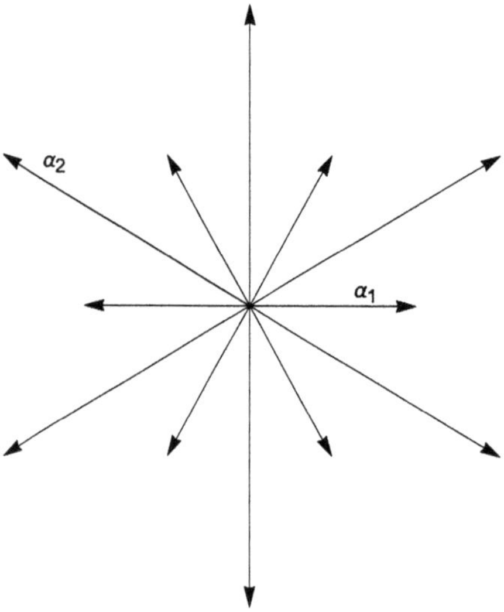

Figure 13.2: The complete root system of the $\mathfrak{g}_2$ Lie algebra.

$\mathfrak{g}_{2(2)}$-generators with the above properties is presented hereby. Denoting by $\{H_1, H_2\}$ the Cartan generators along the two orthonormal directions and adopting the standard Cartan–Weyl normalizations,

$$[E_\alpha, E_{-\alpha}] = \alpha^i H_i, \quad [H_i, E_\alpha] = \alpha^i E_\alpha, \tag{13.5}$$

we have

$$H_1 = \begin{pmatrix} \frac{1}{2} & 0 & 0 & 0 & 0 & 0 & 0 \\ 0 & -\frac{1}{2} & 0 & 0 & 0 & 0 & 0 \\ 0 & 0 & 1 & 0 & 0 & 0 & 0 \\ 0 & 0 & 0 & 0 & 0 & 0 & 0 \\ 0 & 0 & 0 & 0 & -1 & 0 & 0 \\ 0 & 0 & 0 & 0 & 0 & \frac{1}{2} & 0 \\ 0 & 0 & 0 & 0 & 0 & 0 & -\frac{1}{2} \end{pmatrix}; \quad H_2 = \begin{pmatrix} \frac{\sqrt{3}}{2} & 0 & 0 & 0 & 0 & 0 & 0 \\ 0 & \frac{\sqrt{3}}{2} & 0 & 0 & 0 & 0 & 0 \\ 0 & 0 & 0 & 0 & 0 & 0 & 0 \\ 0 & 0 & 0 & 0 & 0 & 0 & 0 \\ 0 & 0 & 0 & 0 & 0 & 0 & 0 \\ 0 & 0 & 0 & 0 & 0 & -\frac{\sqrt{3}}{2} & 0 \\ 0 & 0 & 0 & 0 & 0 & 0 & -\frac{\sqrt{3}}{2} \end{pmatrix};$$

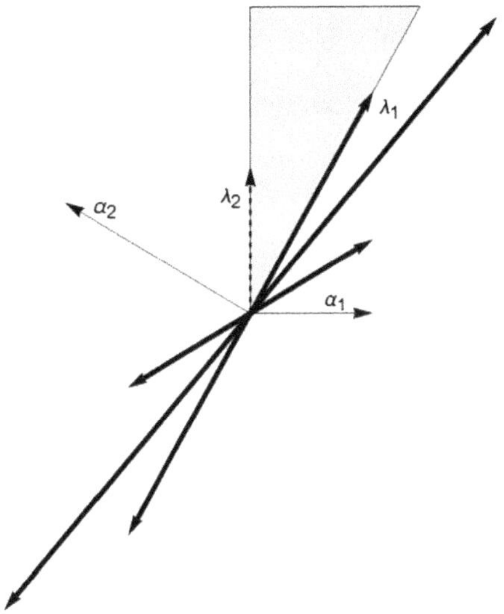

Figure 13.3: The six non-vanishing weights of the fundamental representation of the $\mathfrak{g}_2$ Lie algebra. The fundamental weight λ^1 is the highest weight of this representation.

$$E_{\alpha_1} = \begin{pmatrix} 0 & \frac{1}{\sqrt{2}} & 0 & 0 & 0 & 0 & 0 \\ 0 & 0 & 0 & 0 & 0 & 0 & 0 \\ 0 & 0 & 0 & 1 & 0 & 0 & 0 \\ 0 & 0 & 0 & 0 & 1 & 0 & 0 \\ 0 & 0 & 0 & 0 & 0 & 0 & 0 \\ 0 & 0 & 0 & 0 & 0 & 0 & \frac{1}{\sqrt{2}} \\ 0 & 0 & 0 & 0 & 0 & 0 & 0 \end{pmatrix}; \quad E_{\alpha_2} = \begin{pmatrix} 0 & 0 & 0 & 0 & 0 & 0 & 0 \\ 0 & 0 & \sqrt{\frac{3}{2}} & 0 & 0 & 0 & 0 \\ 0 & 0 & 0 & 0 & 0 & 0 & 0 \\ 0 & 0 & 0 & 0 & 0 & 0 & 0 \\ 0 & 0 & 0 & 0 & 0 & -\sqrt{\frac{3}{2}} & 0 \\ 0 & 0 & 0 & 0 & 0 & 0 & 0 \\ 0 & 0 & 0 & 0 & 0 & 0 & 0 \end{pmatrix};$$

$$E_{\alpha_1+\alpha_2} = \begin{pmatrix} 0 & 0 & \frac{1}{\sqrt{2}} & 0 & 0 & 0 & 0 \\ 0 & 0 & 0 & -1 & 0 & 0 & 0 \\ 0 & 0 & 0 & 0 & 0 & 0 & 0 \\ 0 & 0 & 0 & 0 & 0 & -1 & 0 \\ 0 & 0 & 0 & 0 & 0 & 0 & \frac{1}{\sqrt{2}} \\ 0 & 0 & 0 & 0 & 0 & 0 & 0 \\ 0 & 0 & 0 & 0 & 0 & 0 & 0 \end{pmatrix}; \quad E_{2\alpha_1+\alpha_2} = \begin{pmatrix} 0 & 0 & 0 & -1 & 0 & 0 & 0 \\ 0 & 0 & 0 & 0 & \frac{1}{\sqrt{2}} & 0 & 0 \\ 0 & 0 & 0 & 0 & 0 & -\frac{1}{\sqrt{2}} & 0 \\ 0 & 0 & 0 & 0 & 0 & 0 & 1 \\ 0 & 0 & 0 & 0 & 0 & 0 & 0 \\ 0 & 0 & 0 & 0 & 0 & 0 & 0 \\ 0 & 0 & 0 & 0 & 0 & 0 & 0 \end{pmatrix};$$

$$E_{3\alpha_1+\alpha_2} = \begin{pmatrix} 0 & 0 & 0 & 0 & -\sqrt{\frac{3}{2}} & 0 & 0 \\ 0 & 0 & 0 & 0 & 0 & 0 & 0 \\ 0 & 0 & 0 & 0 & 0 & 0 & -\sqrt{\frac{3}{2}} \\ 0 & 0 & 0 & 0 & 0 & 0 & 0 \\ 0 & 0 & 0 & 0 & 0 & 0 & 0 \\ 0 & 0 & 0 & 0 & 0 & 0 & 0 \\ 0 & 0 & 0 & 0 & 0 & 0 & 0 \end{pmatrix}; \quad E_{3\alpha_1+2\alpha_2} = \begin{pmatrix} 0 & 0 & 0 & 0 & 0 & -\sqrt{\frac{3}{2}} & 0 \\ 0 & 0 & 0 & 0 & 0 & 0 & -\sqrt{\frac{3}{2}} \\ 0 & 0 & 0 & 0 & 0 & 0 & 0 \\ 0 & 0 & 0 & 0 & 0 & 0 & 0 \\ 0 & 0 & 0 & 0 & 0 & 0 & 0 \\ 0 & 0 & 0 & 0 & 0 & 0 & 0 \\ 0 & 0 & 0 & 0 & 0 & 0 & 0 \end{pmatrix}.$$

We refer the reader to [72] for a discussion on a particular splitting of the $\mathfrak{g}_2$ Lie algebra, named *golden splitting*, which was motivated by the discovery of the so-called c-map among homogenous spaces that are special Kählerian and homogenous spaces that are instead quaternionic. All that followed from supersymmetry, yet it is obviously an intrinsic property of certain Lie algebras that might have been discovered by Killing, Cartan, or Weyl if they had searched for it, independently of any supersymmetry or dimensionality reduction of supergravity theories. We do not dwell on these issues since we prefer to touch upon other general aspects of the theory in view of a wider spectrum of its applications.

13.2 The Lie algebra $\mathfrak{f}_4$ and its fundamental representation

The next case that we consider is no longer of rank $r = 2$, and the root system cannot be easily visualized as in previous cases. We discuss the exceptional Lie algebra $\mathfrak{f}_4$, whose rank is 4 and whose structure is codified in the Dynkin diagram presented in Figure 13.4. We show how we can explicitly construct the fundamental and the adjoint representations of this exceptional, non-simply laced Lie algebra. We give some details in this case in order to show how the factors $N_{\alpha\beta}$ appearing in the general form (11.38) of the Lie algebra were determined precisely from the construction of the fundamental representation.

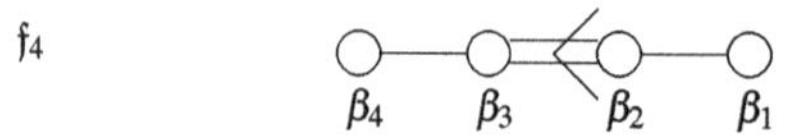

Figure 13.4: The Dynkin diagram of $\mathfrak{f}_4$ and the labeling of simple roots.

Hence, let us proceed to such a construction.

Denoting by $y_{1,2,3,4}$ a basis of orthonormal vectors,

$$y_i \cdot y_j = \delta_{ij}, \tag{13.6}$$

a possible choice of simple roots β_i which reproduces the Cartan matrix encoded in the Dynkin diagram (13.4) is the following:

$$\begin{aligned} \beta_1 &= -y_1 - y_2 - y_3 + y_4, \\ \beta_2 &= 2y_3, \\ \beta_3 &= y_2 - y_3, \\ \beta_4 &= y_1 - y_2. \end{aligned} \tag{13.7}$$

With this basis of simple roots the full root system composed of 48 vectors is given by

$$\Phi_{\mathfrak{f}_4} \equiv \underbrace{\pm y_i \pm y_j;}_{24\text{ roots}} \quad \underbrace{\pm y_i}_{8\text{ roots}}\,; \quad \underbrace{\pm y_1 \pm y_2 \pm y_3 \pm y_4}_{16\text{ roots}} \tag{13.8}$$

and one can order the positive roots by height, as displayed in Table 13.1. Since the con-

Table 13.1: List of the positive roots of the exceptional Lie algebra $\mathfrak{f}_4$. In this table the first column is the name of the root, the second column gives its decomposition in terms of simple roots, and the last column provides the component of the root vector in $\mathbb{R}^4$.

β_1	=	β_1	=	$\{-1,-1,-1,1\}$
β_2	=	β_2	=	$\{0,0,2,0\}$
β_3	=	β_3	=	$\{0,1,-1,0\}$
β_4	=	β_4	=	$\{1,-1,0,0\}$
β_5	=	$\beta_1+\beta_2$	=	$\{-1,-1,1,1\}$
β_6	=	$\beta_2+\beta_3$	=	$\{0,1,1,0\}$
β_7	=	$\beta_3+\beta_4$	=	$\{1,0,-1,0\}$
β_8	=	$\beta_1+\beta_2+\beta_3$	=	$\{-1,0,0,1\}$
β_9	=	$\beta_2+2\beta_3$	=	$\{0,2,0,0\}$
β_{10}	=	$\beta_2+\beta_3+\beta_4$	=	$\{1,0,1,0\}$
β_{11}	=	$\beta_1+\beta_2+2\beta_3$	=	$\{-1,1,-1,1\}$
β_{12}	=	$\beta_1+\beta_2+\beta_3+\beta_4$	=	$\{0,-1,0,1\}$
β_{13}	=	$\beta_2+2\beta_3+\beta_4$	=	$\{1,1,0,0\}$
β_{14}	=	$\beta_1+2\beta_2+2\beta_3$	=	$\{-1,1,1,1\}$
β_{15}	=	$\beta_1+\beta_2+2\beta_3+\beta_4$	=	$\{0,0,-1,1\}$
β_{16}	=	$\beta_2+2\beta_3+2\beta_4$	=	$\{2,0,0,0\}$
β_{17}	=	$\beta_1+2\beta_2+2\beta_3+\beta_4$	=	$\{0,0,1,1\}$
β_{18}	=	$\beta_1+\beta_2+2\beta_3+2\beta_4$	=	$\{1,-1,-1,1\}$
β_{19}	=	$\beta_1+2\beta_2+3\beta_3+\beta_4$	=	$\{0,1,0,1\}$
β_{20}	=	$\beta_1+2\beta_2+2\beta_3+2\beta_4$	=	$\{1,-1,1,1\}$
β_{21}	=	$\beta_1+2\beta_2+3\beta_3+2\beta_4$	=	$\{1,0,0,1\}$
β_{22}	=	$\beta_1+2\beta_2+4\beta_3+2\beta_4$	=	$\{1,1,-1,1\}$
β_{23}	=	$\beta_1+3\beta_2+4\beta_3+2\beta_4$	=	$\{1,1,1,1\}$
β_{24}	=	$2\beta_1+3\beta_2+4\beta_3+2\beta_4$	=	$\{0,0,0,2\}$

sidered Lie algebra is not simply laced, the 24 positive roots split into two subsets of 12 long roots α^ℓ and 12 short roots α^s. They are displayed in Tables 13.2 and 13.3, respectively. Denoting by $\mathbf{\Phi}_\ell$ and $\mathbf{\Phi}_s$ the two subsets, we have the following structure:

$$\begin{aligned} \forall\alpha^\ell,\beta^\ell \in \mathbf{\Phi}_\ell : \quad & \alpha^\ell+\beta^\ell = \left\{\begin{matrix}\text{not a root or}\\ \gamma^\ell \in \mathbf{\Phi}_\ell\end{matrix}\right\}, \\ \forall\alpha^\ell \in \mathbf{\Phi}_\ell \quad \text{and} \quad \forall\beta^s \in \mathbf{\Phi}_s : \quad & \alpha^\ell+\beta^s = \left\{\begin{matrix}\text{not a root or}\\ \gamma^s \in \mathbf{\Phi}_s\end{matrix}\right\}, \\ \forall\alpha^s,\beta^s \in \mathbf{\Phi}_s : \quad & \alpha^s+\beta^s = \left\{\begin{matrix}\text{not a root or}\\ \gamma^s \in \mathbf{\Phi}_s \quad \text{or}\\ \gamma^\ell \in \mathbf{\Phi}_\ell\end{matrix}\right\}. \end{aligned} \tag{13.9}$$

The standard Cartan–Weyl form of the Lie algebra is as follows:

$$[\mathcal{H}_i, E^{\pm\beta}] = \pm\beta^i E^{\pm\beta_I}, \tag{13.10}$$

$$[E^\beta, E^{-\beta}] = \beta\cdot\mathcal{H}, \tag{13.11}$$

Table 13.2: The Φ_ℓ set of the 12 long positive roots in the $\mathfrak{f}_4$ root system.

	$\mathfrak{f}_4$ root labels	$\mathfrak{f}_4$ root in Eucl. basis	root ordered by height
α_1^ℓ	$\{0,1,0,0\}$	$2y_3$	β_2
α_2^ℓ	$\{1,0,0,0\}$	$-y_1-y_2-y_3+y_4$	β_1
α_3^ℓ	$\{1,1,0,0\}$	$-y_1-y_2+y_3+y_4$	β_3
α_4^ℓ	$\{0,1,2,0\}$	$2y_2$	β_9
α_5^ℓ	$\{1,1,2,0\}$	$-y_1+y_2-y_3+y_4$	β_{11}
α_6^ℓ	$\{1,2,2,0\}$	$-y_1+y_2+y_3+y_4$	β_{14}
α_7^ℓ	$\{0,1,2,2\}$	$2y_1$	β_{16}
α_8^ℓ	$\{1,1,2,2\}$	$y_1-y_2-y_3+y_4$	β_{18}
α_9^ℓ	$\{1,2,2,2\}$	$y_1-y_2+y_3+y_4$	β_{20}
α_{10}^ℓ	$\{1,2,4,2\}$	$y_1+y_2-y_3+y_4$	β_{22}
α_{11}^ℓ	$\{1,3,4,2\}$	$y_1+y_2+y_3+y_4$	β_{23}
α_{12}^ℓ	$\{2,3,4,2\}$	$2y_4$	β_{24}

Table 13.3: The Φ_s set of the 12 short positive roots in the $\mathfrak{f}_4$ root system..

	$\mathfrak{f}_4$ root labels	$\mathfrak{f}_4$ root in Eucl. basis	root ordered by height
α_1^s	$\{0,0,0,1\}$	y_1-y_2	β_4
α_2^s	$\{0,0,1,0\}$	y_2-y_3	β_3
α_3^s	$\{0,1,1,0\}$	y_2+y_3	β_6
α_4^s	$\{0,0,1,1\}$	y_1-y_3	β_7
α_5^s	$\{1,1,1,0\}$	$-y_1+y_4$	β_8
α_6^s	$\{0,1,1,1\}$	y_1+y_3	β_{10}
α_7^s	$\{1,1,1,1\}$	$-y_2+y_4$	β_{12}
α_8^s	$\{0,1,2,1\}$	y_1+y_2	β_{13}
α_9^s	$\{1,1,2,1\}$	$-y_3+y_4$	β_{15}
α_{10}^s	$\{1,2,2,1\}$	y_3+y_4	β_{17}
α_{11}^s	$\{1,2,3,1\}$	y_2+y_4	β_{19}
α_{12}^s	$\{1,2,3,2\}$	y_1+y_4	β_{21}

$$[E^\beta, E^\gamma] = \begin{cases} N_{\beta\gamma}\, E^{\beta+\gamma} & \text{if } \beta+\gamma \text{ is a root,} \\ 0 & \text{if } \beta+\gamma \text{ is not a root,} \end{cases} \tag{13.12}$$

where $N_{\beta\gamma}$ are numbers that can be determined by constructing an explicit representation of the Lie algebra.

In the following three tables we exhibit the values of $N_{\beta\gamma}$ for the $\mathfrak{f}_4$ Lie algebra:

α_1^ℓ	α_2^ℓ	α_3^ℓ	α_4^ℓ	α_5^ℓ	α_6^ℓ	α_7^ℓ	α_8^ℓ	α_9^ℓ	α_{10}^ℓ	α_{11}^ℓ	α_{12}^ℓ	
0	$-\sqrt{2}$	0	0	$-\sqrt{2}$	0	0	$-\sqrt{2}$	0	$\sqrt{2}$	0	0	α_1^ℓ
$\sqrt{2}$	0	0	$-\sqrt{2}$	0	0	$-\sqrt{2}$	0	0	0	$-\sqrt{2}$	0	α_2^ℓ
0	0	0	$-\sqrt{2}$	0	0	$-\sqrt{2}$	0	0	$-\sqrt{2}$	0	0	α_3^ℓ
0	$\sqrt{2}$	$\sqrt{2}$	0	0	0	0	$-\sqrt{2}$	$\sqrt{2}$	0	0	0	α_4^ℓ
$\sqrt{2}$	0	0	0	0	0	$\sqrt{2}$	0	$\sqrt{2}$	0	0	0	α_5^ℓ
0	0	0	0	0	0	$-\sqrt{2}$	$-\sqrt{2}$	0	0	0	0	α_6^ℓ
0	$\sqrt{2}$	$\sqrt{2}$	0	$-\sqrt{2}$	$\sqrt{2}$	0	0	0	0	0	0	α_7^ℓ
$\sqrt{2}$	0	0	$\sqrt{2}$	0	$\sqrt{2}$	0	0	0	0	0	0	α_8^ℓ
0	0	0	$-\sqrt{2}$	$-\sqrt{2}$	0	0	0	0	0	0	0	α_9^ℓ
$-\sqrt{2}$	0	$\sqrt{2}$	0	0	0	0	0	0	0	0	0	α_{10}^ℓ
0	$\sqrt{2}$	0	0	0	0	0	0	0	0	0	0	α_{11}^ℓ
0	0	0	0	0	0	0	0	0	0	0	0	α_{12}^ℓ

$\underbrace{\hphantom{xxxxxxxx}}_{N_{\alpha^\ell\beta^\ell}}$ (13.13)

α_1^s	α_2^s	α_3^s	α_4^s	α_5^s	α_6^s	α_7^s	α_8^s	α_9^s	α_{10}^s	α_{11}^s	α_{12}^s	
0	$\sqrt{2}$	0	$-\sqrt{2}$	0	0	0	0	$\sqrt{2}$	0	0	0	α_1^ℓ
0	0	$-\sqrt{2}$	0	0	$-\sqrt{2}$	0	$\sqrt{2}$	0	0	0	0	α_2^ℓ
0	$-\sqrt{2}$	0	$\sqrt{2}$	0	0	0	$-\sqrt{2}$	0	0	0	0	α_3^ℓ
$\sqrt{2}$	0	0	0	0	0	$\sqrt{2}$	0	0	0	0	0	α_4^ℓ
$-\sqrt{2}$	0	0	0	0	$-\sqrt{2}$	0	0	0	0	0	0	α_5^ℓ
$\sqrt{2}$	0	0	$\sqrt{2}$	0	0	0	0	0	0	0	0	α_6^ℓ
0	0	0	0	$-\sqrt{2}$	0	0	0	0	0	0	0	α_7^ℓ
0	0	$\sqrt{2}$	0	0	0	0	0	0	0	0	0	α_8^ℓ
0	$\sqrt{2}$	0	0	0	0	0	0	0	0	0	0	α_9^ℓ
0	0	0	0	0	0	0	0	0	0	0	0	α_{10}^ℓ
0	0	0	0	0	0	0	0	0	0	0	0	α_{11}^ℓ
0	0	0	0	0	0	0	0	0	0	0	0	α_{12}^ℓ

$\underbrace{\hphantom{xxxxxxxx}}_{N_{\alpha^\ell\beta^s}}$ (13.14)

α_1^s	α_2^s	α_3^s	α_4^s	α_5^s	α_6^s	α_7^s	α_8^s	α_9^s	α_{10}^s	α_{11}^s	α_{12}^s	
0	1	−1	0	−1	0	0	$\sqrt{2}$	$-\sqrt{2}$	$\sqrt{2}$	1	0	α_1^s
−1	0	$\sqrt{2}$	0	$\sqrt{2}$	1	−1	0	0	−1	0	$\sqrt{2}$	α_2^s
1	$-\sqrt{2}$	0	1	$-\sqrt{2}$	0	−1	0	1	0	0	$\sqrt{2}$	α_3^s
0	0	−1	0	1	$\sqrt{2}$	$\sqrt{2}$	0	0	1	$-\sqrt{2}$	0	α_4^s
1	$-\sqrt{2}$	$\sqrt{2}$	−1	0	1	0	1	0	0	0	$\sqrt{2}$	α_5^s
0	−1	0	$-\sqrt{2}$	−1	0	$\sqrt{2}$	0	1	0	$\sqrt{2}$	0	α_6^s
0	1	1	$-\sqrt{2}$	0	$-\sqrt{2}$	0	1	0	0	$\sqrt{2}$	0	α_7^s
$-\sqrt{2}$	0	0	0	−1	0	−1	0	$\sqrt{2}$	$\sqrt{2}$	0	0	α_8^s
$\sqrt{2}$	0	−1	0	0	−1	0	$-\sqrt{2}$	0	$-\sqrt{2}$	0	0	α_9^s
$-\sqrt{2}$	1	0	−1	0	0	0	$-\sqrt{2}$	$\sqrt{2}$	0	0	0	α_{10}^s
−1	0	0	$\sqrt{2}$	0	$-\sqrt{2}$	$-\sqrt{2}$	0	0	0	0	0	α_{11}^s
0	$-\sqrt{2}$	$-\sqrt{2}$	0	$-\sqrt{2}$	0	0	0	0	0	0	0	α_{12}^s

$\underbrace{\hphantom{xxxxxxxx}}_{N_{\alpha^s\beta^s}}$

(13.15)

The ordering of long and short roots is that displayed in Tables 13.2 and 13.3. The explicit determination of the tensor $N_{\alpha\beta}$ was performed via the explicit construction of the fundamental 26-dimensional representation of this Lie algebra, which we describe in the next subsection.

13.2.1 Explicit construction of the fundamental and adjoint representation of $\mathfrak{f}_4$

The semi-simple complex Lie algebra $\mathfrak{f}_4$ is defined by the Dynkin diagram in Figure 13.4 and a set of simple roots corresponding to such diagram was provided in equation (13.7). A complete list of the 24 positive roots was given in Table 13.1. The roots were further subdivided into the set of 12 long roots and the set of 12 short roots respectively listed in Tables 13.2 and 13.3. The adjoint representation of $\mathfrak{f}_4$ is 52-dimensional, while its fundamental representation is 26-dimensional. This dimensionality is true for all real sections of the Lie algebra, but the explicit structure of the representation is quite different in each real section. Here we are interested in the maximally split real section $\mathfrak{f}_4$. For such a section we have a maximal, regularly embedded subgroup $\mathfrak{so}(5,4) \subset \mathfrak{f}_{4(4)}$. The decomposition of the representations with respect to this particular subgroup is the essential instrument for their actual construction. For the adjoint representation we have the decomposition

$$\underbrace{\mathbf{52}}_{\text{adj } \mathfrak{f}_{4(4)}} \overset{\mathfrak{so}(5,4)}{\Longrightarrow} \underbrace{\mathbf{36}}_{\text{adj } \mathfrak{so}(5,4)} \oplus \underbrace{\mathbf{16}}_{\text{spinor of } \mathfrak{so}(5,4)} , \tag{13.16}$$

while for the fundamental one we have

$$\underbrace{\mathbf{26}}_{\text{fundamental } \mathfrak{f}_{4(4)}} \overset{\mathfrak{so}(5,4)}{\Longrightarrow} \underbrace{\mathbf{9}}_{\text{vector of } \mathfrak{so}(5,4)} \oplus \underbrace{\mathbf{16}}_{\text{spinor of } \mathfrak{so}(5,4)} \oplus \underbrace{\mathbf{1}}_{\text{singlet of } \mathfrak{so}(5,4)} . \tag{13.17}$$

In view of this, we fix our conventions for the $\mathfrak{so}(5,4)$-invariant metric as

$$\eta_{AB} = \operatorname{diag}\{+,+,+,+,+,-,-,-,-\} \tag{13.18}$$

and we perform an explicit construction of the (16×16)-dimensional gamma-matrices which satisfy the Clifford algebra

$$\{\Gamma_A, \Gamma_B\} = \eta_{AB}\mathbf{1} \tag{13.19}$$

and are *all completely real*. This construction is provided by the following tensor products:

$$\begin{aligned}
\Gamma_1 &= \sigma_1 \otimes \sigma_3 \otimes \mathbf{1} \otimes \mathbf{1}, \\
\Gamma_2 &= \sigma_3 \otimes \sigma_3 \otimes \mathbf{1} \otimes \mathbf{1}, \\
\Gamma_3 &= \mathbf{1} \otimes \sigma_1 \otimes \mathbf{1} \otimes \sigma_1, \\
\Gamma_4 &= \mathbf{1} \otimes \sigma_1 \otimes \sigma_1 \otimes \sigma_3, \\
\Gamma_5 &= \mathbf{1} \otimes \sigma_1 \otimes \sigma_3 \otimes \sigma_3, \\
\Gamma_6 &= \mathbf{1} \otimes i\sigma_2 \otimes \mathbf{1} \otimes \mathbf{1}, \\
\Gamma_7 &= \mathbf{1} \otimes \sigma_1 \otimes i\sigma_2 \otimes \sigma_3, \\
\Gamma_8 &= \mathbf{1} \otimes \sigma_1 \otimes \mathbf{1} \otimes i\sigma_2, \\
\Gamma_9 &= i\sigma_2 \otimes \sigma_3 \otimes \mathbf{1} \otimes \mathbf{1},
\end{aligned} \tag{13.20}$$

where by σ_i we have denoted the standard Pauli matrices:

$$\sigma_1 = \begin{pmatrix} 0 & 1 \\ 1 & 0 \end{pmatrix}; \quad \sigma_2 = \begin{pmatrix} 0 & -i \\ i & 0 \end{pmatrix}; \quad \sigma_3 = \begin{pmatrix} 1 & 0 \\ 0 & -1 \end{pmatrix}. \tag{13.21}$$

Moreover, we introduce the C_+ charge conjugation matrix such that

$$\begin{aligned}
C_+ = (C_+)^T; \quad C_+^2 = \mathbf{1}; \\
C_+ \Gamma_A C_+ = (\Gamma_A)^T.
\end{aligned} \tag{13.22}$$

In the basis of equation (13.20) the explicit form of C_+ is given by

$$C_+ = i\sigma_2 \otimes \sigma_1 \otimes i\sigma_2 \otimes \sigma_1. \tag{13.23}$$

Then we define the usual generators $J_{AB} = -J_{BA}$ of the pseudo-orthogonal algebra $\mathfrak{so}(5,4)$ satisfying the commutation relations

$$[J_{AB}, J_{CD}] = \eta_{BC} J_{AD} - \eta_{AC} J_{BD} - \eta_{BD} J_{AC} + \eta_{AD} J_{BC} \tag{13.24}$$

and we construct the spinor and vector representations by respectively setting

$$J^s_{CD} = \frac{1}{4}[\Gamma_C, \Gamma_D]; \quad (J^v_{CD})_A{}^B = \eta_{CA}\delta^B_D - \eta_{DA}\delta^B_C. \tag{13.25}$$

In this way, if v_A denote the components of a vector, ξ denote the components of a real spinor, and $\epsilon^{AB} = -\epsilon^{BA}$ are the parameters of an infinitesimal $\mathfrak{so}(5,4)$ rotation, we can write the $\mathfrak{so}(5,4)$ transformation as follows:

$$\delta_{\mathfrak{so}(5,4)} v_A = 2\epsilon_{AB} v^B; \quad \delta_{\mathfrak{so}(5,4)} \xi = \frac{1}{2}\epsilon^{AB}\Gamma_{AB}\xi, \tag{13.26}$$

where indices are raised and lowered with the metric (13.18). Furthermore, we introduce the conjugate spinors via the position

$$\overline{\xi} \equiv \xi^T C_+. \tag{13.27}$$

With these preliminaries, we are now in a position to write the explicit form of the 26-dimensional fundamental representation of $\mathfrak{f}_4$ and in this way to construct also its structure constants and hence its adjoint representation, which is our main goal.

According to equation (13.16) the parameters of an $\mathfrak{f}_4$ representation are given by an antisymmetric tensor ϵ_{AB} and a spinor q. On the other hand, a *vector* in the 26-dimensional representation is specified by a collection of three objects, namely, a scalar ϕ, a vector v_A, and a spinor ξ. The representation is constructed if we specify the $\mathfrak{f}_{4(4)}$ transformation of these objects. This is done by writing

$$\delta_{\mathfrak{f}_{4(4)}} \begin{pmatrix} \phi \\ v_A \\ \xi \end{pmatrix} \equiv [\epsilon^{AB} T_{AB} + \overline{q}\, Q] \begin{pmatrix} \phi \\ v_A \\ \xi \end{pmatrix} = \begin{pmatrix} \overline{q}\xi \\ 2\epsilon_{AB} v^B + a\overline{q}\Gamma_A \xi \\ \frac{1}{2}\epsilon^{AB}\Gamma_{AB}\xi - 3\phi q - \frac{1}{a} v^A \Gamma_A q \end{pmatrix}, \tag{13.28}$$

where a is a numerical real arbitrary but non-null parameter. Equation (13.28) defines the generators T_{AB} and Q as 26×26 matrices and therefore completely specifies the fundamental representation of the Lie algebra $\mathfrak{f}_{4(4)}$. Explicitly we have

$$T_{AB} = \left(\begin{array}{c|c|c} 0 & 0 & 0 \\ \hline 0 & J^{v}_{AB} & 0 \\ \hline 0 & 0 & J^{s}_{AB} \end{array}\right) \tag{13.29}$$

and

$$Q_\alpha = \left(\begin{array}{c|c|c} 0 & 0 & \delta^\beta_\alpha \\ \hline 0 & 0 & a(\Gamma_A)^\beta_\alpha \\ \hline -3\delta^\beta_\alpha & -\frac{1}{a}(\Gamma_B)^\beta_\alpha & 0 \end{array}\right), \tag{13.30}$$

and the Lie algebra commutation relations are evaluated to be the following ones:

$$\begin{aligned} [T_{AB}, T_{CD}] &= \eta_{BC} T_{AD} - \eta_{AC} T_{BD} - \eta_{BD} T_{AC} + \eta_{AD} T_{BC}, \\ [T_{AB}, Q] &= \frac{1}{2}\Gamma_{AB} Q, \\ [Q_\alpha, Q_\beta] &= -\frac{1}{12}(C_+ \Gamma^{AB})_{\alpha\beta} T_{AB}. \end{aligned} \tag{13.31}$$

Equation (13.31) together with equations (13.20) and (13.22) provides an explicit numerical construction of the structure constants of the maximally split $\mathfrak{f}_{4(4)}$ Lie algebra. What we still have to do is to identify the relation between the tensorial basis of generators in equation (13.31) and the Cartan–Weyl basis in terms of Cartan generators and step operators. To this effect let us enumerate the 52 generators of $\mathfrak{f}_4$ in the tensorial representation according to the following table:

$$
\begin{array}{|c|c|c|c|}
\hline
\Omega_1 = T_{12} & \Omega_2 = T_{13} & \Omega_3 = T_{14} & \Omega_4 = T_{15} \\
\Omega_5 = T_{16} & \Omega_6 = T_{17} & \Omega_7 = T_{18} & \Omega_8 = T_{19} \\
\Omega_9 = T_{23} & \Omega_{10} = T_{24} & \Omega_{11} = T_{25} & \Omega_{12} = T_{26} \\
\Omega_{13} = T_{27} & \Omega_{14} = T_{28} & \Omega_{15} = T_{29} & \Omega_{16} = T_{34} \\
\Omega_{17} = T_{35} & \Omega_{18} = T_{36} & \Omega_{19} = T_{37} & \Omega_{20} = T_{38} \\
\Omega_{21} = T_{39} & \Omega_{22} = T_{45} & \Omega_{23} = T_{46} & \Omega_{24} = T_{47} \\
\Omega_{25} = T_{48} & \Omega_{26} = T_{49} & \Omega_{27} = T_{56} & \Omega_{28} = T_{57} \\
\Omega_{29} = T_{58} & \Omega_{30} = T_{59} & \Omega_{31} = T_{67} & \Omega_{32} = T_{68} \\
\Omega_{33} = T_{69} & \Omega_{34} = T_{78} & \Omega_{35} = T_{79} & \Omega_{36} = T_{89} \\
\Omega_{37} = Q_1 & \Omega_{38} = Q_2 & \Omega_{39} = Q_3 & \Omega_{40} = Q_4 \\
\Omega_{41} = Q_5 & \Omega_{42} = Q_6 & \Omega_{43} = Q_7 & \Omega_{44} = Q_8 \\
\Omega_{45} = Q_9 & \Omega_{46} = Q_{10} & \Omega_{47} = Q_{11} & \Omega_{48} = Q_{12} \\
\Omega_{49} = Q_{13} & \Omega_{50} = Q_{14} & \Omega_{51} = Q_{15} & \Omega_{52} = Q_{16} \\
\hline
\end{array}
\tag{13.32}
$$

Then, as Cartan subalgebra we take the linear span of the generators,

$$CSA \equiv \text{span}(\Omega_5, \Omega_{13}, \Omega_{20}, \Omega_{26}), \tag{13.33}$$

and furthermore we specify the following basis:

$$
\begin{aligned}
\mathcal{H}_1 &= \Omega_5 + \Omega_{13}; \quad & \mathcal{H}_2 &= \Omega_5 - \Omega_{13}; \\
\mathcal{H}_3 &= \Omega_{20} + \Omega_{26}; \quad & \mathcal{H}_4 &= \Omega_{20} - \Omega_{26}.
\end{aligned}
\tag{13.34}
$$

With respect to this basis the step operators corresponding to the positive roots of $\mathfrak{f}_{4(4)}$ as ordered and displayed in Table 13.1 are those enumerated in Table 13.4. The step operators corresponding to negative roots are obtained from those associated with positive ones via the following relation:

$$E^{-\beta} = -\mathcal{C} E^{\beta} \mathcal{C}, \tag{13.35}$$

where the 26×26 symmetric matrix $\mathcal{C}$ is defined in the following way:

$$
\mathcal{C} = \left(\begin{array}{c|c|c} \mathbf{1} & 0 & 0 \\ \hline 0 & \eta & 0 \\ \hline 0 & 0 & C_+ \end{array} \right).
\tag{13.36}
$$

A further comment is necessary about the normalizations of the step operators E^β which are displayed in Table 13.4. They have been fixed with the following criterion. Once we have constructed the algebra via the generators (13.29) and (13.30), we have the Lie structure constants encoded in equation (13.31) and hence we can diagonalize the adjoint action of the Cartan generators (13.34) to find which linear combinations of the remaining generators correspond to which root. Each root space is one-dimensional and therefore we are left with the task of choosing an absolute normalization for what we want to call the step operators:

Table 13.4: Listing of the step operators corresponding to the positive roots of $\mathfrak{f}_4$.

name	Dynkin lab	comp. root	step operator E^β
$\beta[1]$	$\{1,0,0,0\}$	$-y_1-y_2-y_3+y_4$	$(-\Omega_3-\Omega_8+\Omega_{23}-\Omega_{33})$
$\beta[2]$	$\{0,1,0,0\}$	$2y_3$	$(\Omega_{16}-\Omega_{21}+\Omega_{25}+\Omega_{36})$
$\beta[3]$	$\{0,0,1,0\}$	y_2-y_3	$(\Omega_{37}+\Omega_{39}+\Omega_{41}-\Omega_{43}+\Omega_{45}-\Omega_{47}+\Omega_{49}+\Omega_{51})$
$\beta[4]$	$\{0,0,0,1\}$	y_1-y_2	$(\Omega_{11}+\Omega_{28})$
$\beta[5]$	$\{1,1,0,0\}$	$-y_1-y_2+y_3+y_4$	$-\frac{1}{\sqrt{2}}(-\Omega_2+\Omega_7+\Omega_{18}+\Omega_{32})$
$\beta[6]$	$\{0,1,1,0\}$	y_2+y_3	$-\frac{1}{\sqrt{2}}(-\Omega_{38}-\Omega_{40}+\Omega_{42}-\Omega_{44}+\Omega_{46}-\Omega_{48}-\Omega_{50}-\Omega_{52})$
$\beta[7]$	$\{0,0,1,1\}$	y_1-y_3	$-(-\Omega_{37}-\Omega_{39}+\Omega_{41}-\Omega_{43}+\Omega_{45}-\Omega_{47}-\Omega_{49}-\Omega_{51})$
$\beta[8]$	$\{1,1,1,0\}$	$-y_1+y_4$	$-\frac{1}{2}(\Omega_{38}+\Omega_{40}-\Omega_{42}+\Omega_{44}+\Omega_{46}-\Omega_{48}-\Omega_{50}-\Omega_{52})$
$\beta[9]$	$\{0,1,2,0\}$	$2y_2$	$-\frac{1}{2}(\Omega_1+\Omega_6+\Omega_{12}-\Omega_{31})$
$\beta[10]$	$\{0,1,1,1\}$	y_1+y_3	$-\frac{1}{\sqrt{2}}(\Omega_{38}+\Omega_{40}+\Omega_{42}-\Omega_{44}+\Omega_{46}-\Omega_{48}+\Omega_{50}+\Omega_{52})$
$\beta[11]$	$\{1,1,2,0\}$	$-y_1+y_2-y_3+y_4$	$-\frac{1}{2\sqrt{2}}(\Omega_{10}+\Omega_{15}-\Omega_{24}+\Omega_{35})$
$\beta[12]$	$\{1,1,1,1\}$	$-y_2+y_4$	$-\frac{1}{2}(-\Omega_{38}-\Omega_{40}-\Omega_{42}+\Omega_{44}+\Omega_{46}-\Omega_{48}+\Omega_{50}+\Omega_{52})$
$\beta[13]$	$\{0,1,2,1\}$	y_1+y_2	$-\frac{1}{\sqrt{2}}(\Omega_4+\Omega_{27})$
$\beta[14]$	$\{1,2,2,0\}$	$-y_1+y_2+y_3+y_4$	$-\frac{1}{4}(-\Omega_9+\Omega_{14}+\Omega_{19}+\Omega_{34})$
$\beta[15]$	$\{1,1,2,1\}$	$-y_3+y_4$	$-\frac{1}{2}(\Omega_{22}-\Omega_{30})$
$\beta[16]$	$\{0,1,2,2\}$	$2y_1$	$-\frac{1}{2}(\Omega_1-\Omega_6+\Omega_{12}+\Omega_{31})$
$\beta[17]$	$\{1,2,2,1\}$	y_3+y_4	$-\frac{1}{2\sqrt{2}}(\Omega_{17}+\Omega_{29})$
$\beta[18]$	$\{1,1,2,2\}$	$y_1-y_2-y_3+y_4$	$-\frac{1}{2\sqrt{2}}(\Omega_{10}+\Omega_{15}+\Omega_{24}-\Omega_{35})$
$\beta[19]$	$\{1,2,3,1\}$	y_2+y_4	$-\frac{1}{2\sqrt{2}}(\Omega_{38}-\Omega_{40}+\Omega_{42}+\Omega_{44}+\Omega_{46}+\Omega_{48}+\Omega_{50}-\Omega_{52})$
$\beta[20]$	$\{1,2,2,2\}$	$y_1-y_2+y_3+y_4$	$-\frac{1}{4}(-\Omega_9+\Omega_{14}-\Omega_{19}-\Omega_{34})$
$\beta[21]$	$\{1,2,3,2\}$	y_1+y_4	$-\frac{1}{2\sqrt{2}}(-\Omega_{38}+\Omega_{40}+\Omega_{42}+\Omega_{44}+\Omega_{46}+\Omega_{48}-\Omega_{50}+\Omega_{52})$
$\beta[22]$	$\{1,2,4,2\}$	$y_1+y_2-y_3+y_4$	$-\frac{1}{4}(\Omega_3+\Omega_8+\Omega_{23}-\Omega_{33})$
$\beta[23]$	$\{1,3,4,2\}$	$y_1+y_2+y_3+y_4$	$-\frac{1}{4\sqrt{2}}(\Omega_2-\Omega_7+\Omega_{18}+\Omega_{32})$
$\beta[24]$	$\{2,3,4,2\}$	$2y_4$	$-\frac{1}{8}(\Omega_{16}+\Omega_{21}+\Omega_{25}-\Omega_{36})$

$$E^\beta = \lambda_\beta \quad \text{(linear combination of } \Omega\text{'s)}. \tag{13.37}$$

The values of λ_β are now determined by the following non-trivial conditions:

1. The differences $\mathbb{H}^i = (E^{\beta_i} - E^{-\beta_i})$ should close a subalgebra $\mathbb{H} \subset \mathfrak{f}_{4(4)}$, the maximal compact subalgebra $\mathfrak{su}(2)_R \oplus \mathfrak{usp}(6)$.
2. The sums $\mathbb{K}^i = \frac{1}{\sqrt{2}}(E^{\beta_i} + E^{-\beta_i})$ should span a 28-dimensional representation of $\mathbb{H}$, namely, the aforementioned $\mathfrak{su}(2)_R \oplus \mathfrak{usp}(6)$.

We arbitrarily choose the first four λ_β associated with simple roots and then all the others are determined. The result is displayed in Table 13.4. Using the Cartan generators defined by equation (13.34) and the step operators enumerated in Table 13.4 one can calculate the structure constants of $\mathfrak{f}_4$ in the Cartan–Weyl basis, namely,

$$\begin{aligned}
[\mathscr{H}_i, \mathscr{H}_j] &= 0, \\
[\mathscr{H}_i, E^\beta] &= \beta^i E^\beta, \\
[E^\beta, E^{-\beta}] &= \beta \cdot \mathscr{H}, \\
[E^{\beta_i}, E^{\beta_j}] &= N_{\beta_i,\beta_j} E^{\beta_i+\beta_j}.
\end{aligned} \tag{13.38}$$

In particular one obtains the explicit numerical value of the coefficients $\mathscr{N}_{\beta_i,\beta_j}$, which, as is well known, are the only ones not completely specified by the components of the root vectors in the root system. The result of this computation, following from equation (13.31), is that encoded in equations (13.13), (13.14), and (13.15).

As a last point we can investigate the properties of the maximal compact subalgebra $\mathfrak{su}(2)_R \oplus \mathfrak{usp}(6) \subset \mathfrak{f}_{4(4)}$. As is well known, a basis of generators for these subalgebras is provided by

$$H_i = (E^{\beta_i} - E^{-\beta_i}) \quad (i = 1, \dots, 24), \tag{13.39}$$

but it is not a priori clear which are the generators of $\mathfrak{su}(2)_R$ and which are those of $\mathfrak{usp}(6)$. By choosing a basis of Cartan generators of the compact algebra and diagonalizing their adjoint action this distinction can be established. The generators of $\mathfrak{su}(2)_R$ are the linear combinations

$$\begin{aligned}
J_X &= \frac{1}{4\sqrt{2}}(H_1 - H_{14} + H_{20} - H_{22}), \\
J_Y &= \frac{1}{4\sqrt{2}}(H_5 + H_{11} - H_{18} + H_{23}), \\
J_Z &= \frac{1}{4\sqrt{2}}(-H_2 + H_9 - H_{16} - H_{24}),
\end{aligned} \tag{13.40}$$

close the standard commutation relations

$$[J_i, J_j] = \epsilon_{ijk} J_k, \tag{13.41}$$

and commute with all the generators of $\mathfrak{usp}(6)$. These latter are displayed as follows:

$$\begin{aligned}
\mathscr{H}_1^{(Usp6)} &= -\frac{H_2}{2} - \frac{H_9}{2} + \frac{H_{16}}{2} - \frac{H_{24}}{2}, \\
\mathscr{H}_2^{(Usp6)} &= -\frac{H_2}{2} + \frac{H_9}{2} + \frac{H_{16}}{2} + \frac{H_{24}}{2}, \\
\mathscr{H}_3^{(Usp6)} &= \frac{H_2}{2} + \frac{H_9}{2} + \frac{H_{16}}{2} - \frac{H_{24}}{2},
\end{aligned} \tag{13.42}$$

which are the Cartan generators. On the other hand, the nine pairs of generators which are rotated one into the other by the Cartans with eigenvalues equal to the roots of the compact algebra are the following ones:

$$\begin{array}{|l|l|} \hline W_1 = H_{10} & Z_1 = H_7 \\ W_2 = H_4 & Z_2 = -H_{13} \\ W_3 = H_6 & Z_3 = -H_3 \\ W_4 = -H_1 + H_{14} + H_{20} - H_{22} & Z_4 = -H_5 - H_{11} - H_{18} + H_{23} \\ W_5 = H_{21} & Z_5 = -H_8 \\ W_6 = H_1 + H_{14} + H_{20} + H_{22} & Z_6 = H_5 - H_{11} - H_{18} - H_{23} \\ W_7 = -H_1 - H_{14} + H_{20} + H_{22} & Z_7 = H_5 - H_{11} + H_{18} + H_{23} \\ W_8 = H_{17} & Z_8 = H_{15} \\ W_9 = H_{12} & Z_9 = H_{19} \\ \hline \end{array} \tag{13.43}$$

The construction of the $\mathfrak{f}_4$ Lie algebra presented in this section was published in [100]. All the steps described in the present section are electronically implemented by the MATHEMATICA NoteBook named *F44construzia.nb*, which can be downloaded from the repository https://www.degruyter.com/document/isbn/9783111201535/html.

We provided the details of the $\mathfrak{f}_{4(4)}$ fundamental representation for pedagogical reasons in order to demonstrate how the explicit determination of the matrices for a basis of generators is, for the exceptional Lie algebras, a matter of inventiveness and a case-by-case adventure. This is not surprising since such sporadic mathematical objects do not fit in a general scheme by very definition. In agreement with our strategy we skip the construction of the fundamental 56-dimensional representation of $\mathfrak{e}_{7(7)}$, which is thoroughly described in [72] and is electronically implemented in the MATHEMATICA NoteBook *e77construzia* available from the site https://www.degruyter.com/document/isbn/9783111201535/html.

We rather turn to the case of the $\mathfrak{e}_8$ Lie algebra, which is not covered in [72] and which presents another peculiarity. The fundamental representation coincides in this case with the adjoint representation.

13.3 The exceptional Lie algebra $\mathfrak{e}_8$

The $\mathfrak{e}_8$ algebra is the largest exceptional Lie algebra with dimension 248 and rank 8. The root lattice, spanned by the integer linear combinations $n^i \alpha_i$, has the notable property of being self-dual, that is, the dual root lattice coincides with the root lattice:

$$\Lambda_{\text{weight}}[\mathfrak{e}_8] = \Lambda_{\text{root}}[\mathfrak{e}_8]. \tag{13.44}$$

In line with this the other notable property of this remarkable algebra is the absence of a fundamental representation smaller than the adjoint one. Indeed, the smallest linear representation of $\mathfrak{e}_8$ is just the adjoint one, whose highest weight is the highest positive root of the $\mathfrak{e}_8$ root system:

$$\Pi_{\text{highest}} = \psi \equiv \alpha_{120}. \tag{13.45}$$

The $\mathfrak{e}_8$ Dynkin diagram is displayed in Figure 13.5, and our chosen explicit representation of the simple roots in $\mathbb{R}^8$ is the following one:

$$\begin{aligned}
\alpha_1 &= \{0,1,-1,0,0,0,0,0\},\\
\alpha_2 &= \{0,0,1,-1,0,0,0,0\},\\
\alpha_3 &= \{0,0,0,1,-1,0,0,0\},\\
\alpha_4 &= \{0,0,0,0,1,-1,0,0\},\\
\alpha_5 &= \{0,0,0,0,0,1,-1,0\},\\
\alpha_6 &= \{0,0,0,0,0,1,1,0\},\\
\alpha_7 &= \left\{-\frac{1}{2},-\frac{1}{2},-\frac{1}{2},-\frac{1}{2},-\frac{1}{2},-\frac{1}{2},-\frac{1}{2},-\frac{1}{2}\right\},\\
\alpha_8 &= \{1,-1,0,0,0,0,0,0\}.
\end{aligned} \tag{13.46}$$

The Cartan matrix encoded in the diagram in Figure 13.5 has the following explicit appearance:

$$\mathscr{C}_{ij} \equiv \alpha_i \cdot \alpha_j = \begin{pmatrix}
2 & -1 & 0 & 0 & 0 & 0 & 0 & -1\\
-1 & 2 & -1 & 0 & 0 & 0 & 0 & 0\\
0 & -1 & 2 & -1 & 0 & 0 & 0 & 0\\
0 & 0 & -1 & 2 & -1 & -1 & 0 & 0\\
0 & 0 & 0 & -1 & 2 & 0 & 0 & 0\\
0 & 0 & 0 & -1 & 0 & 2 & -1 & 0\\
0 & 0 & 0 & 0 & 0 & -1 & 2 & 0\\
-1 & 0 & 0 & 0 & 0 & 0 & 0 & 2
\end{pmatrix}. \tag{13.47}$$

The remarkable property (13.44) simply follows from the fact that the determinant of the Cartan matrix is one, Det $\mathscr{C} = 1$, so that the inverse $\mathscr{C}^{-1}$ also has integer valued entries:

$$\mathscr{C}^{-1} = \begin{pmatrix}
6 & 8 & 10 & 12 & 6 & 8 & 4 & 3\\
8 & 12 & 15 & 18 & 9 & 12 & 6 & 4\\
10 & 15 & 20 & 24 & 12 & 16 & 8 & 5\\
12 & 18 & 24 & 30 & 15 & 20 & 10 & 6\\
6 & 9 & 12 & 15 & 8 & 10 & 5 & 3\\
8 & 12 & 16 & 20 & 10 & 14 & 7 & 4\\
4 & 6 & 8 & 10 & 5 & 7 & 4 & 2\\
3 & 4 & 5 & 6 & 3 & 4 & 2 & 2
\end{pmatrix}. \tag{13.48}$$

$\mathfrak{e}_8$

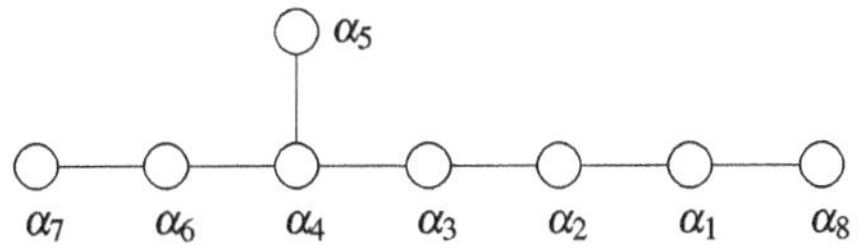

Figure 13.5: The Dynkin diagram of the $\mathfrak{e}_8$ Lie algebra.

This means that the fundamental weights

$$\lambda^i \equiv (\mathscr{C}^{-1})^{ij} \alpha_j \tag{13.49}$$

are linear combinations with integer coefficients of the simple roots and hence lie in the root lattice.

In our chosen basis we have

$$\begin{aligned}
\lambda_1 &= \{1, 1, 0, 0, 0, 0, 0, -2\}, \\
\lambda_2 &= \{1, 1, 1, 0, 0, 0, 0, -3\}, \\
\lambda_3 &= \{1, 1, 1, 1, 0, 0, 0, -4\}, \\
\lambda_4 &= \{1, 1, 1, 1, 1, 0, 0, -5\}, \\
\lambda_5 &= \left\{\frac{1}{2}, \frac{1}{2}, \frac{1}{2}, \frac{1}{2}, \frac{1}{2}, \frac{1}{2}, -\frac{1}{2}, -\frac{5}{2}\right\}, \\
\lambda_6 &= \left\{\frac{1}{2}, \frac{1}{2}, \frac{1}{2}, \frac{1}{2}, \frac{1}{2}, \frac{1}{2}, \frac{1}{2}, -\frac{7}{2}\right\}, \\
\lambda_7 &= \{0, 0, 0, 0, 0, 0, 0, -2\}, \\
\lambda_8 &= \{1, 0, 0, 0, 0, 0, 0, -1\}.
\end{aligned} \tag{13.50}$$

In these conventions the highest weight of the adjoint representation is $\lambda_8 = \alpha_{120}$, and hence it has Dynkin labels $\{0, 0, 0, 0, 0, 0, 0, 1\}$.

13.3.1 Construction of the adjoint representation

The explicit construction of the adjoint representation of $\mathfrak{e}_8$ in its maximally split real form $\mathfrak{e}_{8(8)}$ is done by means of the following steps:

1. Firstly, by means of an algorithm inductive on the height of the roots we construct all the 120 positive roots. It turns out that we have roots of all heights in the interval 1 to 29; their explicit form is displayed in the Mathematica NoteBook *e88construzia23*.
2. Secondly, utilizing the following enumeration for the axes:

$$248 = \underbrace{1, \dots, 8}_{\text{Cartan generators}}, \underbrace{1+8, \dots, 8+120}_{\text{positive roots}}, \underbrace{1+128, \dots, 120+128}_{\text{negative roots}}, \tag{13.51}$$

   we construct the 248×248 matrices representing the Cartan generators and fulfilling the following commutation relations:

$$[\mathscr{H}^i, \mathscr{H}^j] = 0; \tag{13.52}$$

$$[\mathscr{H}^i, \mathscr{E}^\alpha] = \alpha^i \mathscr{E}^\alpha; \quad \forall \alpha \in \text{root system}. \tag{13.53}$$

3. Then we construct the 248 × 248 matrices that represent the step operators $\mathcal{E}^{\alpha_i}$ associated with the simple roots α_i ($i = 1, \dots, 8$) in such a way that equation (13.53) is indeed confirmed,

$$[\mathcal{H}^i, \mathcal{E}^{\alpha_j}] = \alpha^i_j \mathcal{E}^{\alpha_j}; \quad \forall \alpha_j > 0, \tag{13.54}$$

and setting

$$\mathcal{E}^{-\alpha_j} \equiv (\mathcal{E}^{\alpha_j})^T \tag{13.55}$$

we also obtain

$$\begin{aligned} [\mathcal{H}^i, \mathcal{E}^{-\alpha_j}] &= -\alpha^i_j \mathcal{E}^{-\alpha_j}; \quad \forall \alpha_j > 0; \\ [\mathcal{E}^{\alpha_j}, \mathcal{E}^{-\alpha_j}] &= \alpha^i_j \mathcal{H}_i; \quad \forall \alpha_j > 0. \end{aligned} \tag{13.56}$$

4. Finally we obtain all the higher height step operators by defining

$$\mathcal{E}^{\beta} \equiv [\mathcal{E}^{\alpha_i}, \mathcal{E}^{\gamma}], \tag{13.57}$$

where, by definition,

$$\beta = \alpha_i + \gamma \tag{13.58}$$

and the index i of the simple root α_i is the smallest for all pairs $\{\alpha_i, \gamma\}$ for which equation (13.58) is true. This is an algorithmic way of fixing the normalization of the step operators and it is useful for computer aided constructions.

13.3.1.1 The cocycle $N^{\alpha\beta}$

Before we proceed with the illustration of our explicit construction we have to pause and discuss a delicate point. The commutation relations of a Lie algebra in the Cartan–Weyl form are completely fixed once also the cocycle $N^{\alpha,\beta}$ is given in the commutation rules of the step operators:

$$[\mathcal{E}^{\alpha}, \mathcal{E}^{\beta}] = N_{\alpha\beta} \mathcal{E}^{\alpha+\beta}. \tag{13.59}$$

There is no way of representing $N^{\alpha,\beta}$ with an analytic formula, just because it is a cocycle. Since our Lie algebra is simply laced, all the non-vanishing entries of $N^{\alpha,\beta}$ are either +1 or −1, yet there is no better way to specify it than to enumerate the pairs of roots for which it is positive and the pairs of roots for which it is negative. It suffices to enumerate the entries $N^{\alpha,\beta}$ for both α and β positive roots. Indeed, once we know $N^{\alpha,\beta}$ for $\alpha, \beta > 0$ we can extend its definition to the other cases through the following identities, which follow from the Jacobi identities:

$$\begin{aligned} N^{-\alpha,-\beta} &= -N^{\alpha,\beta}, \\ N^{\alpha,-\beta} &= N^{\alpha-\beta,\beta}, \\ N^{\beta,\alpha,} &= -N^{\alpha,\beta}. \end{aligned} \tag{13.60}$$

Our consistent choice of $N^{\alpha\beta}$ is encoded in an array named *Nalfa* which is generated in the library when the MATHEMATICA NoteBook *e88construzia23* is evaluated (download it from the repository https://www.degruyter.com/document/isbn/9783111201535/html). The same NoteBook verifies also the identity

$$N^{\alpha\beta}\, N^{\alpha+\beta,\gamma} + N^{\beta,\gamma}\, N^{\gamma+\beta,\alpha} + N^{\gamma,\alpha}\, N^{\alpha+\gamma,\beta} = 0, \tag{13.61}$$

which follows from the Jacobi identities and which is the cocycle condition. Once this identity is verified, we are sure that the structure constants as given in equation (11.38) by the Cartan–Weyl basis are consistent and we possess the adjoint representation.

13.3.1.2 The Cartan generators

First of all we set the form of the Cartan generators named $\mathcal{H}^i$ $(i = 1, \dots, 8)$. The corresponding 248×248 matrices are diagonal and they are constructed as follows:

$$\begin{aligned} (\mathcal{H}^i)_m{}^{\ell} &= 0; \quad i = 1,\dots,8;\ m,\ell = 1,\dots,8; \\ (\mathcal{H}^i)_{m+8}{}^{\ell+8} &= -\alpha^i_m \delta^\ell_m; \quad i = 1,\dots,8;\ m,\ell = 1,\dots 120; \\ (\mathcal{H}^i)_{m+128}{}^{\ell+128} &= \alpha^i_m \delta^\ell_m; \quad i = 1,\dots,8;\ m,\ell = 1,\dots 120; \\ (\mathcal{H}^i)_{m+128}{}^{\ell} &= 0; \quad i = 1,\dots,8;\ m,\ell = 1,\dots 120; \\ (\mathcal{H}^i)_m{}^{\ell+128} &= 0; \quad i = 1,\dots,8;\ m,\ell = 1,\dots 120, \end{aligned} \tag{13.62}$$

where by α^i_m we denote the i-th component of the m-th root.

13.3.1.3 The step operators associated with simple roots

The simple root step operators are constructed in three steps.

First step

In the first step we fix the components of $\mathcal{E}^{\alpha_i}$ where one leg is in the Cartan subalgebra and the other is along the simple roots:

$$\begin{aligned} (\mathcal{E}^{\alpha_i})_j{}^{i+8} &= -\alpha^j_i; \quad i = 1,\dots,8;\ j = 1,\dots,8; \\ (\mathcal{E}^{\alpha_i})_{i+128}{}^{j} &= \alpha^j_i; \quad i = 1,\dots,8;\ j = 1,\dots,8. \end{aligned} \tag{13.63}$$

Equation (13.63) is obligatory and unambiguous, since it is the transcription into the adjoint representation matrix of the commutation relation (13.54).

Second step

In the second step we fix the components of $\mathcal{E}^{\alpha_i}$ where one leg is along the simple roots and the other is along the roots of height $h = 2$:

$$\begin{aligned}(\mathcal{E}^{\alpha_i})_{k+8}{}^{p+8} &= \epsilon_{ik} \quad (\text{iff } \alpha_p = \alpha_i + \alpha_k); \quad i = 1,\dots,8;\ k = 1,\dots,8;\\ (\mathcal{E}^{\alpha_i})_{k+128}{}^{p+128} &= -\epsilon_{ik} \quad (\text{iff } \alpha_p = \alpha_i + \alpha_k); \quad i = 1,\dots,8;\ k = 1,\dots,8,\end{aligned} \tag{13.64}$$

where by definition $\epsilon_{ik} = 1$ iff $i < k$, $\epsilon_{ik} = -1$ iff $i > k$, and $\epsilon_{ik} = 0$ iff $i = k$.

Third step

In the third step we fix the components of $\mathcal{E}^{\alpha_i}$ where one of the legs is along the roots of height larger than or equal to 3:

$$\begin{aligned}(\mathcal{E}^{\alpha_i})_{k+8}{}^{p+8} &= (-1)^{\frac{i(i+1)}{2}+h[\alpha_p]} \quad (\text{iff } \alpha_p = \alpha_i + \alpha_k),\\ (\mathcal{E}^{\alpha_i})_{k+128}{}^{p+128} &= -(-1)^{\frac{i(i+1)}{2}+h[\alpha_p]} \quad (\text{iff } \alpha_p = \alpha_i + \alpha_k),\\ i &= 1,\dots,8,\ k = 9,\dots,120.\end{aligned}$$

In the above equation $h[\alpha_p]$ denotes the height of the root α_p.

Fourth step

The fourth step is equivalent to determining by trial and error the $N_{\alpha\beta}$ cocycle. As we anticipated, it was through a series of educated guesses that we were able to determine a set of sign changes that was sufficient to guarantee the implementation of equations (13.54), (13.55), and (13.56) and the cocycle condition. This set of sign changes is displayed below:

$$\begin{aligned}(\mathcal{E}^{\alpha_8})_{17,25} &= -1,\\ (\mathcal{E}^{\alpha_8})_{25+120,17+120} &= 1,\\ (\mathcal{E}^{\alpha_4})_{23,29} &= -1,\\ (\mathcal{E}^{\alpha_4})_{29+120,23+120} &= 1,\\ (\mathcal{E}^{\alpha_8})_{24,34} &= 1,\\ (\mathcal{E}^{\alpha_8})_{34+120,24+120} &= -1,\\ (\mathcal{E}^{\alpha_8})_{31,43} &= -1,\\ (\mathcal{E}^{\alpha_8})_{43+120,31+120} &= 1,\\ (\mathcal{E}^{\alpha_8})_{38,50} &= 1,\\ (\mathcal{E}^{\alpha_8})_{50+120,38+120} &= -1,\\ (\mathcal{E}^{\alpha_8})_{39,51} &= 1,\end{aligned}$$

$$\begin{aligned}
(\mathcal{E}^{\alpha_8})_{51+120,39+120} &= -1, \\
(\mathcal{E}^{\alpha_8})_{46,56} &= -1, \\
(\mathcal{E}^{\alpha_8})_{56+120,46+120} &= 1, \\
(\mathcal{E}^{\alpha_8})_{53,63} &= 1, \\
(\mathcal{E}^{\alpha_8})_{63+120,53+120} &= -1, \\
(\mathcal{E}^{\alpha_7})_{51,58} &= -1, \\
(\mathcal{E}^{\alpha_7})_{58+120,51+120} &= 1, \\
(\mathcal{E}^{\alpha_7})_{56,64} &= 1, \\
(\mathcal{E}^{\alpha_7})_{64+120,56+120} &= -1, \\
(\mathcal{E}^{\alpha_7})_{63,70} &= -1, \\
(\mathcal{E}^{\alpha_7})_{70+120,63+120} &= 1, \\
(\mathcal{E}^{\alpha_3})_{63,69} &= -1, \\
(\mathcal{E}^{\alpha_3})_{69+120,63+120} &= 1.
\end{aligned}$$

Implementing next the algorithm specified in equations (13.57) and (13.58) one obtains all the step operators and finally all the 248 generators of the $\mathfrak{e}_8$ Lie algebra given as 248×248 matrices.

13.3.2 Final comments on the $\mathfrak{e}_8$ root systems

From the above displayed table it is evident that the 120 positive roots subdivide into two subsets, one containing 56 elements and the other containing 64 elements. In the first set the Euclidean components of the roots are all integers, in the second set they are all half-integers. The 56 integer valued roots are the positive roots of the D_8 Lie subalgebra, namely, $\mathfrak{so}(16, \mathbb{C})$. The 64 half-integer valued roots are one half of the 128 weights of the spinor representation of $\mathfrak{so}(16)$. Indeed, the $\mathfrak{e}_8$ Lie algebra has a regularly embedded $\mathfrak{so}(16)$ subalgebra with decomposition

$$\mathrm{adj}[\mathfrak{e}_8] = \mathrm{adj}[\mathfrak{so}(16)] \oplus \mathrm{spinor}[\mathfrak{so}(16)]. \tag{13.65}$$

13.3.2.1 The $\mathfrak{e}_{7(7)} \subset \mathfrak{e}_{8(8)}$ simple root basis

In addition to the decomposition (13.65) there is another one which corresponds to the golden splitting of the $\mathfrak{e}_{8(8)}$ Lie algebra, namely,

$$\mathrm{adj}[\mathfrak{e}_8] = (\mathrm{adj}[\mathfrak{e}_{7(7)}], \mathbf{1}) \oplus (\mathbf{1}, \mathrm{adj}[\mathfrak{sl}(2, \mathbb{R})]) \oplus (\mathbf{56}, \mathbf{2}). \tag{13.66}$$

To implement the decomposition (13.66) it is convenient to use a second basis for the root system, which we presently describe. We choose $\alpha_1, \alpha_2, \dots, \alpha_7$ to be the simple roots of $\mathfrak{e}_{7(7)}$ extended by means of an eighth vanishing component. Then α_8 is uniquely determined in order to have the right scalar product with the other simple roots. Hence, we set $\alpha_i = \beta_i$, where

$$
\begin{aligned}
\beta_1 &= \{0, 1, -1, 0, 0, 0, 0, 0\},\\
\beta_2 &= \{0, 0, 1, -1, 0, 0, 0, 0\},\\
\beta_3 &= \{0, 0, 0, 1, -1, 0, 0, 0\},\\
\beta_4 &= \{0, 0, 0, 0, 1, -1, 0, 0\},\\
\beta_5 &= \{0, 0, 0, 0, 0, 1, -1, 0\},\\
\beta_6 &= \{0, 0, 0, 0, 0, 1, 1, 0\},\\
\beta_7 &= \left\{0, -\frac{1}{2}, -\frac{1}{2}, -\frac{1}{2}, -\frac{1}{2}, -\frac{1}{2}, -\frac{1}{2}, \frac{1}{\sqrt{2}}\right\},\\
\beta_8 &= \left\{-\frac{1}{\sqrt{2}}, -1, 0, 0, 0, 0, 0, -\frac{1}{\sqrt{2}}\right\}.
\end{aligned}
\tag{13.67}
$$

The MATHEMATICA NoteBook that implements all the construction algorithms explained above is named *e88construzia23.nb* which can be downloaded from the repository https://www.degruyter.com/document/isbn/9783111201535/html.

13.4 Bibliographical note

The main sources for the constructions of the exceptional Lie algebras presented in this chapter are not textbooks, but rather research literature from the last decade pertaining to the realm of supergravity and its various applications to cosmology, black-hole physics, and other fundamental issues. The electronic implementation of these constructions by means of suitable Mathematica codes was precisely motivated by supergravity applications. The versions of such MATHEMATICA NoteBooks listed in Appendix A are renewed and polished ones that have been made user-friendly for the purpose of distribution and use in connection with the present textbook.

1. The here presented construction of the $\mathfrak{g}_{2(2)}$ Lie algebra has its origin in [101, 102].
2. The here presented construction of the $\mathfrak{f}_{4(4)}$ Lie algebra has its origin in [100].
3. The here presented construction of the $\mathfrak{e}_{7(7)}$ Lie algebra has its origin in [103] and the papers there quoted.
4. The here presented construction of the $\mathfrak{e}_{8(8)}$ Lie algebra has its origin in [104].

Furthermore, some of the above constructions were further revised and refined in [105]. Many more theoretical details are provided in the specialized book [72].

14 In depth study of a simple group

"Simple" does not mean "easy." I have learned that the things that seem the simplest are often the most powerful of all.

Christie Golden, Thrall Twilight of the Aspects

14.1 A simple crystallographic point group in seven dimensions

In Chapter 5 we provided the ADE classification of finite subgroups of the three-dimensional rotation group SO(3), while in Chapter 8 we analyzed the general notion of lattices in n dimensions and displayed the classification of crystallographic lattices in the plane and in the familiar Euclidean three-space ($n = 3$). In Chapter 12 we witnessed the emergence of other important lattices in higher dimensions, namely, *the root and weight lattices* associated with simple Lie algebras. The concept of crystallographic groups was presented in the rigorous Definition 8.1.1, which is dimension-dependent. Namely, the same abstract group is crystallographic in those dimensions where it admits an integer valued representation and not in others.

Summarizing our previous discussions, in $d = 3$ the list of available point groups was given in equation (8.20). In this list there is no simple one which is non-Abelian. They are all either solvable or Abelian and this implies that their irreducible representations can be constructed by means of the induction strategy explained in Section 4.4. A simple group which occurs in the ADE classification is the icosahedral group I_{60} which is isomorphic to the simple alternating group A_5 (the even permutations of five objects). It is barred out by the crystallographic condition because it contains elements of order five.

In many respects this is the analog of what happens with algebraic equations. The algebraic equations of order two, three, or four are always solvable by radicals since their Galois group is solvable. In degree $d \geq 5$ the generic equation is not solvable because the Galois group is generically not solvable.

A natural question arises at this point. Is the condition on the possible orders of the point group elements 2, 3, 4, 6 intrinsic to the crystallographic constraint in any dimension or it is a specific feature of $d = 3$?

The correct answer to the above question is the second option, and in this section we show a counterexample of a crystallographic group in seven dimensions that has group elements of order seven. Not only that. Ours is an example of a simple non-Abelian crystallographic point group!

To the best of our knowledge, the analog of the ADE classification of finite rotation groups in $d > 5$ is so far non-existing, which is quite remarkable. Even less is known about higher-dimensional crystallographic groups.

According to our general strategy the goal we pursue in this section is to familiarize the reader with the distinctive properties of a simple group. At the same time show-

https://doi.org/10.1515/9783111201535-014

ing that it is crystallographic in seven dimensions, we illustrate the orbits of the chosen group in the lattice that it maps into itself. Since group orbits in $d = 3$ are in correspondence with Platonic regular solids, it is interesting to see the rich zoology of such orbits (hence analogs of regular solids) in higher dimensions.

14.1.1 The simple group L_{168}

The finite group

$$\mathrm{L}_{168} \equiv \mathrm{PSL}(2, \mathbb{Z}_7) \tag{14.1}$$

is the second smallest simple group after the alternating group A_5 which has 60 elements and coincides with the symmetry group of the regular icosahedron or dodecahedron. As anticipated by its given name, L_{168} has 168 elements; they can be identified with all the possible 2×2 matrices with determinant one whose entries belong to the finite field $\mathbb{Z}_7$, counting them up to an overall sign. In complete analogy to what we did in Section 9.5.1 and in particular in equation (9.71), the group under consideration is the quotient of the modular group $\Gamma \equiv \mathrm{PSL}(2, \mathbb{Z})$ with respect to the normal subgroup $\Gamma(7)$ defined with 3 replaced by 7 as in equation (9.72). In projective geometry, L_{168} is classified as a *Hurwitz group* since it is the automorphism group of a Hurwitz–Riemann surface, namely, a surface of genus g with a maximum of $84(g-1)$ conformal automorphisms.[1] The Hurwitz surface pertaining to the Hurwitz group L_{168} is the Klein[2] quartic [6], namely, the locus $\mathscr{K}_4$ in $\mathbb{P}_2(\mathbb{C})$ cut out by the following quartic polynomial constraint on the homogeneous coordinates $\{x, y, z\}$:

$$x^3 y + y^3 z + z^3 x = 0. \tag{14.2}$$

Indeed, $\mathscr{K}_4$ is a genus $g = 3$ compact Riemann surface and it can be realized as the quotient of the hyperbolic Poincaré plane $\mathbb{H}_2$ by a certain group Γ that acts freely on $\mathbb{H}_2$ by isometries.

The L_{168} group, which is also isomorphic to $\mathrm{GL}(3, \mathbb{Z}_2)$ [106], has received a lot of attention in mathematics and it has important applications in algebra, geometry, and number theory; for instance, besides being associated with the Klein quartic, L_{168} is the automorphism group of the Fano plane [107, 108].

1 Hurwitz' automorphisms theorem proved in 1893 [5] states that the order $|\mathscr{G}|$ of the group $\mathscr{G}$ of orientation-preserving conformal automorphisms of a compact Riemann surface of genus $g > 1$ admits the following upper bound: $|\mathscr{G}| \leq 84(g-1)$.

2 We already mentioned Felix Klein in Chapter 3. Much more about him and about his role in the development of geometry and group theory in the nineteenth century is presented in [28].

The reason why we consider L_{168} is associated with another property of this finite simple group, which was proved more than 20 years ago in [109], namely,

$$L_{168} \subset G_{2(-14)}. \tag{14.3}$$

This means that L_{168} is a finite subgroup of the compact form of the exceptional Lie group G_2 and the seven-dimensional fundamental representation of the latter is irreducible upon restriction to L_{168}.

The group L_{168} happens to be crystallographic in $d = 7$, the preserved lattice being the root lattice of either the simple Lie algebra $\mathfrak{a}_7$ or, even more inspiringly, of the exceptional Lie algebra $\mathfrak{e}_7$. Actually, L_{168} is a subgroup of the $\mathfrak{e}_7$ Weyl group.

As stated we are interested in its properties in order to illustrate the case of *a simple crystallographic non-Abelian group*.

14.1.2 Structure of the simple group $L_{168} = PSL(2, \mathbb{Z}_7)$

For the reasons outlined above we consider the simple group (14.1) and its crystallographic action in $d = 7$. The Hurwitz simple group L_{168} is abstractly presented as[3]

$$L_{168} = (R, S, T \parallel R^2 = S^3 = T^7 = RST = (TSR)^4 = \mathbf{e}) \tag{14.4}$$

and, as its name implicitly advocates, it has order 168:

$$|L_{168}| = 168. \tag{14.5}$$

The elements of this simple group are organized in six conjugacy classes according to the following scheme:

Conjugacy class	$\mathscr{C}_1$	$\mathscr{C}_2$	$\mathscr{C}_3$	$\mathscr{C}_4$	$\mathscr{C}_5$	$\mathscr{C}_6$
representative of the class	$\mathbf{e}$	R	S	TSR	T	SR
order of the elements in the class	1	2	3	4	7	7
number of elements in the class	1	21	56	42	24	24

(14.6)

As one can see from (14.6), the group contains elements of order two, three, four, and seven and there are two inequivalent conjugacy classes of elements of the highest order. According to the general theory of finite groups, there are six different irreducible representations of dimensions one, six, seven, eight, three, and three, respectively. The character table of the group L_{168} can be found in the mathematical literature, for instance in the book [70]. It reads as follows:

3 In the rest of this section we follow closely the results obtained by the present author in the paper [110].

Representation	$\mathscr{C}_1$	$\mathscr{C}_2$	$\mathscr{C}_3$	$\mathscr{C}_4$	$\mathscr{C}_5$	$\mathscr{C}_6$
$D_1[L_{168}]$	1	1	1	1	1	1
$D_6[L_{168}]$	6	2	0	0	−1	−1
$D_7[L_{168}]$	7	−1	1	−1	0	0
$D_8[L_{168}]$	8	0	−1	0	1	1
$DA_3[L_{168}]$	3	−1	0	1	$\frac{1}{2}(-1+i\sqrt{7})$	$\frac{1}{2}(-1-i\sqrt{7})$
$DB_3[L_{168}]$	3	−1	0	1	$\frac{1}{2}(-1-i\sqrt{7})$	$\frac{1}{2}(-1+i\sqrt{7})$

(14.7)

Soon we will retrieve it by constructing explicitly all the irreducible representations.

14.1.3 The seven-dimensional irreducible representation

For many purposes the most interesting representation is the seven-dimensional one. Indeed, its properties are the very reason to consider the group L_{168} in the present context. The following three statements are true:

1. The seven-dimensional irreducible representation is crystallographic since all elements $\gamma \in L_{168}$ are represented by integer valued matrices $D_7(\gamma)$ in a basis of vectors that span a lattice, namely, the root lattice Λ_{root} of the $\mathfrak{a}_7$ simple Lie algebra.
2. The seven-dimensional irreducible representation provides an immersion $L_{168} \hookrightarrow SO(7)$ since its elements preserve the symmetric Cartan matrix of $\mathfrak{a}_7$:

$$\forall \gamma \in L_{168} : \quad D_7^T(\gamma)\mathscr{C}D_7(\gamma) = \mathscr{C},$$
$$\mathscr{C}_{i,j} = \alpha_i \cdot \alpha_j \quad (i,j = 1\dots,7), \tag{14.8}$$

defined in terms of the simple roots α_i whose standard construction in terms of the unit vectors ϵ_i of $\mathbb{R}^8$ is recalled below:[4]

$$\begin{aligned} &\alpha_1 = \epsilon_1 - \epsilon_2; \quad \alpha_2 = \epsilon_2 - \epsilon_3 =; \quad \alpha_3 = \epsilon_3 - \epsilon_4; \\ &\alpha_4 = \epsilon_4 - \epsilon_5; \quad \alpha_5 = \epsilon_5 - \epsilon_6 =; \quad \alpha_6 = \epsilon_6 - \epsilon_7; \\ &\alpha_7 = \epsilon_7 - \epsilon_8. \end{aligned} \tag{14.9}$$

3. Actually the seven-dimensional representation defines an embedding $L_{168} \hookrightarrow G_2 \subset SO(7)$ since there exists a three-index antisymmetric tensor ϕ_{ijk} satisfying the relations of octonionic structure constants[5] that is preserved by all the matrices $D_7(\gamma)$:

4 We refer the reader to Chapter 11 and in particular to Section 11.5.1 for the explicit form of the Cartan matrices associated with $\mathfrak{a}_\ell$ algebras.

5 For the history of quaternions and octonions we refer the reader to [28].

$$\forall\gamma \in \mathrm{L}_{168}: \quad D_7(\gamma)_{ii'} D_7(\gamma)_{jj'} D_7(\gamma)_{kk'} \phi_{i'j'k'} = \phi_{ijk}. \tag{14.10}$$

Let us prove the above statements. It suffices to write the explicit form of the generators R, S, and T in the crystallographic basis of the considered root lattice:

$$\mathbf{v} \in \Lambda_{\text{root}} \quad \Leftrightarrow \quad \mathbf{v} = n_i \alpha_i, \quad n_i \in \mathbb{Z}. \tag{14.11}$$

Explicitly if we set

$$\begin{aligned}
\mathscr{R} &= \begin{pmatrix} 0 & 0 & 0 & 0 & 0 & 0 & -1 \\ 0 & 0 & 0 & 0 & 0 & -1 & 0 \\ 0 & 0 & -1 & 1 & 0 & -1 & 0 \\ 0 & -1 & 0 & 1 & 0 & -1 & 0 \\ 0 & -1 & 0 & 1 & -1 & 0 & 0 \\ 0 & -1 & 0 & 0 & 0 & 0 & 0 \\ -1 & 0 & 0 & 0 & 0 & 0 & 0 \end{pmatrix}, \\
\mathscr{S} &= \begin{pmatrix} 0 & 0 & 0 & 0 & 0 & 0 & -1 \\ 1 & 0 & 0 & 0 & 0 & 0 & -1 \\ 1 & 0 & 0 & -1 & 1 & 0 & -1 \\ 1 & 0 & -1 & 0 & 1 & 0 & -1 \\ 1 & 0 & -1 & 0 & 1 & -1 & 0 \\ 1 & 0 & -1 & 0 & 0 & 0 & 0 \\ 1 & -1 & 0 & 0 & 0 & 0 & 0 \end{pmatrix}, \\
\mathscr{T} &= \begin{pmatrix} 0 & 0 & 0 & 0 & 0 & -1 & 1 \\ 1 & 0 & 0 & 0 & 0 & -1 & 1 \\ 0 & 1 & 0 & 0 & 0 & -1 & 1 \\ 0 & 0 & 1 & 0 & 0 & -1 & 1 \\ 0 & 0 & 0 & 1 & 0 & -1 & 1 \\ 0 & 0 & 0 & 0 & 1 & -1 & 1 \\ 0 & 0 & 0 & 0 & 0 & 0 & 1 \end{pmatrix},
\end{aligned} \tag{14.12}$$

we find that the defining relations of L_{168} are satisfied:

$$\mathscr{R}^2 = \mathscr{S}^3 = \mathscr{T}^7 = \mathscr{R}\mathscr{S}\mathscr{T} = (\mathscr{T}\mathscr{S}\mathscr{R})^4 = \mathbf{1}_{7\times7}, \tag{14.13}$$

and furthermore we have

$$\mathscr{R}^{\mathrm{T}}\mathscr{C}\mathscr{R} = \mathscr{S}^{\mathrm{T}}\mathscr{C}\mathscr{S} = \mathscr{T}^{\mathrm{T}}\mathscr{C}\mathscr{T} = \mathscr{C}, \tag{14.14}$$

where the explicit form of the $\mathfrak{a}_7$ Cartan matrix is recalled below:

$$\mathscr{C} = \begin{pmatrix} 2 & -1 & 0 & 0 & 0 & 0 & 0 \\ -1 & 2 & -1 & 0 & 0 & 0 & 0 \\ 0 & -1 & 2 & -1 & 0 & 0 & 0 \\ 0 & 0 & -1 & 2 & -1 & 0 & 0 \\ 0 & 0 & 0 & -1 & 2 & -1 & 0 \\ 0 & 0 & 0 & 0 & -1 & 2 & -1 \\ 0 & 0 & 0 & 0 & 0 & -1 & 2 \end{pmatrix}. \tag{14.15}$$

This proves statements 1 and 2.

In order to prove statement 3 we proceed as follows. In $\mathbb{R}^7$ we consider the antisymmetric three-index tensor ϕ_{ABC} that, in the standard orthonormal basis, has the following components:

$$\begin{aligned} \phi_{1,2,6} &= \frac{1}{6}, \\ \phi_{1,3,4} &= -\frac{1}{6}, \\ \phi_{1,5,7} &= -\frac{1}{6}, \\ \phi_{2,3,7} &= \frac{1}{6} \qquad \text{all other components vanish.} \\ \phi_{2,4,5} &= \frac{1}{6}, \\ \phi_{3,5,6} &= -\frac{1}{6}, \\ \phi_{4,6,7} &= -\frac{1}{6}, \end{aligned} \tag{14.16}$$

This tensor satisfies the algebraic relations of octonionic structure constants, namely,[6]

$$\phi_{ABM}\phi_{CDM} = \frac{1}{18}\delta^{AB}_{CD} + \frac{2}{3}\Phi_{ABCD}, \tag{14.17}$$

$$\phi_{ABC} = -\frac{1}{6}\epsilon_{ABCPQRS}\Phi_{ABCD}, \tag{14.18}$$

and the subgroup of SO(7) which leaves ϕ_{ABC} invariant is, by definition, the compact section $G_{(2,-14)}$ of the complex G_2 Lie group (see for instance [111]). A particular matrix that transforms the standard orthonormal basis of $\mathbb{R}^7$ into the basis of simple roots α_i is the following one:

6 In this equation the indices of the G_2-invariant tensor are denoted with uppercase letters of the Latin alphabet, as was the case in the quoted literature on weak G_2-structures. In the following we will use lowercase Latin letters, the uppercase Latin letters being reserved for $d = 8$.

$$\mathfrak{M} = \begin{pmatrix} \sqrt{2} & -\frac{1}{\sqrt{2}} & 0 & 0 & 0 & 0 & 0 \\ 0 & -\frac{1}{\sqrt{2}} & \sqrt{2} & -\frac{1}{\sqrt{2}} & 0 & 0 & 0 \\ 0 & 0 & 0 & -\frac{1}{\sqrt{2}} & \sqrt{2} & -\frac{1}{\sqrt{2}} & 0 \\ 0 & 0 & 0 & 0 & 0 & -\frac{1}{\sqrt{2}} & \sqrt{2} \\ 0 & -\frac{1}{\sqrt{2}} & 0 & \frac{1}{\sqrt{2}} & 0 & -\frac{1}{\sqrt{2}} & 0 \\ 0 & 0 & 0 & -\frac{1}{\sqrt{2}} & 0 & 0 & 0 \\ 0 & \frac{1}{\sqrt{2}} & 0 & 0 & 0 & -\frac{1}{\sqrt{2}} & 0 \end{pmatrix}, \tag{14.19}$$

since

$$\mathfrak{M}^T \mathfrak{M} = \mathscr{C}. \tag{14.20}$$

Defining the transformed tensor

$$\varphi_{ijk} \equiv (\mathfrak{M}^{-1})_i^{\ I} (\mathfrak{M}^{-1})_j^{\ J} (\mathfrak{M}^{-1})_k^{\ K} \phi_{IJK} \tag{14.21}$$

we can explicitly verify that

$$\begin{aligned} \varphi_{ijk} &= (\mathscr{R})_i^{\ p} (\mathscr{R})_j^{\ q} (\mathscr{R})_k^{\ r} \varphi_{pqr}, \\ \varphi_{ijk} &= (\mathscr{S})_i^{\ p} (\mathscr{S})_j^{\ q} (\mathscr{S})_k^{\ r} \varphi_{pqr}, \\ \varphi_{ijk} &= (\mathscr{T})_i^{\ p} (\mathscr{T})_j^{\ q} (\mathscr{T})_k^{\ r} \varphi_{pqr}. \end{aligned} \tag{14.22}$$

Hence, being preserved by the three generators $\mathscr{R}$, $\mathscr{S}$, and $\mathscr{T}$, the antisymmetric tensor φ_{ijk} is preserved by the entire discrete group L_{168} which, henceforth, is a subgroup of $G_{(2,-14)} \subset SO(7)$, as shown by intrinsic group theoretical arguments in [109]. The other representations of the group L_{168} were explicitly constructed about 18 years ago by Pierre Ramond and his younger collaborators in [112]. They are completely specified by giving the matrix form of the three generators R, S, and T satisfying the defining relations (14.4).

14.1.4 The three-dimensional complex representations

The two three-dimensional irreducible representations are complex and they are conjugate to each other. It suffices to give the form of the generators for one of them. The generators of the conjugate representation are the complex conjugates of the same matrices.

Setting

$$\rho \equiv e^{\frac{2i\pi}{7}}, \tag{14.23}$$

we have the following form for the representation **3**:

$$D[R]_3 = \begin{pmatrix} \frac{i(\rho^2-\rho^5)}{\sqrt{7}} & \frac{i(\rho-\rho^6)}{\sqrt{7}} & \frac{i(\rho^4-\rho^3)}{\sqrt{7}} \\ \frac{i(\rho-\rho^6)}{\sqrt{7}} & \frac{i(\rho^4-\rho^3)}{\sqrt{7}} & \frac{i(\rho^2-\rho^5)}{\sqrt{7}} \\ \frac{i(\rho^4-\rho^3)}{\sqrt{7}} & \frac{i(\rho^2-\rho^5)}{\sqrt{7}} & \frac{i(\rho-\rho^6)}{\sqrt{7}} \end{pmatrix},$$

$$D[S]_3 = \begin{pmatrix} \frac{i(\rho^3-\rho^6)}{\sqrt{7}} & \frac{i(\rho^3-\rho)}{\sqrt{7}} & \frac{i(\rho-1)}{\sqrt{7}} \\ \frac{i(\rho^2-1)}{\sqrt{7}} & \frac{i(\rho^6-\rho^5)}{\sqrt{7}} & \frac{i(\rho^6-\rho^2)}{\sqrt{7}} \\ \frac{i(\rho^5-\rho^4)}{\sqrt{7}} & \frac{i(\rho^4-1)}{\sqrt{7}} & \frac{i(\rho^5-\rho^3)}{\sqrt{7}} \end{pmatrix}, \qquad (14.24)$$

$$D[T]_3 = \begin{pmatrix} -ie^{\frac{3i\pi}{14}} & 0 & 0 \\ 0 & -ie^{-\frac{i\pi}{14}} & 0 \\ 0 & 0 & -e^{-\frac{i\pi}{7}} \end{pmatrix}.$$

14.1.5 The six-dimensional representation

We introduce the following shorthand notation:

$$c_n = \cos\left[\frac{2\pi}{7}n\right], \qquad s_n = \sin\left[\frac{2\pi}{7}n\right]. \qquad (14.25)$$

The generators of the group L_{168} in the six-dimensional irreducible representation can be explicitly written as follows:

$$D[R]_6 = \begin{pmatrix} \frac{c_3-1}{\sqrt{2}} & \frac{c_2-1}{\sqrt{2}} & \frac{c_1-1}{\sqrt{2}} & c_3-c_1 & c_1-c_2 & c_2-c_3 \\ \frac{c_2-1}{\sqrt{2}} & \frac{c_1-1}{\sqrt{2}} & \frac{c_3-1}{\sqrt{2}} & c_2-c_3 & c_3-c_1 & c_1-c_2 \\ \frac{c_1-1}{\sqrt{2}} & \frac{c_3-1}{\sqrt{2}} & \frac{c_2-1}{\sqrt{2}} & c_1-c_2 & c_2-c_3 & c_3-c_1 \\ c_3-c_1 & c_2-c_3 & c_1-c_2 & \frac{c_1-1}{\sqrt{2}} & \frac{c_2-1}{\sqrt{2}} & \frac{c_3-1}{\sqrt{2}} \\ c_1-c_2 & c_3-c_1 & c_2-c_3 & \frac{c_2-1}{\sqrt{2}} & \frac{c_3-1}{\sqrt{2}} & \frac{c_1-1}{\sqrt{2}} \\ c_2-c_3 & c_1-c_2 & c_3-c_1 & \frac{c_3-1}{\sqrt{2}} & \frac{c_1-1}{\sqrt{2}} & \frac{c_2-1}{\sqrt{2}} \end{pmatrix},$$

$$D[S]_6 = \begin{pmatrix} \frac{(c_3-1)\rho^2}{\sqrt{2}} & \frac{(c_2-1)\rho^4}{\sqrt{2}} & \frac{(c_1-1)\rho}{\sqrt{2}} & (c_3-c_1)\rho^3 & (c_1-c_2)\rho^5 & (c_2-c_3)\rho^6 \\ \frac{(c_2-1)\rho^2}{\sqrt{2}} & \frac{(c_1-1)\rho^4}{\sqrt{2}} & \frac{(c_3-1)\rho}{\sqrt{2}} & (c_2-c_3)\rho^3 & (c_3-c_1)\rho^5 & (c_1-c_2)\rho^6 \\ \frac{(c_1-1)\rho^2}{\sqrt{2}} & \frac{(c_3-1)\rho^4}{\sqrt{2}} & \frac{(c_2-1)\rho}{\sqrt{2}} & (c_1-c_2)\rho^3 & (c_2-c_3)\rho^5 & (c_3-c_1)\rho^6 \\ (c_3-c_1)\rho^2 & (c_2-c_3)\rho^4 & (c_1-c_2)\rho & \frac{(c_1-1)\rho^3}{\sqrt{2}} & \frac{(c_2-1)\rho^5}{\sqrt{2}} & \frac{(c_3-1)\rho^6}{\sqrt{2}} \\ (c_1-c_2)\rho^2 & (c_3-c_1)\rho^4 & (c_2-c_3)\rho & \frac{(c_2-1)\rho^3}{\sqrt{2}} & \frac{(c_3-1)\rho^5}{\sqrt{2}} & \frac{(c_1-1)\rho^6}{\sqrt{2}} \\ (c_2-c_3)\rho^2 & (c_1-c_2)\rho^4 & (c_3-c_1)\rho & \frac{(c_3-1)\rho^3}{\sqrt{2}} & \frac{(c_1-1)\rho^5}{\sqrt{2}} & \frac{(c_2-1)\rho^6}{\sqrt{2}} \end{pmatrix},$$

$$D[T]_6 = (D[R]_6 \cdot D[S]_6)^{-1}. \qquad (14.26)$$

14.1.6 The eight-dimensional representation

Utilizing the same notations as before we can write the matrix form of the generators also in the irreducible eight-dimensional representation:

$$
\begin{aligned}
&D[R]_8 = \\
&\begin{pmatrix}
2-2c_1 & 0 & 2c_1+2c_2-4c_3 & 2-2c_2 & 0 & 2-2c_3 & 0 & 2\sqrt{3}c_1-2\sqrt{3}c_2\\
0 & -2c_1+4c_2-2 & 0 & 0 & 2c_2-4c_3+2 & 0 & 4c_1-2c_3-2 & 0\\
2c_1+2c_2-4c_3 & 0 & -c_1+2c_2-c_3 & -4c_1+2c_2+2c_3 & 0 & 2c_1-4c_2+2c_3 & 0 & \sqrt{3}c_1-\sqrt{3}c_3\\
2-2c_2 & 0 & -4c_1+2c_2+2c_3 & 2-2c_3 & 0 & 2-2c_1 & 0 & 2\sqrt{3}c_2-2\sqrt{3}c_3\\
0 & 2c_2-4c_3+2 & 0 & 0 & 4c_1-2c_3-2 & 0 & 2c_1-4c_2+2 & 0\\
2-2c_3 & 0 & 2c_1-4c_2+2c_3 & 2-2c_1 & 0 & 2-2c_2 & 0 & 2\sqrt{3}c_3-2\sqrt{3}c_1\\
0 & 4c_1-2c_3-2 & 0 & 0 & 2c_1-4c_2+2 & 0 & -2c_2+4c_3-2 & 0\\
2\sqrt{3}c_1-2\sqrt{3}c_2 & 0 & \sqrt{3}c_1-\sqrt{3}c_3 & 2\sqrt{3}c_2-2\sqrt{3}c_3 & 0 & 2\sqrt{3}c_3-2\sqrt{3}c_1 & 0 & c_1-2c_2+c_3
\end{pmatrix},\\
&D[S]_8 = \begin{pmatrix}
c_1 & s_1 & 0 & 0 & 0 & 0 & 0 & 0\\
-s_1 & c_1 & 0 & 0 & 0 & 0 & 0 & 0\\
0 & 0 & 1 & 0 & 0 & 0 & 0 & 0\\
0 & 0 & 0 & c_3 & s_3 & 0 & 0 & 0\\
0 & 0 & 0 & -s_3 & c_3 & 0 & 0 & 0\\
0 & 0 & 0 & 0 & 0 & c_2 & s_2 & 0\\
0 & 0 & 0 & 0 & 0 & -s_2 & c_2 & 0\\
0 & 0 & 0 & 0 & 0 & 0 & 0 & 1
\end{pmatrix},\\
&D[T]_8 = \left(D[R]_8 \cdot D[S]_8\right)^{-1}.
\end{aligned}
\tag{14.27}
$$

14.1.7 The proper subgroups of L_{168}

From the complexity of the other irreps, compared with the simplicity of the seven-dimensional one, it is already clear that this latter should be considered the natural defining representation. The crystallographic nature of the group in $d = 7$ has already been stressed. We introduce the $\mathfrak{a}_7$ weight lattice which, by definition, is just the dual of the root lattice. Explicitly

$$\Lambda_w \ni \mathbf{w} = n_i \lambda^i : \quad n^i \in \mathbb{Z} \tag{14.28}$$

is spanned by the simple weights that are implicitly defined by the relations

$$\lambda^i \cdot \alpha_j = \delta^i_j \quad \Rightarrow \quad \lambda^i = (\mathscr{C}^{-1})^{ij} \alpha_j. \tag{14.29}$$

Since the group L_{168} is crystallographic on the root lattice, by necessity it is crystallographic also on the weight lattice. Given the generators of the group L_{168} in the basis of simple roots we obtain the same in the basis of simple weights through the following transformation:

$$\mathscr{R}_W = \mathscr{C}\mathscr{R}\mathscr{C}^{-1}; \quad \mathscr{S}_W = \mathscr{C}\mathscr{S}\mathscr{C}^{-1}; \quad \mathscr{T}_W = \mathscr{C}\mathscr{T}\mathscr{C}^{-1}. \tag{14.30}$$

Explicitly we find

$$\mathscr{R}_W = \begin{pmatrix} 0 & 0 & 0 & 0 & 0 & 0 & -1 \\ 0 & 0 & 0 & -1 & -1 & -1 & 0 \\ 0 & 0 & -1 & 0 & 0 & 0 & 0 \\ 0 & 0 & 1 & 1 & 1 & 0 & 0 \\ 0 & 0 & 0 & 0 & -1 & 0 & 0 \\ 0 & -1 & -1 & -1 & 0 & 0 & 0 \\ -1 & 0 & 0 & 0 & 0 & 0 & 0 \end{pmatrix}, \tag{14.31}$$

$$\mathscr{S}_W = \begin{pmatrix} -1 & -1 & -1 & -1 & -1 & -1 & -1 \\ 1 & 1 & 1 & 1 & 0 & 0 & 0 \\ 0 & 0 & 0 & -1 & 0 & 0 & 0 \\ 0 & 0 & 0 & 1 & 1 & 1 & 0 \\ 0 & 0 & 0 & 0 & 0 & -1 & 0 \\ 0 & 0 & -1 & -1 & -1 & 0 & 0 \\ 0 & -1 & 0 & 0 & 0 & 0 & 0 \end{pmatrix}, \tag{14.32}$$

$$\mathscr{T}_W = \begin{pmatrix} -1 & -1 & -1 & -1 & -1 & -1 & 0 \\ 1 & 0 & 0 & 0 & 0 & 0 & 0 \\ 0 & 1 & 0 & 0 & 0 & 0 & 0 \\ 0 & 0 & 1 & 0 & 0 & 0 & 0 \\ 0 & 0 & 0 & 1 & 0 & 0 & 0 \\ 0 & 0 & 0 & 0 & 1 & 0 & 0 \\ 0 & 0 & 0 & 0 & 0 & 1 & 1 \end{pmatrix}. \tag{14.33}$$

Given the weight basis, which is useful in several constructions, let us conclude our survey of the remarkable simple group L_{168} by a brief discussion of its subgroups, none of which, obviously, is normal.

L_{168} contains maximal subgroups only of index 8 and 7, namely, of order 21 and 24. The order 21 subgroup G_{21} is the unique non-Abelian group of that order and abstractly it has the structure of the semi-direct product $\mathbb{Z}_3 \ltimes \mathbb{Z}_7$. We already met this abstract group in Section 8.4.1.2. Up to conjugation there is only one subgroup G_{21}. On the other hand, up to conjugation, there are two different groups of order 24 that are both isomorphic to the octahedral group O_{24}.

14.1.7.1 The maximal subgroup G_{21}

The group G_{21} has two generators $\mathscr{X}$ and $\mathscr{Y}$ that satisfy the following relations:

$$\mathscr{X}^3 = \mathscr{Y}^7 = \mathbf{1}; \quad \mathscr{X}\mathscr{Y} = \mathscr{Y}^2\mathscr{X}. \tag{14.34}$$

Comparing the above presentation (14.34) with equation (8.73) we realize that the subgroup $G_{21} \subset L_{168}$ is indeed isomorphic, as claimed above, with the abstract group studied in Section 8.4.1.2 as a non abelian group extension of $\mathbb{Z}_7$. The organization of the 21 group elements into conjugacy classes is displayed below:

Conjugacy class	C_1	C_2	C_3	C_4	C_5
representative of the class	e	$\mathcal{Y}$	$\mathcal{X}^{-2}\mathcal{Y}\mathcal{X}\mathcal{Y}^2$	$\mathcal{Y}\mathcal{X}^{-2}$	$\mathcal{X}$
order of the elements in the class	1	7	7	3	3
number of elements in the class	1	3	3	7	7

(14.35)

As we see there are five conjugacy classes, which implies that there should be five irreducible representations the square of whose dimensions should sum up to the group order 21. The solution of this problem is

$$21 = 1^2 + 1^2 + 1^2 + 3^2 + 3^2 \tag{14.36}$$

and the corresponding character table is given below:

	e	$\mathcal{Y}$	$\mathcal{X}^2\mathcal{Y}\mathcal{X}\mathcal{Y}^2$	$\mathcal{Y}\mathcal{X}^2$	$\mathcal{X}$
$D_1[G_{21}]$	1	1	1	1	1
$DX_1[G_{21}]$	1	1	1	$-(-1)^{1/3}$	$(-1)^{2/3}$
$DY_1[G_{21}]$	1	1	1	$(-1)^{2/3}$	$-(-1)^{1/3}$
$DA_3[G_{21}]$	3	$\frac{1}{2}i(i+\sqrt{7})$	$-\frac{1}{2}i(-i+\sqrt{7})$	0	0
$DB_3[G_{21}]$	3	$-\frac{1}{2}i(-i+\sqrt{7})$	$\frac{1}{2}i(i+\sqrt{7})$	0	0

(14.37)

In the weight basis the two generators of the G_{21} subgroup of L_{168} can be chosen to be the following matrices, and this fixes our representative of the unique conjugacy class:

$$\begin{aligned}
\mathcal{X} &= \begin{pmatrix} 1 & 1 & 1 & 1 & 1 & 1 & 1 \\ 0 & 0 & 0 & 0 & 0 & 0 & -1 \\ 0 & -1 & -1 & -1 & -1 & -1 & 0 \\ 0 & 1 & 1 & 1 & 0 & 0 & 0 \\ 0 & 0 & -1 & -1 & 0 & 0 & 0 \\ 0 & 0 & 1 & 1 & 1 & 0 & 0 \\ 0 & 0 & 0 & -1 & -1 & 0 & 0 \end{pmatrix}, \\
\mathcal{Y} &= \begin{pmatrix} 0 & 1 & 1 & 0 & 0 & 0 & 0 \\ 0 & 0 & 0 & 1 & 1 & 1 & 1 \\ 0 & 0 & -1 & -1 & -1 & -1 & -1 \\ 0 & 0 & 1 & 1 & 0 & 0 & 0 \\ -1 & -1 & -1 & -1 & 0 & 0 & 0 \\ 1 & 1 & 1 & 1 & 1 & 0 & 0 \\ 0 & 0 & 0 & 0 & 0 & 1 & 0 \end{pmatrix}.
\end{aligned} \tag{14.38}$$

14.1.7.2 The maximal subgroups O_{24A} and O_{24B}

As we know from Section 4.4.2, the octahedral group O_{24} has two generators S and T that satisfy the following relations:

$$S^2 = T^3 = (ST)^4 = \mathbf{1}. \tag{14.39}$$

The 24 elements are organized in five conjugacy classes according to the scheme displayed below:

Conjugacy class	C_1	C_2	C_3	C_4	C_5
representative of the class	e	T	$STST$	S	ST
order of the elements in the class	1	3	2	2	4
number of elements in the class	1	8	3	6	6

(14.40)

The irreducible representations of O_{24} were explicitly constructed in Section 8.2.2. We repeat here the corresponding character table mentioning also a standard representative of each conjugacy class:

	e	T	$STST$	S	ST
$D_1[O_{24}]$	1	1	1	1	1
$D_2[O_{24}]$	1	1	1	−1	−1
$D_3[O_{24}]$	2	−1	2	0	0
$D_4[O_{24}]$	3	0	−1	−1	1
$D_5[O_{24}]$	3	0	−1	1	−1

(14.41)

By computer calculations we have verified that there are just two disjoint conjugacy classes of O_{24} maximal subgroups in L_{168}, which we have named A and B, respectively. We have chosen two standard representatives, one for each conjugacy class, which we have named O_{24A} and O_{24B}, respectively. To fix these subgroups it suffices to mention the explicit form of their generators in the weight basis.

For the group O_{24A}, we choose

$$T_A = \begin{pmatrix} 1 & 1 & 1 & 1 & 1 & 1 & 1 \\ 0 & 0 & 0 & 0 & 0 & 0 & -1 \\ 0 & -1 & -1 & -1 & -1 & -1 & 0 \\ 0 & 1 & 1 & 1 & 0 & 0 & 0 \\ 0 & 0 & -1 & -1 & 0 & 0 & 0 \\ 0 & 0 & 1 & 1 & 1 & 0 & 0 \\ 0 & 0 & 0 & -1 & -1 & 0 & 0 \end{pmatrix}, \quad S_A = \begin{pmatrix} 0 & 0 & 0 & 1 & 1 & 1 & 0 \\ 0 & 0 & 0 & 0 & -1 & -1 & 0 \\ -1 & -1 & -1 & -1 & 0 & 0 & 0 \\ 1 & 1 & 0 & 0 & 0 & 0 & 0 \\ 0 & 0 & 1 & 1 & 1 & 1 & 1 \\ 0 & -1 & -1 & -1 & -1 & -1 & -1 \\ 0 & 1 & 1 & 1 & 1 & 0 & 0 \end{pmatrix}. \tag{14.42}$$

For the group O_{24B}, we choose

$$T_B = \begin{pmatrix} 1 & 1 & 1 & 1 & 0 & 0 & 0 \\ 0 & -1 & -1 & -1 & 0 & 0 & 0 \\ 0 & 1 & 1 & 1 & 1 & 0 & 0 \\ 0 & 0 & -1 & -1 & -1 & 0 & 0 \\ 0 & 0 & 1 & 1 & 1 & 1 & 0 \\ 0 & 0 & 0 & -1 & -1 & -1 & 0 \\ 0 & 0 & 0 & 1 & 1 & 1 & 1 \end{pmatrix},$$
$$S_B = \begin{pmatrix} 0 & 0 & 1 & 1 & 1 & 0 & 0 \\ -1 & -1 & -1 & -1 & -1 & 0 & 0 \\ 1 & 1 & 1 & 1 & 1 & 1 & 1 \\ 0 & 0 & 0 & 0 & 0 & 0 & -1 \\ 0 & -1 & -1 & -1 & -1 & -1 & 0 \\ 0 & 1 & 1 & 1 & 0 & 0 & 0 \\ 0 & 0 & 0 & -1 & 0 & 0 & 0 \end{pmatrix}. \tag{14.43}$$

14.1.7.3 The tetrahedral subgroup $T_{12} \subset O_{24}$

Every octahedral group O_{24} has, up to O_{24}-conjugation, a unique tetrahedral subgroup T_{12} whose order is 12. The abstract description of the tetrahedral group is provided by the following presentation in terms of two generators:

$$T_{12} = (s, t \mid s^2 = t^3 = (st)^3 = 1). \tag{14.44}$$

The 12 elements are organized into four conjugacy classes as displayed below:

Class	C_1	C_2	C_3	C_4
standard representative	1	s	t	t^2s
order of the elements in the conjugacy class	1	2	3	3
number of elements in the conjugacy class	1	3	4	4

(14.45)

The two tetrahedral subgroups $T_{12A} \subset O_{24A}$ and $T_{12B} \subset O_{24B}$ are not conjugate under the big group L_{168}. Hence, we have two conjugacy classes of tetrahedral subgroups of L_{168}.

14.1.7.4 The dihedral subgroup $\mathrm{Dih}_3 \subset O_{24}$

Every octahedral group O_{24} has a dihedral subgroup Dih_3 whose order is six. The abstract description of the dihedral group Dih_3 is provided by the following presentation in terms of two generators:

$$\mathrm{Dih}_3 = (A, B \mid A^3 = B^2 = (BA)^2 = 1). \tag{14.46}$$

The six elements are organized into three conjugacy classes as displayed below:

Conjugacy class	C_1	C_2	C_3
standard representative of the class	1	A	B
order of the elements in the class	1	3	2
number of elements in the class	1	2	3

(14.47)

Differently from the case of the tetrahedral subgroups, the two dihedral subgroups $\mathrm{Dih}_{3A} \subset \mathrm{O}_{24A}$ and $\mathrm{Dih}_{3B} \subset \mathrm{O}_{24B}$ turn out to be conjugate under the big group L_{168}. Actually there is just one L_{168}-conjugacy class of dihedral subgroups Dih_3.

14.1.7.5 Enumeration of the possible subgroups and orbits

In $d = 3$ the orbits of the octahedral group acting on the cubic lattice are the vertices of regular geometrical figures. Since L_{168} has a crystallographic action on the mentioned seven-dimensional weight lattice, its orbits $\mathcal{O}$ in Λ_w correspond to the analog regular geometrical figures in $d = 7$. Every orbit is in correspondence with a coset G/H, where G is the big group and H is one of its possible subgroups. Indeed, H is the stability subgroup of an element of the orbit.

Since the maximal subgroups of L_{168} are of index 7 or 8, we can have subgroups $\mathrm{H} \subset \mathrm{L}_{168}$ that are either G_{21} or O_{24} or subgroups thereof. Furthermore, as is well known, the order $|\mathrm{H}|$ of any subgroup $\mathrm{H} \subset \mathrm{G}$ must be a divisor of $|\mathrm{G}|$. Hence, we conclude that

$$|\mathrm{H}| \in \{1, 2, 3, 4, 6, 7, 8, 12, 21, 24\}. \tag{14.48}$$

Correspondingly we might have L_{168}-orbits $\mathcal{O}$ in the weight lattice Λ_w, whose length is one of the following nine numbers:

$$\ell_\mathcal{O} \in \{168, 84, 56, 42, 28, 24, 21, 14, 8, 7\}. \tag{14.49}$$

Combining the information about the possible group orders in (14.48) with the fact that the maximal subgroups are of index 8 or 7, we arrive at the following list of possible subgroups H (up to conjugation) of the group L_{168}:

Order 24: Either $\mathrm{H} = \mathrm{O}_{24A}$ or $\mathrm{H} = \mathrm{O}_{24B}$.
Order 21: The only possibility is $\mathrm{H} = \mathrm{G}_{21}$.
Order 12: The only possibilities are $\mathrm{H} = \mathrm{T}_{12A}$ or $\mathrm{H} = \mathrm{T}_{12B}$, where T_{12} is the tetrahedral subgroup of the octahedral group O_{24}.
Order 8: Either $\mathrm{H} = \mathbb{Z}_2 \times \mathbb{Z}_2 \times \mathbb{Z}_2$ or $\mathrm{H} = \mathbb{Z}_2 \times \mathbb{Z}_4$.
Order 7: The only possibility is $\mathbb{Z}_7$.
Order 6: Either $\mathrm{H} = \mathbb{Z}_2 \times \mathbb{Z}_3$ or $\mathrm{H} = \mathrm{Dih}_3$, where Dih_3 denotes the dihedral subgroup of index 3 of the octahedral group O_{24}.
Order 4: Either $\mathrm{H} = \mathbb{Z}_2 \times \mathbb{Z}_2$ or $\mathrm{H} = \mathbb{Z}_4$.
Order 3: The only possibility is $\mathrm{H} = \mathbb{Z}_3$
Order 2: The only possibility is $\mathrm{H} = \mathbb{Z}_2$.

14.1.7.6 Synopsis of the L_{168} orbits in the weight lattice Λ_w

In [110], the present author presented his results, obtained by means of computer calculations, on the orbits of the considered simple group acting on the $\mathfrak{a}_7$ weight lattice. They are briefly summarized below:

1. Orbits of length 8 (one parameter **n**; stability subgroup $H^s = G_{21}$).
2. Orbits of length 14 (two types, A and B) (one parameter **n**; stability subgroup $H^s = T_{12A,B}$).
3. Orbits of length 28 (one parameter **n**; stability subgroup $H^s = \mathrm{Dih}_3$).
4. Orbits of length 42 (one parameter **n**; stability subgroup $H^s = \mathbb{Z}_4$).
5. Orbits of length 56 (three parameters **n, m, p**; stability subgroup $H^s = \mathbb{Z}_3$).
6. Orbits of length 84 (three parameters **n, m, p**; stability subgroup $H^s = \mathbb{Z}_2$).
7. Generic orbits of length 168 (seven parameters; stability subgroup $H^s = \mathbf{1}$).

As we already said, the above list is in some sense the seven-dimensional analog of Platonic solids; only in some sense, since it is a complete classification for the group L_{168}, yet we are not aware of a classification of the other crystallographic subgroups of SO(7), if any exist.

Notwithstanding this ignorance, the piece of knowledge we have summarized above is already impressively complicated and demonstrates how even flat geometry becomes more sophisticated in higher dimensions.

The next natural question is why just $d = 7$ should attract our geometrical attention. There are several mathematical reasons for the number 7. They are probably all related to each other:

1. The possible division algebras are $\mathbb{R}$, $\mathbb{C}$, $\mathbb{H}$, and $\mathbb{O}$, that is, the real numbers, the complex numbers, the quaternions, and the octonions. The corresponding number of imaginary units is zero, one, three, and seven, respectively. The automorphism groups of these division algebras are 1, U(1), SU(2), and $G_{2(-14)}$, respectively.
2. The spheres that are globally parallelizable are $\mathbb{S}^1$, $\mathbb{S}^3$, and $\mathbb{S}^7$.
3. The manifolds of restricted holonomy are the complex ones, the Kähler ones, the quaternionic ones, which exist in all dimensions $d = 2n$, respectively $d = 4n$, and then, just in $d = 7$, we have the G_2 manifolds and in $d = 8$ we have the Spin(7) manifolds.
4. Seven are the dimensions that one has to compactify in order to step down from the 11-dimensional M-theory to our $d = 4$ space-time, and many solutions of the theory naturally perform the splitting $11 = 4 + 7$.

14.2 Bibliographical note

The bulk of the material on the group PSL(2, 7) comes from the papers [110, 108].

15 A primary on the theory of connections and metrics

We must admit with humility that, while number is purely a product of our minds, space has a reality outside our minds, so that we cannot completely prescribe its properties a priori . . .

Carl Friedrich Gauss

15.1 Introduction

The present chapter deals with the second stage in the development of modern differential geometry. Once the notions of manifold and fiber bundle have been introduced, the latter requiring Lie groups as fundamental ingredients, one considers differential calculus on these spaces and rather than ordinary derivatives one utilizes *covariant derivatives*. In general relativity one mainly uses the covariant derivative on the tangent bundle, but it is important to realize that we can define covariant derivatives on general fiber bundles. Indeed, the covariant derivative is the physicist's name for the mathematical concept of *connection* that we are going to introduce and illustrate in this chapter. It is also important to stress that even restricting one's attention to the tangent bundle, the connection used in general relativity is just a particular one, the so-called *Levi-Civita connection*, which arises from a more fundamental object, the *metric*. As we are going to see soon, a manifold endowed with a *metric structure* is a space where one can measure lengths, specifically the length of all curves.

We should also stress that manifolds with a metric structure are not necessarily geometrical models of the physical space. For instance in the geometrical approach to thermodynamics quoted in the preface [113, 114, 115, 116, 117, 118, 119, 120, 121, 122, 123], the metric manifold object of study is the space of thermodynamical states singled out by pressure, temperature, and chemical potentials.

In other fields of science, the manifolds or fiber bundles on which we construct a metric or just a connection can have the most diversified interpretations; one noticeable instance of applications is given by the recent and fast developing field of geometric deep learning, also mentioned in the preface [29, 30, 31, 32, 33, 34].

A generic connection on the tangent bundle is named *an affine connection*, and the Levi-Civita connection is a specific affine connection that is uniquely determined by the *metric structure*. As we shall presently illustrate, every connection has associated with it another object (actually a 2-form) that we call its curvature. The curvature of the Levi-Civita connection is what we call the Riemann curvature of a manifold and it is the main concern of general relativity. It encodes the intuitive geometrical notion of curvature of a surface or hypersurface. The field equations of Einstein's theory are statements about the Riemann curvature of space-time, which is related to its energy-matter content. We should be aware that the notion of curvature applies to generic connections

https://doi.org/10.1515/9783111201535-015

on generic fiber bundles, in particular on principal bundles. Physically these connections and curvatures are not less important than the Levi-Civita connection and the Riemann curvature. They constitute the main mathematical objects entering the theory of fundamental non-gravitational interactions, which are all *gauge theories*. The latter constitute the *meta-theory*, which contains the *standard model of fundamental interactions*, namely, our current and completely successful theory of the physical world, yet, as the title of the recent review paper [29] on *geometric deep learning* suggests, the words *metric*, *geodesics*, *connection*, and *gauges* have become very popular in the field of *artificial intelligence*, which is applied to a lot of different technological environments, including *pattern recognition*, *robotic surgery*, *automatic driving*, and *automatic flying*.

Similarly, in other applied fields, the manifold can be the continuous space that encapsulates the possible states of the most diversified complex systems (*mechanical*, *economical*, *environmental*, and so on): as long as there is a reasonable notion of infinitesimal distance and we can define the length of paths inside such an abstract space, we arrive at metrics and Riemannian geometry, or if we just have the action of some groups of transformations mapping the space in itself, we naturally land on the setup of fiber bundles and we come up with the theory of connections.

The term that should be the *sure and winning guide* of the modern scientist, technologist, and engineer is *abstract thinking*. The *material problems* of any sort should be conceptualized and *lifted to the abstract mathematical level*. Once lifted to that level the problems at stake can be addressed in a universal way and the solutions can be projected back to the pragmatic world in a diversified way. Who cares what is the actual interpretation of the points of a manifold? Once the manifold is clearly identified in its proper mathematical environment the pertaining mathematical structures are naturally developed and, at the end of the day, one can ask what, in each field of application, the interpretation of the universal results might be. So, for instance, in the geometric approach to thermodynamics, the scalar curvature associated with the natural metric provided by *equations of state* turns out to provide a measure of the *correlation length* between the gas or liquid particles and hence it provides information on the intermolecular forces. *The same equations have the same solutions* was a famous aphorism of Richard Feynman.

15.2 Connections on principal bundles: the mathematical definition

We come to the contemporary mathematical definition of a connection on a principal fiber bundle. The adopted viewpoint is that of Ehresmann, and what we are going to define is usually named an *Ehresmann connection* in the mathematical literature.

15.2.1 Ehresmann connections on a principal fiber bundle

Let $P(\mathcal{M}, G)$ be a principal fiber bundle with base manifold $\mathcal{M}$ and structural group G. Let us moreover denote by π the projection

$$\pi : P \to \mathcal{M}. \tag{15.1}$$

Consider the action of the Lie group G on the total space P:

$$\forall g \in G, \quad g : G \to G. \tag{15.2}$$

By definition this action is vertical in the sense that

$$\forall u \in P, \forall g \in G : \quad \pi(g(u)) = \pi(u), \tag{15.3}$$

namely, it moves points only along the fibers. Given any element $X \in \mathbb{G}$, where we have denoted by $\mathbb{G}$ the Lie algebra of the structural group, we can consider the one-dimensional subgroup generated by it,

$$g_X(t) = \exp[tX], \quad t \in \mathbb{R}, \tag{15.4}$$

and consider the curve $\mathcal{C}_X(t, u)$ in the manifold P obtained by acting with $g_X(t)$ on some point $u \in P$:

$$\mathcal{C}_X(t, u) \equiv g_X(t)(u). \tag{15.5}$$

The vertical action of the structural group implies that

$$\pi(\mathcal{C}_X(t, u)) = p \in \mathcal{M} \quad \text{if } \pi(u) = p. \tag{15.6}$$

These items allow us to construct a linear map # which associates a vector field $\mathbf{X}^\#$ over P with every element X of the Lie algebra $\mathbb{G}$:

$$\begin{aligned} &\# : \mathbb{G} \ni X \to \mathbf{X}^\# \in \Gamma(TP, P), \\ &\forall f \in C^\infty(P), \quad \mathbf{X}^\# f(u) \equiv \frac{d}{dt} f(\mathcal{C}_X(t, u))\Big|_{t=0}. \end{aligned} \tag{15.7}$$

Focusing on any point $u \in P$ of the bundle, the map (15.7) reduces to a map

$$\#_u : \mathbb{G} \to T_u P \tag{15.8}$$

from the Lie algebra to the tangent space at u. The map $\#_u$ is not surjective. We introduce the following definition.

Definition 15.2.1. At any point $p \in P(M,G)$ of a principal fiber bundle, by vertical subspace $V_u P$ of the tangent space $T_u P$ we mean the image of the map $\#_u$:

$$T_u P \supset V_u \equiv \text{Im}\,\#_u. \tag{15.9}$$

Indeed, any vector $t \in \text{Im}\,\#_u$ lies in the tangent space to the fiber G_p, where $p \equiv \pi(u)$.

The meaning of Definition 15.2.1 becomes clearer if we consider both Figure 15.1 and a local trivialization of the bundle including $u \in P$. In such a local trivialization the bundle point u is identified by a pair,

$$u \overset{\text{loc. triv.}}{\longrightarrow} (p, f), \quad \text{where} \begin{cases} p \in \mathscr{M}, \\ f \in G, \end{cases} \tag{15.10}$$

and the curve $\mathscr{C}_X(t,u)$ takes the following appearance:

$$\mathscr{C}_X(t,u) \overset{\text{loc. triv.}}{\longrightarrow} (p, e^{tX} f), \quad p = \pi(u). \tag{15.11}$$

Correspondingly, calling a^i the group parameters and x^μ the coordinates on the base space $\mathscr{M}$, a generic function $f \in C^\infty(P)$ is just a function $f(x,\alpha)$ of the m coordinates x and of the n parameters α. Comparing equation (15.7) with equation (7.24) we see that, in the local trivialization, the vertical tangent vector $\mathbf{X}^\#$ reduces to nothing else but the right-invariant vector field on the group manifold associated with the same Lie algebra element X:

$$\mathbf{X}^\# \overset{\text{loc. triv.}}{\longrightarrow} \mathbf{X}_R = X_R^i(\alpha) \frac{\overrightarrow{\partial}}{\partial \alpha^i}. \tag{15.12}$$

As we see from equation (15.12), in a local trivialization a vertical vector contains no derivatives with respect to the base space coordinates. This means that its projection onto the base manifold is zero. Indeed, it follows from Definition 15.2.1 that

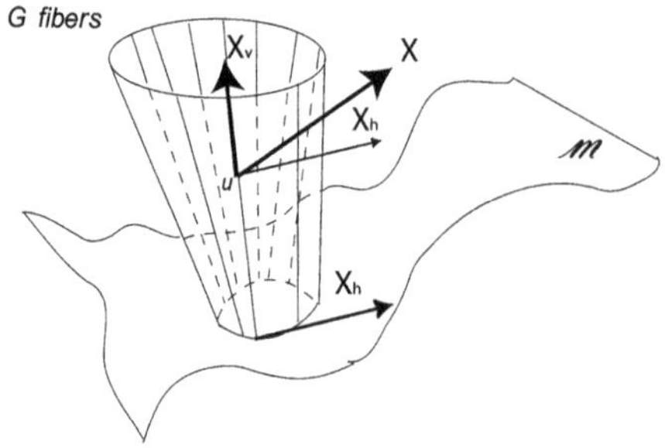

Figure 15.1: The tangent space to a principal bundle P splits at any point $u \in P$ into a vertical subspace along the fibers and a horizontal subspace parallel to the base manifold. This splitting is the intrinsic geometric meaning of a connection.

$$\mathbf{X} \in V_u P \quad \Rightarrow \quad \pi_* \mathbf{X} = 0, \tag{15.13}$$

where π_* is the push-forward of the projection map.

Having defined the vertical subspace of the tangent space one would be interested in giving a definition also of the horizontal subspace $H_u P$, which, intuitively, must be somehow parallel to the tangent space $T_p \mathscr{M}$ to the base manifold at the projection point $p = \pi(u)$. The dimension of the vertical space is the same as the dimension of the fiber, that is, $n = \dim G$. The dimension of the horizontal space $H_u P$ must be the same as the dimension of the base manifold, $m = \dim \mathscr{M}$. Indeed, $H_u P$ should be the orthogonal complement of $V_u P$. Easy to say, but orthogonal with respect to what? This is precisely the point. Is there an a priori intrinsically defined way of defining the orthogonal complement to the vertical subspace $V_u \subset T_u P$? The answer is that there is not. Given a basis $\{\mathbf{v}^\mu\}$ of n vectors for the subspace $V_u P$, there are infinitely many ways of finding m extra vectors $\{\mathbf{h}^i\}$ which complete this basis to a basis of $T_u P$. The span of any such collection of m vectors $\{\mathbf{h}^i\}$ is a possible legitimate definition of the orthogonal complement $H_u P$. This arbitrariness is the root of the mathematical notion of a *connection*. Providing a fiber bundle with a connection precisely means introducing a rule that uniquely defines the orthogonal complement $H_u P$.

Definition 15.2.2. Let $P(M, G)$ be a principal fiber bundle. A *connection* on P is a rule which at any point $u \in P$ defines a unique splitting of the tangent space $T_u P$ into the vertical subspace $V_u P$ and into a horizontal complement $H_u P$ satisfying the following properties:

(i) $T_u P = H_u P \oplus V_u P$.

(ii) Any smooth vector field $\mathbf{X} \in \Gamma(TP, P)$ separates into the sum of two smooth vector fields, $\mathbf{X} = \mathbf{X}^H + \mathbf{X}^V$, such that at any point $u \in P$ we have $\mathbf{X}_u^H \in H_u P$ and $\mathbf{X}_u^V \in V_u P$.

(iii) $\forall g \in G$ we have $H_{gu} P = L_{g*} H_u P$.

The third defining property of the connection states that all the horizontal spaces along the same fiber are related to each other by the push-forward of the left translation on the group manifold.

This beautiful purely geometrical definition of the connection due to Ehresmann emphasizes that it is just an intrinsic attribute of the principal fiber bundle. In a trivial bundle, which is a direct product of manifolds, the splitting between vertical and horizontal spaces is done once for ever. The vertical space is the tangent space to the fiber, the horizontal space is the tangent space to the base. In a non-trivial bundle the splitting between vertical and horizontal directions has to be reconsidered at every next point, and fixing this ambiguity is the task of the connection.

15.2.1.1 The connection 1-form

The algorithmic way to implement the splitting rule advocated by Definition 15.2.2 is provided by introducing a connection 1-form **A** which is just a Lie algebra valued 1-form on the bundle P satisfying two precise requirements.

Definition 15.2.3. Let $P(\mathcal{M}, G)$ be a principal fiber bundle with structural Lie group G, whose Lie algebra we denote by $\mathbb{G}$. A connection 1-form on this bundle is a Lie algebra valued section of the cotangent bundle, $\mathbf{A} \in \mathbb{G} \otimes \Gamma(T^*P, P)$, which satisfies the following defining properties:
(i) $\forall \mathbf{X} \in \mathbb{G} : \mathbf{A}(\mathbf{X}^\#) = \mathbf{X}$.
(ii) $\forall \mathfrak{g} \in G : L^*_{\mathfrak{g}}\mathbf{A} = \mathrm{Adj}_{\mathfrak{g}^{-1}}\mathbf{A} \equiv \mathfrak{g}^{-1}\mathbf{A}\mathfrak{g}$.

Given the connection 1-form **A** the splitting between vertical and horizontal subspaces is performed through the following.

Definition 15.2.4. At any $u \in P$, the horizontal subspace H_uP of the tangent space to the bundle is provided by the kernel of **A**:

$$H_uP \equiv \{\mathbf{X} \in T_uP \mid \mathbf{A}(\mathbf{X}) = 0\}. \tag{15.14}$$

The actual meaning of Definitions 15.2.3 and 15.2.4 and their coherence with Definition 15.2.2 become clear if we consider a local trivialization of the bundle.

Just as before, let x^μ be the coordinates on the base manifold and let α^i be the group parameters. With respect to this local patch, the connection 1-form **A** has the following appearance:

$$\mathbf{A} = \widehat{\mathscr{A}} + \mathrm{A}, \tag{15.15}$$

$$\widehat{\mathscr{A}} = \widehat{\mathscr{A}_\mu}\, dx^\mu = dx^\mu \mathscr{A}^I_\mu T_I, \tag{15.16}$$

$$\mathrm{A} = \mathrm{A}_i\, d\alpha^i = d\alpha^i\, \mathrm{A}^I_i T_I, \tag{15.17}$$

where, a priori, both the components $\widehat{\mathscr{A}_\mu}(x, \alpha)$ and $\mathrm{A}_i(x, \alpha)$ could depend on both x^μ and α^i. The second equality in equations (15.17) and (15.16) is due to the fact that $\mathscr{A}$ and A are Lie algebra valued, and hence they can be expanded along a basis of generators of $\mathbb{G}$. Actually the dependence on the group parameters or better on the group element $\mathfrak{g}(\alpha)$ is completely fixed by the defining axioms. Let us consider first the implications of property (i) in Definition 15.2.3. To this effect let $X \in \mathbb{G}$ be expanded along the same basis of generators, that is, $X = X^A T_A$. As we already remarked in equation (15.12), the image of X through the map # is $\mathbf{X}^\# = X^A \mathbf{T}^{(R)}_A$, where $\mathbf{T}^{(R)}_A = \overset{(r)}{\Sigma}{}^i_A(\alpha)\, \overrightarrow{\partial}/\partial\alpha^i$ are the right-invariant vector fields. Hence, property (i) requires

$$X^A \mathrm{A}^I_i\, \overset{(r)}{\Sigma}{}^i_A(\alpha)\, T_I = X^A T_A, \tag{15.18}$$

which implies

$$\mathrm{A}_i^I \overset{(r)}{\Sigma}{}^i_A(\alpha) = \delta^I_A. \tag{15.19}$$

Therefore, $\mathrm{A}_i^I = \overset{(r)}{\Sigma}{}^A_i(\alpha)$ and the vertical component of the 1-form connection is nothing else but the right-invariant 1-form:

$$\mathrm{A} = \sigma_{(R)} \equiv d\mathfrak{g}\mathfrak{g}^{-1}. \tag{15.20}$$

Consider now the implications of defining property (ii). In the chosen local trivialization the left action of any element γ of the structural group G is $L_\gamma : (x, \mathfrak{g}(\alpha)) \to (x, \gamma \cdot \mathfrak{g}(\alpha))$ and its pull-back acts on the right-invariant 1-form $\sigma_{(R)}$ by adjoint transformations:

$$L_\gamma^* : \sigma_{(R)} \to d(\gamma \cdot \mathfrak{g})(\gamma \cdot \mathfrak{g})^{-1} = \gamma \cdot \sigma_{(R)} \cdot \gamma^{-1}. \tag{15.21}$$

Property (ii) of Definition 15.2.3 requires the same to be true for the complete connection 1-form $\mathbf{A}$. This fixes the $\mathfrak{g}$ dependence of $\widehat{\mathscr{A}}$, that is, we must have

$$\widehat{\mathscr{A}}(x, \mathfrak{g}) = \mathfrak{g} \cdot \mathscr{A} \cdot \mathfrak{g}^{-1}, \tag{15.22}$$

where

$$\mathscr{A} = \mathscr{A}^I_\mu(x)\, dx^\mu\, T_I \tag{15.23}$$

is a $\mathbb{G}$ Lie algebra valued 1-form on the base manifold $\mathscr{M}$. Consequently we can summarize the content of Definition 15.2.3 by stating that in any local trivialization $u = (x, \mathfrak{g})$ the connection 1-form has the following structure:

$$\mathbf{A} = \mathfrak{g} \cdot \mathscr{A} \cdot \mathfrak{g}^{-1} + d\mathfrak{g} \cdot \mathfrak{g}^{-1}. \tag{15.24}$$

15.2.1.2 Gauge transformations

From equation (15.24) we easily work out the action on the 1-form connection of the transition function from one local trivialization to another. Let us recall equations (6.117) and (6.118). In the case of a principal bundle the transition function $\mathfrak{t}_{\alpha\beta}(x)$ is just a map from the intersection of two open neighborhoods of the base space into the structural group,

$$\mathfrak{t}_{\alpha\beta} : \mathscr{M} \supset U_\alpha \cap U_\beta \to G, \tag{15.25}$$

and we have

$$\textit{transition:} \quad (x, \mathfrak{g}_\alpha) \to (x, \mathfrak{g}_\beta) = \big(x, \mathfrak{g}_\alpha \cdot \mathfrak{t}_{\alpha\beta}(x)\big). \tag{15.26}$$

Correspondingly, combining this information with equation (15.24) we conclude that

$$\begin{aligned} &\mathbf{A}_\alpha \to \mathbf{A}_\beta, \\ &\mathbf{A}_\alpha = \mathfrak{g} \cdot \mathscr{A}_\alpha \cdot \mathfrak{g}^{-1} + d\mathfrak{g} \cdot \mathfrak{g}^{-1}, \\ &\mathbf{A}_\beta = \mathfrak{g} \cdot \mathscr{A}_\beta \cdot \mathfrak{g}^{-1} + d\mathfrak{g} \cdot \mathfrak{g}^{-1}, \\ &\mathscr{A}_\beta = \mathfrak{t}_{\alpha\beta}(x) \cdot \mathscr{A}_\alpha \cdot \mathfrak{t}^{-1}_{\alpha\beta}(x) + d\mathfrak{t}_{\alpha\beta}(x) \cdot \mathfrak{t}^{-1}_{\alpha\beta}(x). \end{aligned} \tag{15.27}$$

The last equation of (15.27) is the most significant for the bridge from mathematics to physics. The 1-form $\mathscr{A}$ defined over the base manifold $\mathscr{M}$, which in particular can be space-time, is what physicists name *a gauge potential*, the prototype of which is the gauge potential of electrodynamics; the transformation from one local trivialization to another, encoded in the last equation of (15.27), is what physicists call a *gauge transformation*. The basic idea about gauge transformations is that of *local symmetry*. Some physical system or rather some physical law is invariant with respect to the transformations of a group G, not only globally, but rather also locally, namely, the transformation can be chosen differently from one point of space-time to the next. The basic mathematical structure realizing this physical idea is that of fiber bundle, and the 1-form connection on a fiber bundle is the appropriate mathematical structure which encompasses all mediators of all physical interactions. This is the appropriate place to stress once again that, although *gauge theories* are by now an obligatory piece of study in physics, the *complete geometrization* of modern physics does not mean that the geometry of fiber bundles and all its related subtopics are relevant only for physics. On the contrary, as pointed out in the introductory section to the present chapter, applications of the same concepts in diverse fields become nowadays more and more frequent. The logical error one should not make is to think of these mathematical theories as computational techniques. Just as physicists realized little by little, in the course of a long historical development, the mathematical structure characterizing a physical object is just its true, intrinsic essence. For instance to the question "*what is a particle?*" a modern quantum field theorist would invariably answer "*it is an infinite-dimensional unitary irreducible representation of the non-compact Poincaré group.*" All other descriptions of a particle are just metaphoric and pertain to science popularization. What is true in physics is going to be true in the description of other phenomena investigated in other branches of science of technology. The mathematical structure of any phenomenon is the true essence of the phenomenon or of the device when talking about engineering.

15.2.1.3 Horizontal vector fields and covariant derivatives

Let us come back to Definition 15.2.4 of horizontal vector fields. In a local trivialization, a generic vector field on P is of the form

$$\mathbf{X} = X^\mu \partial_\mu + X^i \partial_i, \tag{15.28}$$

where we have used the shorthand $\partial_\mu = \partial/\partial x^\mu$ and $\partial_i = \partial/\partial \alpha^i$. The horizontality condition in equation (15.14) becomes

$$0 = X^\mu \mathscr{A}_\mu^A \mathfrak{g} \cdot T_A \cdot \mathfrak{g}^{-1} + X^i \partial_i \mathfrak{g} \cdot \mathfrak{g}^{-1}. \tag{15.29}$$

Multiplying on the left by $\mathfrak{g}^{-1}$ and on the right by $\mathfrak{g}$, equation (15.29) can be easily solved for X^i, yielding in this way the general form of a horizontal vector field on a principal fiber bundle $P(\mathscr{M}, G)$ endowed with a connection 1-form $\mathbf{A} \leftrightarrow \mathscr{A}$. We get

$$\mathbf{X}^{(H)} = X^\mu(\vec{\partial}_\mu - \mathscr{A}_\mu^A \mathbf{T}_A^{(L)}) = X^\mu(\vec{\partial}_\mu - \mathscr{A}_\mu^A \overset{(\ell)}{\Sigma_A^i} \vec{\partial}_i). \tag{15.30}$$

To obtain the above result we made use of the following identities:

$$\mathfrak{g}^{-1} \cdot \partial_i \mathfrak{g} = \overset{(\ell)}{\Sigma_i^A}; \quad \mathbf{T}_A^{(L)} = \overset{(\ell)}{\Sigma_A^i} \partial_i; \quad \overset{(\ell)}{\Sigma_i^A} \overset{(\ell)}{\Sigma_B^i} = \delta_B^A, \tag{15.31}$$

which follow from the definitions of left-invariant 1-forms and vector fields discussed in Chapter 6.

Equation (15.30) is as relevant for the bridge between mathematics and physics as equation (15.27). Indeed, from (15.30) we realize that the mathematical notion of horizontal vector fields coincides with the physical notion of *covariant derivative*. Expressions like that in the parentheses of (15.30) first appeared in the early decades of the twentieth century, when classical electrodynamics was rewritten in the language of special relativity and the 4-potential $\mathscr{A}_\mu$ was introduced. Indeed, with the development first of the Dirac equation and then of quantum electrodynamics, the covariant derivative of the electron wave function made its entrance in the language of theoretical physics:

$$\nabla_\mu \psi = (\partial_\mu - i\mathscr{A}_\mu)\psi. \tag{15.32}$$

The operator ∇_μ might be assimilated with the action of a horizontal vector field (15.30) if we just had a one-dimensional structural group U(1) with a single generator and we were acting on functions over $P(\mathscr{M}, \mathrm{U}(1))$ that are eigenstates of the unique left-invariant vector field $\mathbf{T}^{(L)}$ with eigenvalue i. This is indeed the case in electrodynamics. The group U(1) is composed of all unimodular complex numbers $\exp[i\alpha]$ and the left-invariant vector field is simply $\mathbf{T}^{(L)} = \vec{\partial}/\partial\alpha$. The wave function $\psi(x)$ is actually a section of a rank 1 complex vector bundle associated with the principal bundle $P(\mathscr{M}_4, \mathrm{U}(1))$. Indeed, $\psi(x)$ is a complex spinor whose overall phase factor $\exp[i\alpha(x)]$ takes values in the U(1)-fiber over $x \in \mathscr{M}_4$. Since $\mathbf{T}^{(L)} \exp[i\alpha] = i\exp[i\alpha]$, the covariant derivative (15.32) is of the form (15.30).

This shows that the 1-form connection coincides, at least in the case of electrodynamics, with the physical notion of gauge potential, which upon quantization describes the photon field, that is, the mediator of electromagnetic interactions. The 1954 invention of non-Abelian gauge theories by Yang and Mills started from the generalization of the covariant derivative (15.32) to the case of the $\mathfrak{su}(2)$ Lie algebra. Introducing three gauge potentials $\mathscr{A}_\mu^A$ corresponding to the three standard generators $\mathrm{J_A}$ of $\mathfrak{su}(2)$,

$$[\mathrm{J_A}, \mathrm{J_B}] = \epsilon_{ABC}\mathrm{J_C}; \quad \mathrm{J_A} = i\sigma^A; \quad (A, B, C = 1, 2, 3), \tag{15.33}$$

where σ^A denote the 2×2 Pauli matrices

$$\sigma^1 = \begin{pmatrix} 0 & 1 \\ 1 & 0 \end{pmatrix}; \quad \sigma^2 = \begin{pmatrix} 0 & -i \\ i & 0 \end{pmatrix}; \quad \sigma^3 = \begin{pmatrix} 1 & 0 \\ 0 & -1 \end{pmatrix}, \tag{15.34}$$

they wrote

$$\nabla_\mu \begin{pmatrix} \psi_1 \\ \psi_2 \end{pmatrix} = (\partial_\mu - i\mathscr{A}_\mu^A \mathrm{J_A}) \begin{pmatrix} \psi_1 \\ \psi_2 \end{pmatrix}. \tag{15.35}$$

From the $\mathfrak{su}(2) \times \mathfrak{u}(1)$ standard model of electro-weak interactions of Glashow, Salam, and Weinberg, we know nowadays that, in appropriate combinations with $\mathscr{A}_\mu$, the three gauge fields $\mathscr{A}_\mu^A$ describe the $W^\pm$ and Z_0 particles discovered at CERN in 1983 by the UA1 experiment of Carlo Rubbia. They have spin one and mediate the weak interactions.

The mentioned examples show that the Lie algebra valued connection 1-form defined by Ehresmann on principal bundles is clearly related to the gauge fields of particle physics since it enters the construction of covariant derivatives via the notion of horizontal vector fields. On the other hand, the same 1-form must be related to gravitation as well, since also there one deals with covariant derivatives,

$$\nabla_\mu t_{\lambda_1 \ldots \lambda_n} \equiv \partial_\mu t_{\lambda_1 \ldots \lambda_n} - \begin{Bmatrix} \rho \\ \mu\lambda_1 \end{Bmatrix} t_{\rho\lambda_2 \ldots \lambda_n} - \begin{Bmatrix} \rho \\ \mu\lambda_2 \end{Bmatrix} t_{\lambda_1 \rho \ldots \lambda_n} \cdots - \begin{Bmatrix} \rho \\ \mu\lambda_n \end{Bmatrix} t_{\lambda_1 \ldots \lambda_{n-1}\rho}, \tag{15.36}$$

sustained by the Levi-Civita connection and the Christoffel symbols

$$\begin{Bmatrix} \lambda \\ \mu\nu \end{Bmatrix} \equiv \frac{1}{2} g^{\lambda\sigma} (\partial_\mu g_{\nu\sigma} + \partial_\nu g_{\mu\sigma} - \partial_\sigma g_{\mu\nu}). \tag{15.37}$$

The key to understand the unifying point of view offered by the notion of Ehresmann connection on a principal bundle is obtained by recalling what we emphasized in Chapter 6 when we introduced the very notion of fiber bundles. Following Definition 6.7.4, let us recall that with each principal bundle $P(\mathscr{M}, G)$ one can associate as many *associated vector bundles* as there are linear representations of the structural group G, namely, infinitely many. It suffices to use as transition functions the corresponding linear representations of the transition functions of the principal bundle as displayed in equation (6.129). In all such associated bundles the fibers are vector spaces of dimension r (the rank of the bundle) and the transition functions are $r \times r$ matrices. Every linear representation of a Lie group G of dimension r induces a linear representation of its

Lie algebra $\mathbb{G}$, where the left (right)-invariant vector fields $\mathbf{T}_A^{(L/R)}$ are mapped into $r \times r$ matrices

$$\mathbf{T}_A^{(L/R)} \to D(T_A) \tag{15.38}$$

satisfying the same commutation relations:

$$[D(T_A), D(T_B)] = C_{AB}{}^C D(T_C). \tag{15.39}$$

It is therefore tempting to assume that given a 1-form connection $\mathbf{A} \leftrightarrow \mathscr{A}$ on a principal bundle one can define a 1-form connection on every associated vector bundle by taking its $r \times r$ matrix representation $D(\mathbf{A}) \leftrightarrow D(\mathscr{A})$. In which sense does this matrix valued 1-form define a connection on the considered vector bundle? To answer such a question we obviously need first to define connections on generic vector bundles, which is what we do in the next section.

15.3 Connections on a vector bundle

Let us now consider a generic vector bundle $E \overset{\pi}{\Longrightarrow} \mathscr{M}$ of rank r.[1] Its standard fiber is an r-dimensional vector space V and the transition functions from one local trivialization to another are maps:

$$\psi_{\alpha\beta} : U_\alpha \cap U_\beta \mapsto \mathrm{Hom(V, V)} \sim \mathrm{GL(r, \mathbb{R})}. \tag{15.40}$$

In other words, without further restrictions, the structural group of a generic vector bundle is just GL(r, ℝ).

The notion of a connection on generic vector bundles is formulated according to the following definition.

Definition 15.3.1. Let $E \overset{\pi}{\Longrightarrow} \mathscr{M}$ be a vector bundle, let $T\mathscr{M} \overset{\pi}{\Longrightarrow} \mathscr{M}$ be the tangent bundle to the base manifold $\mathscr{M}$ ($\dim \mathscr{M} = m$), and let $\Gamma(E, \mathscr{M})$ be the space of sections of E. A *connection* ∇ on E is a rule that with each vector field $\mathbf{X} \in \Gamma(T\mathscr{M}, \mathscr{M})$ associates a map

$$\nabla_{\mathbf{X}} : \Gamma(E, \mathscr{M}) \mapsto \Gamma(E, \mathscr{M}) \tag{15.41}$$

satisfying the following defining properties:
(a) $\nabla_{\mathbf{X}}(a_1 s_1 + b_1 s_2) = a_1 \nabla_{\mathbf{X}} s_1 + a_2 \nabla_{\mathbf{X}} s_2$.
(b) $\nabla_{a_1 \mathbf{X}_1 + a_2 \mathbf{X}_2} s = a_1 \nabla_{\mathbf{X}_1} s + a_2 \nabla_{\mathbf{X}_2} s$.
(c) $\nabla_{\mathbf{X}}(f \cdot s) = \mathbf{X}(f) \cdot s + f \cdot \nabla_{\mathbf{X}} s$.
(d) $\nabla_{f \cdot \mathbf{X}} s = f \cdot \nabla_{\mathbf{X}} s$.

---

1 Let us remind the reader that in the mathematical literature the rank of a vector bundle is just the dimension of its standard fiber (see equation (15.45) and the following lines).

where $a_{1,2} \in \mathbb{R}$ are real numbers, $s_{1,2} \in \Gamma(E, \mathcal{M})$ are sections of the vector bundle, and $f \in C^{\infty}(\mathcal{M})$ is a smooth function.

The abstract description of a connection provided by Definition 15.3.1 becomes more explicit if we consider bases of sections for the two involved vector bundles. To this effect let $\{s_i(p)\}$ be a basis of sections of the vector bundle $E \overset{\pi}{\Longrightarrow} \mathcal{M}$ $(i = 1, 2, \ldots, r = \operatorname{rank} E)$ and let $\{\mathbf{e}_\mu(p)\}$ $(\mu = 1, 2, \ldots, m = \dim \mathcal{M})$ be a basis of sections for the tangent bundle to the base manifold $T\mathcal{M} \overset{\pi}{\Longrightarrow} \mathcal{M}$. Then we can write

$$\nabla_{\mathbf{e}_\mu} s_i \equiv \nabla_\mu s_i = \Theta_{\mu i}{}^{j} s_j, \tag{15.42}$$

where the local functions $\Theta_{\mu i}{}^{j}(p)$ $(\forall p \in \mathcal{M})$ are called the *coefficients of the connection.* They are necessary and sufficient to specify the connection on an arbitrary section of the vector bundle. To this effect let $s \in \Gamma(E, \mathcal{M})$. By the definition of basis of sections we can write

$$s = c^i(p) s_i(p) \tag{15.43}$$

and correspondingly we obtain

$$\begin{aligned} \nabla_\mu s &= (\nabla_\mu c^i) \cdot s_i, \\ \nabla_\mu c^i &\equiv \partial_\mu c^i + \Theta_{\mu j}{}^{i} c^j, \end{aligned} \tag{15.44}$$

having defined $\partial_\mu c^i \equiv e_\mu(c^i)$. Comparing this with the discussion at the end of Section 15.2.1.1, we see that in the language of physicists $\nabla_\mu c^i$ can be identified as the covariant derivatives of the vector components c^i. Let us now observe that for any vector bundle $E \overset{\pi}{\Longrightarrow} \mathcal{M}$, if

$$r \equiv \operatorname{rank}(E) \tag{15.45}$$

is the rank, namely, the dimension of the standard fiber, and

$$m \equiv \dim \mathcal{M}$$

is the dimension of the base manifold, the connection coefficients can be viewed as a set of m matrices, each of them $r \times r$:

$$\forall e_\mu(p): \quad \Theta_{\mu j}{}^{i} = (\Theta_{e_\mu})_j{}^{i}. \tag{15.46}$$

Hence, more abstractly we can say that the connection on a vector bundle associates with each vector field defined on the base manifold a matrix of dimension equal to the rank of the bundle:

$$\forall X \in \Gamma(T\mathcal{M},\mathcal{M}): \quad \nabla : X \mapsto \Theta_X = r\times r \text{ matrix depending on the base point } p \in \mathcal{M}. \tag{15.47}$$

The relevant question is the following: can we relate the matrix Θ_X to a 1-form and make a bridge between the above definition of a connection and that provided by the Ehresmann approach? The answer is yes, and it is encoded in the following equivalent definition.

Definition 15.3.2. Let $E \overset{\pi}{\Longrightarrow} \mathcal{M}$ be a vector bundle, let $T^*\mathcal{M} \overset{\pi}{\Longrightarrow} \mathcal{M}$ be the cotangent bundle to the base manifold $\mathcal{M}$ ($\dim \mathcal{M} = m$), and let $\Gamma(E,\mathcal{M})$ be the space of sections of E. A *connection* ∇ on E is a map

$$\nabla : \Gamma(E,\mathcal{M}) \mapsto \Gamma(E,\mathcal{M}) \otimes T^*\mathcal{M} \tag{15.48}$$

which with each section $s \in \Gamma(E,\mathcal{M})$ associates a 1-form ∇s with values in $\Gamma(E,\mathcal{M})$ such that the following defining properties are satisfied:
a) $\nabla(a_1 s_1 + b_1 s_2) = a_1 \nabla s_1 + a_2 \nabla s_2$;
b) $\nabla(f \cdot s) = df \cdot s + f \cdot \nabla s$;

where $a_{1,2} \in \mathbb{R}$ are real numbers, $s_{1,2} \in \Gamma(E,\mathcal{M})$ are sections of the vector bundle, and $f \in C^\infty(\mathcal{M})$ is a smooth function.

The relation between the two definitions of a connection on a vector bundle is now easily obtained by stating that for each section $s \in \Gamma(E,\mathcal{M})$ and for each vector field $\mathbf{X} \in \Gamma(T\mathcal{M},\mathcal{M})$ we have

$$\nabla_{\mathbf{X}} s = \nabla s(\mathbf{X}). \tag{15.49}$$

The reader can easily see that using (15.49) and relying on the properties of ∇s established by Definition 15.3.2, those of $\nabla_X s$ required by Definition 15.3.1 are all satisfied.

Consider now, just as before, a basis of sections of the vector bundle $\{s_i(p)\}$. According to its second definition, Definition 15.3.2, a connection ∇ singles out an $r \times r$ matrix valued 1-form $\Theta \in \Gamma(T^*\mathcal{M},\mathcal{M})$ through the following equation:

$$\nabla s_i = \Theta_i^j s_j. \tag{15.50}$$

Clearly the connection coefficients introduced in equation (15.47) are nothing but the values of Θ on each vector field $\mathbf{X}$:

$$\Theta_{\mathbf{X}} = \Theta(\mathbf{X}). \tag{15.51}$$

A natural question which arises at this point is the following. In view of our comments at the end of Section 15.2.1.1, could we identify Θ with the matrix representation of $\mathcal{A}$, the principal connection on the corresponding principal bundle? In other words, could we write

$$\Theta = D(\mathscr{A})? \tag{15.52}$$

For a single generic vector bundle this question seems tautological. Indeed, the structural group is simply GL(r, ℝ) and with the corresponding Lie algebra $\mathfrak{gl}(r, \mathbb{R})$ being made up of the space of generic matrices it seems obvious that Θ could be viewed as Lie algebra valued. The real question however is another. The identification (15.52) is legitimate if and only if Θ transforms as the matrix representation of $\mathscr{A}$. This is precisely the case. Let us demonstrate it. Consider the intersection of two local trivializations. On $U_\alpha \cap U_\beta$ the relation between two bases of sections is given by

$$s_i^{(\alpha)} = D(\mathfrak{t}_{(\alpha\beta)})_i^{\ j} s_j^{(\beta)}, \tag{15.53}$$

where $\mathfrak{t}_{(\alpha\beta)}$ is the transition function seen as an abstract group element. Equation (15.53) implies

$$\Theta^{(\alpha)} = D(d\mathfrak{t}_{(\alpha\beta)}) \cdot D(\mathfrak{t}_{(\alpha\beta)})^{-1} + D(\mathfrak{t}_{(\alpha\beta)}) \cdot \Theta^{(\beta)} \cdot D(\mathfrak{t}_{(\alpha\beta)})^{-1}, \tag{15.54}$$

which is consistent with the gauge transformation rule (15.27) if we identify Θ as in equation (15.52).

The outcome of the above discussion is that once we have defined a connection 1-form $\mathscr{A}$ on a principal bundle $P(\mathscr{M}, G)$ a connection is induced on any associated vector bundle $E \overset{\pi}{\Longrightarrow} \mathscr{M}$. It is simply defined as follows:

$$\forall s \in \Gamma(E, \mathscr{M}): \quad \nabla s = ds + D(\mathscr{A})s, \tag{15.55}$$

where $D()$ denotes the linear representation of both G and $\mathbb{G}$, which define the associated bundle.

15.4 An illustrative example of fiber bundle and connection

In this section we discuss a simple example that illustrates both the general definition of fiber bundle and that of principal connection. The chosen example has an intrinsic physical interest since it corresponds to the mathematical description of a magnetic monopole in the standard electromagnetic theory. It is also geometrically relevant since it shows how a differentiable manifold can sometimes be reinterpreted as the total space of a fiber bundle constructed on a smaller manifold.

15.4.1 The magnetic monopole and the Hopf fibration of $\mathbb{S}^3$

We introduce our case study defining a family of principal bundles that depend on an integer $n \in \mathbb{Z}$. Explicitly let $P(\mathbb{S}^2, \mathrm{U}(1))$ be a principal bundle where the base manifold is a 2-sphere $\mathcal{M} = \mathbb{S}^2$, the structural group is a circle $\mathbb{S}^1 \sim \mathrm{U}(1)$, and by the definition of principal bundle, the standard fiber coincides with the structural group $F = \mathbb{S}^1$. An element $f \in F$ of this fiber is just a complex number of modulus 1, $f = \exp[i\theta]$. To describe this bundle we need an atlas of local trivializations and hence, to begin with, an atlas of open charts of the base manifold $\mathbb{S}^2$. We use the two charts provided by the stereographic projection. Defining $\mathbb{S}^2$ via the standard quadratic equation in $\mathbb{R}^3$,

$$\mathbb{S}^2 \subset \mathbb{R}^3 : \{x_1, x_2, x_3\} \in \mathbb{S}^2 \leftrightarrow x_1^2 + x_2^2 + x_3^2 = 1, \tag{15.56}$$

we define the two open neighborhoods of the North and the South Pole, respectively named H^+ and H^-, which correspond in $\mathbb{R}^3$ language to the exclusion of either the point $\{0, 0, -1\}$ or the point $\{0, 0, 1\}$. These neighborhoods are shown in Figure 15.2. As we already know from Chapter 6, the stereographic projections s_N and s_S map H^- and H^+ onto the complex plane $\mathbb{C}$ as

$$\begin{aligned} s_N \quad H^- \to \mathbb{C} \quad z_N \equiv s_N(x_1, x_2, x_3) &= \frac{x_1 + ix_2}{1 - x_3}, \\ s_S \quad H^+ \to \mathbb{C} \quad z_S \equiv s_S(x_1, x_2, x_3) &= \frac{x_1 - ix_2}{1 + x_3}, \end{aligned} \tag{15.57}$$

and on the intersection $H^- \cap H^+$ we have the transition function $z_N = 1/z_S$ (see equation (6.20)). We construct our fiber bundle introducing two local trivializations respectively associated with the two open charts H^- and H^+:

$$\begin{aligned} \phi_N^{-1} : \quad \pi^{-1}(H^-) \to H^- \otimes U(1) : \quad \phi_N^{-1}(\{x_1, x_2, x_3\}) &= (z_N, \exp[i\psi_N]), \\ \phi_S^{-1} : \quad \pi^{-1}(H^+) \to H^+ \otimes U(1) : \quad \phi_S^{-1}(\{x_1, x_2, x_3\}) &= (z_S, \exp[i\psi_S]). \end{aligned} \tag{15.58}$$

To complete the construction of the bundle we still need to define a transition function between the two local trivializations, namely, a map

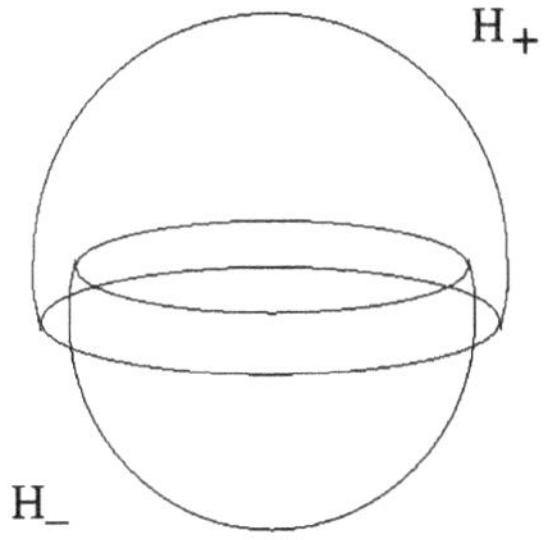

Figure 15.2: The two open charts H^+ and H^- covering the 2-sphere.

$$t_{SN}: \quad H^- \cap H^+ \rightarrow \mathrm{U}(1), \tag{15.59}$$

such that

$$(z_S, \exp[i\psi_S]) = \left(\frac{1}{z_N}, t_{SN}(z_N)\exp[i\psi_N]\right). \tag{15.60}$$

In order to illustrate the geometrical interpretation of the transition function we are going to write, it is convenient to make use of polar coordinates on the 2-sphere. Following standard conventions, we parameterize the points of $\mathbb{S}^2$ by means of the two angles $\theta \in [0, \pi]$ and $\phi \in [0, 2\pi]$:

$$x_1 = \sin\theta\cos\phi; \quad x_2 = \sin\theta\sin\phi; \quad x_3 = \cos\theta. \tag{15.61}$$

On the intersection $H^+ \cap H^-$ we obtain

$$z_N = \frac{\sin\theta}{1-\cos\theta}\exp[i\phi]; \quad z_S = \frac{\sin\theta}{1+\cos\theta}\exp[-i\phi], \tag{15.62}$$

and we see that the azimuth angle ϕ parameterizes the points of the equator identified by the condition $\theta = \pi/2 \Rightarrow x_3 = 0$. Then we write the transition function as follows:

$$t_{SN}(z_N) = \exp[in\phi] = \left(\frac{z_N}{|z_N|}\right)^n = \left(\frac{|z_S|}{z_S}\right)^n; \quad n \in \mathbb{Z}. \tag{15.63}$$

The catch in the above equation is that each choice of the integer number n identifies a different inequivalent principal bundle. The difference is that while going around the equator the transition function can cover the structural group $\mathbb{S}^1$, once, twice, or an integer number of times. The transition function can also go clockwise or counterclockwise and this accounts for the sign of n. We will see that the integer n characterizing the bundle can be interpreted as the magnetic charge of a magnetic monopole. Before going into that, let us first consider the special properties of the case $n = 1$.

15.4.1.1 The case $n = 1$ and the Hopf fibration of $\mathbb{S}^3$

Let us call $P_{(n)}(\mathbb{S}^2, \mathrm{U}(1))$ the principal bundles we have constructed in the previous section. Here we want to show that the total space of the bundle $P_{(1)}(\mathbb{S}^2, \mathrm{U}(1))$ is actually the 3-sphere $\mathbb{S}^3$.

To this effect we define $\mathbb{S}^3$ as the locus in $\mathbb{R}^4$ of the standard quadric,

$$\mathbb{S}^3 \subset \mathbb{R}^4 : \{X_1, X_2, X_3, X_4\} \in \mathbb{S}^3 \leftrightarrow X_1^2 + X_2^2 + X_3^2 + X_4^2 = 1, \tag{15.64}$$

and we introduce the *Hopf map*

$$\pi : \mathbb{S}^3 \rightarrow \mathbb{S}^2 \tag{15.65}$$

which is explicitly given by

$$\begin{aligned} x_1 &= 2(X_1X_3 + X_2X_4), \\ x_2 &= 2(X_2X_3 - X_1X_4), \\ x_3 &= X_1^2 + X_2^2 - X_3^2 - X_4^2. \end{aligned} \tag{15.66}$$

It is immediately verified that x_i $(i = 1, 2, 3)$ as given in equation (15.66) satisfy (15.56) if X_i $(i = 1, 2, 3, 4)$ satisfy equation (15.64). Hence, this is indeed a projection map from $\mathbb{S}^3$ to $\mathbb{S}^2$ as claimed in (15.65). We want to show that π can be interpreted as the projection map in the principal fiber bundle $P_{(1)}(\mathbb{S}^2, \mathrm{U}(1))$. To obtain this result we need first to show that for each $p \in \mathbb{S}^3$ the fiber over p is diffeomorphic to a circle, namely, $\pi^{-1}(p) \sim \mathrm{U}(1)$. To this effect we begin by complexifying the coordinates of $\mathbb{R}^4$,

$$Z_1 = X_1 + iX_2; \quad Z_2 = X_3 + iX_4, \tag{15.67}$$

so that the equation defining $\mathbb{S}^3$ (15.64) becomes the following equation in $\mathbb{C}^2$:

$$|Z_1|^2 + |Z_2|^2 = 1. \tag{15.68}$$

Then we call $U^- \subset \mathbb{S}^3$ and $U^+ \subset \mathbb{S}^3$ the two open neighborhoods where we respectively have $Z_2 \neq 0$ and $Z_1 \neq 0$. The Hopf map (15.66) can be combined with the stereographic projection of the North Pole in the neighborhood U^- and with the stereographic projection of the South Pole in the neighborhood U^+, yielding

$$\begin{aligned} \forall \{Z_1, Z_2\} \in U^- \subset \mathbb{S}^3; \quad s_N \circ \pi(Z_1, Z_2) &= \frac{Z_1}{Z_2}, \\ \forall \{Z_1, Z_2\} \in U^+ \subset \mathbb{S}^3; \quad s_S \circ \pi(Z_1, Z_2) &= \frac{Z_2}{Z_1}. \end{aligned} \tag{15.69}$$

From both local descriptions of the Hopf map we see that

$$\forall \lambda \in \mathbb{C} \quad \pi(\lambda Z_1, \lambda Z_2) = \pi(Z_1, Z_2). \tag{15.70}$$

On the other hand, if $(Z_1, Z_2) \in \mathbb{S}^3$ we have $(\lambda Z_1, \lambda Z_2) \in \mathbb{S}^3$ if and only if $\lambda \in \mathrm{U}(1)$, that is, if $|\lambda|^2 = 1$. This proves the desired result that $\pi^{-1}(p) \sim \mathrm{U}(1)$ for all $p \in \mathbb{S}^3$. The two equations in (15.69) can also be interpreted as two local trivializations of a principal U(1) bundle with $\mathbb{S}^2$ base manifold whose total space is the 3-sphere $\mathbb{S}^3$. Indeed, we can rewrite them in the opposite direction as follows:

$$\begin{aligned} \phi_N^{-1} : U^- \equiv \pi^{-1}(H^-) \to H^- \otimes U(1); \quad [Z_1, Z_2] &\to \left[\frac{Z_1}{Z_2}, \frac{Z_2}{|Z_2|}\right], \\ \phi_S^{-1} : U^+ \equiv \pi^{-1}(H^+) \to H^+ \otimes U(1); \quad [Z_1, Z_2] &\to \left[\frac{Z_2}{Z_1}, \frac{Z_1}{|Z_1|}\right]. \end{aligned} \tag{15.71}$$

This time the transition function does not have to be invented; rather, it is decided a priori by the Hopf map. Indeed, we immediately read it from (15.71):

$$t_{SN} = \frac{Z_1}{|Z_1|}\frac{|Z_2|}{Z_2} = \frac{Z_1/Z_2}{|Z_1/Z_2|} = \frac{z_N}{|z_N|}, \tag{15.72}$$

which coincides with equation (15.63) for $n = 1$. This is the result we wanted to prove: the 3-sphere is the total space for the principal fiber $P_{(1)}(\mathbb{S}^2, \mathrm{U}(1))$.

15.4.1.2 The U(1)-connection of the Dirac magnetic monopole

The Dirac monopole [124] of magnetic charge n corresponds to introducing a vector potential $\mathscr{A} = \mathscr{A}_i(x)\, dx^i$ that is a well-defined connection on the principal U(1) bundle $P_{(n)}(\mathbb{S}^2, \mathrm{U}(1))$ we discussed above. Physically the discussion goes as follows. Let us work in ordinary three-dimensional space $\mathbb{R}^3$ and assume that in the origin of our coordinate system there is a magnetically charged particle.

Then we observe a magnetic field of the form

$$\mathbf{H} = \frac{m}{r^3}\mathbf{r} \quad \Leftrightarrow \quad H_i = \frac{m}{r^3}x_i. \tag{15.73}$$

From the relation $F_{ij} = \epsilon_{ijk}H_k$ we conclude that the electromagnetic field strength associated with such a magnetic monopole is the following 2-form on $\mathbb{R}^3 - 0$:

$$\mathfrak{F} = \frac{m}{r^3}\epsilon_{ijk}x^i\, dx^j \wedge dx^k. \tag{15.74}$$

Indeed, equation (15.74) makes sense everywhere in $\mathbb{R}^3$ except at the origin $\{0, 0, 0\}$. Using polar coordinates rather than Cartesian ones the 2-form $\mathfrak{F}$ assumes a very simple expression:

$$\mathfrak{F} = \frac{g}{2}\sin\theta\, d\theta \wedge d\phi, \tag{15.75}$$

that is, it is proportional to the volume form on the 2-sphere $\mathrm{Vol}(\mathbb{S}^2) \equiv \sin\theta\, d\theta \wedge d\phi$. We leave it as an exercise for the reader to calculate the relation between the constant g appearing in (15.75) and the constant m appearing in (15.74). Hence, the 2-form F can be integrated on any 2-sphere of arbitrary radius and the result, according to the Gauss law, is proportional to the charge of the magnetic monopole. Explicitly we have

$$\int_{\mathbb{S}^2}\mathfrak{F} = \frac{g}{2}\int_0^{\pi}\sin\theta\, d\theta\int_0^{2\pi} d\phi = 2\pi g. \tag{15.76}$$

Consider now the problem of writing a vector potential for this magnetic field. By definition, in each local trivialization (U_α, ϕ_α) of the underlying U(1) bundle we must have

$$\mathfrak{F} = d\mathscr{A}_\alpha, \tag{15.77}$$

where $\mathscr{A}_\alpha$ is a 1-form on U_α. Yet we have just observed that $\mathfrak{F}$ is proportional to the volume form on $\mathbb{S}^2$, which is a closed but not exact form. Indeed, if the volume form were exact, namely, if there existed a global section ω of $T^*\mathbb{S}^2$ such that Vol $= d\omega$, then using the Stokes theorem we would arrive at the absurd conclusion that the volume of the sphere vanishes: indeed, $\int_{\mathbb{S}^2} \mathrm{Vol} = \int_{\partial\mathbb{S}^2} \omega = 0$ since the 2-sphere has no boundary. Hence, there cannot be a vector potential globally defined on the 2-sphere. The best we can do is to have two different definitions of the vector potential, one in the North Pole patch and one in the South Pole patch. Recalling that the vector potential is the component of a connection on a U(1) bundle we multiply it by the Lie algebra generator of U(1) and we write the ansatz for the full connection in the two patches:

$$\begin{aligned} \text{on } U^-; \quad \mathscr{A} &= \mathscr{A}_N = i\frac{1}{2}g(1 - \cos\theta)\, d\phi, \\ \text{on } U^+; \quad \mathscr{A} &= \mathscr{A}_S = -i\frac{1}{2}g(1 + \cos\theta)\, d\phi. \end{aligned} \tag{15.78}$$

As one sees, the first definition of the connection does not vanish as long as $\theta \neq 0$, that is, as long as we stay away from the North Pole, while the second definition does not vanish as long as $\theta \neq \pi$, that is, as long as we stay away from the South Pole. On the other hand, if we calculate the exterior derivative of either $\mathscr{A}_N$ or $\mathscr{A}_S$ we obtain the same result, namely, i times the 2-form (15.75):

$$\mathfrak{F} = d\mathscr{A}_S = d\mathscr{A}_N = i \times \frac{g}{2} \sin\theta\, d\theta \wedge d\phi. \tag{15.79}$$

The transition function between the two local trivializations can now be reconstructed from

$$\mathscr{A}_S = \mathscr{A}_N + t_{SN}^{-1} dt_{NS} = -i g\, d\phi, \tag{15.80}$$

which implies

$$t_{NS} = \exp[i g \phi], \tag{15.81}$$

and by comparison with equation (15.63) we see that the magnetic charge g is the integer n classifying the principal U(1) bundles. The reason why it has to be an integer was already noted. On the equator of the 2-sphere the transition function realizes a map of the circle $\mathbb{S}^1$ into itself. Such a map *winds* a certain number of times but cannot wind a non-integer number of times, because its image would no longer be a closed circle.

In this way we see that the quantization of the magnetic charge is a topological condition.

15.5 Riemannian and pseudo-Riemannian metrics: the mathematical definition

We have discussed at length the notion of connections on fiber bundles. From the historical review presented at the beginning of the chapter we know that connections appeared first in relation with the metric geometry introduced by Gauss and Riemann and developed by Ricci, Levi-Civita, and Bianchi.[2] It is now time to come to the rigorous modern mathematical definition of metrics.

The mathematical environment in which we will work is that provided by a differentiable manifold $\mathcal{M}$ of dimension $\dim \mathcal{M} = m$. As several times emphasized in Chapter 6, we can construct many different fiber bundles on the same base manifold $\mathcal{M}$, but there are two which are intrinsically defined by the differentiable structure of $\mathcal{M}$, namely, the tangent bundle $T\mathcal{M} \overset{\pi}{\Longrightarrow} \mathcal{M}$ and its dual, the cotangent bundle $T^*\mathcal{M} \overset{\pi}{\Longrightarrow} \mathcal{M}$. Of these bundles we can take the tensor powers (symmetric, antisymmetric, or none) which are bundles whose sections transform with the corresponding tensor powers of the transition functions of the original bundle. The sections of these tensor powers are the tensors introduced by Ricci and Levi-Civita in their fundamental paper *Méthodes de calcul differential absolu et leurs applications*, for whose differentiation Christoffel had invented his symbols (see equations (15.36) and (15.37)).

For brevity let us use a special notation for the space of sections of the tangent bundle, namely, for the algebra of vector fields: $\mathfrak{X}(\mathcal{M}) \equiv \Gamma(T\mathcal{M}, \mathcal{M})$. Then we have the following definition.

Definition 15.5.1. A *pseudo-Riemannian* metric on a manifold $\mathcal{M}$ is a $C^\infty(\mathcal{M})$, symmetric, bilinear, non-degenerate form on $\mathfrak{X}(\mathcal{M})$, namely, it is a map

$$g(,):\mathfrak{X}(\mathcal{M}) \otimes \mathfrak{X}(\mathcal{M}) \longrightarrow C^\infty(\mathcal{M})$$

satisfying the following properties:
(i) $\forall \mathbf{X}, \mathbf{Y} \in \mathfrak{X}(\mathcal{M}) : g(\mathbf{X}, \mathbf{Y}) = g(\mathbf{Y}, \mathbf{X}) \in C^\infty(\mathcal{M})$.
(ii) $\forall \mathbf{X}, \mathbf{Y}, \mathbf{Z} \in \mathfrak{X}(\mathcal{M}), \forall f, g \in C^\infty(\mathcal{M}) : g(f\mathbf{X} + h\mathbf{Y}, \mathbf{Z}) = fg(\mathbf{X}, \mathbf{Z}) + hg(\mathbf{Y}, \mathbf{Z})$.
(iii) $\{\forall X \in \mathfrak{X}(\mathcal{M}), g(\mathbf{X}, \mathbf{Y}) = 0\} \Rightarrow Y \equiv 0$.

Definition 15.5.2. A *Riemannian* metric on a manifold $\mathcal{M}$ is a pseudo-Riemannian metric which in addition satisfies the following two properties:
(iv) $\forall \mathbf{X} \in \mathfrak{X}(\mathcal{M}) : g(\mathbf{X}, \mathbf{X}) \geq 0$,
(v) $g(\mathbf{X}, \mathbf{X}) = 0 \Rightarrow \mathbf{X} \equiv 0$.

The two additional properties satisfied by a Riemannian metric state that it should be not only a symmetric non-degenerate form but also a positive definite one.

2 For a comprehensive historical account see [28].

The definition of pseudo-Riemannian and Riemannian metrics can be summarized by saying that a metric is a section of the second symmetric tensor power of the cotangent bundle, that is,

$$g \in \Gamma\Big(\bigotimes_{\text{symm}}{}^2 T^* \mathcal{M}, \mathcal{M}\Big). \tag{15.82}$$

In a coordinate patch the metric is described by a symmetric rank 2 tensor $g_{\mu\nu}(x)$. Indeed, we can write

$$g = g_{\mu\nu}(x) dx^\mu \otimes dx^\nu, \tag{15.83}$$

and we have

$$\forall \mathbf{X}, \mathbf{Y} \in \mathfrak{X}(\mathcal{M}) : \quad g(\mathbf{X}, \mathbf{Y}) = g_{\mu\nu}(x) X^\mu(x) Y^\nu(x) \tag{15.84}$$

if

$$\mathbf{X} = X^\mu(x) \overrightarrow{\partial}_\mu; \quad \mathbf{Y} = Y^\nu(x) \overrightarrow{\partial}_\nu \tag{15.85}$$

are the local expressions of the two vector fields in the considered coordinate patch. The essential point in the definition of the metric is the statement that $g(X, Y) \in C^\infty(\mathcal{M})$, that is, that $g(X, Y)$ is a scalar. This implies that the coefficients $g_{\mu\nu}(x)$ should transform from a coordinate patch to the next one with the inverse of the transition functions with which the vector field components transform. This is what equation (15.82) implies. On the other hand, according to the viewpoint conceived by Gauss and further developed by Riemann, Ricci, and Levi-Civita, the metric is the mathematical structure which allows to define infinitesimal lengths and, by integration, the length of any curve. Indeed, given a curve $\mathcal{C} : [0, 1] \longrightarrow \mathcal{M}$ let $\mathbf{t}(s)$ be its tangent vector at any point of the curve $\mathcal{C}$, parameterized by $s \in [0, 1]$. We define the length of the curve as

$$\mathfrak{s}(\mathcal{C}) \equiv \int_0^1 g(\mathbf{t}(s), \mathbf{t}(s)) \, ds. \tag{15.86}$$

15.5.1 Signatures

At each point of the manifold $p \in \mathcal{M}$, the coefficients $g_{\mu\nu}(p)$ of a metric g constitute a symmetric non-degenerate real matrix. Such a matrix can always be diagonalized by means of an orthogonal transformation $\mathcal{O}(p) \in \mathrm{SO(N)}$, which obviously varies from point to point, namely, we can write

$$g_{\mu\nu}(p) = \mathcal{O}^T(p) \begin{pmatrix} \lambda_1(p) & 0 & \cdots & 0 & 0 \\ 0 & \lambda_2(p) & 0 & \cdots & 0 \\ \vdots & \vdots & \vdots & \vdots & \vdots \\ 0 & \cdots & \cdots & \lambda_{N-1}(p) & 0 \\ 0 & \cdots & \cdots & \cdots & \lambda_N(p) \end{pmatrix} \mathcal{O}(p), \tag{15.87}$$

where the real numbers $\lambda_i(p)$ are the eigenvalues. Each of them depends on the point $p \in \mathcal{M}$, but none of them can vanish, because then the determinant of the metric would be zero, which contradicts one of the defining axioms. Consider next the diagonal matrix

$$\mathcal{L}(p) = \begin{pmatrix} \frac{1}{\sqrt{|\lambda_1(p)|}} & 0 & \cdots & 0 & 0 \\ 0 & \frac{1}{\sqrt{|\lambda_2(p)|}} & 0 & \cdots & 0 \\ \vdots & \vdots & \vdots & \vdots & \vdots \\ 0 & \cdots & \cdots & \frac{1}{\sqrt{|\lambda_{N-1}(p)|}} & 0 \\ 0 & \cdots & \cdots & \cdots & \frac{1}{\sqrt{|\lambda_{N-1}(p)|}} \end{pmatrix}. \tag{15.88}$$

The matrix $M = \mathcal{O}\cdot\mathcal{L}$, where for brevity the dependence on the point p has been omitted, is such that

$$g_{\mu\nu} = M^T \begin{pmatrix} \text{sign}[\lambda_1] & 0 & \cdots & 0 & 0 \\ 0 & \text{sign}[\lambda_2] & 0 & \cdots & 0 \\ \vdots & \vdots & \vdots & \vdots & \vdots \\ 0 & \cdots & \cdots & \text{sign}[\lambda_{N-1}] & 0 \\ 0 & \cdots & \cdots & \cdots & \text{sign}[\lambda_N] \end{pmatrix} M, \tag{15.89}$$

having denoted by $\text{sign}[x]$ the function which takes the value 1 if $x > 0$ and the value -1 if $x < 0$. Hence, we arrive at the following conclusion and definition.

Definition 15.5.3. Let $\mathcal{M}$ be a differentiable manifold of dimension m endowed with a metric g. At every point $p \in \mathcal{M}$, by means of a transformation $g \mapsto S^T \cdot g \cdot S$ the metric tensor $g_{\mu\nu}(p)$ can be reduced to a diagonal matrix with p entries equal to 1 and $m - p$ entries equal to -1. The pair of integers $(p, m - p)$ is called the *signature* of the metric g.

The rationale of the above definition is that the signature of a metric is an intrinsic property of g, independent of both the chosen coordinate patch and the chosen point.

This issue was already discussed in Section 3.5.6 (see in particular equation (3.39)), where it was recalled that it is the content of a theorem proved in 1852 by James Sylvester. According to Sylvester's theorem a symmetric non-degenerate matrix A can always be transformed into a diagonal one with ±1 entries by means of a substitution $A \mapsto B^T \cdot A \cdot B$. On the other hand, no such transformation can alter the signature $(p, m - p)$, which is intrinsic to the matrix A. This is what happens for a single matrix. Consider now a point-dependent matrix like the metric tensor g, whose entries are smooth functions. Defining by $s = 2p - m$ the difference between the number of positive and negative eigenvalues of g, it follows that also s is a smooth function. Yet s is an integer by definition. Hence, it has to be a constant.

The metrics on a differentiable manifold are therefore intrinsically characterized by their signatures. Riemannian are the positive definite metrics with signature $(m, 0)$. Lorentzian are the metrics with signature $(1, m-1)$, just as the flat metric of Minkowski space. There are also metrics with more elaborate signatures which appear in many mathematical constructions related to supergravity.

15.6 The Levi-Civita connection

Having established the rigorous mathematical notion of both a metric and a connection we come back to the ideas of Riemannian curvature and torsion, which were heuristically touched upon in the course of our historical outline. In particular, we are now in a position to derive from clear-cut mathematical principles the Christoffel symbols anticipated in equation (15.37), the Riemann tensor, and the torsion tensor. The starting point for the implementation of this plan is provided by a careful consideration of the special properties of affine connections.

15.6.1 Affine connections

In Definitions 15.3.1 and 15.3.2 we fixed the notion of a connection on a generic vector bundle $E \overset{\pi}{\Longrightarrow} \mathcal{M}$. In particular, we can consider the tangent bundle $T\mathcal{M} \overset{\pi}{\Longrightarrow} \mathcal{M}$. A connection on $T\mathcal{M}$ is called *affine*. It follows that we can give the following definition.

Definition 15.6.1. Let $\mathcal{M}$ be an m-dimensional differentiable manifold. An affine connection on $\mathcal{M}$ is a map

$$\nabla : \mathfrak{X}(\mathcal{M}) \times \mathfrak{X}(\mathcal{M}) \to \mathfrak{X}(\mathcal{M})$$

which satisfies the following properties:
(i) $\forall \mathbf{X}, \mathbf{Y}, \mathbf{Z} \in \mathfrak{X}(\mathcal{M}) : \nabla_{\mathbf{X}}(\mathbf{Y} + \mathbf{Z}) = \nabla_{\mathbf{X}}\mathbf{Y} + \nabla_{\mathbf{X}}\mathbf{Z}$.
(ii) $\forall \mathbf{X}, \mathbf{Y}, \mathbf{Z} \in \mathfrak{X}(\mathcal{M}) : \nabla_{(\mathbf{X}+\mathbf{Y})}\mathbf{Z} = \nabla_{\mathbf{X}}\mathbf{Z} + \nabla_{\mathbf{Y}}\mathbf{Z}$.
(iii) $\forall \mathbf{X}, \mathbf{Y} \in \mathfrak{X}(\mathcal{M}), \forall f \in C^{\infty}(\mathcal{M}) : \nabla_{f\mathbf{X}}\mathbf{Y} = f\nabla_{\mathbf{X}}\mathbf{Y}$.
(iv) $\forall \mathbf{X}, \mathbf{Y} \in \mathfrak{X}(\mathcal{M}), \forall f \in C^{\infty}(\mathcal{M}) : \nabla_{\mathbf{X}}(f\mathbf{Y}) = \mathbf{X}[f]\mathbf{Y} + f\nabla_{\mathbf{X}}\mathbf{Y}$.

Clearly also affine connections are encoded into corresponding connection 1-forms, which are traditionally denoted by the symbol Γ. In the affine case Γ is $\mathfrak{gl}(m, \mathbb{R})$-Lie algebra valued since the structural group of $T\mathcal{M}$ is $\mathrm{GL}(m, \mathbb{R})$. Let $\{\mathbf{e}_\mu\}$ be a basis of sections of the tangent bundle so that any vector field $\mathbf{X} \in \mathfrak{X}(\mathcal{M})$ can be written as follows:

$$\mathbf{X} = X^{\mu}(x)\mathbf{e}_{\mu}. \tag{15.90}$$

The connection 1-form is defined by calculating the *covariant differentials* of the basis elements:

$$\nabla \mathbf{e}_{\nu} = \Gamma_{\nu}{}^{\rho}\mathbf{e}_{\rho}. \tag{15.91}$$

Let us introduce the dual basis of $T^*\mathscr{M}$, namely, the set of 1-forms ω^{μ} such that

$$\omega^{\mu}(\mathbf{e}_{\nu}) = \delta^{\mu}_{\nu}. \tag{15.92}$$

The matrix valued 1-form Γ can be expanded along such a basis, yielding

$$\Gamma = \omega^{\mu}\Gamma_{\mu} \quad \Rightarrow \quad \Gamma_{\nu}{}^{\rho} = \omega^{\mu}\Gamma^{\rho}_{\mu\nu}. \tag{15.93}$$

The three-index symbols $\Gamma^{\rho}_{\mu\nu}$ encode, patch by patch, the considered affine connection. According to Definition 15.6.1 these connection coefficients are equivalently defined by setting

$$\nabla_{\mathbf{e}_{\mu}}\mathbf{e}_{\nu} \equiv \nabla\mathbf{e}_{\nu}(\mathbf{e}_{\mu}) = \Gamma_{\mu\nu}{}^{\rho}\mathbf{e}_{\rho}. \tag{15.94}$$

15.6.2 Curvature and torsion of an affine connection

With every connection 1-form $\mathscr{A}$ on a principal bundle $P(\mathscr{M}, G)$ we can associate a curvature 2-form:

$$\begin{aligned} \mathfrak{F} &\equiv d\mathscr{A} + \mathscr{A}\wedge\mathscr{A} \\ &\equiv \mathfrak{F}^{I}T_{I} = \left(d\mathscr{A}^{I} + \frac{1}{2}f_{JK}{}^{I}\mathscr{A}^{J}\wedge\mathscr{A}^{K}\right)T_{I}, \end{aligned} \tag{15.95}$$

which is $\mathbb{G}$ Lie algebra valued. We note that, evaluated on any associated vector bundle, $E \overset{\pi}{\Longrightarrow} \mathscr{M}$, the connection $\mathscr{A}$ becomes a matrix, and the same is true of the curvature $\mathscr{F}$. In that case the first line of equation (15.95) is to be understood in the sense both of matrix multiplication and of wedge product, namely, the element (i,j) of $\mathscr{A}\wedge\mathscr{A}$ is calculated as $\mathscr{A}_{i}{}^{k}\wedge\mathscr{A}_{k}{}^{j}$, with summation over the dummy index k.

We can apply the general formula (15.95) to the case of an affine connection. In that case the curvature 2-form is traditionally denoted with the letter $\mathfrak{R}$ in honor of Riemann. We obtain

$$\mathfrak{R} \equiv d\Gamma + \Gamma\wedge\Gamma, \tag{15.96}$$

which, using the basis $\{\mathbf{e}_{\mu}\}$ for the tangent bundle and its dual $\{\omega^{\nu}\}$ for the cotangent bundle, becomes

$$\begin{aligned}\mathfrak{R}_\mu{}^\nu &= d\Gamma_\mu{}^\nu + \Gamma_\mu{}^\rho \wedge \Gamma_\rho{}^\nu \\ &= \omega^\lambda \wedge \omega^\sigma \frac{1}{2} \mathscr{R}_{\lambda\sigma\mu}{}^\nu,\end{aligned} \tag{15.97}$$

the four-index symbols $\mathscr{R}_{\lambda\sigma\mu}{}^\nu$ being, by definition, twice the components of the 2-form along the basis $\{\omega^\nu\}$. In particular, in an open chart $U \subset \mathscr{M}$, whose coordinates we denote by x^μ, we can choose the holonomic basis of sections $\mathbf{e}_\mu = \partial_\mu \equiv \partial/\partial x^\mu$, whose dual is provided by the differentials $\omega^\nu = dx^\nu$, and we get

$$\mathscr{R}_{\lambda\sigma\mu}{}^\nu = \partial_\lambda \Gamma_{\sigma\mu}^{\nu} - \partial_\sigma \Gamma_{\lambda\mu}^{\nu} + \Gamma_{\lambda\mu}^{\rho}\Gamma_{\sigma\rho}^{\nu} - \Gamma_{\sigma\mu}^{\rho}\Gamma_{\lambda\rho}^{\nu}. \tag{15.98}$$

Comparing equation (15.98) with the Riemann–Christoffel symbols of equation (15.37), we see that the latter could be identified with the components of the curvature 2-form of an affine connection Γ if the Christoffel symbols introduced in equation (15.37) were the coefficients of such a connection. Which connection is the one described by the Christoffel symbols and how is it defined? The answer is the *Levi-Civita connection*. Its definition is given in the next paragraph.

15.6.2.1 Torsion and torsionless connections

The notion of torsion applies to affine connections and distinguishes them from general connections on generic fiber bundles. Intuitively torsion has to do with the fact that when we parallel transport vectors along a loop the transported vector can differ from the original one not only through a rotation but also through a displacement. While the infinitesimal rotation angle is related to the curvature tensor, the infinitesimal displacement is related to the torsion tensor. Rigorously we have the following definition.

Definition 15.6.2. Let $\mathscr{M}$ be an m-dimensional manifold and let ∇ denote an affine connection on its tangent bundle. The torsion T_∇ is a map

$$\mathrm{T}_\nabla : \mathfrak{X}(\mathscr{M}) \times \mathfrak{X}(\mathscr{M}) \to \mathfrak{X}(\mathscr{M})$$

defined as follows:

$$\forall \mathbf{X}, \mathbf{Y} \in \mathfrak{X}(\mathscr{M}) : \quad \mathrm{T}_\nabla(\mathbf{X}, \mathbf{Y}) = -\mathrm{T}_\nabla(\mathbf{Y}, \mathbf{X}) \equiv \nabla_{\mathbf{X}}\mathbf{Y} - \nabla_{\mathbf{Y}}\mathbf{X} - [\mathbf{X}, \mathbf{Y}] \in \mathfrak{X}(\mathscr{M}).$$

Given a basis of sections of the tangent bundle $\{\mathbf{e}_\mu\}$ we can calculate their commutators:

$$[\mathbf{e}_\mu, \mathbf{e}_\nu] = K_{\mu\nu}{}^\rho(p)\mathbf{e}_\rho, \tag{15.99}$$

where the point-dependent coefficients $K_{\mu\nu}{}^\rho(p)$ are called the *contorsion* coefficients. They do not form a tensor, since they depend on the choice of basis. For instance in the holonomic basis $\mathbf{e}_\mu = \partial_\mu$ the contorsion coefficients are zero, while they do not vanish in other bases. Notwithstanding their non-tensorial character they can be calculated in any basis and once this is done we obtain a true tensor, namely, the torsion from Definition 15.6.2. Explicitly we have

$$\begin{aligned} T_\nabla(\mathbf{e}_\mu, \mathbf{e}_\nu) &= \mathcal{T}^\rho_{\mu\nu}\mathbf{e}_\rho, \\ \mathcal{T}^\rho_{\mu\nu} &= \Gamma_{\mu\nu}{}^\rho - \Gamma_{\nu\mu}{}^\rho - K_{\mu\nu}{}^\rho. \end{aligned} \tag{15.100}$$

Definition 15.6.3. An affine connection ∇ is named torsionless if its torsion tensor vanishes identically, that is, if $T_\nabla(\mathbf{X}, \mathbf{Y}) = 0$, $\forall \mathbf{X}, \mathbf{Y} \in \mathfrak{X}(\mathcal{M})$.

It follows from equation (15.100) that the coefficients of a torsionless affine connection are symmetric in the lower indices in the holonomic basis. Indeed, if the contorsion vanishes, imposing zero torsion reduces to the condition

$$\Gamma_{\mu\nu}{}^\rho = \Gamma_{\nu\mu}{}^\rho. \tag{15.101}$$

15.6.2.2 The Levi-Civita metric connection

Consider now the case where the manifold $\mathcal{M}$ is endowed with a metric g. Independently of the signature of the latter (Riemannian or pseudo-Riemannian) we can define a unique affine connection which preserves the scalar products defined by g and is torsionless. That affine connection is the Levi-Civita connection. Explicitly we have the following definition.

Definition 15.6.4. Let $(\mathcal{M}, g)$ be a (pseudo-)Riemannian manifold. The associated Levi-Civita connection ∇^g is that unique affine connection which satisfies the following two conditions:

(i) ∇^g is torsionless, namely, $T_{\nabla^g}(\,,\,) = 0$.

(ii) The metric is covariantly constant under the transport defined by ∇^g, that is, $\forall \mathbf{Z}, \mathbf{X}, \mathbf{Y} \in \mathfrak{X}(\mathcal{M}) : \mathbf{Z}g(\mathbf{X}, \mathbf{Y}) = g(\nabla^g_{\mathbf{Z}}\mathbf{X}, \mathbf{Y}) + g(\mathbf{X}, \nabla^g_{\mathbf{Z}}\mathbf{Y})$.

The idea behind such a definition is very simple and intuitive. Consider two vector fields $\mathbf{X}, \mathbf{Y}$. We can measure their scalar product and hence the angle they form by evaluating $g(\mathbf{X}, \mathbf{Y})$. Consider now a third vector field $\mathbf{Z}$ and let us parallel transport the previously given vectors in the direction defined by $\mathbf{Z}$. For an infinitesimal displacement we have $\mathbf{X} \mapsto \mathbf{X} + \nabla_{\mathbf{Z}}\mathbf{X}$ and $\mathbf{Y} \mapsto \mathbf{Y} + \nabla_{\mathbf{Z}}\mathbf{Y}$. We can compare the scalar product of the parallel transported vectors with that of the original ones. Imposing the second condition listed in Definition 15.6.4 corresponds to stating that the scalar product of the parallel transported vectors should just be the increment along Z of the scalar product of the original ones. This is the very intuitive notion of parallelism.

It is now very easy to verify that the Christoffel symbols defined in equation (15.37) are just the coefficients of the Levi-Civita connection in the holonomic basis $\mathbf{e}_\mu = \partial_\mu \equiv \partial/\partial x^\mu$. As we already remarked, in this case the contorsion vanishes and a torsionless connection has symmetric coefficients according to equation (15.101). On the other hand, the second condition of Definition 15.6.4 translates into

$$\partial_\lambda g_{\mu\nu} = \Gamma_{\lambda\mu}{}^{\sigma} g_{\sigma\nu} + \Gamma_{\lambda\nu}{}^{\sigma} g_{\mu\sigma}, \tag{15.102}$$

which admits the Christoffel symbols in (15.37) as a unique solution. There is a standard trick to see this and solve equation (15.102) for Γ. Just write three copies of the same equation with cyclically permuted indices:

$$\partial_\lambda g_{\mu\nu} = \Gamma_{\lambda\mu}{}^{\sigma} g_{\sigma\nu} + \Gamma_{\lambda\nu}{}^{\sigma} g_{\mu\sigma}, \tag{15.103}$$

$$\partial_\mu g_{\nu\lambda} = \Gamma_{\mu\nu}{}^{\sigma} g_{\sigma\lambda} + \Gamma_{\mu\lambda}{}^{\sigma} g_{\nu\sigma}, \tag{15.104}$$

$$\partial_\nu g_{\lambda\mu} = \Gamma_{\nu\lambda}{}^{\sigma} g_{\sigma\mu} + \Gamma_{\nu\mu}{}^{\sigma} g_{\lambda\sigma}. \tag{15.105}$$

Next sum equation (15.103) with equation (15.104) and subtract equation (15.105). In the result of this linear combination use the symmetry of $\Gamma_{\lambda\mu}{}^{\sigma}$ in its lower indices. With this procedure you obtain that $\Gamma_{\lambda\mu}{}^{\sigma}$ is equal to the Christoffel symbols.

Recalling equation (15.36), which defines the covariant derivative of a generic tensor field according to the tensor calculus of Ricci and Levi-Civita, we discover the interpretation of equation (15.102). It just states that the covariant derivative of the metric tensor should be zero:

$$\nabla_\lambda g_{\mu\nu} = 0. \tag{15.106}$$

Hence, the Levi-Civita connection is that affine torsionless connection with respect to which the metric tensor is covariantly constant.

15.7 Geodesics

Once we have an affine connection we can answer the question that was at the root of the whole development of differential geometry, namely, *which lines are straight in a curved space?* To use a car driving analogy, the straight lines are obviously those that imply no turning of the steering wheel. In geometric terms steering the wheel corresponds to changing one's direction while proceeding along the curve, and such a change is precisely measured by the parallel transport of the tangent vector to the curve along itself.

Let $\mathscr{C}(\lambda)$ be a curve $[0,1] \mapsto \mathscr{M}$ in a manifold of dimension m and let $\mathbf{t}$ be its tangent vector. In each coordinate patch the considered curve is represented as $x^\mu = x^\mu(\lambda)$ and the tangent vector has the following components:

$$t^\mu(\lambda) = \frac{d}{d\lambda} x^\mu(\lambda). \tag{15.107}$$

According to the above discussion we can rightly say that a curve is straight if we have

$$\nabla_{\mathbf{t}} \mathbf{t} = 0. \tag{15.108}$$

The above condition immediately translates into a set of m differential equations of the second order for the functions $x^\mu(\lambda)$. Observing that

$$\frac{d}{d\lambda} x^\rho(\lambda) \partial_\rho \left[\frac{d}{d\lambda} x^\mu(\lambda) \right] = \frac{d^2 x^\mu}{d\lambda^2}, \tag{15.109}$$

we conclude that equation (15.108) just coincides with

$$\frac{d^2 x^\mu}{d\lambda^2} + \frac{dx^\rho}{d\lambda} \frac{dx^\sigma}{d\lambda} \Gamma_{\rho\sigma}{}^{\mu} = 0, \tag{15.110}$$

which is named the geodesic equation. The solutions of these differential equations are the straight lines of the considered manifold and are named the *geodesics*. A solution is completely determined by the initial conditions, which, given the order of the differential system, are $2m$. These correspond to giving the values $x^\mu(0)$, namely, the initial point of the curve, and the values of $\frac{dx^\rho}{d\lambda}(0)$, namely, the initial tangent vector $\mathbf{t}(0)$. So we can conclude that at every point $p \in \mathcal{M}$ there is a geodesic departing along any chosen direction in the tangent space $T_p\mathcal{M}$.

We can define geodesics with respect to any affine connection Γ, yet nothing guarantees a priori that such straight lines should also be the shortest routes from one point to another of the considered manifold $\mathcal{M}$. In the case we have a metric structure, lengths are defined and we can consider the variational problem of calculating extremal curves for which any variation makes them longer. It suffices to implement standard variational calculus to the length functional (see equation (15.86)):[3]

$$\begin{aligned} \mathfrak{s} &= \int d\tau \equiv \int \sqrt{2\mathcal{L}}\, d\lambda, \\ \mathcal{L} &\equiv \frac{1}{2} g_{\mu\nu}(x) \frac{dx^\mu}{d\lambda} \frac{dx^\nu}{d\lambda}. \end{aligned} \tag{15.111}$$

Performing a variational calculation we find that the length is extremal if

$$\delta \mathfrak{s} = \int \frac{1}{\sqrt{2\mathcal{L}}} \delta \mathcal{L}\, d\lambda = 0. \tag{15.112}$$

We are free to use any parameter λ to parameterize the curves. Let us use the affine parameter $\lambda = \tau$ defined by the condition

$$2\mathcal{L} = g_{\mu\nu}(x) \frac{dx^\mu}{d\tau} \frac{dx^\nu}{d\tau} = 1. \tag{15.113}$$

3 For a more detailed discussion of the use of the proper length parameter versus a generic affine parameter we refer the reader to Section 3.8 of the first volume of [79].

In this case equation (15.112) reduces to $\delta\mathscr{L} = 0$, which is the standard variational equation for a Lagrangian $\mathscr{L}$, where the affine parameter τ is the time and x^μ are the Lagrangian coordinates q^μ. It is a straightforward exercise to verify that the Euler–Lagrange equation of this system

$$\frac{d}{d\tau}\frac{\partial\mathscr{L}}{\partial\dot{x}^\mu} - \frac{\partial\mathscr{L}}{\partial x^\mu} = 0 \tag{15.114}$$

coincides with the geodesic equation (15.110) where for Γ we use the Christoffel symbols (15.37).

In this way we reach a very important conclusion. The Levi-Civita connection is that unique affine connection for which also in curved space the curves of extremal length (typically the shortest ones) are straight, just as in flat space. This being true, the geodesics can be directly obtained from the variational principle, which is the easiest and fastest way.

15.8 Geodesics in Lorentzian and Riemannian manifolds: two simple examples

Let us now illustrate geodesics using some simple examples which will also be useful to emphasize the difference between Riemannian and Lorentzian manifolds. In a Riemannian manifold the metric is positive definite and there is only one type of geodesics. Indeed, the norm of the tangent vector is always positive and the auxiliary condition (15.113) defining the affine parameter is unique. In a Lorentzian case, on the other hand, we have three kinds of geodesics, depending on the sign of the norm of the tangent vector. *Time-like geodesics* are those where $g(\mathbf{t},\mathbf{t}) > 0$ and the auxiliary condition is precisely stated as in equation (15.113). However, we also have *space-like geodesics*, where $g(\mathbf{t},\mathbf{t}) < 0$, and *null-like geodesics*, where $g(\mathbf{t},\mathbf{t}) = 0$. In these cases the auxiliary condition defining the affine parameter is reformulated as $2\mathscr{L} = -1$ and $2\mathscr{L} = 0$, respectively.

In general relativity, time-like geodesics are the world lines traced in space-time by massive particles that move at a speed less than that of light. Null-like geodesics are the world lines traced by massless particles moving at the speed of light, while space-like geodesics, corresponding to superluminal velocities, violate causality and cannot be traveled by any physical particle.

15.8.1 The Lorentzian example of dS_2

An interesting toy example that can be used to illustrate in a pedagogical way many aspects of the so far developed theory is given by two-dimensional de Sitter space. We can describe this pseudo-Riemannian manifold as an algebraic locus in $\mathbb{R}^3$, writing the following quadratic equation:

$$\mathbb{R}^3 \supset \mathrm{dS}_2 \quad : -X^2 + Y^2 + Z^2 = 1. \tag{15.115}$$

A parametric solution of the defining locus equation (15.115) is easily obtained:

$$X = \sinh t; \quad Y = \cosh t \sin\theta; \quad Z = \cosh t \cos\theta. \tag{15.116}$$

An overall picture of the manifold is given in Figure 15.3. The parameters t and θ can be taken as coordinates on the dS_2 surface on which we can define a Lorentzian metric by means of the pull-back of the standard $\mathrm{SO}(1,2)$ metric on three-dimensional Minkowski space, namely,

$$\begin{aligned} ds^2_{\mathrm{dS}_2} &= -dX^2 + dY^2 + dZ^2 \\ &= -dt^2 + \cosh^2 t\, d\theta^2. \end{aligned} \tag{15.117}$$

The first thing to note about the above metric is that it describes an expanding two-dimensional universe where the spatial sections at constant time t = const are circles $\mathbb{S}^1$. Indeed, the angle θ can be regarded as the coordinate on $\mathbb{S}^1$ and $d\theta^2 = ds^2_{\mathbb{S}^1}$ is the corresponding metric, so we can write

$$ds^2_{\mathrm{dS}_2} = -dt^2 + a^2(t)\, ds^2_{\mathbb{S}^1}, \tag{15.118}$$

where

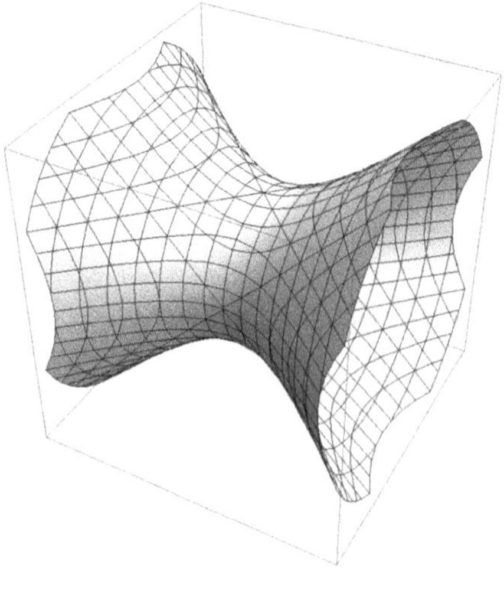

Figure 15.3: Two-dimensional de Sitter space is a hyperbolic rotational surface that can be visualized in three dimensions.

$$a(t) = \cosh t. \tag{15.119}$$

Equation (15.118) provides a paradigm. This is precisely the structure one meets in the discussion of relativistic cosmology (see Chapter 5 of Volume Two of [79]). The second important thing to note about the metric (15.117) is that it has Lorentzian signature. Hence, we are not supposed to find just one type of geodesics; rather, we have to discuss three types of them:

1. The null geodesics for which the tangent vector is light-like.
2. The time-like geodesics for which the tangent vector is time-like.
3. The space-like geodesics for which the tangent vector is space-like.

According to our general discussion, the proper length of any curve on dS_2 is given by the value of the following integral:

$$\mathfrak{s} = \int \sqrt{-\left(\frac{dt}{d\lambda}\right)^2 + \cosh^2 t \left(\frac{d\theta}{d\lambda}\right)^2}\, d\lambda \equiv \int \sqrt{2\mathscr{L}}\, d\lambda, \tag{15.120}$$

where λ is any parameter labeling the points along the curve. Performing a variational calculation we find that the length is extremal if

$$\delta\mathfrak{s} = \int \frac{1}{\sqrt{2\mathscr{L}}} \delta\mathscr{L}\, d\lambda = 0 \quad \Rightarrow \quad \delta\mathscr{L} = 0. \tag{15.121}$$

Hence, as long as we use for λ an *affine parameter*, defined by the auxiliary condition

$$g_{\mu\nu}\frac{dx^\mu}{d\lambda}\frac{dx^\nu}{d\lambda} = -\dot{t}^2 + \cosh^2 t\,\dot{\theta}^2 = k = \begin{cases} 1; & \text{space-like,} \\ 0; & \text{null-like,} \\ -1; & \text{time-like,} \end{cases} \tag{15.122}$$

we can just treat

$$\mathscr{L} = -\dot{t}^2 + \cosh^2 t\,\dot{\theta}^2 \tag{15.123}$$

as the Lagrangian of an ordinary mechanical problem. The corresponding Euler–Lagrange equations of motion are

$$\begin{aligned} 0 &= \partial_\lambda \frac{\partial\mathscr{L}}{\partial\dot{\theta}} - \frac{\partial\mathscr{L}}{\partial\theta} = \partial_\lambda(\cosh^2 t\,\dot{\theta}), \\ 0 &= \partial_\lambda \frac{\partial\mathscr{L}}{\partial\dot{t}} - \frac{\partial\mathscr{L}}{\partial t} = (\partial_\lambda \dot{t} - \cosh t \sinh t\,\dot{\theta}^2). \end{aligned} \tag{15.124}$$

The first of the above equations shows that θ is a cyclic variable and hence we have a first integral of the motion:

$$\text{const} = \ell \equiv \cosh^2 t\dot{\theta}, \tag{15.125}$$

which deserves the name of *angular momentum*. Indeed, the existence of this first integral follows from the SO(2) rotational symmetry of the metric (15.117). For the concept of symmetry of a metric, namely, isometry, we refer the reader to Chapter 16. Thanks to ℓ and the auxiliary condition (15.122), the geodesic equations are immediately reduced to quadratures. Let us discuss the resulting three types of geodesics separately.

15.8.1.1 Null geodesics

For null geodesics we have

$$0 = -\dot{t}^2 + \cosh^2 t\dot{\theta}^2. \tag{15.126}$$

Combining this information with (15.125) we immediately get

$$\begin{aligned} \dot{t} &= \pm\frac{\ell}{\cosh t}, \\ \dot{\theta} &= \frac{\ell}{\cosh^2 t}. \end{aligned} \tag{15.127}$$

The ratio of the above two equations yields the differential equation of the null orbits,

$$\frac{d\theta}{dt} = \pm\frac{1}{\cosh t}, \tag{15.128}$$

which is immediately integrated in the following form:

$$\tan\frac{\theta+\alpha}{2} = \pm\tanh\frac{t}{2}, \tag{15.129}$$

where the arbitrary angle α is the integration constant that parameterizes the family of all possible null-like curves on dS_2. In order to visualize the structure of such curves in the ambient three-dimensional space, it is convenient to use the following elliptic and hyperbolic trigonometric identities:

$$\begin{aligned} \sinh t &= \frac{2\tanh\frac{t}{2}}{1-\tanh^2\frac{t}{2}}; \quad & \cosh t &= \frac{1+\tanh^2\frac{t}{2}}{1-\tanh^2\frac{t}{2}}; \\ \sin\phi &= \frac{2\tan\frac{\phi}{2}}{1+\tan^2\frac{\phi}{2}}; \quad & \cos t &= \frac{1-\tan^2\frac{\phi}{2}}{1+\tan^2\frac{\phi}{2}}. \end{aligned} \tag{15.130}$$

Setting $y = \tanh\frac{t}{2} = \tan\frac{\theta+\alpha}{2}$, utilizing the parametric solution of the locus equations (15.116) and also (15.130), we obtain the form of the null geodesics in $\mathbb{R}^3$:

$$X = x; \quad Y = x\cos\alpha - \sin\alpha; \quad Z = \cos\alpha + x\sin\alpha;$$
$$x \equiv \frac{2y}{1-y^2}. \tag{15.131}$$

It is evident from (15.131) that null geodesics are straight lines that lie on the hyperbolic dS_2 surface (see Figure 15.4).

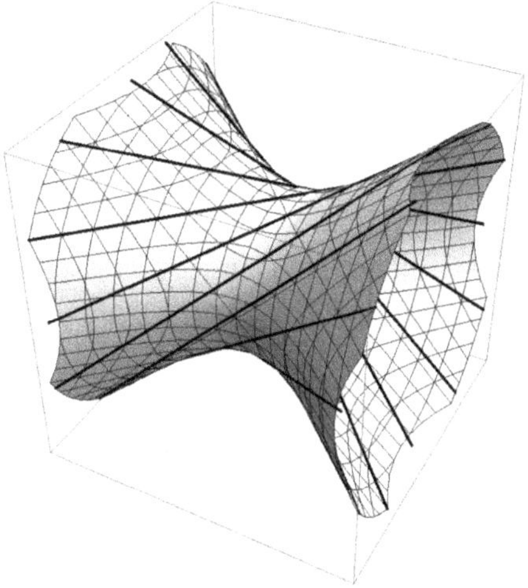

Figure 15.4: The null geodesics on dS_2 are straight lines lying on the hyperbolic surface. In this figure we show a family of these straight lines parameterized by the angle α.

15.8.1.2 Time-like geodesics

For time-like geodesics we have

$$-1 = -\dot{t}^2 + \cosh^2 t\,\dot{\theta}^2, \tag{15.132}$$

and following the same steps as in the previous case we obtain the following differential equation for time-like orbits:

$$\frac{dt}{d\theta} = \pm\frac{1}{\ell}\cosh t\sqrt{\ell^2 + \cosh^2 t}, \tag{15.133}$$

which is immediately reduced to quadratures and integrated as follows:

$$\tan\frac{\theta+\alpha}{2} = \frac{\ell\sinh(t)}{\sqrt{4\ell^2 + 2\cosh(2t) + 2}}. \tag{15.134}$$

Equation (15.134) provides the analytic form of all time-like geodesics in dS_2. The two integration constants are ℓ (the angular momentum) and the angle α.

It is also instructive to visualize the time geodesics as three-dimensional curves that lie on the hyperbolic surface. To this effect we set once again $y = \tan\frac{\theta+\alpha}{2}$, and using both the orbit equation (15.134) and the identities in (15.130) we obtain

$$\begin{aligned}\sin(\theta+\alpha) &= \frac{4\sqrt{2}\ell\sqrt{2\ell^2+\cosh(2t)+1}\sinh(t)}{7\ell^2+(\ell^2+4)\cosh(2t)+4},\\ \cos(\theta+\alpha) &= \frac{4}{\frac{\ell^2\sinh^2(t)}{2\ell^2+\cosh(2t)+1}+2}-1.\end{aligned} \tag{15.135}$$

Changing variables,

$$t = \operatorname{arcsinh} x, \tag{15.136}$$

equation (15.135) combined with the parametric description of the surface (15.116) yields the parametric form of the time-like geodesics in three-dimensional space:

$$\begin{aligned} X &= x,\\ Y &= \frac{4x\sqrt{x^2+1}\ell\sqrt{x^2+\ell^2+1}\cos(\alpha)-\sqrt{x^2+1}(4(\ell^2+1)-x^2(\ell^2-4))\sin(\alpha)}{(\ell^2+4)x^2+4(\ell^2+1)},\\ Z &= \frac{\sqrt{x^2+1}(4(\ell^2+1)-x^2(\ell^2-4))\cos(\alpha)+4x\sqrt{x^2+1}\ell\sqrt{x^2+\ell^2+1}\sin(\alpha)}{(\ell^2+4)x^2+4(\ell^2+1)}.\end{aligned} \tag{15.137}$$

See Figure 15.5.

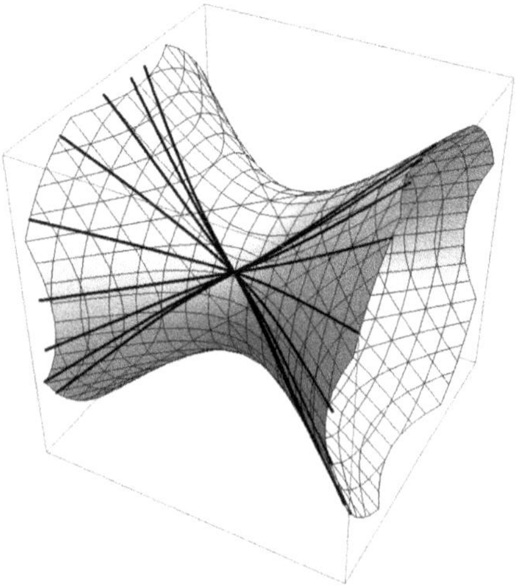

Figure 15.5: The time-like geodesics on dS_2. In this figure we show a family of geodesics parameterized by the value of the angular momentum ℓ and passing through the same point.

15.8.1.3 Space-like geodesics

For space-like geodesics we have

$$1 = -\dot{t}^2 + \cosh^2 t\dot{\theta}^2, \tag{15.138}$$

and we obtain the following differential equation for space-like orbits:

$$\frac{dt}{d\theta} = \pm\frac{1}{\ell}\cosh t\sqrt{\ell^2 - \cosh^2 t}, \tag{15.139}$$

which is integrated as follows:

$$\tan\frac{\theta + \alpha}{2} = \frac{\ell\sinh(t)}{\sqrt{4\ell^2 - 2\cosh(2t) - 2}}. \tag{15.140}$$

As one sees, the difference with respect to the equation describing time-like orbits resides only in two signs. By means of algebraic substitutions completely analogous to those used in the previous case we obtain the parameterization of the space-like geodesics as three-dimensional curves. We find

$$\begin{aligned}
X &= x,\\
Y &= \frac{4x\sqrt{x^2+1}\ell\sqrt{-x^2+\ell^2-1}\cos(\alpha) + \sqrt{x^2+1}((\ell^2+4)x^2 - 4\ell^2 + 4)\sin(\alpha)}{(x^2+4)\ell^2 - 4(x^2+1)},\\
Z &= \frac{4x\sqrt{x^2+1}\ell\sqrt{-x^2+\ell^2-1}\sin(\alpha) - \sqrt{x^2+1}((\ell^2+4)x^2 - 4\ell^2 + 4)\cos(\alpha)}{(x^2+4)\ell^2 - 4(x^2+1)}.
\end{aligned} \tag{15.141}$$

The sign changes with respect to the time-like case have significant consequences. For a given value ℓ of the angular momentum the range of the X coordinate and hence of the x parameter is limited by

$$-\sqrt{\ell^2 - 1} < x < \sqrt{\ell^2 - 1}. \tag{15.142}$$

Outside of this range coordinates become imaginary, as is evident from equation (15.141). In Figure 15.6 we display the shape of a family of space-like geodesics.

15.8.2 The Riemannian example of the Lobachevsky–Poincaré plane

The second example of geodesic calculation we present is that related to the hyperbolic upper plane model of Lobachevsky geometry found by Poincaré.

As many readers already know, the question of whether non-Euclidean geometries did or did not exist was a central issue in mathematical and philosophical thought for almost 2000 years. The crucial question was whether the fifth postulate of Euclid about

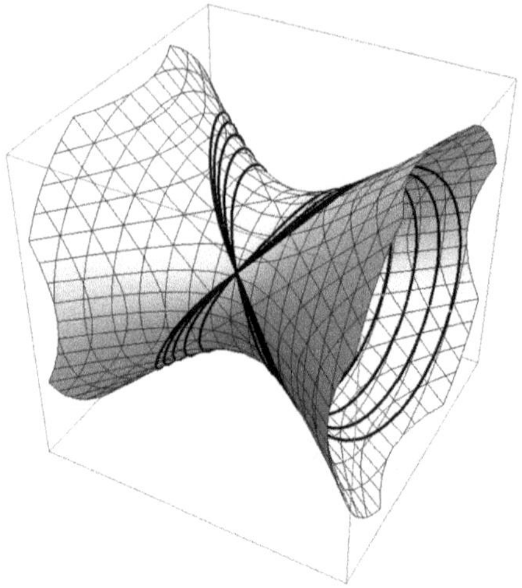

Figure 15.6: The space-like geodesics on dS_2. In this figure we show a family of space-time geodesics all going through the same point.

parallel lines was independent of the previous ones or not. Many mathematicians tried to demonstrate the fifth postulate and typically came to erroneous or tautological conclusions since, as we now know, the fifth postulate is indeed independent and distinguishes Euclidean from other equally self-consistent geometries. The first attempt of a proof, by *Posidonius of Rhodes* (135–51 BC), dates back to as early as the first century BC. This encyclopedic scholar, acclaimed as one of the most erudite men of his epoch, tried to modify the definition of parallelism in order to prove the postulate, but came to inconclusive and contradictory statements. In the modern era the most interesting and deepest attempt at the proof of the postulate is that of the Italian Jesuit *Giovanni Girolamo Saccheri* (1667–1733). In his book *Euclides ab omni naevo vindicatus*, Saccheri tried to demonstrate the postulate with a *reductio ad absurdum* argument. So doing, he actually proved a series of theorems in non-Euclidean geometry whose implications seemed so unnatural and remote from sensorial experience that Saccheri considered them absurd and flattered himself with the presumption of having proved the fifth postulate. The first to discover a consistent model of non-Euclidean geometry was probably Gauss around 1828. However, he refrained from publishing his result since he did not wish to hear the *screams of Boeotians*, referring to the German philosophers of the time who, following Kant, considered Euclidean geometry an *a priori truth* of human thought. Less influenced by post-Kantian philosophy in the remote town of Kazan of whose university he was for many years the rector, the Russian mathematician *Nicolai Ivanovich Lobachevsky* (1793–1856) discovered and formulated a consistent axiomatic setup of non-Euclidean geometry where the fifth postulate did not hold true and the sum of internal angles of a triangle was less than π. An explicit model of Lobachevsky geometry was first created by *Eugenio Beltrami* (1836–1900) by means of lines drawn on the hyperbolic surface known as the pseudo-sphere and then analytically realized by *Henri Poincaré* (1854–1912) some years later.

In 1882 Poincaré defined the following two-dimensional Riemannian manifold $(\mathcal{M}, g)$, where $\mathcal{M}$ is the upper plane

$$\mathbb{R}^2 \supset \mathcal{M} : \quad (x, y) \in \mathcal{M} \quad \Leftrightarrow \quad y > 0 \tag{15.143}$$

and the metric g is defined by the following infinitesimal line element:

$$ds^2 = \frac{dx^2 + dy^2}{y^2}. \tag{15.144}$$

Lobachevsky geometry is realized by all polygons in the upper plane $\mathcal{M}$ whose sides are arcs of geodesics with respect to the Poincaré metric (15.144). Let us derive the general form of such geodesics.

This time the metric has Euclidean signature and there is just one type of geodesic curves. Following our variational method the effective Lagrangian is

$$\mathcal{L} = \frac{1}{2}\frac{\dot{x}^2 + \dot{y}^2}{y^2}, \tag{15.145}$$

where the dot denotes derivatives with respect to the length parameter s. The Lagrangian variable x is cyclic (namely, it appears only under derivatives) and from this fact we immediately obtain a first order integral of motion:

$$\frac{\dot{x}}{y^2} = \frac{1}{R} = \text{const.} \tag{15.146}$$

The name R given to this conserved quantity follows from its geometrical interpretation that we will next discover. Using the information from (15.146) in the auxiliary condition

$$2\mathcal{L} = 1, \tag{15.147}$$

which defines the affine length parameter we obtain

$$\dot{y}^2 = \left(1 - \frac{1}{R^2}y^2\right)y^2, \tag{15.148}$$

and by eliminating ds between equations (15.146) and (15.148) we obtain

$$\frac{1}{R}dx = \frac{y\,dy}{\sqrt{1 - \frac{y^2}{R^2}}}, \tag{15.149}$$

which upon integration yields

$$\frac{1}{R}(x - x_0) = \sqrt{1 - \frac{y^2}{R^2}}, \tag{15.150}$$

where x_0 is the integration constant. Squaring the above relation we get

$$(x - x_0)^2 + y^2 = R^2, \tag{15.151}$$

which has an immediate interpretation. A geodesic is just the arc lying in the upper plane of any circle of radius R having center in $(x_0, 0)$, that is, some point lying on the real axis.

With this result Lobachevsky geometry is easily visualized. Examples of planar figures with geodesic sides are presented in Figure 15.7.

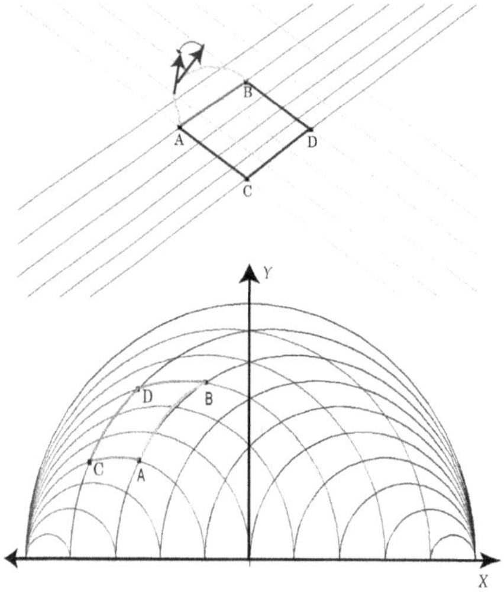

Figure 15.7: The geodesics of the Poincaré metric in the upper plane compared to the geodesics of the Euclidean metric, namely, the straight lines.

15.8.3 Another Riemannian example: the catenoid

As a further Riemannian example we consider the catenoid revolution surface introduced in 1744 by Leonhard Euler [125].

The catenoid is parametrically described as follows:

$$\begin{aligned} X &= \frac{1}{\sqrt{2}}\cos(\theta)\cosh(C), \\ Y &= \frac{1}{\sqrt{2}}\sin(\theta)\cosh(C), \\ Z &= \frac{1}{\sqrt{2}}C, \end{aligned} \tag{15.152}$$

where $\theta \in [0, 2\pi]$ is an angle and $C \in [-\infty, +\infty]$ is a real parameter. Its appearance in three-dimensional space is displayed in Figure 15.8.

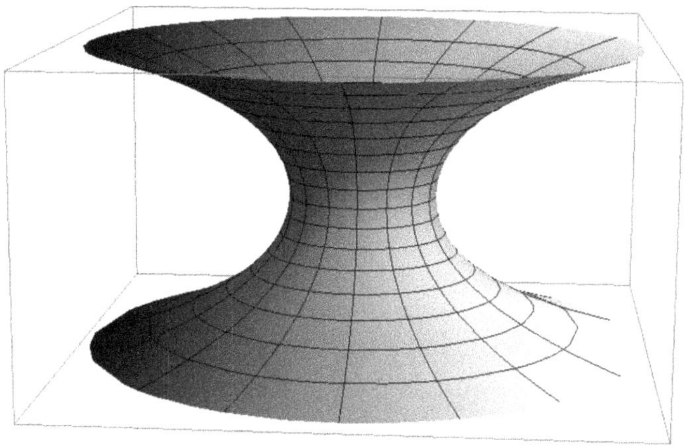

Figure 15.8: A representation of the catenoid as a surface immersed in three-dimensional space.

The pull-back of the ambient flat metric $ds^2_{\mathbb{E}^3} = dX^2 + dY^2 + dZ^2$ on the locus (15.152) yields the following two-dimensional Riemannian metric:

$$ds^2_{\text{catenoid}} = \frac{1}{2}\cosh^2 C(dC^2 + d\theta^2). \tag{15.153}$$

We derive the geodesics of the above metric.

As in the previous cases we realize that there is a cyclic variable which appears in the metric only through its own differential, namely, the angle θ. This tells us that every geodesic is characterized by a first integral of motion, which we can call the *angular momentum* of the *particle* moving on the catenoid surface:

$$J = \cosh^2 C\frac{d\theta}{d\tau}. \tag{15.154}$$

Substituting back the condition (15.154) in the metric (15.153), we obtain

$$\left(\frac{ds}{d\tau}\right)^2 = \frac{1}{2}\operatorname{sech}^2(C)\left(\left(\frac{dC}{d\tau}\right)^2\cosh^4(C) + J^2\right). \tag{15.155}$$

Utilizing as parameter along the geodesic the arc length, namely, setting $\tau = s$, we obtain

$$\frac{dC}{ds} = \pm\sqrt{\operatorname{sech}^4(C)(\cosh(2C) - J^2 + 1)}, \tag{15.156}$$

and eliminating the variable s between equation (15.156) and equation (15.154) we get the orbit equation

$$\frac{\mathrm{d}\theta}{\mathrm{d}C} = \frac{J}{\sqrt{\cosh(2C) - J^2 + 1}}, \tag{15.157}$$

which by integration yields

$$\begin{aligned} \theta &= \mathfrak{f}(C,J) + \theta_0, \\ \mathfrak{f}(C,J) &\equiv -\frac{iJ\sqrt{\frac{-\cosh(2C)+J^2-1}{J^2-2}}\,F(iC| - \frac{2}{J^2-2})}{\sqrt{\cosh(2C) - J^2 + 1}}, \end{aligned} \tag{15.158}$$

where $F(\phi\,|m) \equiv \int_0^\phi (1 - m\sin^2\theta)^{-1/2}\,d\theta$ denotes the elliptic integral of specified parameters.

In this way we have solved the geodesic problem completely. We have two integration constants, namely, J and the reference angle θ_0, and the three-dimensional image of the geodesic curves is parametrically provided by

$$\begin{aligned} X &= \frac{1}{\sqrt{2}}\cos(\mathfrak{f}(C,J) + \theta_0)\cosh(C), \\ Y &= \frac{1}{\sqrt{2}}\sin(\mathfrak{f}(C,J) + \theta_0)\cosh(C), \\ Z &= \frac{1}{\sqrt{2}}C; \quad C \in [-\infty, +\infty]. \end{aligned} \tag{15.159}$$

A few of them are displayed in Figure 15.9.

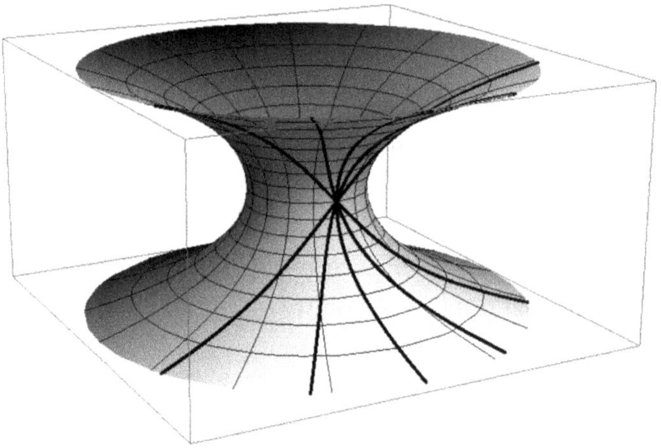

Figure 15.9: A family of geodesics on the catenoid surface all passing through the same point.

15.9 Bibliographical note

Bibliographical sources of the material presented in this chapter are:
1. Volume 1 of [79].
2. Volume 1 of the three-volume monograph on supergravity [126].
3. The textbook [38].
4. The textbook [42].
5. The textbook [76].

16 Isometries and the geometry of coset manifolds

The art of doing mathematics consists in finding that special case which contains all the germs of generality.

David Hilbert

16.1 Conceptual and historical introduction

The word isometry comes from the ancient Greek word *ισομετρία*, which means equality of measures.

The origin of the modern concept of isometry is rooted in that of *congruence* of geometrical figures, which Euclid never introduced explicitly, yet implicitly assumed when he proceeded to identify those triangles that can be superimposed onto one another.

It was indeed the question which transformations define such congruences that led Felix Klein to the famous Erlangen Program.[1] Klein understood that Euclidean congruences are based on the transformations of the Euclidean group, and he came to the idea that other geometries are based on different groups of transformations with respect to which we consider congruences.

Such a concept, however, would have been essentially empty without the additional element which was the subject of Chapter 15, namely, the *metric*. The area and the volume of geometrical figures, the length of sides, and the relative angles have to be measured in order to compare them. These measurements can be performed if and only if we have a metric g, in other words, if the *substratum* of the considered geometry is a *Riemannian* or a *pseudo-Riemannian* manifold $(\mathcal{M}, g)$.

Therefore, the group of transformations which according to the vision of the Erlangen Program defines a geometry is the *group of isometries* $\mathrm{G}_{\mathrm{iso}}$ of a given Riemannian space $(\mathcal{M}, g)$, the elements of this group being diffeomorphisms,

$$\phi : \mathcal{M} \to \mathcal{M}, \tag{16.1}$$

such that their pull-back on the metric form leaves it invariant:

$$\forall \phi \in \mathrm{G}_{\mathrm{iso}} : \quad \phi^\star \left[g_{\mu\nu}(x)\, dx^\mu\, dx^\nu \right] = g_{\mu\nu}(x)\, dx^\mu\, dx^\nu. \tag{16.2}$$

Quite intuitively it becomes clear that the structure of $\mathrm{G}_{\mathrm{iso}}$ is determined by the manifold $\mathcal{M}$ and by its metric g, so that the concept of geometries is now identified with that of Riemannian spaces $(\mathcal{M}, g)$.

1 For a historical account of the conceptual developments linked with the Erlangen Program see the book [28] by the present author.

https://doi.org/10.1515/9783111201535-016

A generic metric g has no isometries and hence there are no congruences to study. (Pseudo-)Riemannian manifolds with no isometry, or with few isometries, are relevant to several different problems pertaining to physics and also to other sciences, yet they are not in the vein of the Erlangen Program, aiming at the classification of geometries in terms of groups. Hence, we can legitimately ask ourselves the question whether such a program can be ultimately saved, notwithstanding our discovery that a geometry is necessarily based on a (pseudo-)Riemannian manifold $(\mathcal{M}, g)$. The answer is obviously yes if we can invert the relation between the metric g and its isometry group $\mathrm{G}_{\mathrm{iso}}$. Given a Lie group G, can we construct the Riemannian manifold $(\mathcal{M}, g)$ which admits G as its own isometry group $\mathrm{G}_{\mathrm{iso}}$? Indeed we can; the answers are also exhaustive if we add an additional requirement, namely, that of transitivity.

Definition 16.1.1. A group G acting on a manifold $\mathcal{M}$ by means of diffeomorphisms

$$\forall \gamma \in \mathrm{G}, \quad \gamma : \mathcal{M} \to \mathcal{M} \tag{16.3}$$

has a transitive action if and only if

$$\forall p, q \in \mathcal{M}, \quad \exists \gamma \in \mathrm{G} / \gamma(q) = p. \tag{16.4}$$

If the Riemannian manifold $(\mathcal{M}, g)$ admits a transitive group of isometries, this means that any point of $\mathcal{M}$ can be mapped into any other by means of a transformation that is an isometry. In this case the very manifold $\mathcal{M}$ and its metric g are completely determined by group theory: $\mathcal{M}$ is necessarily a *coset manifold* G/H, namely, the space of equivalence classes of elements of G with respect to multiplication (either on the right or on the left) by elements of a subgroup $\mathrm{H} \subset \mathrm{G}$ (see Section 4.2.2). The metric g is induced on the equivalence classes by the Killing metric of the Lie algebra, defined on $\mathbb{G}$.

The present chapter, after a study of Killing vector fields, namely, of the infinitesimal generators that realize the Lie algebra $\mathbb{G}$ of the isometry group, will be devoted to the geometry of coset manifolds. Among them particular attention will be given to the so-called *symmetric spaces* characterized by an additional reflection symmetry whose nature will become clear to the reader in the following sections.

16.1.1 Symmetric spaces and Élie Cartan

The full-fledged classification of all symmetric spaces was the gigantic achievement of Élie Cartan. The classification of symmetric spaces is at the same time a classification of the real forms of the complex Lie algebras and it is the conclusive step in the path initiated with the classification of complex Lie algebras, to which Chapter 11 was devoted. At the same time, the geometries of non-compact symmetric spaces can be formulated in terms of other quite interesting algebraic structures, the *normed solvable Lie algebras*. The class of the latter is wider than that of symmetric spaces, and this provides a generalization path leading to a wider class of geometries, all of them under firm algebraic control. This will be the topic of the last two sections of the present chapter, which is propaedeutic to the most advanced topics in Lie algebra theory currently utilized in modern theoretical physics.[2]

16.1.2 Where and how do coset manifolds come into play?

By now it should be clear to the reader that, just as we have the whole spectrum of linear representations of a Lie algebra $\mathbb{G}$ and of its corresponding Lie group G, in the same way we have the set of *non-linear representations* of the same Lie algebra $\mathbb{G}$ and of the same Lie group G. These are encoded in all possible coset manifolds G/H with their associated G-invariant metrics.

Where and how do these geometries pop up?

The answer is that they appear at several different levels of analysis and in connection with different aspects of both physical theories and other branches of science. Let us enumerate them.

(A) A first context of utilization of coset manifolds G/H is in the quest for solutions of Einstein equations in $d = 4$ or in higher dimensions. One is typically interested in space-times with a prescribed *isometry* and one tries to fit into the equations G/H metrics whose parameters depend on some residual coordinate like the time t in cosmology or the radius r in black-hole physics. The field equations of the theory reduce to few-parameter differential equations in the residual space.

(B) Another instance of utilization of coset manifolds is in the context of σ-models. In physical theories that include scalar fields $\phi^I(x)$ the kinetic term is necessarily of the following form:

$$\mathcal{L}_{\text{kin}} = \frac{1}{2}\gamma_{IJ}(\phi)\partial_\mu\phi^I(x)\partial_\nu\phi^J(x)g^{\mu\nu}(x), \tag{16.5}$$

where $g^{\mu\nu}(x)$ is the metric of space-time, while $\gamma_{IJ}(\phi)$ can be interpreted as the metric of some manifold $\mathcal{M}_{\text{target}}$ of which the fields ϕ^I are the coordinates and whose dimension is just equal to the number of scalar fields present in the theory. If we require the field theory to have some Lie group symmetry G, we have either linear representations or non-linear ones. In the first case the metric γ_{IJ} is constant and invariant under the linear transformations of G acting on $\phi^I(x)$. In the second case the manifold $\mathcal{M}_{\text{target}}$ = G/H is some coset of the considered group and $\gamma_{IJ}(\phi)$ is the corresponding G-invariant metric.

2 For an account of the advances in Lie algebra and geometry streaming from modern theoretical physics see the book [72] by the present author.

(C) In mathematics and sometimes in physics you can consider structures that depend on a continuous set of parameters, for instance the solutions of certain differential equations, like the self-duality constraint for gauge field strengths or the Ricci-flat metrics on certain manifolds, or the algebraic surfaces of a certain degree in some projective spaces. The parameters corresponding to all the possible deformations of the considered structure constitute themselves a manifold $\mathcal{M}$ which typically has some symmetries and in many cases is actually a coset manifold. A typical example is provided by the so-called Kummer surface K3, whose Ricci-flat metric no one has so far constructed, yet we know a priori that it depends on 3×19 parameters that span the homogeneous space $\frac{\mathrm{SO}(3,19)}{\mathrm{SO}(3)\times\mathrm{SO}(19)}$.

(D) In many instances of field theories that include scalar fields there is a scalar potential term $\mathrm{V}(\phi)$ which has a certain group of symmetries G. The vacua of the theory, namely, the set of extrema of the potential, usually fill up a coset manifold G/H, where $\mathrm{H} \subset \mathrm{G}$ is the residual symmetry of the vacuum configuration $\phi = \phi_0$.

(E) In condensed matter theories quite often it happens that the *order parameter* takes values in a G/H symmetric space that encodes the symmetries of the physical system.

(F) In the last few years, as we remarked several times in the course of previous chapters, a lot of attention has been attracted by new approaches to *convolutionary neural networks* (CNNs) utilized in *deep learning* that include principles of symmetry by means of imposing *equivariance* with respect to some group G. As shown for instance in [33] and [34], the geometry of coset manifolds G/H is just the mathematical environment where the structure of such CNNs can be formulated.

16.1.3 The deep insight of supersymmetry

In supersymmetric field theories, in particular in supergravities that are supersymmetric extensions of Einstein gravity coupled to matter multiplets, all the uses listed above of coset manifolds do occur, but there is an additional ingredient whose consequences are very deep and far reaching for geometry: supersymmetry itself. Consistency with supersymmetry introduces further restrictions on the geometry of target manifolds $\mathcal{M}_{\text{target}}$ that are required to fall in specialized categories like *Kähler manifolds*, *special Kähler manifolds*, *quaternionic Kähler manifolds*, and so on. These geometries, which we collectively dub *special geometries*, require the existence of complex structures and encompass both manifolds that do not have transitive groups of isometries and homogeneous manifolds G/H. In the second case, which is one of the main focuses of interest in the essay [72], the combination of the special structures with the theory of Lie algebras produces new insights in homogenous geometries that would have been inconceivable outside the framework of supergravity. This is what we call the deep geometrical insight of supersymmetry. As several examples manifest (for instance supersymmetry in condensed matter conformal field theories, application of supergravity to the theory of graphene [127], and more), the new geometric structures introduced by supergravity have a potentially much larger field of applications than fundamental physical theories that can pop up unexpectedly here and there in connection with different problems in different disciplines.

In the present book, which is introductory although comprehensive with respect to aspects of group theory, we just provide a primary on the notions of G/H geometries.

16.2 Isometries and Killing vector fields

The existence of continuous isometries is related with the existence of Killing vector fields. Here we explain the underlying mathematical theory which leads to the study of coset manifolds and symmetric spaces.

Suppose that the diffeomorphism considered in equation (16.1) is infinitesimally close to the identity

$$x^\mu \to \phi^\mu(x) \simeq x^\mu + k^\mu(x). \tag{16.6}$$

The condition for this diffeomorphism to be an isometry is a differential equation for the components of the vector field $\mathbf{k} = k^\mu \partial_\mu$ which immediately follows from (16.2):

$$\nabla_\mu k_\nu + \nabla_\nu k_\mu = 0. \tag{16.7}$$

Hence, given a metric one can investigate the nature of its isometries by trying to solve the linear homogeneous equation (16.7) determining its general integral. The important point is that, if we have two Killing vectors $\mathbf{k}$ and $\mathbf{w}$, also their commutator $[\mathbf{k}, \mathbf{w}]$ is a Killing vector. This follows from the fact that the product of two finite isometries is also an isometry. Hence, Killing vector fields form a finite-dimensional Lie algebra $\mathbb{G}_{\text{iso}}$ and one can turn the question around. Rather than calculating the isometries of a given metric, one can address the problem of constructing (pseudo-)Riemannian manifolds that have a prescribed isometry algebra. Due to the well-established classification of semi-simple Lie algebras this becomes a very fruitful point of view.

In particular, one is interested in homogeneous spaces, namely, in (pseudo-)Riemannian manifolds where each point of the manifold can be reached from a reference one by the action of an isometry.

Homogeneous spaces are identified with coset manifolds, whose differential geometry can be thoroughly described and calculated in pure Lie algebra terms.

16.3 Coset manifolds

Coset manifolds are a natural generalization of group manifolds and play a very important ubiquitous role in mathematics, physics, and several other branches of science, last but not least deep learning and artificial intelligence.

In group theory (irrespectively of whether the group G is finite or infinite and continuous or discrete) we have the concept of *coset space* G/H, which is just the set of equivalence classes of elements $g \in \mathrm{G}$, where the equivalence is defined by right multiplication with elements $h \in \mathrm{H} \subset \mathrm{G}$ of a subgroup:

$$\forall g, g' \in \mathrm{G}: \quad g \sim g' \quad \text{iff} \quad \exists h \in \mathrm{H} \backslash gh = g'. \tag{16.8}$$

That is, two group elements are equivalent if and only if they can be mapped into each other by means of some element of the subgroup. The equivalence classes which constitute the elements of G/H are usually denoted $g\mathrm{H}$, where g is any representative of the class, that is, any one of the equivalent G-group elements the class is composed of. The definition we have just provided by means of right multiplication can be obviously replaced by an analogous one based on left multiplication. In this case we construct the coset H\G composed of *right lateral classes* Hg, while gH are named the *left lateral classes*. For non-Abelian groups G and generic subgroups H the left G/H and right H\G coset spaces have different not coinciding elements. Working with one or with the other definition is just a matter of conventions. We choose to work with *left classes*.

Coset manifolds arise in the context of Lie group theory when G is a Lie group and H is a Lie subgroup thereof. In that case the set of lateral classes gH can be endowed with a manifold structure inherited from the manifold structure of the parent group G. Furthermore, on G/H we can construct *invariant metrics* such that all elements of the original group G are isometries of the constructed metric. As we show below, the curvature tensor of invariant metrics on coset manifolds can be constructed in purely algebraic terms starting from the structure constants of the Lie algebra $\mathbb{G}$, bypassing all analytic differential calculations.

Coset manifolds are easily identified with *homogeneous spaces* which we presently define.

Definition 16.3.1. A Riemannian or pseudo-Riemannian manifold $\mathscr{M}_g$ is said to be homogeneous if it admits as an isometry the transitive action of a group G. A group acts transitively if any point of the manifold can be reached from any other by means of the group action.

A notable and very common example of such homogeneous manifolds is provided by the spheres $\mathbb{S}^n$ and their non-compact generalizations, the pseudo-spheres $\mathbb{H}_{\pm}^{(n+1-m,m)}$. Let x^I denote the Cartesian coordinates in $\mathbb{R}^{n+1}$ and let

$$\eta_{IJ} = \mathrm{diag}(\underbrace{+, + \cdots, +}_{n+1-m}, \underbrace{-, -, \ldots, -}_{m}) \tag{16.9}$$

be the coefficient of a non-degenerate quadratic form with signature $(n+1-m, m)$:

$$\langle \mathbf{x}, \mathbf{x} \rangle_\eta \equiv x^I x^J \eta_{IJ}. \tag{16.10}$$

We obtain a pseudo-sphere $\mathbb{H}_{\pm}^{(n+1-m,m)}$ by defining the algebraic locus

$$\mathbf{x} \in \mathbb{H}_{\pm}^{(n+1-m,m)} \quad \Leftrightarrow \quad \langle \mathbf{x}, \mathbf{x} \rangle_{\eta} \equiv \pm 1, \tag{16.11}$$

which is a manifold of dimension n. The spheres $\mathbb{S}^n$ correspond to the particular case $\mathbb{H}_{+}^{n+1,0}$, where the quadratic form is positive definite and the sign in the right hand side of equation (16.11) is positive. Obviously with a positive definite quadratic form this is the only possibility.

All these algebraic loci are invariant under the transitive action of the group $\mathrm{SO}(n+1, n+1-m)$ realized by matrix multiplication on the vector $\mathbf{x}$ since

$$\forall g \in \mathrm{G}: \quad \langle \mathbf{x}, \mathbf{x} \rangle_{\eta} = \pm 1 \quad \Leftrightarrow \quad \langle g\mathbf{x}, g\mathbf{x} \rangle_{\eta} = \pm 1, \tag{16.12}$$

that is, the group maps solutions of the constraint (16.11) into solutions of the same and, furthermore, all solutions can be generated starting from a standard reference vector:

$$\langle \mathbf{x}, \mathbf{x} \rangle_{\eta} = 0 \quad \Rightarrow \quad \exists g \in \mathrm{G} \backslash \mathbf{x} = g\mathbf{x}_0^{\pm}, \tag{16.13}$$

where

$$x_0^{+} = \begin{pmatrix} 1 \\ 0 \\ \vdots \\ 0 \\ \hline 0 \\ 0 \\ \vdots \\ 0 \end{pmatrix} ; \quad x_0^{-} = \begin{pmatrix} 0 \\ 0 \\ \vdots \\ 0 \\ \hline 1 \\ 0 \\ \vdots \\ 0 \end{pmatrix}, \tag{16.14}$$

the line separating the first $n+1-m$ entries from the last m. Equation (16.13) guarantees that the locus is invariant under the action of G, while equation (16.14) states that G is transitive.

Definition 16.3.2. In a homogeneous space $\mathcal{M}_g$, the subgroup $\mathrm{H}_p \subset \mathrm{G}$ which leaves a point $p \in \mathcal{M}_g$ fixed ($\forall h \in \mathrm{H}_p$, $hp = p$) is named the *isotropy subgroup* of the point.[3] Because of the transitive action of G, any other point $p' = gp$ has an isotropy subgroup $\mathrm{H}_{p'} = g\mathrm{H}_p g^{-1}$ which is conjugate to H_p and therefore isomorphic to it.

3 Let us remark that the notion of isotropy subgroup utilized while discussing coset manifolds abstractly coincides with the notion of stability subgroup generically used when a group G acts as a transformation group on a set (see Section 4.2.12).

It follows that, up to conjugation, the isotropy group of a homogeneous manifold $\mathcal{M}_g$ is unique and corresponds to an intrinsic property of such a space. It suffices to calculate the isotropy group H_0 of a conventional properly chosen reference point p_0; all other isotropy groups will immediately follow. For brevity H_0 will be just renamed H.

In our example of the spaces $\mathbb{H}_{\pm}^{(n+1-m,m)}$ the isotropy group is immediately derived by looking at the form of the vectors $\mathbf{x}_0^{\pm}$: all elements of G which rotate the vanishing entries of these vectors among themselves are clearly elements of the isotropy group. Hence, we find

$$\begin{aligned} H &= SO(n, m) \quad \text{for } \mathbb{H}_{+}^{(n+1-m,m)}, \\ H &= SO(n+1, m-1) \quad \text{for } \mathbb{H}_{-}^{(n+1-m,m)}. \end{aligned} \tag{16.15}$$

It is natural to label any point p of a homogeneous space by the parameters describing the G-group element which carries a conventional point p_0 into p. These parameters, however, are redundant: because of the H-isotropy there are infinitely many ways to reach p from p_0. Indeed, if g does that job, any other element of the lateral class gH does the same. It follows by this simple discussion that the homogeneous manifold $\mathcal{M}_g$ can be identified with the coset manifold G/H defined by the transitive group G divided by the isotropy group H.

Focusing once again on our example we find

$$\mathbb{H}_{+}^{(n+1-m,m)} = \frac{SO(n+1-m, m)}{SO(n-m, m)}; \quad \mathbb{H}_{-}^{(n+1-m,m)} = \frac{SO(n+1-m, m)}{SO(n+1-m, m-1)}. \tag{16.16}$$

In particular the spheres correspond to

$$\mathbb{S}^n = \mathbb{H}_{+}^{(n+1,0)} = \frac{SO(n+1)}{SO(n)}. \tag{16.17}$$

Other important examples are

$$\mathbb{H}_{+}^{(n+1,1)} = \frac{SO(n+1, 1)}{SO(n, 1)}; \quad \mathbb{H}_{-}^{(n+1,1)} = \frac{SO(n+1, 1)}{SO(n+1)}. \tag{16.18}$$

The general classification of homogeneous (pseudo-)Riemannian spaces corresponds therefore to the classification of the coset manifolds G/H for all Lie groups G and their closed Lie subgroups H ⊂ G.

The equivalence classes constituting the points of the coset manifold can be labeled by a set of d coordinates $y \equiv \{y^1, \dots, y^d\}$, where

$$d = \dim \frac{G}{H} \equiv \dim G - \dim H. \tag{16.19}$$

There are of course many different ways of choosing the y-parameters since, just as in any other manifold, there are many possible coordinate systems. What is specific to coset manifolds is that, given any coordinate system y by means of which we label the equivalence classes, within each equivalence class we can choose a representative group element $\mathbb{L}(y) \in \mathrm{G}$. The choice must be made in such a way that $\mathbb{L}(y)$ should be a smooth function of the parameters y. Furthermore, for different values y and y', the group elements $\mathbb{L}(y)$ and $\mathbb{L}(y')$ should never be equivalent; in other words, no $h \in \mathrm{H}$ should exist such that $\mathbb{L}(y) = \mathbb{L}(y')h$. Under left multiplication by $g \in \mathrm{G}$, $\mathbb{L}(y)$ is in general carried into another equivalence class with coset representative $\mathbb{L}(y')$. Yet the g image of $\mathbb{L}(y)$ is not necessarily $\mathbb{L}(y')$: it is typically some other element of the same class, so that we can write

$$\forall g \in G: \quad g\mathbb{L}(y) = \mathbb{L}(y')h(g,y); \quad h(g,y) \in \mathrm{H}, \tag{16.20}$$

where we emphasized that the H-element necessary to map $\mathbb{L}(y')$ into the g-image of $\mathbb{L}(y)$ depends in general both on the point y and on the chosen transformation g. Equation (16.20) is pictorially described in Figure 16.1. For the spheres a possible set of coordinates y can be obtained by means of the stereographic projection whose concept is recalled here. Considering the $\mathbb{S}^n$ sphere immersed in $\mathbb{R}^{n+1}$, from the North Pole $\{1, 0, 0, \dots, 0\}$ one draws the line that goes through the point $p \in \mathbb{S}^n$ and considers the point $\pi(p) \in \mathbb{R}^n$ where such a line intersects the $\mathbb{R}^n$ plane tangent to the sphere in the South Pole and orthogonal to the line that joins the North and the South Pole. The n coordinates $\{y^1, \dots, y^n\}$ of $\pi(p)$ can be taken as labels of an open chart in $\mathbb{S}^n$ (see Figure 6.3).

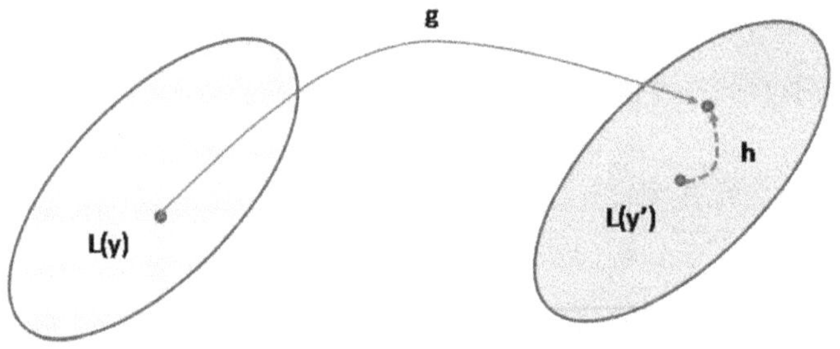

Figure 16.1: Pictorial description of the action of the group G on the coset representatives.

As another explicit example, we consider the case of the Euclidean hyperbolic spaces $\mathbb{H}_{-}^{(n,1)}$ identified as coset manifolds in equation (16.18). In this case, to introduce a coset parameterization means to write a family of SO(n, 1) matrices $\mathbb{L}(\mathbf{y})$ depending smoothly on an n-component vector $\mathbf{y}$ in such a way that for different values of $\mathbf{y}$ such matrices cannot be mapped into one another by means of right multiplication with any element h of the subgroup $\mathrm{SO(n)} \subset \mathrm{SO(n,1)}$:

$$\mathrm{SO(n,1)} \supset \mathrm{SO(n)} \ni h = \left(\begin{array}{c|c} \mathcal{O} & 0 \\ \hline 0 & 1 \end{array}\right); \quad \mathcal{O}^T\mathcal{O} = \mathbf{1}_{n\times n}. \tag{16.21}$$

An explicit parameterization of this type can be written as follows:

$$\mathbb{L}(\mathbf{y}) = \left(\begin{array}{c|c} \mathbf{1}_{n\times n} + 2\frac{\mathbf{y}\mathbf{y}^T}{1-\mathbf{y}^2} & -2\frac{\mathbf{y}}{1-\mathbf{y}^2} \\ \hline -2\frac{\mathbf{y}^T}{1-\mathbf{y}^2} & \frac{1+\mathbf{y}^2}{1-\mathbf{y}^2} \end{array} \right), \tag{16.22}$$

where $\mathbf{y}^2 \equiv \mathbf{y} \cdot \mathbf{y}$ denotes the standard SO(n) invariant scalar product in $\mathbb{R}^n$. Why do the matrices $\mathbb{L}(\mathbf{y})$ form a good parameterization of the coset? The reason is simple. First of all, observe that

$$\mathbb{L}(\mathbf{y})^T \eta \mathbb{L}(\mathbf{y}) = \eta, \tag{16.23}$$

where

$$\eta = \mathrm{diag}(+, +, \ldots, +, -). \tag{16.24}$$

This guarantees that $\mathbb{L}(\mathbf{y})$ are elements of SO(n, 1). Secondly, observe that the image $\mathbf{x}(\mathbf{y})$ of the standard vector $\mathbf{x}_0$ through $\mathbb{L}(\mathbf{y})$,

$$\mathbf{x}(\mathbf{y}) \equiv \mathbb{L}(\mathbf{y})\mathbf{x}_0 = \mathbb{L}(\mathbf{y}) \left(\begin{array}{c} 0 \\ \vdots \\ 0 \\ \hline 1 \end{array} \right) = \frac{1}{1-\mathbf{y}^2} \left(\begin{array}{c} 2y^1 \\ \vdots \\ 2y^n \\ \hline \frac{1+\mathbf{y}^2}{1-\mathbf{y}^2} \end{array} \right), \tag{16.25}$$

lies, as it should, in the algebraic locus $\mathbb{H}_-^{(n,1)}$,

$$\mathbf{x}(\mathbf{y})^T \eta \mathbf{x}(\mathbf{y}) = -1, \tag{16.26}$$

and has n linearly independent entries (the first n) parameterized by $\mathbf{y}$. Hence, the lateral classes can be labeled by $\mathbf{y}$ and this concludes our argument to show that (16.22) is a good coset parameterization. The coset representative $\mathbb{L}(0) = \mathbf{1}_{(n+1)\times(n+1)}$ corresponds to the identity class which is usually named the *origin* of the coset.

16.3.1 The geometry of coset manifolds

In order to study the geometry of a coset manifold G/H, the first important step is provided by the orthogonal decomposition of the corresponding Lie algebra, namely, by

$$\mathbb{G} = \mathbb{H} \oplus \mathbb{K}, \tag{16.27}$$

where $\mathbb{G}$ is the Lie algebra of G, the subalgebra $\mathbb{H} \subset \mathbb{G}$ is the Lie algebra of the subgroup H, and $\mathbb{K}$ denotes a vector space orthogonal to $\mathbb{H}$ with respect to the Cartan–Killing metric of $\mathbb{G}$. By the definition of subalgebra we always have

$$[\mathbb{H}, \mathbb{H}] \subset \mathbb{H}, \tag{16.28}$$

while in general one has

$$[\mathbb{H}, \mathbb{K}] \subset \mathbb{H} \oplus \mathbb{K}. \tag{16.29}$$

Definition 16.3.3. Let G/H be a Lie coset manifold and let the orthogonal decomposition of the corresponding Lie algebra be as in equation (16.27). If the condition

$$[\mathbb{H}, \mathbb{K}] \subset \mathbb{K} \tag{16.30}$$

applies, the coset G/H is called *reductive*.

Equation (16.30) has an obvious and immediate interpretation. The complementary space $\mathbb{K}$ forms a linear representation of the subalgebra $\mathbb{H}$ under its adjoint action within the ambient algebra $\mathbb{G}$.

Almost all of the "reasonable" coset manifolds which occur in various fields of different branches of science are reductive. Violation of *reductivity* is a sort of pathology whose study we can disregard in the scope of this book. We will consider only reductive coset manifolds.

Definition 16.3.4. Let G/H be a reductive coset manifold. If in addition to (16.30) also the condition

$$[\mathbb{K}, \mathbb{K}] \subset \mathbb{H} \tag{16.31}$$

applies, then the coset manifold G/H is called a *symmetric space*.

Let T_A $(A = 1, \dots, n)$ denote a complete basis of generators for the Lie algebra $\mathbb{G}$,

$$[T_A, T_B] = C^C{}_{AB} T_C, \tag{16.32}$$

and let T_i $(i = 1, \dots, m)$ denote a complete basis for the subalgebra $\mathbb{H} \subset \mathbb{G}$. We also introduce the notation T_a $(a = 1, \dots, n-m)$ for a set of generators that provide a basis of the complementary subspace $\mathbb{K}$ in the orthogonal decomposition (16.27). We nickname T_a the *coset generators*. Using such notations, equation (16.32) splits into the following three ones:

$$[T_j, T_k] = C^i{}_{jk} T_i, \tag{16.33}$$
$$[T_i, T_b] = C^a{}_{ib} T_a, \tag{16.34}$$
$$[T_b, T_c] = C^i{}_{bc} T_i + C^a{}_{bc} T_a. \tag{16.35}$$

Equation (16.33) encodes the property of $\mathbb{H}$ of being a subalgebra. Equation (16.34) encodes the property of the considered coset of being reductive. Finally, if in equation (16.35) we have $C^a{}_{bc} = 0$, the coset is not only reductive but also symmetric.

We are able to provide explicit formulae for the Riemann tensor of reductive coset manifolds equipped with G-invariant metrics in terms of such structure constants. Prior to that we consider the infinitesimal transformation and the very definition of the Killing vectors with respect to which the metric has to be invariant.

16.3.1.1 Infinitesimal transformations and Killing vectors

Let us consider the transformation law (16.20) of the coset representative. For a group element g infinitesimally close to the identity, we have

$$g \simeq 1 + \epsilon^A T_A, \tag{16.36}$$
$$h(y,g) \simeq 1 - \epsilon_A W^i_A(y) T_i, \tag{16.37}$$
$$y'^{\alpha} \simeq y^{\alpha} + \epsilon^A k^{\alpha}_A. \tag{16.38}$$

The induced h transformation in equation (16.20) depends in general on the infinitesimal G-parameters ϵ^A and on the point in the coset manifold y, as shown in equation (16.37). The y-dependent rectangular matrix $W^i_A(y)$ is usually named the $\mathbb{H}$-compensator. The shift in the coordinates y^{α} is also proportional to ϵ^A, and the vector fields

$$\mathbf{k}_A = k^{\alpha}_A(y)\frac{\partial}{\partial y^{\alpha}} \tag{16.39}$$

are called the *Killing vectors of the coset*. Such a name will be justified when we show that on G/H we can construct a (pseudo-)Riemannian metric which admits the vector fields (16.39) as generators of infinitesimal isometries. For the time being those in (16.39) are just a set of vector fields that, as we prove below, close the Lie algebra of the group G.

Inserting equations (16.36)–(16.38) into the transformation law (16.20) we obtain

$$T_A \mathbb{L}(y) = \mathbf{k}_A \mathbb{L}(y) - W^i_A(y)\mathbb{L}(y)T_i. \tag{16.40}$$

Consider now the commutator $g_2^{-1} g_1^{-1} g_2 g_1$ acting on $\mathbb{L}(y)$. If both group elements $g_{1,2}$ are infinitesimally close to the identity in the sense of equation (16.36), then we obtain

$$g_2^{-1} g_1^{-1} g_2 g_1 \mathbb{L}(y) \simeq (1 - \epsilon^A_1 \epsilon^B_2 [T_A, T_B])\mathbb{L}(y). \tag{16.41}$$

By explicit calculation we find

$$\begin{aligned}[T_A, T_B]\mathbb{L}(y) &= T_A T_B \mathbb{L}(y) - T_B T_A \mathbb{L}(y) \\ &= [\mathbf{k}_A, \mathbf{k}_B]\mathbb{L}(y) - (\mathbf{k}_A W^i_B - \mathbf{k}_B W^i_A + 2C^i{}_{jk} W^j_A W^k_B)\mathbb{L}(y)T_i.\end{aligned} \tag{16.42}$$

On the other hand, using the Lie algebra commutation relations we obtain

$$[T_A, T_B]\mathbb{L}(y) = C^C{}_{AB} T_C \mathbb{L}(y) = C^C{}_{AB}(\mathbf{k}_C \mathbb{L}(y) - W^i_C \mathbb{L}(y) T_i). \tag{16.43}$$

By equating the right hand sides of equation (16.42) and equation (16.43) we conclude that

$$[\mathbf{k}_A, \mathbf{k}_B] = C^C{}_{AB}\mathbf{k}_C, \tag{16.44}$$

$$\mathbf{k}_A W^i_B - \mathbf{k}_B W^i_A + 2C^i{}_{jk} W^j_A W^k_B = C^C{}_{AB} W^i_C, \tag{16.45}$$

where we separately compared the terms with and without W's, since the decomposition of a group element into $\mathbb{L}(y)h$ is unique.

Equation (16.44) shows that the Killing vector fields defined above close the commutation relations of the $\mathbb{G}$ algebra.

Instead, equation (16.45) will be used to construct a consistent $\mathbb{H}$-covariant Lie derivative.

In the case of the spaces $\mathbb{H}^{(n,1)}_{-}$, which we choose as illustrative example, the Killing vectors can be easily calculated by following the above described procedure step by step. For pedagogical purposes we find it convenient to present such a calculation in a slightly more general setup by introducing the following coset representative that depends on a discrete parameter $\kappa = \pm 1$:

$$\mathbb{L}_\kappa(\mathbf{y}) = \left(\begin{array}{c|c} \mathbf{1}_{n\times n} + 2\mathbf{y}\mathbf{y}^T\frac{\kappa}{1+\kappa\mathbf{y}^2} & -2\frac{\mathbf{y}}{1+\kappa\mathbf{y}^2} \\ \hline 2\kappa\frac{\mathbf{y}^T}{1+\kappa\mathbf{y}^2} & \frac{1-\kappa\mathbf{y}^2}{1+\kappa\mathbf{y}^2} \end{array}\right). \tag{16.46}$$

An explicit calculation shows that

$$\mathbb{L}_\kappa(\mathbf{y})^T \underbrace{\left(\begin{array}{cccc|c} 1 & 0 & \cdots & 0 & 0 \\ 0 & 1 & \cdots & 0 & 0 \\ \vdots & \vdots & \vdots & \vdots & \vdots \\ 0 & \cdots & 0 & 1 & 0 \\ \hline 0 & \cdots & 0 & 0 & \kappa \end{array}\right)}_{\eta_\kappa} \mathbb{L}_\kappa(\mathbf{y}) = \underbrace{\left(\begin{array}{cccc|c} 1 & 0 & \cdots & 0 & 0 \\ 0 & 1 & \cdots & 0 & 0 \\ \vdots & \vdots & \vdots & \vdots & \vdots \\ 0 & \cdots & 0 & 1 & 0 \\ \hline 0 & \cdots & 0 & 0 & \kappa \end{array}\right)}_{\eta_\kappa}. \tag{16.47}$$

Namely, $\mathbb{L}_{-1}(\mathbf{y})$ is an $\mathrm{SO}(n,1)$ matrix, while $\mathbb{L}_1(\mathbf{y})$ is an $\mathrm{SO}(n+1)$ group element. Furthermore, defining, as in equation (16.25),

$$\mathbf{x}_\kappa(\mathbf{y}) \equiv \mathbb{L}_\kappa(\mathbf{y}) \left(\begin{array}{c} 0 \\ \vdots \\ 0 \\ \hline\hline 1 \end{array}\right), \tag{16.48}$$

we find that

$$\mathbf{x}_\kappa(\mathbf{y})^T \eta_\kappa \mathbf{x}_\kappa(\mathbf{y}) = \kappa. \tag{16.49}$$

Hence, by means of $\mathbb{L}_1(\mathbf{y})$ we parameterize the points of the n-sphere $\mathbb{S}^n$, while by means of $\mathbb{L}_{-1}(\mathbf{y})$ we parameterize the points of $\mathbb{H}_{-}^{(n,1)}$, also called the n-pseudo-sphere or the n-hyperboloid. In both cases the stability subalgebra is $\mathfrak{so}(n)$, for which a basis of generators is provided by the following matrices:

$$J_{ij} = \mathscr{I}_{ij} - \mathscr{I}_{ji}; \quad i,j = 1,\dots,n, \tag{16.50}$$

having denoted by

$$\mathscr{I}_{ij} = \left(\begin{array}{cccc|c} 0 & \cdots & \cdots & 0 & 0 \\ 0 & \cdots & 1 & 0 & 0 \;\}i\text{-th row} \\ 0 & \cdots & \cdots & 0 & 0 \\ \hline 0 & \cdots & \underbrace{\cdots}_{j\text{-th column}} & 0 & 0 \end{array}\right) \tag{16.51}$$

the $(n+1) \times (n+1)$ matrices whose only non-vanishing entry is the ij-th one, equal to 1.

The commutation relations of the $\mathfrak{so}(n)$ generators are very simple. We have

$$[J_{ij}, J_{k\ell}] = -\delta_{ik} J_{j\ell} + \delta_{jk} J_{i\ell} - \delta_{j\ell} J_{ik} + \delta_{i\ell} J_{jk}. \tag{16.52}$$

The coset generators can instead be chosen as the matrices

$$P_i = \left(\begin{array}{cccc|c} 0 & \cdots & \cdots & 0 & 0 \\ 0 & \cdots & 0 & 0 & 1 \;\}i\text{-th row} \\ 0 & \cdots & \cdots & 0 & 0 \\ \hline 0 & \cdots & \underbrace{-\kappa}_{i\text{-th column}} & 0 & 0 \end{array}\right) \tag{16.53}$$

and satisfy the following commutation relations:

$$[J_{ij}, P_k] = -\delta_{ik} P_j + \delta_{jk} P_i, \tag{16.54}$$

$$[P_i, P_j] = -\kappa J_{ij}. \tag{16.55}$$

Equation (16.54) states that the generators P_i transform as an n-vector under $\mathfrak{so}(n)$ rotations (reductivity), while equation (16.55) shows that for both signs $\kappa = \pm 1$ the considered coset manifold is a symmetric space. Correspondingly, we denote by $\mathbf{k}_{ij} = k_{ij}^{\ell}(y)\frac{\partial}{\partial y^{\ell}}$ the Killing vector fields associated with the action of the generators J_{ij},

$$J_{ij}\mathbb{L}_{\kappa}(\mathbf{y}) = \mathbf{k}_{ij}\mathbb{L}_{\kappa}(\mathbf{y}) + \mathbb{L}_{\kappa}(\mathbf{y}) J_{pq} W_{ij}^{pq}(y), \tag{16.56}$$

while we denote by $\mathbf{k}_i = k_i^{\ell}(y)\frac{\partial}{\partial y^{\ell}}$ the Killing vector fields associated with the action of the generators P_i,

$$P_i\mathbb{L}_{\kappa}(\mathbf{y}) = \mathbf{k}_i\mathbb{L}_{\kappa}(\mathbf{y}) + \mathbb{L}_{\kappa}(\mathbf{y}) J_{pq} W_i^{pq}(y). \tag{16.57}$$

Resolving conditions (16.56) and (16.57) we obtain

$$\mathbf{k}_{ij} = y_i\partial_j - y_j\partial_i, \tag{16.58}$$

$$\mathbf{k}_i = \frac{1}{2}(1 - \kappa\mathbf{y}^2)\partial_i + \kappa y_i \quad \mathbf{y}\cdot\vec{\partial}. \tag{16.59}$$

The $\mathbb{H}$-compensators W_i^{pq} and W_{ij}^{pq} can also be extracted from the same calculation but since their explicit form is not essential in this context we skip them.

16.3.1.2 Vielbeins, connections, and metrics on G/H

Consider next the following 1-form defined over the reductive coset manifold G/H:

$$\Sigma(y) = \mathbb{L}^{-1}(y)\,\mathrm{d}\mathbb{L}(y), \tag{16.60}$$

which generalizes the Maurer–Cartan form defined over the group manifold G. As a consequence of its own definition the 1-form Σ satisfies the equation

$$0 = \mathrm{d}\Sigma + \Sigma \wedge \Sigma, \tag{16.61}$$

which provides the clue to the entire (pseudo-)Riemannian geometry of the coset manifold. To work out the latter we start by decomposing Σ along a complete set of generators of the Lie algebra $\mathbb{G}$. According to the notations introduced in the previous subsection, we put

$$\Sigma = V^a T_a + \omega^i T_i. \tag{16.62}$$

The set of $(n-m)$ 1-forms $V^a = V^a_\alpha(y)\,dy^\alpha$ provides a covariant frame for the cotangent bundle $\mathrm{T}^\star(\mathrm{G/H})$, namely, a complete basis of sections of this vector bundle that transform in a proper way under the action of the group G. On the other hand, $\omega = \omega^i T_i = \omega^i_\alpha(y)\,dy^\alpha T_i$ is called the $\mathbb{H}$-connection. Indeed, according to the theory exposed in Section 15.2, ω turns out to be the 1-form of a bona fide principal connection on the principal fiber bundle

$$\mathscr{P}\left(\frac{\mathrm{G}}{\mathrm{H}},\mathrm{H}\right) : \mathrm{G} \xrightarrow{\pi} \frac{\mathrm{G}}{\mathrm{H}}, \tag{16.63}$$

which has the Lie group G as total space, the coset manifold G/H as base space, and the closed Lie subgroup H ⊂ G as structural group. The bundle $\mathscr{P}(\frac{\mathrm{G}}{\mathrm{H}},\mathrm{H})$ is uniquely defined by the projection that associates with each group element $g \in \mathrm{G}$ the equivalence class $g\mathrm{H}$ it belongs to.

The above setup is standard in all applications of G/H manifolds to general relativity (see for instance the textbook [79]), Kaluza–Klein compactifications of higher-dimensional supergravity (see for instance [126] for a review), and other sectors of mathematical physics. It is appropriate to stress that it has been adopted entirely by the new trends in *geometrical deep learning* that we have repeatedly emphasized (see [29] for a recent review).

Introducing the decomposition (16.62) into the Maurer–Cartan equation (16.61), the latter can be rewritten as the following pair of equations:

$$dV^a + C^a{}_{ib}\omega^i \wedge V^b = -\frac{1}{2}C^a{}_{bc}V^b \wedge V^c, \tag{16.64}$$

$$d\omega^i + \frac{1}{2}C^i{}_{jk}\omega^j \wedge \omega^k = -\frac{1}{2}C^i{}_{bc}V^b \wedge V^c, \tag{16.65}$$

where we have used the Lie algebra structure constants organized as in equations (16.33)–(16.35).

Let us now consider the transformations of the 1-forms we have introduced.

Under left multiplication by a constant group element $g \in \mathrm{G}$ the 1-form $\Sigma(y)$ transforms as follows:

$$\begin{aligned}\Sigma(y') &= h(y,g)\mathbb{L}^{-1}(y)g^{-1}\,d(g\mathbb{L}(y)h^{-1}) \\ &= h(y,g)^{-1}\Sigma(y)h(y,g) + h(y,g)^{-1}\,dh(y,g),\end{aligned} \tag{16.66}$$

where $y' = g.y$ is the new point in the manifold G/H whereto y is moved by the action of g. Projecting the above equation on the coset generators T_a we obtain

$$V^a(y') = V^b(y)\mathscr{D}_b{}^a(h(y,g)), \tag{16.67}$$

where $\mathscr{D} = \exp[\mathfrak{D}_{\mathbb{H}}]$, having denoted by $\mathfrak{D}_{\mathbb{H}}$ the $(n-m)$-dimensional representation of the subalgebra $\mathbb{H}$ which occurs in the decomposition of the adjoint representation of $\mathbb{G}$:

$$\mathrm{adj}\,\mathbb{G} = \underbrace{\mathrm{adj}\,\mathbb{H}}_{=\mathfrak{A}_{\mathbb{H}}} \oplus \mathfrak{D}_{\mathbb{H}}. \tag{16.68}$$

Projecting on the other hand on the $\mathbb{H}$-subalgebra generators T_i we get

$$\omega(y') = \mathscr{A}[h(y,g)]\omega(y)\mathscr{A}^{-1}[h(y,g)] + \mathscr{A}[h(y,g)]\,d\mathscr{A}^{-1}[h(y,g)], \tag{16.69}$$

where we have set

$$\mathscr{A} = \exp[\mathfrak{A}_{\mathbb{H}}]. \tag{16.70}$$

Considering a complete basis T_A of generators for the full Lie algebra $\mathbb{G}$, the adjoint representation is defined as follows:

$$\forall g \in \mathrm{G}: \quad g^{-1}T_A g \equiv \mathrm{adj}(g)_A{}^B T_B. \tag{16.71}$$

In the explicit basis of T_A generators the decomposition (16.68) means that, once restricted to the elements of the subgroup $\mathrm{H} \subset \mathrm{G}$, the adjoint representation becomes block-diagonal:

$$\forall h \in \mathrm{H}: \quad \mathrm{adj}(h) = \left(\begin{array}{c|c} \mathscr{D}(h) & 0 \\ \hline 0 & \mathscr{A}(h) \end{array} \right). \tag{16.72}$$

Note that for such decomposition to hold true the coset manifold has to be reductive according to definition (16.30).

The infinitesimal form of equation (16.67) is

$$V^a(y + \delta y) - V^a(y) = -\epsilon^A W_A^i(y) C^a{}_{ib} V^b(y), \tag{16.73}$$

$$\delta y^a = \epsilon^A k_A^a(y) \tag{16.74}$$

for a group element $g \in \mathrm{G}$ very close to the identity as in equation (16.36).

Similarly, the infinitesimal form of equation (16.69) is

$$\omega^i(y + \delta y) - \omega^i(y) = -\epsilon^A (C^i{}_{kj} W_A^k \omega^j + \mathrm{d}W_A^i). \tag{16.75}$$

16.3.1.3 Lie derivatives

The Lie derivative of a tensor $T_{\alpha_1 \dots \alpha_p}$ along a vector field v^μ provides the change in shape of that tensor under an infinitesimal diffeomorphism:

$$y^\mu \mapsto y^\mu + v^\mu(y). \tag{16.76}$$

Explicitly one sets

$$\ell_{\mathbf{v}} T_{\alpha_1 \dots \alpha_p}(y) = v^\mu \partial_\mu T_{\alpha_1 \dots \alpha_p} + (\partial_{\alpha_1} v^\gamma) T_{\gamma \alpha_2 \dots \alpha_p} + \cdots + (\partial_{\alpha_p} v^\gamma) T_{\alpha_1 \alpha_2 \dots \gamma}. \tag{16.77}$$

In the case of p-forms, namely, of antisymmetric tensors, the definition (16.77) of Lie derivative can be recast into a more intrinsic form using both the exterior differential d and the *contraction operator*.

Definition 16.3.5. Let $\mathscr{M}$ be a differentiable manifold and let $\Lambda_k(\mathscr{M})$ be the vector bundles of differential k-forms on $\mathscr{M}$. Let $\mathbf{v} \in \Gamma(T\mathscr{M}, \mathscr{M})$ be a vector field. The contraction $\mathrm{i}_{\mathbf{v}}$ is a linear map

$$\mathrm{i}_{\mathbf{v}} : \Lambda_k(\mathscr{M}) \to \Lambda_{k-1}(\mathscr{M}) \tag{16.78}$$

such that for any $\omega^{(k)} \in \Lambda_k(\mathscr{M})$ and for any set of $k-1$ vector fields $\mathbf{w}_1, \dots, \mathbf{w}_{k-1}$ we have

$$\mathrm{i}_{\mathbf{v}} \omega^{(k)}(\mathbf{w}_1, \dots, \mathbf{w}_{k-1}) \equiv k\omega^{(k)}(\mathbf{v}, \mathbf{w}_1, \dots, \mathbf{w}_{k-1}). \tag{16.79}$$

Then by going to components we can verify that the tensor definition (16.77) is equivalent to the following one.

Definition 16.3.6. Let $\mathscr{M}$ be a differentiable manifold and let $\Lambda_k(\mathscr{M})$ be the vector bundles of differential k-forms on $\mathscr{M}$. Let $\mathbf{v} \in \Gamma(T\mathscr{M}, \mathscr{M})$ be a vector field. The Lie derivative $\ell_{\mathbf{v}}$ is a linear map

$$\ell_{\mathbf{v}} : \Lambda_k(\mathcal{M}) \to \Lambda_k(\mathcal{M}) \tag{16.80}$$

such that for any $\omega^{(k)} \in \Lambda_k(\mathcal{M})$ we have

$$\ell_{\mathbf{v}}\omega^{(k)} \equiv \mathrm{i}_{\mathbf{v}}\, \mathrm{d}\omega^{(k)} + \mathrm{d}\mathrm{i}_{\mathbf{v}}\omega^{(k)}. \tag{16.81}$$

On the other hand, for vector fields the tensor definition (16.77) is equivalent to the following one.

Definition 16.3.7. Let $\mathcal{M}$ be a differentiable manifold and let $T\mathcal{M} \to \mathcal{M}$ be the tangent bundle whose sections are the vector fields. Let $\mathbf{v} \in \Gamma(T\mathcal{M}, \mathcal{M})$ be a vector field. The Lie derivative $\ell_{\mathbf{v}}$ is a linear map

$$\ell_{\mathbf{v}} : \Gamma(T\mathcal{M}, \mathcal{M}) \to \Gamma(T\mathcal{M}, \mathcal{M}) \tag{16.82}$$

such that for any $\mathbf{w} \in \Gamma(T\mathcal{M}, \mathcal{M})$ we have

$$\ell_{\mathbf{v}}\mathbf{w} \equiv [\mathbf{v}, \mathbf{w}]. \tag{16.83}$$

The most important properties of the Lie derivative, which immediately follow from its definition, are the following ones:

$$\begin{aligned} [\ell_{\mathbf{v}}, \mathrm{d}] &= 0, \\ [\ell_{\mathbf{v}}, \ell_{\mathbf{w}}] &= \ell_{[\mathbf{v},\mathbf{w}]}. \end{aligned} \tag{16.84}$$

The first of the above equations states that the Lie derivative commutes with exterior derivative. This is just a consequence of the invariance of the exterior algebra of k-forms with respect to diffeomorphisms. On the other hand, the second equation states that the Lie derivative provides an explicit representation of the Lie algebra of vector fields on tensors.

The Lie derivatives along the Killing vectors of the frames V^a and the $\mathbb{H}$-connection ω^i introduced in the previous subsection are

$$\ell_{\mathbf{v}_A} V^a = W^i_A C^a{}_{ib} V^b, \tag{16.85}$$

$$\ell_{\mathbf{v}_A} \omega^i = -(\mathrm{d}W^i_A + C^i{}_{kj} W^k_A \omega^j). \tag{16.86}$$

This result can be interpreted by saying that associated with every Killing vector $\mathbf{k}_A$ there is an infinitesimal $\mathbb{H}$-gauge transformation

$$\mathbf{W}_A = W^i_A(y) T_i \tag{16.87}$$

and that the Lie derivative of both V^a and ω^i along the Killing vectors is just such local gauge transformation pertaining to their respective geometrical type. The frame V^a is a section of an H-vector bundle and transforms as such, while ω^i is a connection and it transforms as a connection should do.

16.3.1.4 Invariant metrics on coset manifolds

The result (16.85)–(16.86) has a very important consequence which constitutes the fundamental motivation to consider coset manifolds. Indeed, this result instructs us to construct G-invariant metrics on G/H, namely, metrics that admit all the above discussed Killing vectors as generators of true isometries.

The argument is quite simple. We saw that the 1-forms V^a transform in a linear representation $\mathfrak{D}_{\mathbb{H}}$ of the isotropy subalgebra $\mathbb{H}$ (and group H). Hence, if τ_{ab} is a symmetric H-invariant constant 2-tensor, by setting

$$ds^2 = \tau_{ab} V^a \otimes V^b = \underbrace{\tau_{ab} V^a_\alpha(y) V^b_\beta(y)}_{g_{\alpha\beta}(y)} \, dy^\alpha \otimes dy^\beta \tag{16.88}$$

we obtain a metric for which all the above constructed Killing vectors are indeed Killing vectors, that is,

$$\begin{aligned}\ell_{\mathbf{k}_A} ds^2 &= \tau_{ab}(\ell_{\mathbf{k}_A} V^a \otimes V^b + V^a \otimes \ell_{\mathbf{k}_A} V^b) \\ &= \underbrace{\tau_{ab}([\mathfrak{D}_{\mathbb{H}}(W_A)]^a{}_c \delta^b_d + [\mathfrak{D}_{\mathbb{H}}(W_A)]^b{}_c \delta^a_d)}_{=0 \quad \text{by invariance}} V^c \otimes V^d \end{aligned} \tag{16.89}$$

$$= 0. \tag{16.90}$$

The key point in order to utilize the above construction is the decomposition of the representation $\mathfrak{D}_{\mathbb{H}}$ into irreducible representations. Typically, for most common cosets, $\mathfrak{D}_{\mathbb{H}}$ is already irreducible. In this case there is just one invariant H-tensor τ and the only free parameter in the definition of the metric (16.88) is an overall scale constant. Indeed, if τ_{ab} is an invariant tensor, any multiple thereof $\tau'_{ab} = \lambda \tau_{ab}$ is also invariant. In the case $\mathfrak{D}_{\mathbb{H}}$ splits into $\mathfrak{r}$ irreducible representations,

$$\mathfrak{D}_{\mathbb{H}} = \left(\begin{array}{c|c|c|c|c} \mathfrak{D}_1 & 0 & \dots & 0 & 0 \\ \hline 0 & \mathfrak{D}_2 & 0 & \dots & 0 \\ \hline \vdots & \vdots & \vdots & \vdots & \vdots \\ \hline 0 & \dots & 0 & \mathfrak{D}_{\mathfrak{r}-1} & 0 \\ \hline 0 & 0 & \dots & 0 & \mathfrak{D}_{\mathfrak{r}} \end{array}\right), \tag{16.91}$$

we have $\mathfrak{r}$ irreducible invariant tensors $\tau^{(i)}_{a_i b_i}$ in correspondence with such irreducible blocks and we can introduce $\mathfrak{r}$ independent scale factors:

$$\tau = \left(\begin{array}{c|c|c|c|c} \lambda_1 \tau^{(1)} & 0 & \dots & 0 & 0 \\ \hline 0 & \lambda_2 \tau^{(2)} & 0 & \dots & 0 \\ \hline \vdots & \vdots & \vdots & \vdots & \vdots \\ \hline 0 & \dots & 0 & \lambda_{p-1} \tau^{(p-1)} & 0 \\ \hline 0 & 0 & \dots & 0 & \lambda_p \tau^{(p)} \end{array}\right). \tag{16.92}$$

Correspondingly we arrive at a continuous family of G-invariant metrics on G/H depending on $\mathfrak{r}$ parameters, or, as it is customary to say in this context, $\mathfrak{r}$ *moduli*. The number $\mathfrak{r}$ defined by equation (16.91) is called the *rank of the coset manifold* G/H.

In this section we confine ourselves to the most common case of rank 1 cosets ($\mathfrak{r} = 1$), assuming, furthermore, that the algebras $\mathbb{G}$ and $\mathbb{H}$ are both semi-simple. By an appropriate choice of basis for the coset generators T^a, the invariant tensor τ_{ab} can always be reduced to the form

$$\tau_{ab} = \eta_{ab} = \text{diag}(\underbrace{+,+,\dots,+}_{n_+},\underbrace{-,-,\dots,-}_{n_-}), \tag{16.93}$$

where the two numbers n_+ and n_- sum up to the dimension of the coset,

$$n_+ + n_- = \dim \frac{\mathrm{G}}{\mathrm{H}} = \dim \mathbb{K}, \tag{16.94}$$

and provide the dimensions of the two eigenspaces, $\mathbb{K}_\pm \subset \mathbb{K}$, respectively corresponding to real and purely imaginary eigenvalues of the matrix $\mathfrak{D}_{\mathbb{H}}(W)$, which represents a generic element W of the isotropy subalgebra $\mathbb{H}$.

Focusing on our example (16.46), which encompasses both the spheres and the pseudo-spheres, depending on the sign of κ, we find that

$$n_+ = 0; \quad n_- = n, \tag{16.95}$$

so that in both cases ($\kappa = \pm 1$) the invariant tensor is proportional to a Kronecker delta:

$$\eta_{ab} = \delta_{ab}. \tag{16.96}$$

The reason is that the subalgebra $\mathbb{H}$ is the compact $\mathfrak{so}(n)$; hence, the matrix $\mathfrak{D}_{\mathfrak{H}}(W)$ is antisymmetric and all of its eigenvalues are purely imaginary.

If we consider cosets with non-compact isotropy groups, then the invariant tensor τ_{ab} develops a non-trivial Lorentzian signature η_{ab}. In any case, if we restrict ourselves to rank 1 cosets, the general form of the metric is

$$ds^2 = \lambda^2 \eta_{ab} V^a \otimes V^b, \tag{16.97}$$

where λ is a scale factor. This allows us to introduce the *Vielbein*

$$E^a = \lambda V^a \tag{16.98}$$

and calculate the *spin connection* from the vanishing torsion equation:

$$0 = \mathrm{d}E^a - \omega^{ab} \wedge E^c \eta_{bc}. \tag{16.99}$$

Using the Maurer–Cartan equations (16.64)–(16.65), equation (16.99) can be immediately solved by

$$\omega^{ab}\eta_{bc} \equiv \omega^a{}_c = \frac{1}{2\lambda} C^a{}_{cd} E^d + C^a{}_{ci}\omega^i. \tag{16.100}$$

Inserting this in the definition of the curvature 2-form

$$\mathfrak{R}^a{}_b = \mathrm{d}\omega^a{}_b - \omega^a{}_c \wedge \omega^c{}_b \tag{16.101}$$

allows to calculate the Riemann tensor defined by

$$\mathfrak{R}^a{}_b = \mathscr{R}^a{}_{bcd} E^c \wedge E^d. \tag{16.102}$$

Using once again the Maurer–Cartan equations (16.64)–(16.65), we obtain

$$\mathscr{R}^a{}_{bcd} = \frac{1}{\lambda^2}\left(-\frac{1}{4} C^a{}_{be} C^e{}_{cd} - \frac{1}{8} C^a{}_{ec} C^e{}_{bd} + \frac{1}{8} C^a{}_{ed} C^e{}_{bc} - \frac{1}{2} C^a{}_{bi} C^i{}_{cd}\right), \tag{16.103}$$

which, as previously announced, provides the expression of the Riemann tensor in terms of structure constants.

In the case of symmetric spaces, $C^a{}_{be} = 0$, formula (16.103) simplifies to

$$\mathscr{R}^a{}_{bcd} = -\frac{1}{2\lambda^2} C^a{}_{bi} C^i{}_{cd}. \tag{16.104}$$

16.3.1.5 Spheres and pseudo-spheres

In order to illustrate the structures presented in the previous section we consider the explicit example of the spheres and pseudo-spheres. Applying the outlined procedure to this case we immediately get

$$\begin{aligned} E^a &= -\frac{2}{\lambda}\frac{\mathrm{d}y^a}{1+\kappa y^2}, \\ \mathfrak{R}^{ab} &= 2\frac{\kappa}{\lambda^2} E^a \wedge E^b. \end{aligned} \tag{16.105}$$

This means that for spheres and pseudo-spheres the Riemann tensor is proportional to an antisymmetrized Kronecker delta:

$$R^{ab}{}_{cd} = \frac{\kappa}{\lambda^2}\delta^{[a}_{[c}\delta^{b]}_{d]}. \tag{16.106}$$

16.4 The real sections of a complex Lie algebra and symmetric spaces

In the context of coset manifolds a very interesting class that finds important applications in various sectors of science and in particular in supergravity and superstring theory is the following one:

$$\mathscr{M}_{\mathbb{G}_R} = \frac{\mathrm{G}_R}{\mathrm{H}_c}, \tag{16.107}$$

where G_R is some semi-simple Lie group and $\mathrm{H}_c \subset \mathrm{G}_R$ is its maximal compact subgroup. The Lie algebra $\mathbb{H}_c$ of the denominator H_c is the maximal compact subalgebra $\mathbb{H} \subset \mathbb{G}_R$ which has typically rank $r_{\text{compact}} > r$. Denoting as usual by $\mathbb{K}$ the orthogonal complement of $\mathbb{H}_c$ in $\mathbb{G}_R$,

$$\mathbb{G}_R = \mathbb{H}_c \oplus \mathbb{K}, \tag{16.108}$$

and defining as non-compact rank or rank of the coset $\mathrm{G}_\mathrm{R}/\mathrm{H}$ the dimension of the non-compact Cartan subalgebra,

$$r_{nc} = \operatorname{rank}(\mathrm{G}_\mathrm{R}/\mathrm{H}) \equiv \dim \mathscr{H}^{n.c.}; \quad \mathscr{H}^{n.c.} \equiv \mathrm{CSA}_{\mathbb{G}(\mathbb{C})} \cap \mathbb{K}, \tag{16.109}$$

we obtain $r_{nc} < r$.

By definition the Lie algebra $\mathbb{G}_R$ is a real section of a complex semi-simple Lie algebra. In Section 12.1.1.1 we met the first two universal instances of real sections of a simple complex Lie algebra $\mathbb{G}(\mathbb{C})$, namely, the *maximally split* and the *maximally compact real sections*.

All other possible real sections are obtained by studying the available Cartan involutions of the complex Lie algebra. Consider the following definition.

Definition 16.4.1. Let

$$\theta : \mathfrak{g} \to \mathfrak{g} \tag{16.110}$$

be a linear automorphism of the compact Lie algebra $\mathfrak{g} = \mathbb{G}_c$, where $\mathbb{G}_c$ is the maximal compact real section of a complex semi-simple Lie algebra $\mathbb{G}(\mathbb{C})$. By definition we have

$$\forall \alpha, \beta \in \mathbb{R}, \quad \forall \mathbf{X}, \mathbf{Y} \in \mathfrak{g} : \quad \begin{cases} \theta(\alpha\mathbf{X} + \beta\mathbf{Y}) = \alpha\theta(\mathbf{X}) + \beta\theta(\mathbf{Y}), \\ \theta([\mathbf{X}, \mathbf{Y}]) = [\theta(\mathbf{X}), \theta(\mathbf{Y})]. \end{cases} \tag{16.111}$$

If $\theta^2 = \mathbf{Id}$, then θ is called a Cartan involution of the Lie algebra $\mathfrak{g}$.

For any Cartan involution θ the possible eigenvalues are ± 1. This allows us to split the entire Lie algebra $\mathfrak{g}$ in two subspaces corresponding to the eigenvalues 1 and -1 respectively:

$$\mathfrak{g} = \mathfrak{H}_\theta \oplus \mathfrak{p}_\theta. \tag{16.112}$$

One immediately realizes that

$$\begin{aligned} [\mathfrak{H}_\theta, \mathfrak{H}_\theta] &\subset \mathfrak{H}_\theta, \\ [\mathfrak{H}_\theta, \mathfrak{p}_\theta] &\subset \mathfrak{p}_\theta, \\ [\mathfrak{p}_\theta, \mathfrak{p}_\theta] &\subset \mathfrak{H}_\theta. \end{aligned} \tag{16.113}$$

Hence, for any Cartan involution $\mathfrak{H}_\theta$ is a subalgebra and θ singles out a symmetric homogeneous compact coset manifold:

$$\mathcal{M}_\theta = \frac{\mathrm{G_c}}{\mathrm{H}_\theta}, \quad \text{where } \mathrm{H}_\theta \equiv \exp[\mathfrak{H}_\theta];\ \mathrm{G_c} \equiv \exp[\mathfrak{g}]. \tag{16.114}$$

The structure in (16.113) has also another important consequence. If we define the vector space

$$\mathfrak{g}_\theta^\star = \mathfrak{H}_\theta \oplus \mathfrak{p}_\theta^\star; \quad \mathfrak{p}_\theta^\star \equiv i\,\mathfrak{p}_\theta, \tag{16.115}$$

we see that $\mathfrak{g}_\theta^\star$ is closed under the Lie bracket and hence it is a Lie algebra. It is some real section of the complex Lie algebra $\mathbb{G}(\mathbb{C})$ and we can consider a new, generally noncompact coset manifold:

$$\mathcal{M}_\theta^\star = \frac{\mathrm{G}_\theta^\star}{\mathrm{H}_\theta}; \quad \mathrm{H}_\theta \equiv \exp[\mathfrak{H}_\theta]; \quad \mathrm{G}_\theta^\star \equiv \exp[\mathfrak{g}_\theta^\star]. \tag{16.116}$$

An important theorem for which we refer the reader to classical textbooks [37, 38, 128][4] states that all real forms of a Lie algebra, up to isomorphism, are obtained in this way. Furthermore, according to the same theorem, θ can always be chosen in such a way that it maps the compact Cartan subalgebra into itself:

$$\theta : \mathcal{H}_c \to \mathcal{H}_c. \tag{16.117}$$

This short discussion reveals that the classification of real forms of a complex Lie algebra $\mathbb{G}(\mathbb{C})$ is in one-to-one correspondence with the classification of symmetric spaces, the complexification of whose Lie algebra of isometries is $\mathbb{G}(\mathbb{C})$. For this reason we have postponed the discussion of the real forms to the present chapter devoted to homogeneous coset manifolds.

Let us now consider the action of the Cartan involution on the Cartan subalgebra $\mathcal{H}_c = \mathrm{span}\{iH_i\}$ of the maximal compact section $\mathbb{G}_c$. Choosing a basis of $\mathcal{H}_c$ aligned with the simple roots

$$\mathcal{H}_c = \mathrm{span}\{iH_{\alpha_i}\}, \tag{16.118}$$

we see that the action of the Cartan involution θ is by duality transferred to the simple roots α_i and hence to the entire root lattice. As a consequence we can introduce the notion of real and imaginary roots. One argues as follows.

4 The proof is also summarized in Appendix B of [129].

We split the Cartan subalgebra into its compact and non-compact subalgebras:

$$\begin{aligned} \mathrm{CSA}_{\mathbb{G}_R} &= i\,\mathcal{H}^{\mathrm{comp}} \oplus \mathcal{H}^{n.c.}, \\ &\quad\ \Updownarrow \qquad\quad \Updownarrow \\ \mathrm{CSA}_{\mathbb{G}_{\max}} &= \mathcal{H}^{\mathrm{comp}} \oplus \mathcal{H}^{n.c.}, \end{aligned} \tag{16.119}$$

defining

$$\begin{aligned} h \in \mathcal{H}^{\mathrm{comp}} \quad &\Leftrightarrow \quad \theta(h) = h, \\ h \in \mathcal{H}^{n.c} \quad &\Leftrightarrow \quad \theta(h) \neq h. \end{aligned} \tag{16.120}$$

Then every vector in the dual of the full Cartan subalgebra, in particular every root α, can be decomposed into its parallel and its transverse part to $\mathcal{H}^{n.c.}$:

$$\alpha = \alpha_{||} \oplus \alpha_{\perp}. \tag{16.121}$$

A root α is called *imaginary* if $\alpha_{||} = 0$. On the contrary, a root α is called real if $\alpha_{\perp} = 0$. Generically a root is complex.

Given the original Dynkin diagram of a complex Lie algebra we can characterize a real section by mentioning which of the simple roots are imaginary. We do this by painting the imaginary roots black. The result is a Tits–Satake diagram like that in Figure 16.2 which corresponds to the real Lie algebra $\mathfrak{so}(p, 2\ell - p + 1)$ for $p > 2$, $\ell > 2$.

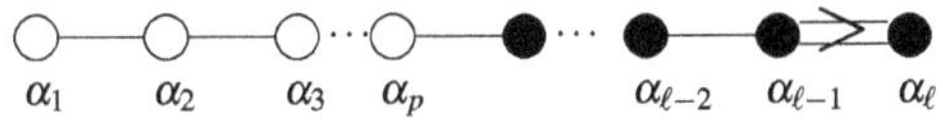

Figure 16.2: The Tits–Satake diagram representing the real form $\mathfrak{so}(p, 2\ell - p + 1)$ of the complex $\mathfrak{so}(2\ell + 1)$ Lie algebra.

16.5 The solvable group representation of non-compact coset manifolds

Definition 16.5.1. A Riemannian space $(\mathcal{M}, g)$ is called *normal* if it admits a completely solvable[5] Lie group $\exp[\mathrm{Solv}(\mathcal{M})]$ of isometries that acts on the manifold in a simply transitive manner (i. e., for every two points in the manifold there is one and only one group element connecting them). The group $\exp[\mathrm{Solv}(\mathcal{M})]$ is generated by a so-called *normal metric Lie algebra*, $\mathrm{Solv}(\mathcal{M})$, which is a completely solvable Lie algebra endowed with a Euclidean metric.

5 A solvable Lie algebra s is completely solvable if the adjoint operation ad_X for all generators $X \in s$ has only real eigenvalues. The nomenclature of the Lie algebra is carried over to the corresponding Lie group in general in this chapter.

The main tool to classify and study homogeneous spaces of the type (16.116) is provided by a theorem [38] that states that if a Riemannian manifold $(\mathcal{M}, g)$ is normal, according to Definition 16.5.1, then it is metrically equivalent to the solvable group manifold

$$\begin{aligned} \mathcal{M} &\simeq \exp[\mathrm{Solv}(\mathcal{M})], \\ g|_{e\in\mathcal{M}} &= \langle , \rangle, \end{aligned} \tag{16.122}$$

where $\langle , \rangle$ is a Euclidean metric defined on the normal solvable Lie algebra $\mathrm{Solv}(\mathcal{M})$. The key point is that non-compact coset manifolds of the form (16.116) are all normal. This is so because there is always, *for all real forms except the maximally compact one*, a *solvable subalgebra* with the following features:

$$\begin{aligned} \forall\, \mathrm{Solv}\left(\frac{\mathrm{G}_R}{\mathrm{H}_c}\right) &\subset \mathrm{G}_R, \\ \dim\left[\mathrm{Solv}\left(\frac{\mathrm{G}_R}{\mathrm{H}_c}\right)\right] &= \dim\left(\frac{\mathrm{G}_R}{\mathrm{H}_c}\right), \\ \exp\left[\mathrm{Solv}\left(\frac{\mathrm{G}_R}{\mathrm{H}_c}\right)\right] &= \text{transitive on } \frac{\mathrm{G}_R}{\mathrm{H}_c}. \end{aligned} \tag{16.123}$$

It is very easy to single out the appropriate solvable algebra in the case of the maximally split real form $\mathbb{G}_{\mathrm{max}}$. In that case, as is well known, the maximal compact subalgebra has the following form:

$$\mathbb{H}_c = \mathrm{span}\{(E^\alpha - E^{-\alpha})\}; \quad \forall \alpha \in \mathbf{\Phi}_+. \tag{16.124}$$

The solvable algebra that does the required job is the Borelian subalgebra

$$\mathrm{Bor}(\mathbb{G}_{\mathrm{max}}) \equiv \mathscr{H} \oplus \mathrm{span}(E^\alpha); \quad \forall \alpha \in \mathbf{\Phi}_+, \tag{16.125}$$

where $\mathscr{H}$ is the complete Cartan subalgebra and E^α are the step operators associated with the positive roots. That $\mathrm{Bor}(\mathbb{G}_{\mathrm{max}})$ is a solvable Lie algebra follows from the canonical structure of Lie algebras in (11.37) presented in Section 11.2.1. If you exclude the negative roots, you immediately see that the Cartan generators are not in the first derivative of the algebra. The second derivative excludes all the simple roots, the third derivative excludes the roots of height 2, and so on, until you end up in a derivative that makes zero. Hence, the Lie algebra is solvable. Furthermore, it is obvious that any equivalence class of $\frac{\mathrm{G}_R}{\mathrm{H}_c}$ has a representative that is an element of the solvable Lie group $\exp[\mathrm{Bor}(\mathbb{G}_{\mathrm{max}})]$. This is intuitive at the infinitesimal level due to the fact that each element of the complementary space

$$\mathbb{K} = \mathscr{H} \oplus \mathrm{span}[(E^\alpha + E^{-\alpha})] \tag{16.126}$$

which generates the coset can be uniquely rewritten as an element of $\mathrm{Bor}(\mathbb{G}_{max})$ plus an element of the subalgebra $\mathbb{H}_c$. At the finite level an exact formula which connects the solvable representative $\exp[\mathbf{s}]$ (with $\mathbf{s} \in \mathrm{Bor}$) to the orthogonal representative $\exp[\mathbf{k}]$ (with $\mathbf{k} \in \mathbb{K}$) of the same equivalence class was derived by P. Fré and A. Sorin and is presented in [72]. Here it suffices for us to understand that the action of the Borel group is transitive on the coset manifold, so that the coset manifold $\mathrm{G_R}/\mathrm{H_c}$ is indeed normal and its metric can be obtained from the non-degenerate Euclidean metric $\langle , \rangle$ defined over $\mathrm{Bor}(\mathrm{G}_{max}) = \mathrm{Solv}(\frac{\mathrm{G}_{max}}{\mathrm{H}_c})$.

The example of the maximally split case clearly suggests what is the required solvable algebra for other normal forms. We have

$$\mathrm{Solv}\left(\frac{\mathrm{G}_R}{\mathrm{H}_c}\right) = \mathscr{H}^{n.c.} \oplus \mathrm{span}(\mathscr{E}^\alpha); \quad \forall \alpha \in \mathbf{\Phi}_+/\alpha_\parallel \neq 0, \tag{16.127}$$

where $\mathscr{H}^{n.c.}$ is the non-compact part of the Cartan subalgebra and $\mathscr{E}^\alpha$ denotes the combination of step operators pertaining to the positive roots α that appear in the real form $\mathbb{G}_R$ for which the sum is extended only to those roots that are not purely imaginary. Indeed, the step operators pertaining to imaginary roots are included in the maximal compact subalgebra, which now is larger than the number of positive roots.

For any solvable group manifold with a non-degenerate invariant metric[6] the differential geometry of the manifold is completely rephrased in algebraic language through the relation of the Levi-Civita connection and the *Nomizu operator* acting on the solvable Lie algebra. The latter is defined as

$$\mathbb{L}: \quad \mathrm{Solv}(\mathscr{M}) \otimes \mathrm{Solv}(\mathscr{M}) \to \mathrm{Solv}(\mathscr{M}), \tag{16.128}$$

$$\forall X, Y, Z \in \mathrm{Solv}(\mathscr{M}): \quad 2\langle \mathbb{L}_X Y, Z\rangle = \langle [X,Y], Z\rangle - \langle X, [Y,Z]\rangle - \langle Y, [X,Z]\rangle. \tag{16.129}$$

The *Riemann curvature operator* on this group manifold can be expressed as

$$\mathrm{Riem}(X,Y) = [\mathbb{L}_X, \mathbb{L}_Y] - \mathbb{L}_{[X,Y]}. \tag{16.130}$$

This implies that the covariant derivative explicitly reads

$$\mathbb{L}_X Y = \Gamma^Z_{XY} Z, \tag{16.131}$$

where

$$\Gamma^Z_{XY} = \frac{1}{2}(\langle Z, [X,Y]\rangle - \langle X, [Y,Z]\rangle - \langle Y, [X,Z]\rangle)\frac{1}{\langle Z, Z\rangle} \quad \forall X, Y, Z \in \mathrm{Solv}. \tag{16.132}$$

6 See [130, 131, 132, 133, 134] for reviews on the solvable Lie algebra approach to supergravity scalar manifolds and the use of the Nomizu operator.

Equation (16.132) is true for any solvable Lie algebra, but in the case of *maximally non-compact, split algebras* we can write a general form for Γ^Z_{XY}, namely,

$$\begin{aligned}
\Gamma^i_{jk} &= 0,\\
\Gamma^i_{\alpha\beta} &= \frac{1}{2}(-\langle E_\alpha, [E_\beta, H^i]\rangle - \langle E_\beta, [E_\alpha, H^i]\rangle) = \frac{1}{2}\alpha^i\delta_{\alpha\beta},\\
\Gamma^\alpha_{ij} &= \Gamma^\alpha_{i\beta} = \Gamma^i_{j\alpha} = 0,\\
\Gamma^\alpha_{\beta i} &= \frac{1}{2}(\langle E^\alpha, [E_\beta, H_i]\rangle - \langle E_\beta, [H_i, E^\alpha]\rangle) = -\alpha_i\delta^\alpha_\beta,\\
\Gamma^{\alpha+\beta}_{\alpha\beta} &= -\Gamma^{\alpha+\beta}_{\beta\alpha} = \frac{1}{2}N_{\alpha\beta},\\
\Gamma^\alpha_{\alpha+\beta\beta} &= \Gamma^\alpha_{\beta\alpha+\beta} = \frac{1}{2}N_{\alpha\beta},
\end{aligned} \tag{16.133}$$

where $N^{\alpha\beta}$ is defined by the commutator

$$[E_\alpha, E_\beta] = N_{\alpha\beta}E_{\alpha+\beta}. \tag{16.134}$$

The explicit form (16.133) follows from the following choice of the non-degenerate metric:

$$\begin{aligned}
\langle \mathscr{H}_i, \mathscr{H}_j\rangle &= 2\delta_{ij},\\
\langle \mathscr{H}_i, E_\alpha\rangle &= 0,\\
\langle E_\alpha, E_\beta\rangle &= \delta_{\alpha,\beta},
\end{aligned} \tag{16.135}$$

where $\mathscr{H}_i \in$ CSA and E_α are the step operators associated with positive roots $\alpha \in \Phi_+$. For any other *non-split case*, the Nomizu connection exists nonetheless although it does not take the form (16.133). It follows from equation (16.132) upon the choice of an invariant positive metric on Solv and the use of the structure constants of Solv.

16.5.1 The Tits–Satake projection: just a flash

Let us now come back to equation (16.121). Setting all $\alpha_\perp = 0$ corresponds to a projection

$$\Pi_{TS} : \Phi_{\mathbb{G}} \mapsto \overline{\Phi} \tag{16.136}$$

of the original root system $\Phi_{\mathbb{G}}$ onto a new system of vectors living in a Euclidean space of dimension equal to the non-compact rank r_{nc}. A priori this is not obvious, but it is nonetheless true that $\overline{\Phi}$, with only one exception, is by itself the root system of a simple Lie algebra $\mathbb{G}_{TS}$, the Tits–Satake subalgebra of $\mathbb{G}_R$:

$$\overline{\Phi} = \text{root system of } \mathbb{G}_{TS} \subset \mathbb{G}_R. \tag{16.137}$$

The Tits–Satake subalgebra $\mathbb{G}_{TS} \subset \mathbb{G}_R$ is always the maximally non-compact real section of its own complexification. For this reason, considering its maximal compact subalgebra $\mathbb{H}_{TS} \subset \mathbb{G}_{TS}$, we have a new smaller coset $\frac{G_{TS}}{H_{TS}}$, which is maximally split. A natural question arises: what is the relation between the two solvable Lie algebras $\mathrm{Solv}(\frac{G}{H})$ and $\mathrm{Solv}(\frac{G_{TS}}{H_{TS}})$? The explicit answer to this question and the systematic illustration of the geometrical relevance of the Tits–Satake projection is a topic extensively discussed in [72]. Indeed, this projection plays a very significant role in supergravity theories and in the search for black-hole solutions, cosmological potentials, and gaugings of supergravity models. The Tits–Satake projection finds its deepest and most useful interpretation in the context of special geometries and of the c-map, these items being the most conspicuous contributions of supergravity to modern geometry. We refer the reader to the book [72] for the development of such topics.

16.6 Bibliographical note

For the topics discussed in the present chapter, the main references are the second volume of [79], [126], [38], and [72].

17 Functional spaces and non-compact Lie algebras

No other question has ever moved so profoundly the spirit of man; no other idea has so fruitfully stimulated his intellect; yet no other concept stands in greater need of clarification than that of the infinite.

David Hilbert

17.1 Introduction to an introduction

The last developed chapter of this book happens to be an introduction, a very short and scant one, to *functional analysis*, that is, to an entire subdiscipline of modern mathematics. Functional analysis, whose birth we can identify with some fundamental contributions of Hilbert dating back to 1909, is a major field of mathematical analysis of high relevance in all pure and applied sciences.

Conceptually there are several different mathematical issues that come together and cross each other's way in this context. Let us enumerate them in some logical order.

(1) To begin with, we have the issue of integration and of integrable functions. We would like to treat functions $f(x_1, \ldots, x_n)$ over a closed subset of $\mathbb{R}^n$ (an interval $[a, b] \subset \mathbb{R}$ in the one-variable case) as if they were points in a suitable continuous space $\mathscr{H}$ that we are able to organize in some suitable way, introducing at the same time some notion of distance among its points. This is the proper setting in order to evaluate approximations and decide how close an approximant is to a given function. Therefore, one focuses on continuous functions and looks for their integrability. In order to obtain that the limit of a succession of integrable continuous functions is also integrable one has to appropriately revise the notion of integral, and this was done by Lebesgue.

(2) Secondly, one has the issue of infinite-dimensional vector spaces. The notion of a vector space does not imply that it should be finite-dimensional. If we allow the possibility that a vector space V admits an *arbitrary large number* of linearly independent vectors we arrive at the notion of an infinite-dimensional vector space. Every n-dimensional real vector space $V_n(\mathbb{R})$ is isomorphic to $\mathbb{R}^n$. If we introduce a positive definite scalar product, the norm squared of a vector $\mathbf{v} \in V_n(\mathbb{R})$ is $N^2(\mathbf{v}) = \sum_{i=1}^n v_i^2$, where $v_i \in \mathbb{R}$. On the other hand, every n-dimensional vector space $V_n(\mathbb{C})$ over the complex number field is isomorphic to $\mathbb{C}^n$. Introducing a Hermitian scalar product, the norm squared of a vector $\mathbf{v} \in V_n(\mathbb{C})$ is $N^2(\mathbf{v}) = \sum_{i=1}^n |v_i|^2$, where $v_i \in \mathbb{C}$. Then it is natural to assume that an infinite-dimensional complex vector space $V_\infty(\mathbb{C})$ is isomorphic to the space of infinite successions $(v_1, v_2, \ldots, v_n, \ldots)$ of complex numbers; in other words, an infinite-dimensional vector space is, in a formal sense, $\mathbb{C}^\infty$. In this case, however, in order to give sense to such successions of complex numbers we have to assume that they define convergent series, that is,

https://doi.org/10.1515/9783111201535-017

$$\sum_{n=1}^{\infty} |v_n|^2 < \infty. \tag{17.1}$$

(3) Issues (1) and (2) are brought together by the issue of orthogonal functions. Indeed, the available functional spaces turn out to be the spaces of square integrable functions $L^2_w(\Sigma)$ that are maps from a closed domain $\Sigma \subset \mathbb{R}^n$ into $\mathbb{C}$, satisfying the following condition:

$$\begin{aligned} & L^2_w(\Sigma) \ni \mathbf{f} : \Sigma \longrightarrow \mathbb{C}, \\ & \Downarrow \\ & N^2(\mathbf{f}) = \int_\Sigma |\mathbf{f}(\mathbf{x})|^2 w(\mathbf{x})\, d^n\mathbf{x} < \infty, \end{aligned} \tag{17.2}$$

where $w(\mathbf{x})$ is a suitable positive definite real function that is called the *measure*. Such spaces have a naturally defined Hermitian scalar product,

$$\begin{aligned} \forall f(\mathbf{x}), g(\mathbf{x}) \in L^2_w(\Sigma); \quad (\mathbf{f}, \mathbf{g}) &\equiv \int_\Sigma \overline{f(\mathbf{x})} g(\mathbf{x}) w(\mathbf{x})\, d^n\mathbf{x}, \\ (f, g) &= (g, f)^\star, \end{aligned} \tag{17.3}$$

and the problem is raised of finding complete orthonormal bases of functions $e_m(\mathbf{x})$ such that:
- $(e_n, e_m) = \delta_{n,m}$, and
- $\forall f(\mathbf{x}) \in L^2_w(\Sigma)$ we can write $f(\mathbf{x}) = \sum_{n=1}^{\infty} a_n e_m(\mathbf{x})$ with $a_n \in \mathbb{C}$

When such bases are found we see that

$$N^2(\mathbf{f}) = \sum_{n=1}^{\infty} |f_m|^2 < \infty \tag{17.4}$$

and we can identify the vector $\mathbf{f}$ with its succession of coefficients a_m. The issue of orthogonal functions in spaces $L^2_w(\Sigma)$, which we shall collectively name *Hilbert spaces*, is the appropriate way to reorganize a lot of nineteenth-century results about so-called orthogonal polynomials and other classical special functions.

(4) The issue of differential equations and self-adjoint (or Hermitian) operators in a Hilbert space $\mathscr{H} = L^2_w(\Sigma)$ is the fourth main item in the present list. Typically an operator in a functional space is a linear differential operator,[1]

$$\forall f \in \mathscr{H} : \quad \mathscr{L}f = \sum_{m=0}^{p} c_m(x) \frac{d^m f}{dx^m} \in \mathscr{L}, \tag{17.5}$$

1 In the definition below, for simplicity we confine ourselves to the case where the functional space is composed of functions of only one variable x.

and one is led to define differential operators $\mathscr{L}$ that are self-adjoint with respect to the Hermitian scalar product that exists on that functional space, that is, such that

$$\forall f,g \in \mathscr{H}: \quad (g,\mathscr{L}f) = (\mathscr{L}g,f). \tag{17.6}$$

The spectrum of eigenvalues of such operators is composed of real numbers λ_n:

$$\mathscr{L}f_n(x) = \lambda_n f_n(x); \quad \lambda_n \in R. \tag{17.7}$$

As it happens in the case of finite-dimensional vector spaces, the eigenfunctions of an operator corresponding to different eigenvalues are orthogonal among themselves and this provides a connection between this problem and the problem of orthonormal basis for functional Hilbert spaces, in particular orthogonal polynomials. This issue is strongly related with the concepts of quantum mechanics. Indeed, in quantum mechanics the Hilbert space $\mathscr{H}$ is considered as the space of possible quantum states of some physical system and one searches for bases of $\mathscr{H}$ composed of eigenstates of a maximal set of commuting self-adjoint operators $\mathscr{O}_i$ called the *observables*. The eigenvalues of the latter constitute the labels λ_i of any given quantum state. Apart from their relevance in quantum mechanics and quantum field theory, Hilbert spaces are an important tool in many other applied sciences.

(5) The last issue in this list, which is strongly related with the above issue and issue (2), is that of unitary representations of non-compact groups.
We discussed groups, emphasizing finite, discrete, and continuous groups. In previous chapters we met groups in various capacities as conformal groups, monodromy groups, homotopy groups, homology groups, and isometry groups, and we considered their representations.
Let us now focus on the issue of unitarity.

Definition 17.1.1. A representation D of a group G is called *unitary* if the images $D(g)$ of all group elements are unitary operators, $D(g)^\dagger = D(g)^{-1} = D(g^{-1})$.

To this effect, let us recall the definition of the Hermitian conjugate $\mathscr{A}^\dagger$ or adjoint of a linear operator $\mathscr{A}$ in a generic vector space V endowed with a scalar product $(\,,\,)$:

$$\forall f,g \in V; \quad \overline{(g,Af)} = (A^\dagger g,f), \tag{17.8}$$

which applies equally well to the case of finite- and infinite-dimensional vector spaces.
Let us now consider the case of continuous groups, in particular Lie groups G, which have a fundamental direct connection with the notion of Lie algebras.
In an intuitive way the connection between Lie groups and Lie algebras is provided by the consideration of group elements infinitesimally close to the identity $\mathbf{e}$. Developing in power series of a suitable parameter λ, every group element $g \in G$ can be written as follows:

$$g = \exp[\lambda \mathbf{X}] \simeq e + \lambda \mathbf{X} + \mathcal{O}(\lambda^2), \tag{17.9}$$

where $\mathbf{X}$ is an element of a Lie algebra $\mathbb{G}$ whose dimension is equal to the dimension of the Lie group G, which is determined by $\mathbb{G}$.

Any linear representation of the Lie group G induces a linear representation of its Lie algebra $\mathbb{G}$, and vice versa.

In the case of a unitary representation of the group G, the corresponding representation of the Lie algebra $\mathbb{G}$ is provided by anti-Hermitian operators,

$$\forall \mathbf{X} \in \mathbb{G}; \quad D(\mathbf{X})^\dagger = -D(\mathbf{X}), \tag{17.10}$$

which, multiplying by just an i, provides a relation with self-adjoint operators.

The key point in connection with the above discussion is provided by the following general theorem, which we state without proof.

Theorem 17.1.1. *Let G be a non-compact Lie group. All unitary representations of G are necessarily infinite-dimensional, and for this reason they are provided by suitable L^2_w functional spaces. The generators T_I of the Lie algebra $\mathbb{G}$ are represented by $i \times \mathcal{L}_I$, where $\mathcal{L}_I$ are self-adjoint operators in the Hilbert space L^2_w.*

Hence, most of the interesting self-adjoint differential operators in Hilbert spaces are actually generators of suitable Lie groups and the solutions of the corresponding differential equations are eigenvectors of such generators.

17.2 The idea of functional spaces

The first important book written by Hermann Weyl was titled *The Idea of a Riemann Surface*. In Weyl's spirit the present section is titled *The idea of functional spaces*. Our goal is to outline the logic that leads to Hilbert spaces, whose elements are, in a sense to be explained, *functions*. The connection of these mathematical developments with the conceptual development of physics in the twentieth century is very strong. Furthermore, the conceptual development of physics is a *paradigm* for the conceptual development of all other fundamental and technical sciences that quite often follow the steps of mathematicians and physicists with some delay, sometimes rediscovering the various steps of the paradigm under different names and in different environments. Physics, being a mature science, is tied up with mathematics in an almost rigid fashion. Most of the other sciences become harmonized with their mathematical roots in their process of ripening. Such considerations apply to chemistry, molecular biology, complex and dynamical system theory, and recently also computer science and artificial intelligence.

Classical physics, as it evolved from the eighteenth to the nineteenth century, finally led to the notion of *phase space*. All possible states of a physical system are provided by the points of an even-dimensional manifold $\mathbf{\Phi}$, whose coordinates we denote

$\{p^i, q_i\}$ $(i = 1, \ldots, n)$, that is endowed with a symplectic structure encoded in a closed 2-form $\mathbf{\Omega}$ defined over it. The dynamical evolution of a physical system is described by curves in phase space, determined by $2n$ first order differential equations, the *Hamilton equations*. With the advent of quantum mechanics and the Schrödinger equation this mathematical picture changed completely. The state of a system is no longer a point in a manifold; rather, it is described by a function $\Psi(q)$ of the generalized coordinates q_i or the momenta p^i. This object, called the *wave function*, has its own support on half of the phase space $\mathbf{\Phi}$, while the other half of the canonical variables become a set of operators, typically differential, acting on the possible wave functions. Hence, quantum mechanics posed the problem of characterizing the *space of states* of a physical system as a space whose points are functions. The superposition principle, or, differently stated, the linearity of the Schrödinger equation, implies that quantum states form a vector space. In this way the concept of infinite-dimensional vector spaces whose elements (vectors) are "*functions*" emerged almost at the same time as the logical development of mathematical and physical theories.

Here we outline the mathematical logical path.

(a) The first problem to be solved arises from the observation that there are infinite successions of continuous functions defined over an interval of the real line $\mathbb{R}$ or over some closed subset of $\mathbb{R}^n$ whose limit is a discontinuous function. It follows that if we were to consider a functional space defined as the space of continuous function or of differentiable functions, it might be not complete, namely, it might miss the limit of sequences of its own points. One needs some different definition.

(b) Another mathematical discomfort with spaces of functions is that the Riemann definition of an integral is not apt to cope with the case of discontinuous, yet bounded functions, which, as we have seen, can be the limit of successions of Riemann integrable continuous functions.

(c) The French mathematician Lebesgue provided the solution to both above problems introducing a generalized definition of the integral coinciding for continuous functions with the Riemann definition yet allowing for the calculation of the integral of discontinuous functions. Lebesgue integration theory is based on the notion of *measure* $\mu(\mathcal{S})$ of subsets of the real line $\mathcal{S} \subset \mathbb{R}$ and introduces the notion of *measure zero sets*.

(d) Relying on the Lebesgue integral one arrives at a good definition of functional spaces that are complete by considering the spaces $L^2_w(\Sigma)$ of those complex valued functions defined over a region $\Sigma \subset \mathbb{R}^n$ (we will mainly focus on the case $n = 1$) whose squared norm $|f(\mathbf{x})|^2$ is integrable over Σ yielding a finite result. The fundamental theorem named after Fischer and Riesz states that the space $L^2_w(\Sigma)$ is complete. Since the addition of a function that is non-zero only on measure zero sets does not change the value of any Lebesgue integral, it follows that the elements of $L^2_w(\Sigma)$ are not exactly functions $f(x)$; rather, they are equivalence classes of functions with respect to the equivalence relation $f(x) \sim g(x)$ if the difference $f(x) - g(x)$ does not vanish only on measure zero sets.

(e) Having established that $L^2_w(\Sigma)$ is a complete vector space one looks for bases of orthonormal functions $\mathbf{e}_n(x)$ such that any element $f \in L^2_w(\Sigma)$ can be expanded as $f(x) = \sum_{n=0}^{\infty} a_n \mathbf{e}_n(x)$. Once a basis is singled out, the space $L^2_w(\Sigma)$ becomes isomorphic to an abstract Hilbert space $\mathscr{H}$ whose elements are, by definition, successions of complex numbers a_i the series of whose norms is convergent, $\sum_{i=0}^{\infty} |a_i|^2 < \infty$.

(e) The hint where to look for bases of functions is provided by the *Weierstrass theorem*, which states that a continuous bounded function can be approximated with arbitrary accuracy by a polynomial of sufficiently large degree. Since every element of $f \in L^2_w(\Sigma)$ can be seen as the limit of a succession of continuous functions, it follows that the polynomials and in particular the monomials x^n provide a basis of functions. Typically such a basis is not orthonormal.

(f) The Gram–Schmidt orthonormalization algorithm utilized in finite-dimensional vector spaces is iterative and inductive. Hence, it can be easily extended to the infinite-dimensional case. This shows that for $L^2_w(\Sigma)$ one can find orthonormal bases composed of orthogonal polynomials.

(g) The nineteenth-century French mathematician Rodrigues developed an ingenious algorithm for the direct construction of families of orthogonal polynomials associated with suitable weight functions $w(x)$.

17.2.1 Limits of successions of continuous functions

We come to the first point of our logical path and we consider an example of a succession of continuous functions whose limit is a discontinuous one.

We introduce the following functions, which are all defined over the interval $[-1, 1]$ and take values in the interval $[0, 1]$:

$$f_k : \quad \mathbb{R} \supset [-1,1] \longrightarrow [0,1] \subset \mathbb{R},$$

$$f_k(x) = \begin{cases} 0 & \text{if } -1 \le x < -\frac{1}{k}, \\ \frac{kx+1}{2} & \text{if } -\frac{1}{k} \le x < \frac{1}{k}, \\ 1 & \text{if } \frac{1}{k} \le x \le 1. \end{cases} \tag{17.11}$$

For any finite value $k < \infty$, the function $f_k(x)$ is continuous, although its first derivative is discontinuous. In the limit $k \to \infty$ the function $f_\infty(x)$ becomes discontinuous and it is actually the step function $\Theta(x)$ that is zero for $x < 0$ and equal to 1 for $x > 0$. A graphic representation of the functions $f_k(x)$ is provided in Figure 17.1.

The main purpose of functional analysis is that of constructing spaces which contain the limit of any convergent succession of their elements. It is evident from the quoted example that a suitable definition of such complete spaces cannot be provided by considering only continuous functions. The definition has to be wider and based on some other property that encompasses both continuous and discontinuous functions.

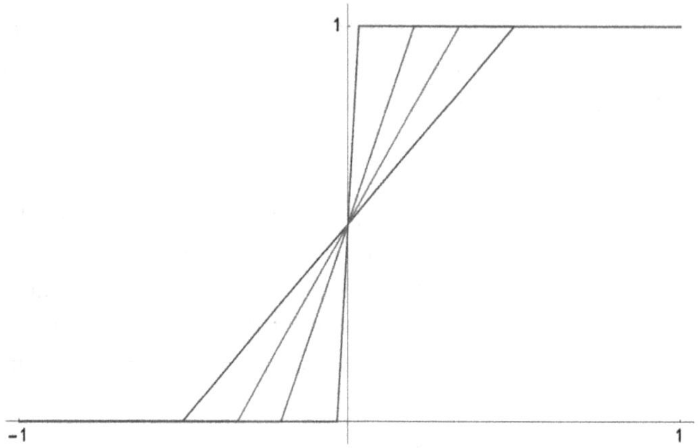

Figure 17.1: The example of the step function $\Theta(x)$, seen as the limit of a succession of continuous functions.

17.2.2 A very short introduction to measure theory and the Lebesgue integral

According to our logical path we come to its third point, namely, to the generalization of the integral definition due to Lebesgue. The first step towards the construction of the latter is provided by the notion of *Lebesgue measure*. In one dimension, that is, for subsets $E \subset \mathbb{R}$, we argue as follows.

If $E = [a, b]$ is a closed interval we set $\mu(E) = b - a$. We do the same for open or semi-open intervals, $\mu([a, b[) = \mu(]a, b]) = \mu(]a, b[) = b - a$. In the case where $E = \bigcup_{i=1}^{n}[a_i, b_i]$ is the union of several disjoint intervals $b_i < a_{i+1}$ we set $\mu(E) = \sum_{i=1}^{n}(n_i - a_i)$. If the intervals are not all disjoint we need to subtract the lengths of the intersections. For instance, if $a_1 = 1$, $a_2 = 3$, $a_3 = 7$ and $b_1 = 5$, $b_2 = 6$, $b_3 = 10$, we have

$$\mu(E) = (5 - 1) + (6 - 3) + (10 - 7) - (5 - 3) = 8. \tag{17.12}$$

If E is made up of a denumerable infinity of intervals, we have

$$\mu(E) = \sum_{i=1}^{\infty}(b_i - a_i), \tag{17.13}$$

which does not necessarily converge to a finite number, although convergence might occur.

As long as the sets to be measured are intervals, every thing is clear, yet the set E might include other types of subsets, for instance isolated points. Can we extend the notion of measure to such subsets? The answer is positive, and it is obtained through the following chain of arguments. One introduces the notion of *external covering* of the set E. An external covering is any system of intervals $C_{\text{ext}}(E) = \{[x_n, y_n], n \in J\}$ such that E is contained in it:

$$C_{\text{ext}}(E) \supset E. \tag{17.14}$$

Suppose for instance that

$$E = [1,3] \cup]2,\pi[\cup \{11,35\}. \tag{17.15}$$

A possible covering is provided for instance by $C^{(1)}_{\text{ext}}(E) = [1,4] \cup [10,40]$. Obviously there are many other possibilities, for instance $C^{(2)}_{\text{ext}}(E) = [1,\frac{7}{2}] \cup [11,35]$. Obviously we have

$$\mu(C^{(2)}_{\text{ext}}(E)) = \frac{53}{2} < \mu(C^{(1)}_{\text{ext}}(E)) = 33. \tag{17.16}$$

Definition 17.2.1. Given a subset $E \subset \mathbb{R}$, by *exterior measure* of E we mean the following:

$$\mu_{\text{ext}}(E) \equiv \inf\{\mu(C_{\text{ext}}(E))\}, \tag{17.17}$$

that is, the measure of the smallest exterior covering of the considered set.

In a fully analogous way one introduces the notion of internal covering of the set $C_{\text{int}}(E) = \{[x_n, y_n], n \in J\}$ such that

$$C_{\text{int}}(E) \subset E. \tag{17.18}$$

Definition 17.2.2. Given a subset $E \subset \mathbb{R}$, by *interior measure* of E we mean the following:

$$\mu_{\text{int}}(E) \equiv \sup\{\mu(C_{\text{int}}(E))\}, \tag{17.19}$$

that is, the measure of the largest possible interior covering of the considered set.

It is obvious that

$$\mu_{\text{int}}(E) \leq \mu_{\text{ext}}(E). \tag{17.20}$$

Definition 17.2.3. A subset $E \subset R$ is L-measurable when the internal and external measures coincide. In this case one sets

$$\mu(E) = \mu_{\text{int}}(E) = \mu_{\text{ext}}(E). \tag{17.21}$$

We say that a set E has Lebesgue measure equal to zero if $\forall \epsilon > 0$ we can find an exterior covering $C_{\text{ext}}(E)$ such that $\mu_{\text{ext}}(C_{\text{ext}}(E)) < \epsilon$. One immediately sees that any collection of isolated points is a measure zero set.

Having established the notion of measure, one can define the Lebesgue integral in the following way.

Definition 17.2.4. Let

$$f : \Sigma \longrightarrow \mathbb{C} \tag{17.22}$$

be a complex valued bounded function with support on $\Sigma \subset \mathbb{R}$. By image of f, denoted by $\mathrm{Im} f \subset \mathbb{C}$, we mean the subset of complex numbers that have a preimage in Σ through f, that is,

$$\mathrm{Im} f = \{z \in \mathbb{C} \mid \exists x \in \Sigma \backslash f(x) = z\}. \tag{17.23}$$

For each $z \in \mathrm{Im} f$, we denote by $f^{-1}(z)$ the set of preimages of z in Σ.

Then the Lebesgue integral of f on Σ is defined as follows:

$$\int_{\Sigma} f(x)\, dx = \sum_{z \in \mathrm{Im} f} z \times \mu(f^{-1}(z)). \tag{17.24}$$

17.2.3 Space of square summable functions

The answer to the problem posed at the end of Section 17.2.1 is provided by the spaces of square summable functions whose idea was anticipated in equation (17.2). We come back to that definition and from now on we specialize it to the case of functions of a single variable x defined over an interval $[a, b]$ of the real line. Furthermore, the key point is that integrations should be performed according to Lebesgue's wider definition of an integral.

Definition 17.2.5. Let $[a, b] \subset \mathbb{R}$ be an interval (finite or infinite) of the real line and let $f(x)$ be a complex valued function defined over the same interval. Furthermore, let $w(x)$ be a positive real valued function defined over that same interval. We say that f belongs to the functional space $L^2_w(a, b)$ if and only if the following integral is convergent and finite:

$$\int_a^b |f(x)|^2 w(x)\, dx < \infty. \tag{17.25}$$

From a conceptual point of view, the spaces $L^2_w(a, b)$ provide the answer to the problems we have outlined in the previous subsection because of their completeness, which is established by an important theorem that we quote without proof.

Theorem 17.2.1 (Fischer–Riesz). *If there exists a sequence of functions $f_\ell(x)$ that represent elements of $L^2_w(a, b)$ and which satisfy the condition*

$$\lim_{k,\ell \to \infty} \int_a^b |f_\ell(x) - f_k(x)|^2 w(x)\, dx = 0, \tag{17.26}$$

then there exists a function $f(x)$ which also represents an element of $L^2_w(a, b)$ such that

$$\lim_{k\to\infty}\int_a^b \left|f(x)-f_k(x)\right|^2 w(x)\,dx = 0, \tag{17.27}$$

that is, $L^2_w(a,b)$ is complete.

Obviously, since the integral involved both in the definition of $L^2_w(a,b)$ and in the statement of the theorem is the Lebesgue integral, it follows that the function $f(x)$ is defined up to functions $g(x)$ that are different from zero on a set of measure zero.

17.2.4 Hilbert space

In a finite m-dimensional vector space $\mathbb{V}_m$ over the complex field $\mathbb{C}$ we associate a set of m complex numbers v_i with every vector of the space. This is simply the isomorphism $\mathbb{V}_m \sim \mathbb{C}^m$, which is explicitly realized any time we have an explicit basis of vectors $|e_i\rangle$:

$$\begin{aligned} \forall|v\rangle \in \mathbb{V}_m : |v\rangle \quad &\Leftrightarrow \quad \{v_1, v_2, \ldots, v_m\}, \\ \mathbb{C} \ni v_i &= \langle v|e_i\rangle, \\ |v\rangle &= \sum_{i=1}^{m} v_i|e_i\rangle. \end{aligned} \tag{17.28}$$

If the basis is orthonormal,

$$\langle e_j|e_i\rangle = \delta_{ij}, \tag{17.29}$$

for any two vectors $|u\rangle, |v\rangle \in \mathbb{V}_m$ we have

$$\langle u|v\rangle = \sum_{i=1}^{m} \bar{u}_i v_i \tag{17.30}$$

and the Cauchy–Schwarz inequality becomes

$$\left|\sum_{i=1}^{m} \bar{u}_i v_i\right| \leq \sqrt{\sum_{i=1}^{m} |u_i|^2}\sqrt{\sum_{i=1}^{m} |v_i|^2}. \tag{17.31}$$

The most straightforward generalization of a complex finite-dimensional vector space $\mathbb{C}^m$ consists of introducing a space $\mathscr{H}$ whose elements are infinite successions of complex numbers:

$$|\mathbf{v}\rangle \in \mathscr{H} \quad \Leftrightarrow \quad |\mathbf{v}\rangle \sim \{v_1, v_2, \ldots, v_k, \ldots\}. \tag{17.32}$$

In this case, differently from the finite-dimensional case, we have an additional constraint that is necessary in order to give a meaning to the considered vectors, i. e., their norms should be finite. This translates into the requirement that the following series must be convergent:

$$|\mathbf{v}|^2 = \langle \mathbf{v}|\mathbf{v}\rangle = \sum_{i=1}^{m} \overline{v}_i v_i \leq \infty. \tag{17.33}$$

Clearly the infinite series $\{v_1, v_2, \dots, v_k, \dots\}$ form a vector space over $\mathbb{C}$ since they can be linearly combined:

$$\alpha\{u_1, u_2, \dots, u_k, \dots\} + \beta\{v_1, v_2, \dots, v_k, \dots\} = \{\alpha u_1 + \beta v_1, \alpha u_2 + \beta v_2 \dots, \alpha u_k + \beta v_k, \dots\}. \tag{17.34}$$

The adjoint of a vector $|\mathbf{v}\rangle$ is given by the complex conjugate succession

$$\langle v| = \{\overline{v}_1, \overline{v}_2, \dots, \overline{v}_k, \dots\}, \tag{17.35}$$

so that the scalar product becomes

$$\langle u|v\rangle = \sum_{i=1}^{\infty} \overline{u}_i v_i. \tag{17.36}$$

This equation is meaningful because, on account of the Cauchy–Schwarz inequality (17.31) and the assumed convergence of the series (17.33), the series on the right hand side of equation (17.36) converges.

Definition 17.2.6. The vector space of the convergent infinite successions

$$\{v_1, v_2, \dots, v_k, \dots\}$$

that are square summable, that is, satisfy equation (17.33), is named the *Hilbert space* $\mathcal{H}$.

Comparing with previous sections we come to the conclusion that provided we can find an orthonormal basis of square integrable functions $f_n(x)$, the space $L^2_w(a,b)$ of *mod square* integrable functions over an interval $[a,b]$ with weight $w(x)$ is just a Hilbert space. Indeed, any vector $|a\rangle \in L^2_w(a,b)$ defines its own *Fourier coefficients* a_n with respect to an orthonormal basis of square integrable functions $f_n(x)$ and their square moduli define a convergent series:

$$\sum_{i=0}^{\infty} |a_i|^2 < \infty. \tag{17.37}$$

Conversely, given any convergent series of coefficients a_n and an orthonormal basis of square integrable functions $f_n(x)$ we can define an element:

$$|f\rangle \equiv \sum_{n=0} a_n f_n(x) \in L^2_w(a,b). \tag{17.38}$$

Hence, the Hilbert space $\mathscr{H}$ is isomorphic to any of the $L^2_w(a,b)$ spaces, which are consequently all isomorphic among themselves.

17.2.5 Infinite orthonormal bases and the Weierstrass theorem

Once we have established the result that every Hilbert space $L^2_w(a,b)$ should admit bases of orthonormal functions $f_n(x)$, our next errand is to construct suitable instances of such bases. Where to look is indicated by a general theorem due to Weierstrass, stating that any continuous bounded function over a closed interval can be approximated, with arbitrary accuracy, by polynomials of a suitable high degree.

To get a feeling about the issue at stake, let us consider some function, constructed with elementary transcendental and algebraic functions, whose graph is sufficiently complicated and twisted. For instance we can consider the following one, defined over the interval $[-1,1]$,

$$g(x) = \frac{(x-\frac{1}{3})\sin(3\pi x) + (x+\frac{1}{5})\sin(5\pi x) + (x-\frac{1}{7})\cos(7\pi x)}{x+3}, \tag{17.39}$$

whose plot is displayed in Figure 17.2.

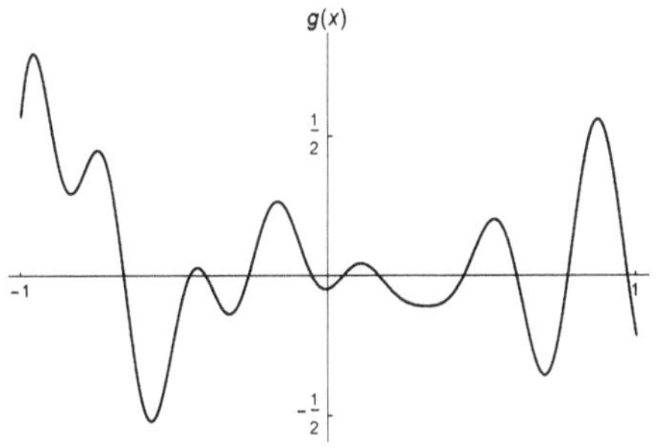

Figure 17.2: Plot of the continuous function (17.39).

In Figure 17.3 we show graphically how the function $g(x)$ can be approximated to a quite high degree of accuracy by a polynomial of degree 25, while lower order polynomials, for instance of degree respectively 15 and 5, just capture the general behavior of the considered function, yet are not able to reproduce all of its oscillations. We do not display the 26 coefficients of the polynomial $\mathscr{P}_{25}(x)$ that have been determined by numerical integration using a basis of Legendre polynomials to be explained precisely in this chapter. It suffices that the reader gets a feeling of the level of accuracy that can be reached by increasing the degree of the polynomial approximant.

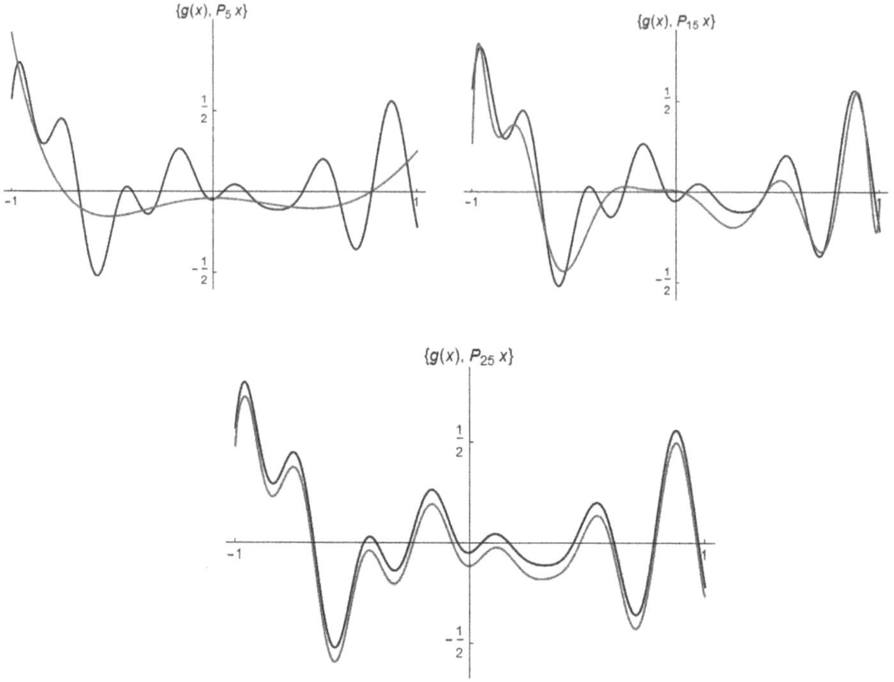

Figure 17.3: In this figure we compare the plot of the function $g(x)$ defined in equation (17.39) with the plots of polynomial approximants respectively of degree 5, 15, and 25. The bluish curve is the exact function, while the yellowish ones are the polynomial approximants.

Theorem 17.2.2 (Weierstrass). *Let the function $f(x)$ be continuous on the finite, closed interval $[a,b] \subset \mathbb{R}$. For any $\epsilon > 0$, there exist a positive integer $n \in \mathbb{N}$ and a corresponding polynomial $\mathscr{P}_n(x) \in \mathbb{C}[x]$ of the n-th degree such that*

$$|f(x) - \mathscr{P}_n(x)| < \epsilon, \quad \forall x \in [a,b]. \tag{17.40}$$

We omit the proof of this classical theorem, which can be found in many textbooks of functional analysis, and we rather focus on its consequences.

17.2.6 Consequences of the Weierstrass theorem

The Weierstrass theorem guarantees that for any continuous function $f(x)$ defined over an interval $[a,b]$ there is a sequence of polynomials $\mathscr{P}_n(x)$ that converges uniformly to it. Obviously uniform convergence implies the weaker mean convergence. Let us call

$$\mathfrak{m}_n(x) = x^n \tag{17.41}$$

the monomials of degree n and let us consider them as a sequence of linear independent vectors $|\mathfrak{m}_j\rangle$ in $L^2_w(a,b)$. A polynomial of degree n is just a linear combination of the first n of these vectors:

$$|\mathscr{P}_n\rangle = \sum_{j=0}^{n} a_j|\mathfrak{m}_j\rangle. \tag{17.42}$$

From the Weierstrass theorem we know that for sufficiently large n we get

$$\int_a^b |f(x) - \mathscr{P}_n(x)|^2 w(x)\,dx < \epsilon^2 \int_a^b w(x)\,dx, \tag{17.43}$$

that is, the distance of the vector $|f\rangle$ from the vector $|\mathscr{P}_n\rangle$ goes to zero as $n \to \infty$:

$$\lim_{n\to\infty} \mathfrak{d}(f, \mathscr{P}_n) = 0. \tag{17.44}$$

Hence, it is sufficient to establish that any square integrable function $f(x) \in L^2_w(a,b)$ is the limit of a sequence of continuous functions $f_j(x)$, and this fact joined with the Weierstrass theorem leads to the conclusion that the monomials in (17.41) constitute a basis of linear independent vectors for $L^2_w(a,b)$. We do not provide the formal proof of the above statement, yet it should be intuitively evident from the example of the step function $\theta(x)$ shown in Figure 17.1.

17.2.7 The Schmidt orthogonalization algorithm

Let us now take a step back and consider a finite-dimensional vector space $\mathbb{V}_m$ over the complex field $\mathbb{C}$ with a Hermitian scalar product which, as usual, we denote $\langle,\rangle$:

$$\begin{aligned} \langle,\rangle : &\quad \mathbb{V}_m \otimes \mathbb{V}_m \longrightarrow \mathbb{C}, \\ \forall|a\rangle, |b\rangle, |c\rangle \in \mathbb{V}_m, \quad \alpha, \beta \in \mathbb{C} : &\quad \langle c|\alpha a + \beta b\rangle = \alpha\langle c|a\rangle + \beta\langle c|b\rangle, \\ \forall|a\rangle, |b\rangle, \in \mathbb{V}_m : &\quad \langle a|b\rangle = \overline{\langle b|a\rangle}. \end{aligned} \tag{17.45}$$

Let us now assume that in the vector space $\mathbb{V}_m$ we have found a set of m linearly independent vectors

$$|v_1\rangle, |v_2\rangle, \dots, |v_m\rangle,$$

where linear independence means that the only solution to the equation

$$\sum_{i=1}^{m} a_i|v_i\rangle = |0\rangle \tag{17.46}$$

is provided by

$$a_1 = a_2 = \cdots = a_m = 0. \tag{17.47}$$

In equation (17.46) the vector $|0\rangle$ is the null element of $\mathbb{V}_m$, such that for any vector $|v\rangle$ we have $|v\rangle + |0\rangle = |v\rangle$.

We have the following theorem.

Theorem 17.2.3 (Gram–Schmidt). *In a vector space $\mathbb{V}_m$ of dimension m, given any set of m linearly independent vectors $|v_i\rangle$ one can always construct out of them, by means of linear combinations, a set of m orthonormal vectors $|e_i\rangle$:*

$$\langle e_j|e_i\rangle = \delta_{ij}. \tag{17.48}$$

The proof of the theorem, which we omit, is by induction and it is also constructive since it outlines a precise algorithm for the construction of the orthonormal basis. This algorithm is named after Gram and Schmidt and it is typically implemented by suitable routines in all symbolic manipulation computer languages. According to such an algorithm starting from a set of linearly independent vectors $|v_i\rangle$ the corresponding orthonormal basis $|e_i\rangle$ is obtained with a linear transformation,

$$|e_i\rangle = \boldsymbol{a}_{ij}|v_j\rangle, \tag{17.49}$$

where the matrix $\boldsymbol{a}_{ij}$ is lower triangular:

$$\boldsymbol{a} = \begin{pmatrix} \star & 0 & \cdots & \cdots & \cdots & 0 \\ \star & \star & 0 & \cdots & 0 & \vdots \\ \vdots & \vdots & \ddots & \ddots & \vdots & \vdots \\ \vdots & \vdots & \cdots & \ddots & 0 & 0 \\ \vdots & \star & \cdots & \cdots & \star & 0 \\ \star & \cdots & \cdots & \cdots & \star & \star \end{pmatrix}. \tag{17.50}$$

The notable thing in the Gram–Schmidt procedure is that it can be easily extended to the infinite-dimensional case if we have a natural ordering of the functions forming a basis. This is certainly the case when we deal with the monomials $\mathfrak{m}_n(x)$. The order is provided by their degree. In this case given the scalar product of $L^2_w(a,b)$ we can easily apply the orthonormalization procedure just as in the finite-dimensional case. We denote by $\mathscr{P}^{(n)}_w(x)$ the elements of the orthonormal basis. Assuming that we have already constructed the first n of such functions, we easily get the $(n+1)$-th element of the basis by setting

$$\begin{aligned} \mathscr{P}^{(n+1)}_w(x) &= \frac{1}{\sqrt{N}} P(x), \\ P(x) &\equiv x^{n+1} - \sum_{j=1}^{n} \int_a^b x^{n+1} \mathscr{P}^{(j)}_w(x) w(x)\, dx, \\ N &\equiv \int_a^b (P(x))^2 w(x)\, dx. \end{aligned} \tag{17.51}$$

Clearly all constructed functions $\mathscr{P}_w^{(n)}(x)$ are polynomials of degree n. In this way we see that convenient bases of $L_w^2(a,b)$ spaces are provided by orthogonal polynomials.

17.3 Orthogonal polynomials

Thanks to the work of several mathematicians of the nineteenth century, in particular Benjamin Rodrigues, we do not have to go through the Gram–Schmidt orthogonalization algorithm. Orthogonal polynomials relative to most interesting weight functions are already constructed through a synthetic formula that gives all of them in one stroke.

Let us consider the space of square integrable functions $L_w^2[a,b]$.

Formula 17.3.1 (Rodrigues). We introduce an infinite set of functions defined by the formula

$$C_{(n)}(x) \equiv \frac{1}{w(x)} \frac{\mathrm{d}^n}{\mathrm{d}x^n}\left(w(x)[s(x)]^n\right); \quad n = 0,1,2,3,\ldots, \tag{17.52}$$

and we impose the following conditions:

(1) $s(x)$ is a polynomial of degree 0, 1, or 2.

(2) The weight function $w(x)$ is real, positive, and integrable in the interval $[a,b]$ and it satisfies the boundary conditions

$$w(a)s(a) = w(b)s(b) = 0. \tag{17.53}$$

(3) $C_1(x)$ is a first order polynomial in x.

We have the following theorem.

Theorem 17.3.1. *The function $C_{(n)}(x)$ constructed by formula* 17.3.1 *is a polynomial of degree n which is orthogonal in $L_w^2[a,b]$ to any polynomial $\mathscr{P}_m$ with degree $m < n$.*

We omit the proof of this theorem, for which we refer the reader to [81] (Chapter III, p. 204 and the following ones).

In the next subsection we examine the possible choices of $w(x)$ that fulfill the required hypotheses.

17.3.1 The classical orthogonal polynomials

Summarizing, thanks to the generalized Rodrigues formula (17.52) we can introduce three classes of Hilbert spaces $L_w^2[a,b]$ where the weight function $w(x)$ and the interval $[a,b]$ are essentially determined by the choice of a polynomial $s(x)$ of degree 0, 1, or 2. Inside the Hilbert space $\mathscr{H} \equiv L^2([a,b],w(x))$ every vector $|f\rangle \in \mathscr{H}$ can be developed into an infinite series of the corresponding orthogonal polynomials. We briefly scan the list of these classical polynomials and the Hilbert spaces they correspond to.

17.3.1.1 Hermite polynomials

The first case corresponds to the choice $s(x) = 1$ and the interval $[-\infty, \infty]$. Indeed, the equation

$$\frac{1}{w}\frac{\mathrm{d}w}{\mathrm{d}x} \equiv C_{(1)}(x) \equiv \alpha x + \beta \tag{17.54}$$

by means of a linear change of variables can always be reduced to

$$\frac{1}{w}\frac{\mathrm{d}w}{\mathrm{d}x} \equiv C_{(1)}(x) \equiv 2x, \tag{17.55}$$

which has the unique solution

$$w(x) = \exp[-x^2] \tag{17.56}$$

and vanishes only at $\pm\infty$.

Hence, the classical Hermite polynomials can be defined as follows:

$$H_n(x) = (-1)^n \exp[x^2]\frac{\mathrm{d}^m}{\mathrm{d}x^m}\exp[-x^2]. \tag{17.57}$$

Relying on the obvious identity

$$2ax\exp[-ax^2] + \frac{\mathrm{d}w(x)}{\mathrm{d}x} = 0, \tag{17.58}$$

one can write an equivalent formula for the same Hermite polynomials:

$$H_n(x) = \sqrt{2^n}\exp\left[\frac{1}{2}x^2\right]\left(x - \frac{d}{dx}\right)^n \exp\left[-\frac{1}{2}x^2\right]. \tag{17.59}$$

In Section 17.4 we show the deep interpretation of formula (17.59) in terms of unitary representations of the Heisenberg group and the quantum mechanical treatment of the harmonic oscillator.

17.3.1.2 Laguerre polynomials

The next case is provided by Laguerre polynomials, which correspond to $s(x) = x$ and to the weight function

$$w(x) = x^\nu \exp[-x]; \quad \nu > -1. \tag{17.60}$$

The interval is $[0, +\infty]$. Hence, we define

$$L_n^\nu(x) = \frac{x^{-\nu}e^x}{n!}\frac{\mathrm{d}^n}{\mathrm{d}x^n}(x^{n+\nu}e^{-x}). \tag{17.61}$$

The Laguerre polynomials are met in the solutions of the Schrödinger equation for the hydrogen atom.

17.3.1.3 Jacobi polynomials

For Jacobi orthogonal polynomials the standard support interval is $[-1,1]$, while the weight function is

$$w(x) = (1-x)^\nu(1+x)^\mu, \quad \nu,\mu > -1, \tag{17.62}$$

and the standard polynomial $s(x)$ is quadratic. Precisely we have

$$s(x) = (1-x^2). \tag{17.63}$$

Hence, the Rodrigues formula for the Jacobi polynomials is

$$P_n^{(\nu,\mu)}(x) = \frac{(-1)^n}{2^n n!(1-x)^\nu(1+x)^\mu}\frac{\mathrm{d}}{\mathrm{d}x}\left[(1-x)^{\nu+n}(1+x)^{\mu+n}\right]. \tag{17.64}$$

17.3.2 The differential equation satisfied by orthogonal polynomials

The classical orthogonal polynomials derived from the Rodrigues formula satisfy, by construction, a second order ordinary linear differential equation that can be written in general terms.

The argument is the following one. According to what is shown in the proof of Theorem 17.3.1 in [81], the expression

$$A_n(x) = \frac{1}{w}\frac{\mathrm{d}}{\mathrm{d}x}\left[s[x]w(x)\frac{\mathrm{d}C_{(n)}}{\mathrm{d}x}\right] \tag{17.65}$$

is a polynomial of degree $\leq n$; hence, it can be decomposed in the basis of the orthogonal polynomials $C_{(i)}(x)$, namely, we can write

$$A_n(x) = -\sum_{i=0}^{n}\lambda^i C_{(i)}(x). \tag{17.66}$$

We calculate the coefficients λ^i evaluating the scalar products:

$$\lambda_n^i = \frac{(A_n, C_{(i)})}{(C_{(i)}, C_{(i)})}. \tag{17.67}$$

We have

$$(A_n, C_{(i)}) = \int_a^b C_{(i)}(x)\frac{\mathrm{d}}{\mathrm{d}x}\left[s(x)w(x)\frac{\mathrm{d}C_{(n)}}{\mathrm{d}x}\right] dx. \tag{17.68}$$

Integrating by parts twice and discarding the boundary terms which vanish under the hypotheses associated with Formula 17.3.1, we obtain

$$\begin{aligned}(A_n, C_{(i)}) &= \int_a^b dx\, w(x)C_{(n)}(x)\left\{\frac{1}{w(x)}\frac{\mathrm{d}}{\mathrm{d}x}\left[s(x)w(x)\frac{\mathrm{d}C_{(i)}}{\mathrm{d}x}\right]\right\}\\ &= 0 \quad \mathit{if} \quad i < n.\end{aligned} \tag{17.69}$$

In other words, only $\lambda_n^n \neq 0$. Hence, the expression $A_n(x)$, which is a linear combination of the first and second derivatives of the polynomial $C_{(n)}(x)$ with x-dependent coefficients, is proportional to $C_{(n)}(x)$. Such a statement is the differential equation satisfied by the orthogonal polynomials, and the previous discussion provides a precise algorithm to derive it.

Applying the above method we obtain the following results.

17.3.2.1 The differential equation satisfied by Hermite polynomials

We have

$$H_n''(x) - 2xH_n'(x) + 2nH(x) = 0. \tag{17.70}$$

17.3.2.2 The differential equation satisfied by Laguerre polynomials

We have

$$x\frac{\mathrm{d}^2L_n^\nu(x)}{\mathrm{d}x^2} + (\nu + 1 - x)\frac{\mathrm{d}L_n^\nu(x)}{\mathrm{d}x} + nL_n^\nu(x) = 0. \tag{17.71}$$

17.3.2.3 The differential equation satisfied by Jacobi polynomials

We have

$$(1-x^2)\frac{\mathrm{d}^2P_n^{(\nu,\mu)}(x)}{\mathrm{d}x^2} + (\mu - \nu - (\mu + \nu + 2)x)\frac{\mathrm{d}P_n^{(\nu,\mu)}(x)}{\mathrm{d}x} + n(n + \mu + \nu + 1)P_n^{(\nu,\mu)}(x) = 0. \tag{17.72}$$

Typically the above differential equations have important physical interpretations in classical or quantum mechanics. As already stressed the Hermite equation is related with the quantum harmonic oscillator, while the Laguerre equation is related to the Schrödinger equation of the hydrogen atom. In his various incarnations, provided by special values of the parameters ν and μ, the Jacobi equation, which is a hypergeometric equation in disguise, is related with spherical harmonics and other highly notable physical issues. In the next section we discuss in detail the relation of Hermite polynomials with the quantum harmonic oscillator and we stress the tight connection of these functions with the issue of *unitary representations* of a non-compact group, namely, the Heisenberg group.

17.4 The Heisenberg group, Hermite polynomials, and the harmonic oscillator

In this section we illustrate by means of an explicit example the point raised in the introduction to the present chapter, namely, the deep relation between Hilbert spaces and the issue of unitary representations of non-compact Lie groups.

17.4.1 The Heisenberg group and its Lie algebra

Let us consider the following set of 4 × 4 matrices depending on three real parameters:

$$\begin{aligned} \mathbf{Heis} &\equiv \{\mathcal{H}(\lambda,\mu,c) \mid \forall \lambda,\mu,c \in \mathbb{R}\}, \\ \mathcal{H}(\lambda,\mu,c) &= \begin{pmatrix} 1 & 0 & 0 & \lambda \\ 0 & 1 & 0 & \mu \\ \mu & -\lambda & 1 & -2c \\ 0 & 0 & 0 & 1 \end{pmatrix}. \end{aligned} \tag{17.73}$$

This set of matrices **Heis** constitutes a continuous group, since:

(1) The product of any two matrices belonging to the set **Heis** belongs to the same set. Indeed, we can easily verify that

$$\begin{aligned} &\mathcal{H}(\lambda_1,\mu_1,c_1)\cdot\mathcal{H}(\lambda_2,\mu_2,c_2) = \mathcal{H}(\lambda_3,\mu_3,c_3), \\ &\{\lambda_3,\mu_3,c_3\} = \left\{\lambda_1+\lambda_2,\mu_1+\mu_2,c_1+c_2+\frac{1}{2}(\lambda_1\mu_2-\lambda_2\mu_1)\right\}. \end{aligned} \tag{17.74}$$

(2) The identity matrix belongs to the set **Heis**:

$$\mathbf{Id}_{4\times 4} = \mathcal{H}(0,0,0) \in \mathbf{Heis}. \tag{17.75}$$

(3) The inverse of any matrix $\mathcal{H}(\lambda,\mu,c) \in$ **Heis** is in the same set since

$$[\mathcal{H}(\lambda,\mu,c)]^{-1} = \mathcal{H}(-\lambda,-\mu,-c) \in \mathbf{Heis}. \tag{17.76}$$

The group **Heis** is named the Heisenberg group, and it is an example of a Lie group although it is none of the classical or exceptional semi-simple groups that we classified in previous chapters. Indeed, *Heis* is not semi-simple.

Inside the Heisenberg group, we easily identify three one-parameter subgroups respectively given by

$$\begin{aligned}\mathbf{Heis} \supset \mathbf{H}_+(\lambda) &= \mathscr{H}(\lambda, 0, 0),\\ \mathbf{Heis} \supset \mathbf{H}_-(\mu) &= \mathscr{H}(0, \mu, 0),\\ \mathbf{Heis} \supset \mathbf{H_c}(c) &= \mathscr{H}(0, 0, c).\end{aligned} \tag{17.77}$$

Expanding in power series of the respective parameters we obtain

$$\begin{aligned}\mathscr{H}_+(\lambda, 0, 0) &= \mathbf{Id} + \lambda\mathbf{X} + \mathscr{O}(\lambda^2),\\ \mathscr{H}_-(0, \mu, 0) &= \mathbf{Id} + \mu\mathbf{P} + \mathscr{O}(\mu^2),\\ \mathscr{H}_\mathbf{c}(0, 0, c) &= \mathbf{Id} + c\mathbf{Z} + \mathscr{O}(c^2),\end{aligned} \tag{17.78}$$

where

$$\begin{aligned}\mathbf{X} &= \begin{pmatrix} 0 & 0 & 0 & 1\\ 0 & 0 & 0 & 0\\ 0 & -1 & 0 & 0\\ 0 & 0 & 0 & 0\end{pmatrix},\\ \mathbf{P} &= \begin{pmatrix} 0 & 0 & 0 & 0\\ 0 & 0 & 0 & 1\\ 1 & 0 & 0 & 0\\ 0 & 0 & 0 & 0\end{pmatrix},\\ \mathbf{Z} &= \begin{pmatrix} 0 & 0 & 0 & 0\\ 0 & 0 & 0 & 0\\ 0 & 0 & 0 & -2\\ 0 & 0 & 0 & 0\end{pmatrix}.\end{aligned} \tag{17.79}$$

The operators **X**, **P**, **Z** form a basis of generators for the Lie algebra of the Heisenberg group that we denote $\mathbb{H}$eis and is given by all the linear combinations with real coefficients of these three objects:

$$\mathbb{H}\mathrm{eis} = \mathrm{span}_{\mathbb{R}}(\mathbf{X}, \mathbf{P}, \mathbf{Z}). \tag{17.80}$$

The commutation relations of these generators are the following ones:

$$[\mathbf{X}, \mathbf{P}] = \mathbf{Z}; \quad [\mathbf{Z}, \mathbf{P}] = [\mathbf{Z}, \mathbf{X}] = \mathbf{0}. \tag{17.81}$$

The above commutation relations are just isomorphic to the Heisenberg quantization rules of quantum mechanics:

$$[\mathbf{x}, \mathbf{p}] = i\hbar\,\mathbf{Id}; \quad [i\hbar\,\mathbf{Id}, \mathbf{x}] = [i\hbar\,\mathbf{Id}, \mathbf{p}] = \mathbf{0} \tag{17.82}$$

under the identification $\mathbf{X} \leftrightarrow i\,\mathbf{x}$, $\mathbf{P} \leftrightarrow i\,\mathbf{p}$, $\mathbf{Z} \leftrightarrow i\hbar\,\mathbf{Id}$. This isomorphism is precisely the reason for the name given to the Lie algebra defined by equation (17.81).

Let us now observe that we can introduce the complex operators

$$\begin{aligned} \mathbf{A}_+ &= -i\frac{1}{2}(\mathbf{X} + i\,\mathbf{P}), \\ \mathbf{A}_- &= -i\frac{1}{2}(\mathbf{X} - i\,\mathbf{P}), \\ \mathfrak{c} &= i\frac{1}{2}\mathbf{Z}, \end{aligned} \tag{17.83}$$

and the Heisenberg–Lie algebra can be rewritten as

$$[\mathbf{A}_+, \mathbf{A}_-] = \mathfrak{c}; \quad [\mathfrak{c}, \mathbf{A}_\pm] = \mathbf{0}. \tag{17.84}$$

The above equation can be taken as a fully equivalent definition of the Heisenberg–Lie algebra of which by means of equation (17.79) we have constructed an explicit four-dimensional linear representation. Such a representation automatically provides a four-dimensional representation of the corresponding Lie group which is encoded in equation (17.73).

The point we want to stress is the following one. The finite-dimensional representation of our group, which can also be considered the defining one, is *not unitary*! It could not be different since the Heisenberg group is *non-compact* and the general Theorem 17.1.1 states that a non-compact group G has no finite-dimensional unitary representation. In order to give a precise meaning to the above statement, we have to clarify the notion of compactness for Lie groups and of unitarity for Lie group and Lie algebra representations.

Definition 17.4.1. A Lie group G is compact if all of its one-parameter subgroups are isomorphic to a circle $\mathbb{S}^1$, that is, to the group of complex numbers of unit modulus.

As we see the Heisenberg group is non-compact since its principal one-parameter subgroups mentioned in equation (17.78) are isomorphic to the group of real numbers $\mathbb{R}$.

As for unitarity, the characterization of unitary representations of Lie algebras was already advocated in formulae (17.9) and (17.10). When the representation of a Lie algebra is unitary in the sense specified above, the induced representation of the corresponding Lie group, obtained by means of the exponential map, is unitary in the standard sense that the matrix representing any group element $\exp[\mathbf{x}]$ is unitary:

$$\mathscr{D}(\exp[\mathbf{x}])^\dagger = \mathscr{D}(\exp[\mathbf{x}])^{-1} = \mathscr{D}(\exp[-\mathbf{x}]); \quad \forall \mathbf{x} \in \mathbb{G}. \tag{17.85}$$

The defining representation of the Heisenberg group would be unitary if the matrices $\mathbf{X}$, $\mathbf{P}$, $\mathbf{Z}$ were anti-Hermitian, implying that the matrices $\mathbf{A}_\pm$ are the Hermitian conjugate of each other, while $\mathfrak{c}$ should be Hermitian. This is far from being true, in agreement with the quoted Theorem 17.1.1.

In any case the elements of the Heisenberg group are given by

$$\exp[c_+\mathbf{A}_+ + c_-\mathbf{A}_-] \in \mathbf{Heis} \quad \textit{iff} \quad c_+^\star = -c_- \tag{17.86}$$

and a unitary representation could be obtained if we were able to construct a triplet of operators

$$\mathbf{a}_\pm = \mathscr{D}(\mathbf{A}_\pm); \quad \omega\mathbf{1} = \mathscr{D}(\mathfrak{c}) \tag{17.87}$$

such that

$$\begin{aligned} [\mathbf{a}_+, \mathbf{a}_-] &= \omega\mathbf{1}; \quad \omega \in \mathbb{R}; \\ [\mathbf{1}, \mathbf{a}_\pm] &= \mathbf{0}; \\ \mathbf{a}_+^\dagger &= \mathbf{a}_-; \quad \mathbf{1}^\dagger = \mathbf{1}. \end{aligned} \tag{17.88}$$

In view of the mentioned general theorem, the operators mentioned in equation (17.87) must necessarily be linear maps in an infinite-dimensional vector space with a Hermitian scalar product, that is, in a functional Hilbert space. This observation provides the clue that links the theory of Hilbert space and quantum mechanics to the unitary representations of the Heisenberg group. In particular the physical system whose quantum states fill the unitary irreducible representation of the Heisenberg group is the *one-dimensional harmonic oscillator*.

Classically a one-dimensional harmonic oscillator is described by the standard Hamiltonian:

$$\mathfrak{H} = \frac{1}{2m}p^2 + \kappa^2 x^2, \tag{17.89}$$

where $x \in [-\infty, \infty]$ is the spatial coordinate and $p = m\frac{dx}{dt}$ is the particle momentum (κ is the coupling constant in the harmonic potential). Upon quantization, the momentum p is replaced by the differential operator $\hat{p} = -i\hbar\frac{d}{d\hat{x}}$, both $\hat{x}$ and $\hat{p}$ operating on the candidate wave functions $\hat{\Psi}(\hat{x})$ that belong to the Hilbert space of square integrable functions:[2]

$$\hat{\Psi}(\hat{x}) \in L_1^2[-\infty, \infty]. \tag{17.90}$$

Following a standard procedure one can introduce operators

$$\hat{\mathbf{a}}_\pm = \frac{1}{\sqrt{2}}\left(i\frac{\hat{p}}{\sqrt{2m}} \pm \kappa\hat{x}\right) \tag{17.91}$$

---

2 Without entering too much into the details we mention that the square integrable wave functions represent the stationary states of the harmonic oscillator, namely, they are eigenstates of the Hamiltonian (17.89).

that fulfill the relations

$$[\hat{\mathbf{a}}_+, \hat{\mathbf{a}}_-] = \omega \mathbf{1}; \quad \omega = \frac{\kappa}{\sqrt{2m}} \hbar. \tag{17.92}$$

Rescaling the variables as

$$\begin{aligned} \hat{x} &= \frac{x}{\sqrt[4]{2m\kappa^2}}, \\ \hat{\mathbf{a}}_\pm &= \sqrt{\omega}\mathbf{a}_\pm, \end{aligned} \tag{17.93}$$

we obtain

$$\begin{aligned} \mathbf{a}_\pm &= \left(x \mp \frac{\mathrm{d}}{\mathrm{d}x} \right), \\ [\mathbf{a}_+, \mathbf{a}_-] &= 1, \\ \mathfrak{H} &= \omega(\mathbf{a}_+\mathbf{a}_- + 1). \end{aligned} \tag{17.94}$$

Now we easily show that in the Hilbert space $L_1^2[-\infty, \infty]$ the operators $\mathbf{a}_\pm$ are indeed adjoint to each other with respect to the scalar product:

$$\forall f(x), g(x) \in L_1^2[-\infty, \infty] : \quad (g, f) \equiv \int_{-\infty}^{\infty} \overline{g(x)} f(x)\, dx. \tag{17.95}$$

By definition the adjoint $A^\dagger$ of an operator A is that one for which the following relation holds true:

$$(A^\dagger g, f) = (g, Af). \tag{17.96}$$

We easily see from equation (17.96) that the adjoint of the operator $\partial \equiv \frac{\mathrm{d}}{\mathrm{d}x}$ is $\partial^\dagger = -\partial$. Indeed,

$$(-\partial g, f) \equiv \int_{-\infty}^{\infty} -\frac{\mathrm{d}\overline{g(x)}}{\mathrm{d}x} f(x)\, dx = -\underbrace{\int_{-\infty}^{\infty} \frac{\mathrm{d}}{\mathrm{d}x}(\overline{g(x)} f(x))\, dx}_{=0} + \int_{-\infty}^{\infty} \overline{g(x)} \frac{\mathrm{d}f(x)}{\mathrm{d}x} = (g, \partial f). \tag{17.97}$$

Furthermore, looking at the Hamiltonian (17.94) we see that its eigenstate of minimal energy is that which is annihilated by the operator $\mathbf{a}_-$. Let us call such minimal eigenfunction $\Psi_0(x)$. By definition it satisfies the differential equation

$$x\Psi_0(x) = \frac{\mathrm{d}}{\mathrm{d}x}\Psi_0(x), \tag{17.98}$$

which has the unique normalizable to one solution

$$\Psi_0(x) = \frac{e^{-\frac{x^2}{2}}}{\sqrt[4]{\pi}}; \quad \int_{-\infty}^{\infty} |\Psi_0(x)|^2 \, dx = 1. \tag{17.99}$$

All the other states, which form a complete orthonormal basis of $L^2_1[-\infty, \infty]$, are obtained by applying n times the creation operator $\mathbf{a}_+$ to the minimal energy eigenfunction, in particular setting

$$\begin{aligned}\Psi_n(x) &\equiv \frac{1}{\sqrt{n!}}(a_+)^n \Psi_0(x) = \frac{1}{\sqrt{n!2^n}}\left(x - \frac{\mathrm{d}}{\mathrm{d}x}\right)^n \Psi_0(x) \\ &= \frac{1}{\sqrt{n!2^n}} H_n(x)\Psi_0,\end{aligned} \tag{17.100}$$

where $H_n(x)$ are polynomials of degree n that can be iteratively calculated from their own definition and are the previously discussed *Hermite polynomials* (see Section 17.3.1.1).

By construction, the functions $\Psi_n(x)$ are eigenstates of the Hamiltonian with eigenvalue $E = \omega(n+1)$:

$$\mathfrak{H}\Psi_n(x) = \omega(n+1)\Psi_n(x); \quad \int_{-\infty}^{\infty} |\Psi_n(x)|^2 \, dx = 1. \tag{17.101}$$

In this way we have discovered that the orthogonal Hermite polynomials $H_n(x)$ that belong to the Hilbert space $L^2_w[-\infty, \infty]$ with weight function

$$w(x) = \Psi_0^2(x) = \frac{e^{-x^2}}{\sqrt{\pi}} \tag{17.102}$$

are related with a unitary, infinite-dimensional representation of the Heisenberg group and Lie algebra. On the other hand, considering the eigenvalue problem for the harmonic oscillator Hamiltonian

$$\mathbf{a}_+\mathbf{a}_-\Psi_{\mathscr{E}}(x) = \mathscr{E}\Psi_{\mathscr{E}}(x) \tag{17.103}$$

and writing the looked for eigenfunction as

$$\Psi_{\mathscr{E}}(x) = H(x)\Psi_0(x), \tag{17.104}$$

from the relation

$$\mathbf{a}_+\mathbf{a}_- + \mathbf{a}_-\mathbf{a}_+ = 2\mathbf{a}_+\mathbf{a}_- + 1 \tag{17.105}$$

we obtain a differential equation for the function $H(x)$:

$$H''(x) - 2xH'(x) + (2\mathscr{E} + 1)H(x) - H(x) = 0. \tag{17.106}$$

The above equation can be solved in the neighborhood of $x = 0$ in terms of a convergent power series for all real values of the parameter $\mathcal{E}$. The corresponding function $H_{\mathcal{E}}(x)$ is called a Hermite function. Yet if $\mathcal{E} \notin \mathbb{N}$, the Hermite function $H_{\mathcal{E}}(x) \notin L^2_w[-\infty, \infty]$ since the integral $\int_{-\infty}^{\infty} |H_{\mathcal{E}}(x)|^2 w(x)\, dx = \infty$ is divergent. On the contrary, if $\mathcal{E} \in \mathbb{N}$, the recurrence relation for the coefficients in the series solution of the Hermite equation (17.106) stops at a certain iteration, yielding a polynomial solution, i. e., the Hermite polynomial $H_n(x)$ that is square integrable with the weight function (17.102).

17.5 Self-adjoint operators

Let us briefly comment on self-adjoint operators in functional spaces $L^2_w(a, b)$ with particular attention to the second order ones. Let $f(x) \in L^2_w(a, b)$ be a representative of an equivalence class of square integrable functions. A second order linear differential operator acting on $f(x)$ is of the form

$$\mathscr{L}f(x) = \frac{d^2 f(x)}{dx^2} + p(x)\frac{df(x)}{dx} + q(x)f(x), \tag{17.107}$$

where $p(x)$ and $q(x)$ are some functions of the variable x that are real when x is real. As one sees, the solutions to such differential equations can be seen as the null vectors of certain differential operators of the second order. In this perspective we are interested in calculating the adjoint of an operator such as that in equation (17.107). As we recalled already a few times, the adjoint of the operator $\mathscr{L}$ is defined by the statement

$$\forall f(x), g(x) \in L^2_w(a, b): \quad (\mathscr{L}^\dagger g, f) = \overline{(g, \mathscr{L}f)}. \tag{17.108}$$

Integrating by parts under the integral and discarding boundary terms since $w(x)$ is supposed to vanish at the boundary we obtain

$$L^\dagger g(x) = \frac{d^2 g(x)}{dx^2} + p^\dagger(x)\frac{dg(x)}{dx} + q^\dagger(x)g(x), \tag{17.109}$$

where

$$\begin{aligned} p^\dagger(x) &= 2\frac{w'(x)}{w(x)} - p(x), \\ q^\dagger(x) &= \frac{w''(x)}{w(x)} - p'(x) - p(x)\frac{w'(x)}{w(x)} + q(x). \end{aligned} \tag{17.110}$$

If we impose that the considered operator must be self-adjoint, that is, $\mathscr{L}^\dagger = \mathscr{L}$, we get the conditions

$$p^\dagger(x) = p(x); \quad q^\dagger(x) = q(x), \tag{17.111}$$

which are uniquely solved by

$$p(x) = \frac{w'(x)}{w(x)}. \tag{17.112}$$

Both in the theory of group representations and in quantum mechanics one is frequently confronted with the problem of determining the spectrum of eigenvalues of self-adjoint operators of the second order in a functional Hilbert space. Typically this is what is required to determine the complete set of pure quantum states, the self-adjoint operator being the Hamiltonian of the system and its eigenvalues being the energy levels. For the harmonic oscillator associated with the unitary representations of the Heisenberg group this is just what we did when we identified

$$\mathscr{L}^{\text{Herm}} f(x) = \frac{d^2 f}{dx^2} - 2x \frac{df(x)}{dx}, \tag{17.113}$$

which indeed represents the Hamiltonian. Note that $p(x) = -2x = \frac{w'(x)}{w(x)}$, the weight function being $w(x) = \text{const} \times \exp[-x^2]$.

18 Harmonic analysis and conclusive remarks

In Chapter 17, which is the last developed one of this book, we showed how functional spaces enter group theory when we consider the unitary irreducible representations of non-compact groups. Actually, apart from its specific interest and value, the material discussed in Chapter 17 was introductory to the ideas that we simply highlight in the present conclusive chapter in order to indicate the way how group theory merges with geometry and functional analysis in multifaceted applications in physics, chemistry, engineering, economics, IT sciences, and other sciences. Whenever the space of *configurations or data* one is dealing with possesses some kind of partial or full symmetry one is bound to ask oneself the question *How should I correctly take such symmetries into account?* Typically the dynamical equations governing the behavior of the system under investigation are differential equations for certain *functions* (or *sections of suitable bundles*) and the working arena is provided by the functional space $\mathfrak{F}$ to which such functions of m variables (x^μ) are supposed to belong. The symmetries are operations on the space $\mathcal{M}$ of these variables that map a point $p \in \mathcal{M}$ into another one p' that is considered completely equivalent to the previous one, i. e., the system in point p has identical properties and behavior as in point p'. Then it happens that the actual number of relevant coordinates is much less than m. In other words, the relevant number of truly relevant differential equations to be solved is much smaller than you might think a priori. The number of relevant dynamical degrees of freedom depends on the cohomogeneity of the space $\mathcal{M}$, typically a differentiable manifold, but in certain cases a discrete, finite, or infinite set. The symmetry is necessarily a group G acting as a group of transformations on the space $\mathcal{M}$. Focusing on the continuous manifold case, which is the most frequent one, we consider the action of G on the manifold $\mathcal{M}$: this action is transitive if any point $p \in \mathcal{M}$ can be reached from any reference point $p_0 \in \mathcal{M}$ by means of a suitable element $g \in$ G. In this case the cohomogeneity is zero and $\mathcal{M}$ is necessarily a coset manifold G/H. The geometry of G/H coset manifolds was discussed in Chapter 16, and we showed that all geometric entities are reduced to algebraic ones. Here in the next subsection we highlight the principles of so-called *harmonic analysis*, namely, the recipe to construct suitable bases on the functional spaces of functions (or sections of bundles) having support on G/H. In the harmonic basis differential calculus becomes algebraic because all derivative operators are replaced by algebraic operators (matrices, finite or infinite ones). Hence, when the symmetry is complete and our data fill a G/H manifold, dynamical equations become trivial and are completely solved. If the symmetry is not complete, this means that from any given point $p \in \mathcal{M}$ we can reach only a submanifold of smaller dimension, let us say $n < m$. This submanifold is named the G-orbit G.p of p. Then the cohomogeneity of $\mathcal{M}$ is $coh = m - n$ and the number of relevant variables is only $m - n$. Indeed, each G orbit is necessarily a copy of the same homogeneous coset manifold G/H and the entire original manifold $\mathcal{M}$ is foliated into G/H coset manifolds. We divide the m coordinates into two sets: n coordinates y^i ($i = 1, \dots, n$) span G/H, while

https://doi.org/10.1515/9783111201535-018

coh coordinates r^a ($a = 1, \dots, coh$) label the various leaves of the foliation. This is precisely what happens in the case of spherical symmetry in gravity, electrodynamics, or quantum mechanics for the hydrogenoid atoms. The three-dimensional space is foliated into spheres of increasing radius r. Each sphere is an identical copy of the coset manifold $\mathbb{S}^2 \equiv \mathrm{SO}(3)/\mathrm{SO}(2)$ and according to the general theory outlined in the next section the functional space in which the wave function $\psi(\mathbf{x})$ lives can be provided with an infinite basis constructed as follows. First we consider a basis for the functional space of functions $\mathrm{Y}(y)$ on $\mathbb{S}^2$. As we show in the next section the basis for functions on G/H is given by the matrix elements of irreducible representations of the group G, in this case SO(3). The irreps of SO(3) are in one-to-one correspondence with the integer $J = \ell \in \mathbb{Z}$ or half-integer numbers $J = \frac{\ell}{n}, \ell \in \mathbb{Z}$. In quantum mechanics the wave function $\psi(\mathbf{x})$ is a scalar function with third component of the angular momentum $m = 0$. In view of the rule that we outline below (see equation (18.14)), the representations containing $m = 0$ are only the integer ones. So the most general function on $\mathbb{S}^2$ is

$$f(y) = \sum_{\ell=0}^{\infty} \sum_{m=-\ell}^{\ell} c_\ell^m \mathrm{Y}_m^{(\ell)}(y), \tag{18.1}$$

where the spherical harmonic functions

$$\mathrm{Y}_m^{(\ell)}(y) = D_{0,m}^{\ell}(\mathbb{L}(y)) \tag{18.2}$$

are given by the matrix element $D_{0,m}^{\ell}(g)$ of indices $\{0, m\}$ in the $(2\ell+1)$-dimensional representation of integer angular momentum $J = \ell$ for the SO(3) group element $g = \mathbb{L}(y)$ corresponding to a coset representative parameterization of the sphere, like for instance that in equation (16.17). This familiar case is the application to SO(3) of the general formula (18.7) discussed below.

Since the familiar Euclidean space has cohomogeneity $coh = 1$, the coordinate labeling the spherical leaves being the radius r the final ansatz for the wave function is obtained by allowing the constant coefficients $c_\ell^m = \psi_{\ell,m}(r)$ to depend on r. In this way we introduce the radial wave functions and the entire structure of the Mendeleev table of elements is decoded in terms of group representations plus the yields of the Schrödinger differential equation for the radial functions. Actually there is a hidden symmetry O(4) and this is the very reason why we happen to be able to integrate the dynamical equation. Such a tale, however, we do not tell here.

18.1 A few highlights on harmonic analysis

Here we give few a highlights of the general theory of harmonic analysis on coset spaces.

18.1.1 Harmonics on coset spaces

Let us consider as a first step a group manifold G. A complete functional basis on G is given by the matrix elements of the G unitary irreducible representations (UIRs): any function

$$\Phi(g) \quad g \in G \tag{18.3}$$

can be expanded as

$$\Phi(g) = \sum_{(\mu)} \sum_{m,n=1}^{\dim(\mu)} c_{mn}^{(\mu)} D_{mn}^{(\mu)}(g), \tag{18.4}$$

where (μ) are the UIRs of G, m, n run in these representations, and D are the elements of these representations. In fact, the $D_{mn}^{(\mu)}(g)$ satisfy orthogonality and completeness relations that for continuous groups are the obvious generalization of equation (4.68):

$$\begin{aligned} \int_G dg\, D_{mn}^{(\mu)}(g) D_{sr}^{(\nu)}(g^{-1}) &= \frac{\mathrm{vol}(G)}{\mathrm{vol}(\mu)} \delta_{mr} \delta_{ns} \delta^{(\mu)(\nu)}, \\ \sum_{(\mu)} D_{mn}^{(\mu)}(g) D_{nm}^{(\mu)}(g'^{-1}) \dim(\mu) &= \delta(g - g')\, \mathrm{vol}(G). \end{aligned} \tag{18.5}$$

If $\Phi(g)$ transforms in an irreducible representation (μ) of G, for example under left multiplication, namely,

$$\Phi_m^{(\mu)}(g'g) = D_{mn}^{(\mu)}(g')\Phi_n^{(\mu)}(g), \tag{18.6}$$

then only a subset of the complete functional basis is present, the D's that transform in the same way, that is,

$$D_{mn}^{(\mu)}, \quad \mu, m \text{ fixed}. \tag{18.7}$$

So in this case the expansion is shorter:

$$\Phi_m^{(\mu)}(g) = \sum_n c_n^{(\mu)} D_{mn}^{(\mu)}(g). \tag{18.8}$$

Let us now consider the functions $\Phi(y)$ on a coset manifold G/H. The matrix elements

$$D_{mn}^{(\mu)}(\mathbb{L}(y)),$$

where $\mathbb{L}(y)$ is a coset representative of the coset space (see equation (16.20) and previous lines) and y are the coset manifold coordinates, constitute a complete functional basis on G/H:

$$\Phi(y) = \sum_{(\mu)} \sum_{m,n=1}^{\dim(\mu)} c_{mn}^{(\mu)} D_{mn}^{(\mu)}(\mathbb{L}(y)), \tag{18.9}$$

satisfying

$$\begin{aligned} \int_{G/H} dm(y) D_{mn}^{(\mu)}(\mathbb{L}(y)) D_{sr}^{(\nu)}(\mathbb{L}(y)^{-1}) &= \frac{\mathrm{vol}(G/H)}{\mathrm{vol}(\mu)} \delta_{mr}\delta_{ns}\delta^{(\mu)(\nu)}, \\ \sum_{(\mu)} D_{mn}^{(\mu)}(\mathbb{L}(y)) D_{nm}^{(\mu)}(\mathbb{L}(y')^{-1}) \dim(\mu) &= \delta(y-y')\,\mathrm{vol}(G/H), \end{aligned} \tag{18.10}$$

where $dm(y)$ is the invariant measure on G/H.

We are interested in functions $\Phi(\mathbb{L}(y))$ on which a linear action of the subgroup $H \subset G$ is well-defined, namely, that transform within an irreducible linear representation (ρ) of H,

$$\forall h \in H: \quad h \cdot \Phi_i^{(\rho)}(\mathbb{L}(y)) \equiv D_{ij}(h)^{(\rho)} \Phi_j^{(\rho)}(\mathbb{L}(y)), \tag{18.11}$$

the index i running in (ρ).

Recalling, as in equation (16.63), that a coset manifold G/H can be interpreted as the base manifold of a principal fiber bundle having the subgroup H as structural group, the functions $\Phi_i^{(\rho)}(\mathbb{L}(y))$ advocated in equation (18.11) are actually sections of an associated vector bundle in the representation (ρ) of the principal bundle (16.63). Such objects are relevant in both classical and quantum field theories and in particular in Kaluza–Klein compactifications of higher-dimensional theories. They are also considered the main ingredients of recent *geometrical deep learning constructions*, as we already remarked.

Hence, how do we determine such vector bundle sections constructed in terms of matrix element representations

$$D_{mn}^{(\mu)}, \quad m, n \text{ running in } (\mu) \text{ of } G? \tag{18.12}$$

The answer is simple. We have to choose matrix elements of the following type:

$$D_{in}^{(\mu)}, \quad i \text{ running in } (\rho) \text{ of } H,\ n \text{ running in } (\mu) \text{ of } G. \tag{18.13}$$

Hence, we have to choose only those irreps (μ) that satisfy the following condition: the decomposition of (μ) with respect to $H \subset G$ must contain the H irreducible representation (ρ):

$$(\mu) \xrightarrow{H} \cdots \oplus (\rho) \oplus \cdots. \tag{18.14}$$

Only in this case $D_{mn}^{(\mu)}$ decomposes in $(\ldots, D_{in}^{(\mu)}, D_{i'n}^{(\mu)}, \ldots)$ and $D_{in}^{(\mu)}$ actually exists. The functions constructed with matrix elements of the representations satisfying (18.14) are named H-*harmonics* on G/H and constitute a complete basis for the sections of the ρ-vector bundle on G/H: $\Phi_i^{(\rho)}(\mathbb{L}(y))$. Its generic expansion is

$$\Phi_i^{(\rho)}(y) = \sum_{(\mu)}{}' \sum_n c_n^{(\mu)} D_{in}^{(\mu)}(\mathbb{L}(y)), \tag{18.15}$$

where $\sum'$ means a sum only on the representations (μ) satisfying the property (18.14). Notice that the H-harmonics have both an index running in an irreducible representation of G (on the right) and an index running in an irreducible representation of H (on the left). The coefficients of the expansion $c_n^{(\mu)}$ have an index of a representation of G present in the expansion $\sum'$.

18.1.2 Differential operators on H-harmonics

H-harmonics have a very powerful property: the action on them of differential operators can be expressed in a completely algebraic way. According to equations (16.62), (16.92), and (16.100) the H-covariant external differential

$$\mathscr{D}^{\mathrm{H}}\mathbb{L}^{-1}(y) \equiv (d + \omega^i T_i)\mathbb{L}^{-1}(y) \tag{18.16}$$

becomes

$$\mathscr{D}^{\mathrm{H}}\mathbb{L}^{-1}(y) = -V^a T_a \mathbb{L}^{-1}(y), \tag{18.17}$$

or, expanding on the Vielbein $\mathscr{D}^{\mathrm{H}} = E^a \mathscr{D}_a^{\mathrm{H}}$,

$$\mathscr{D}_a^{\mathrm{H}}\mathbb{L}^{-1}(y) = -\lambda_a T_a \mathbb{L}^{-1}(y), \tag{18.18}$$

where T_a is the generator of the subspace $\mathbb{K}$ defined by the orthogonal decomposition $\mathbb{G} = \mathbb{K} \oplus \mathbb{H}$ in the representation in which the inverse coset representative is expressed and λ_a is the rescaling of the Vielbein as introduced in equation (16.92).

Since the harmonics *are* the inverse coset representatives in the representation (μ),

$$D^{(\mu)}(\mathbb{L}^{-1}(y))^i{}_n = \mathbb{L}^{-1}(y)^i{}_n. \tag{18.19}$$

More precisely, the harmonic in (ρ, μ), representations $D^{(\mu)i}{}_n$, is obtained performing the decomposition (18.14) of the first index of $D^{(\mu)m}{}_n = \mathbb{L}^{-1m}{}_n$ and taking the (ρ) term. We consider the *inverse* coset representative because for simplicity of notation we want H to act on their left, while it acts on the right of coset representatives.

Hence the action of the H-covariant derivative on the harmonic $D^i{}_n(y) \equiv D^i{}_n(\mathbb{L}(y))$ is

$$\mathscr{D}_a^{\mathrm{H}} D^i{}_n(y) = -\lambda_a \big(T_a D(y)\big)^i{}_n = -\lambda_a (T_a)^i_m D^m{}_n, \tag{18.20}$$

where $(T_a)^i{}_m$ is defined as the (ρ) term in the decomposition (18.14) of the index n in $(T_a)^n{}_m$, namely,

$$(T_a)^n{}_m = \{\dots, (T_a)^i{}_m, \dots\}. \tag{18.21}$$

Utilizing equation (16.100) which expresses the spin $\mathfrak{so}(n)$-connection $\omega^a{}_b$ of the n-dimensional coset manifold G/H in terms of the G-group structure constants and of the H-connection ω^i, it is evident that once the H-covariant differential is reduced to an algebraic action on the harmonics the same will be true of the standard $\mathfrak{so}(n)$-covariant differential:

$$\mathscr{D} \equiv \mathscr{D}^{\mathfrak{so}(n)} = \mathrm{d} + \omega^{ab} J_{ab}, \tag{18.22}$$

where J_{ab} are the $\mathfrak{so}(n)$ generators.

It follows that the Laplace–Beltrami invariant operators of differential geometry, which are all $\mathfrak{so}(n)$ invariant operators quadratic in the $\mathfrak{so}(n)$ derivatives, will have on harmonics an algebraic action and will take on them a numerical eigenvalue that can be calculated a priori in terms of the generators of the G Lie algebra and depend only on the G irreducible representation (μ). In conclusion, when we consider sections of the various vector bundles associated with the principal bundle (16.63), the harmonics constitute a complete basis for the functional space over G/H and all differential operations in this functional space are reduced to algebraic ones.

Thanks to group theory analysis is reduced to algebra.

18.2 Conclusive remarks

The goals pursued in this book were all spelled out in the introduction. At the end of his work, the author confesses his hope that they have been, at least partially, attained. He is completely conscious of the unconventional character of his approach; actually he relied just on these unconventional features in order to attain his goals. This book is not a conventional textbook for math or physics majors, yet it should be regarded as a possible textbook for a course that in the author's view by now should be delivered to mathematics students, physics students, complex systems students, dynamical systems students, chemistry students, crystallographers, and IT science students. Groups being the correct mathematical phrasing of symmetry are transversal to all sciences and even to the different subdisciplines of mathematics. Thinking in terms of groups and symmetries should become the conceptual habit of all types of scientists. Group theory provides a very powerful grammar to unify different languages under a unique intellectual code. As the author quite frequently stressed, so-called *artificial intelligence* is the last corner in the arena where symmetry and group theory fueled the most substantial advances in many scientific directions. We will see which advances can be propelled by group thinking also in this new field. The personal viewpoint of the author is that although machines can do at high speed what humans are able to do only slowly, handling numbers of data many orders of magnitudes smaller than those processed by machines, *yet there is no basis for talking about machine intelligence*. Even in the neural networks case the principle holds true that *what you put in you get out*, you meaning the humans. Inserting symmetry in the convolutional neural networks is a great idea but it is also the proof that *what the machine learns is just what it is instructed to learn by its human architect*. The abstract thinking inherent to group theory and representation theory is a gigantic progress of the human race that was somehow already hidden in platonic philosophy, yet it took more than 2000 years to become crystal clear and correctly formalized. Abstract thinking, this author thinks, will never be realized in machines and talking about artificial intelligence is improper, notwithstanding the spectacular progress made with deep neural networks in many areas.

A Available MATHEMATICA NoteBooks written by the author

Group theory and differential geometry calculations can benefit from the utilization of computer codes. A very appropriate, efficient, and user-friendly language to write such codes is Wolfram MATHEMATICA. The reader and user of the present monograph, in order to improve his/her understanding, to practice with the material illustrated in the book, and also to perform new original calculations, experiments, and checks, might utilize the library of NoteBooks that is put at his/her free disposal in the following repository: https://www.degruyter.com/document/isbn/9783111201535/html.

These NoteBooks, all written and developed by this book's author in the course of many years, were tested in many situations, constantly upgraded, and improved. A complete list is given below:

(1) **GenFinitGroup23.nb**. This is a general NoteBook able to reconstruct the entire set of group elements in the form of $n \times n$ matrices starting from some few elements that are generators. The routines in the NoteBook work both with pure matrices and with name-labeled matrices, organize them in conjugacy classes, verify whether a subset is a subgroup and whether a subgroup is normal, and do more of such operations.

(2) **Octagroup23.nb**. This is a NoteBook entirely dedicated to the octahedral group in three dimensions, its irreducible representations and character table, and other associated issues.

(3) **Tetragroup23.nb**. This is a short NoteBook dedicated to the tetrahedral group and to its application to the vibrational modes of molecules of type XY_4 with tetrahedral symmetry. Emphasis is placed on the generation of pictures.

(4) **L168Group23**. This is a very complete NoteBook that deals with all aspects of the simple group $\mathrm{PSL}(2, \mathbb{Z}_7)$ of order 168, creates its representations and character table, studies its orbits on a regular seven-dimensional lattice, classifies all of its subgroups, and does much more.

(5) **metricgrav23**. This NoteBook is a general tool for calculations of the Riemannian geometry of a generic manifold in n dimensions with an arbitrary signature. The adopted formalism is the standard one of Riemannian tensor calculus. The only required input is the explicit form of the metric in a set of local coordinates that must be specified by the user.

(6) **Vielbgrav23**. Similarly to the previous NoteBook this is a general tool for calculations of the Riemannian geometry of a generic manifold in n dimensions with an arbitrary signature; in this case, however, the adopted formalism is the Vielbein one, and all the calculated tensors are provided in the intrinsic Vielbein basis. It is very useful in formal calculations.

(7) **e77construzia23**. The last four NoteBooks are specialized to the explicit construction of the root space and the explicit matrix form of the fundamental representa-

https://doi.org/10.1515/9783111201535-019

tion of each of four out of five of the exceptional Lie algebras in their maximally split real realization. The first of these NoteBooks is devoted to the Lie algebra $\mathfrak{e}_{7(7)}$. In the main text the $\mathfrak{e}_{7(7)}$ Lie algebra is discussed in Section 13.3.2.1.

(8) **e88construzia23**. This NoteBook is devoted to the Lie algebra $\mathfrak{e}_{8(8)}$. Since for this algebra the smallest representation is the adjoint of 248 dimensions the NoteBook creates a library space where various items and the large matrices of the representation are stored on the hard disk of the user. The steps followed in the NoteBook are the same as those illustrated in Section 13.3.

(9) **f44construzia23**. This NoteBook is devoted to the Lie algebra $\mathfrak{f}_{4(4)}$, whose complicated 26-dimensional fundamental representation is step by step constructed as in Section 13.2. The reader of that section might be interested in playing with the objects dealt with in that section by him/herself. One just has to download the NoteBook *f44construzia23* and run it on his/her computer.

(10) **g22construzia23**. The last NoteBook is devoted to the explicit construction of the $g_{2(2)}$ Lie algebra and its fundamental seven-dimensional representation. In the main text this algebra is discussed in Section 13.1.

The next few pages provide a glimpse of some pages of some of the 10 NoteBooks as they appear exported to pdf. These pages are meant to give the reader a preview of what he/she can download and utilize.

Elaborated in Pietramarazzi March 12th 2023, collecting and improving various subroutines dispersed in various previous NoteBooks

General Discrete Group routines

This Notebook is a background programme. It has to be evaluated. Then its routines can be used in a separate Execution Notebook. It is recommended to read chapter one about the introduced mathematical items and about the available commands and routines. It is advisable that before EVALUATING this NoteBook you QUIT KERNEL.

Pietro G. Fré

Mathematica Code

What the routines contained in this NoteBook do

The routines contained in this NoteBook provide general instruments to construct finite groups in terms of matrices. Some routines generate all the elements of a group, others organize it in conjugacy classes

Group theoretical routines available to the user

1. **generone**. Given a set of matrices named Allgroup the routine generone generates the set of all their products. Generone arrives at a set that closes under multiplication if the original matrices were elements of a finite group.
2. **generoneName**. Given a set of matrices named AllgroupN , labeled, each of them with a name in a pair ***{name, matrix}***, the routine generone generates the set of all their products keeping track of the non--commutative product of names. **generoneName** arrives at a set that closes under multiplication if the original matrices were elements of a finite group.
3. **generonePrec** is identical to **generone** but it is slower because it uses FullSimplify instead of Simplify. It is necessary when the original matrices one starts from have radicals.
4. **coniugatoL** . If you give a set of matrices forming a finite group and you name it **gruppone**, **coniugatoL** produces the set of conjugacy classes into which the finite group is organized. The output of this calculation is named orgclas. After that coniugatoL elaborates a name for each group element that is n_p where n is the equivalence class #n in the order provided in orgclas and p enumerates the elements in the same class as they are arranged in orgclas. The array containing the so named group elements is denoted groupNamed. It is an array of pairs ***{name, matrix}***.
5. **verifiosub**. Given a set of matrices that form a finite group, named **gruppone** and a subset named settino, **verifiosub** verifies whether settino is a subgroup and moreover it verifies whether it is a normal subgroup.
6. **verifiosubName** works exactly like **verifiosub** but it operates on two sets (**gruppone**,settino) not matrices but of labeled matrices namely of pairs ***{name, matrix}***.
7. **quozientus**. Given a set of matrices forming a finite group, named **gruppone** and a normal subgroup named **gruppino**, **quozientus** constructs the equivalence classes G/H namely the quotient group. The output of this calculation is named equaclass.
8. **quozientusName** works exactly like **quozientus** but it operates on two sets (**gruppone**,**gruppino**) not matrices but of labeled matrices namely of pairs ***{name, matrix}***.

9. **stabilione.** You have to call **gruppone** the set of nxn matrices representing a given group G. You have to name **veicolo** an n-vector. The routine **stabilione** constructs the subgroup G that leaves it invariant. The stability group is named **stab.** The orbit of vectors is named **orbitano.**

Example

The best way to explain how the routines work is by means of an example. We treat the case of the octahedral group realized by means of its natural action in 3 dimension. The abstract group has two generators S and T that satisfy the following relations:

$$S^2 = T^3 = (\mathrm{TS})^4 = \boldsymbol{e} \tag{0.1}$$

where $\boldsymbol{e}$ is the neutral element. A possible choice for these generators is the following one

$$S = \begin{pmatrix} 0 & 1 & 0 \\ 1 & 0 & 0 \\ 0 & 0 & -1 \end{pmatrix}; \; T = \begin{pmatrix} 0 & 1 & 0 \\ 0 & 0 & 1 \\ 1 & 0 & 0 \end{pmatrix} \tag{0.2}$$

In the next sections we show how from these two matrices we reconstruct the whole group and its properties

Reconstruction of the Octahedral Group from generators

In this section we utilize both the routine generone and the routine generoneName to generate all the elements of the octahedral group starting from two generators

Matrix form of the generators

In[ ]:= $TT = \begin{pmatrix} 0 & 1 & 0 \\ 0 & 0 & 1 \\ 1 & 0 & 0 \end{pmatrix}; SS = \begin{pmatrix} 0 & 1 & 0 \\ 1 & 0 & 0 \\ 0 & 0 & -1 \end{pmatrix};$

Generation of the group keeping track of the abstract form of the elements

In this section we generate all the 24 elements keeping track of their abstract form as a word. We use the routine generoneName

```
In[ ]:= AllgroupN = {{𝒯, TT}, {𝒮, SS}};
```

```
In[ ]:= generoneName
```

We save the result as a file OctoName

```
In[ ]:= OctoName = AllgroupN;
```

We investigate as a check one of the elements

```
In[ ]:= OctoName[[24]]
```

```
Out[ ]= {𝒯.𝒮.𝒯.𝒯.𝒮.𝒯.𝒮, {{0, -1, 0}, {-1, 0, 0}, {0, 0, -1}}}
```

Generation of the group elements without names

Organization in conjugacy classes of the octahedral group

In this section we use the file containing all the group elements without names to organize them in conjugacy classes. At the end of the calculation the computer will name the elements according to conjugacy classes. We use the routine coiugatoL

```
In[ ]:= gruppone = Octo;
        coniugatoL
```

We save the result

```
In[ ]:= OctoN = groupNamed;
```

We investigate for illustrative goals a couple of elements

```
In[ ]:= OctoN[[1]]
```

Out[ ]= {1_1, {{1, 0, 0}, {0, 1, 0}, {0, 0, 1}}}

```
In[ ]:= OctoN[[18]]
```

Out[ ]= {5_2, {{0, -1, 0}, {0, 0, 1}, {-1, 0, 0}}}

Started in Moscow, Home, May 14th 2016, updated 2018 in Pietramarazzi
Last update March 12th 2023

The simple group $L_{168} \simeq \mathrm{PSL}(2,Z_7)$ and its maximal subgroups.

This Notebook is a background programme. It has to be evaluated. Then its routines can be used in a separate Execution Notebook. It is recommended to read chapters one and two that contain all the explanations about the introduced mathematical items and about the available commands and routines. It is advisable that before EVALUATING this NoteBook you QUIT KERNEL.

Pietro G. Fré
Professor Emeritus
University of Torino, Physics Department, Via. P. Giuria 1, 10125, Torino Italy
Home address Prof. Pietro G. Fré
Via Sant'Ilario, 4/2
15040 Pietramarazzi (AL)
Italy
email: pietro.fre@unito.it

1 Theory and items

WARNING

In this programme the symbols ϵ, ρ,σ,τ, $\mathcal{Y}$,$\mathcal{X}$ are protected and you cannot use them. You can only make substitutions

Description of the items and of their names

The group L168 ≃ PSL(2,Z_7)

Generators

The group L168 has three generators S,T,R with the following relations:

$$R^2 = S^3 = T^7 = RST = TSR^4 = 1$$

Abstractly we use the Greek letter for the generators of L168 and we set:

$$\rho^2 = \sigma^3 = \tau^7 = \rho\sigma\tau = (\tau\sigma\rho)^4 = 1$$

In the programme there are two 7-dimensional realizations of these generators:

1. R,S,T are 7×7 matrices in the orthonormal basis
2. RL,SL,TL are 7×7 matrices in the basis of A7 simple roots where the metric is the A7 Cartan matrix (ROOT LATTICE)

3. RW,SW,TW are 7×7 matrices in the basis of A7 simple weights **where the metric is the A7 Cartan matrix (WEIGHT LATTICE)**

Conjugacy classes

The group L168 has 6 conjugacy classes organized in the following way

Conjugacy Class	C_1	C_2	C_3	C_4	C_5	C_6
representative of the class	e	R	S	TSR	T	SR
order of the elements in the class	1	2	3	4	7	7
number of elements in the class	1	21	56	42	24	24

The abstract form of the group elements as **words** in the generator symbols ρ,σ,τ is given in the file **formL168clas** which has the following appearance:

$$\text{formL168clas[[2]]} = \{\mathcal{E},\ 1,\ \{\epsilon\}\} \tag{0.1}$$

$$\begin{aligned}\text{formL168clas[[2]]} = \{&\mathcal{R}, 21, \{\rho.\sigma.\rho.\sigma.\rho.\tau.\tau,\ \tau.\tau.\sigma.\rho.\tau.\tau.\tau,\ \sigma.\rho.\sigma.\sigma,\ \sigma.\rho.\tau.\tau.\sigma.\rho.\tau.\tau,\ \sigma.\rho.\tau.\sigma.\rho.\tau,\\ &\tau.\sigma,\ \rho.\sigma.\rho.\tau.\sigma.\rho.\tau,\ \tau.\tau.\sigma.\rho.\tau.\sigma.\sigma,\ \rho.\tau.\tau.\sigma.\rho.\sigma.\rho,\ \tau.\sigma.\rho.\tau.\sigma,\ \sigma.\rho.\sigma.\rho.\tau.\sigma.\sigma,\ \rho.\tau.\tau.\sigma.\rho.\tau.\tau,\ \rho.\sigma.\rho.\tau,\\ &\rho.\tau.\sigma.\rho.\tau.\sigma,\ \tau.\tau.\sigma.\rho.\tau.\tau.\sigma.\rho,\ \tau.\tau.\tau.\sigma.\rho.\tau.\tau,\ \tau.\tau.\sigma.\rho.\sigma,\ \rho.\tau.\sigma.\rho,\ \tau.\sigma.\rho.\tau.\tau.\sigma.\rho.\tau,\ \tau.\sigma.\rho.\sigma.\rho.\tau.\tau.\sigma,\ \rho\}\}\end{aligned} \tag{0.2}$$

$$\ldots\ldots\ldots\ldots\ldots\ldots\ldots\ldots\ldots\ldots\ldots \tag{0.3}$$

$$\ldots\ldots\ldots\ldots\ldots\ldots\ldots\ldots\ldots\ldots\ldots \tag{0.4}$$

$$\begin{aligned}&\text{formL168clas[[6]]} =\\ &\{\mathcal{SR}, 24, \{\rho.\sigma,\ \rho.\tau.\sigma.\rho.\sigma.\rho.\tau.\tau,\ \rho.\sigma.\rho.\tau.\tau.\sigma.\rho,\ \rho.\sigma.\rho.\sigma,\ \sigma.\rho.\tau.\sigma.\rho.\sigma.\rho,\ \rho.\tau.\tau.\sigma.\rho.\sigma.\rho.\tau,\ \tau.\sigma.\rho.\sigma.\sigma,\ \tau.\tau.\sigma.\rho.\tau.\sigma,\\ &\quad\rho.\sigma.\rho.\tau.\tau.\tau,\ \tau.\sigma.\sigma,\ \tau.\tau.\tau,\ \rho.\sigma.\rho.\sigma.\rho.\tau.\sigma,\ \sigma.\rho.\tau.\tau.\sigma,\ \rho.\tau.\sigma.\rho.\tau,\ \sigma.\rho.\tau.\tau.\sigma.\sigma,\ \rho.\tau.\tau.\tau.\sigma.\rho,\ \rho.\sigma.\rho.\tau.\sigma.\rho.\sigma,\\ &\quad\tau.\tau.\tau.\sigma,\ \rho.\tau.\tau,\ \tau.\sigma.\rho.\sigma.\rho.\tau.\sigma.\sigma,\ \tau.\sigma.\rho.\tau.\tau.\tau.\sigma,\ \rho.\tau.\tau.\sigma.\sigma,\ \sigma.\rho.\sigma.\rho,\ \sigma.\rho\}\}\end{aligned} \tag{0.5}$$

This file is very important in the construction of irreducible representations in order to provide in each case the precise form of the representation of each invidual group element.

In the programme, after evaluation of this notebook the explicit matrices of the group organized by conjugacy classes are stored in two different files:

1. **L168strutclas** contains the 7×7 matrices in the simple root basis, namely those acting on the configuration space of the root lattice.
2. **L168clasW** contains the 7×7 matrices in the simple root basis, namely those acting on the configuration space of the root lattice.

The standard realization of the 7 simple roots as vectors in 8-dimensions orthogonal to the vector {1,1,1,1,1,1,1,1} is encoded in the object **alp** while the simple weights are encoded in the object **lammi**

finally the Cartarn matrix is encoded in the object CC.

The set of all the group elements is provided in two files:

1. **grupL168** contains the 168 matrices 7×7 in the configuration space (root lattice)
2. **L168W** contains the 168 matrices 7×7 in momentum space (weight lattice)

Characters

The 6 irriducible representations of the group L168 are of dimensions **1,6,7,8,3 and 3**, respectively. The character table is the following one:

0	e	R	S	TRS	T	SR
D_1	1	1	1	1	1	1
D_6	6	2	0	0	−1	−1
D_7	7	−1	1	−1	0	0
D_8	8	0	−1	0	1	1
DA_3	3	−1	0	1	$\frac{1}{2}\left(-1+i\sqrt{7}\right)$	$\frac{1}{2}\left(-1-i\sqrt{7}\right)$
DB_3	3	−1	0	1	$\frac{1}{2}\left(-1-i\sqrt{7}\right)$	$\frac{1}{2}\left(-1+i\sqrt{7}\right)$

The character table is encoded in the object **PchiL168**. The populations of conjugacy classes are encoded in the object **PgiL168**. The names of the representations are encoded in the object **namesD168**.

Embedding into G2 and τ-matrices

The embedding of the group L168 into G2 is demonstrated by showing that there is a G2-invariant 3-tensor that is invariant under L168. The tensor which satisfies the G2 - relations and which is L168 invariant is named ϕ in the orthonormal basis and φ in the root basis. The basis of 8×8 gamma matrices in 7-dimensions such that the G2 invariant tensor coincides with $\phi_{ijk} = \eta\,\tau_{ijk}\,\eta$ are constructed and encoded in the file $\tau\tau$

The maximal subgroups G21 and O24A, O24B

The simple group L168 contains maximal subgroups only of index 8 and 7, namely of order 21 and 24. The order 21 subgroup G21 is the unique non-abelian group of that order and abstractly it has the structure of the semidirect product $\mathbb{Z}_3 \propto \mathbb{Z}_7$. Up to conjugation there is only one subgroup 21 as we explicitly verified with the computer. On the other hand, up to conjugation there are two different groups of order 24 that are both isomorphic to the octahedral group O_{24}.

They are named O24A, O24B.

The group G21

The group G21 has two generators $\mathcal{X}$ and $\mathcal{Y}$ that satisfy the following relations:

$$\mathcal{X}^3 = \mathcal{Y}^7 = 1\ ;\ \mathcal{X}\mathcal{Y} = \mathcal{Y}^2\mathcal{X}$$

Conjugacy classes of G21

Conjugacy Class	C_1	C_2	C_3	C_4	C_5
representative of the class	e	$\mathcal{Y}$	$\mathcal{X}^2\,\mathcal{Y}\mathcal{X}\mathcal{Y}^2$	$\mathcal{Y}\mathcal{X}^2$	$\mathcal{X}$
order of the elements in the class	1	7	7	3	3
number of elements in the class	1	3	3	7	7

Characters

The 5 irriducible representations of the group G21 are of dimensions 1,1,1,3, and 3, respectively. The character table is the following one:

0	e	$\mathcal{Y}$	$\mathcal{X}^2\,\mathcal{Y}\mathcal{X}\mathcal{Y}^2$	$\mathcal{Y}\mathcal{X}^2$	$\mathcal{X}$
D_1	1	1	1	1	1
DX_1	1	1	1	$-(-1)^{1/3}$	$(-1)^{2/3}$
DY_1	1	1	1	$(-1)^{2/3}$	$-(-1)^{1/3}$
DA_3	3	$\frac{1}{2}\,i\left(i+\sqrt{7}\right)$	$-\frac{1}{2}\,i\left(-i+\sqrt{7}\right)$	0	0
DB_3	3	$-\frac{1}{2}\,i\left(-i+\sqrt{7}\right)$	$\frac{1}{2}\,i\left(i+\sqrt{7}\right)$	0	0

The character table is encoded in the object **Pchi21**. The populations of conjugacy classes are encoded in the object **Pgi21**. The names of the representations are encoded in the object **names21**.

Our choice of the representative for the entire conjugacy class of G21 maximal subgroups is given, abstractly, by setting the following generators:

$$\mathcal{Y} = \rho.\tau.\tau.\tau.\sigma.\rho \quad ;\ \mathcal{X} = \sigma.\rho.\sigma.\rho.\tau.\tau$$

In the programme the entire group G21 is encoded in the following files.

1. **G21W** contains all the 21 elements as 7×7 matrices in the weight basis
2. **G21clasW** contains all the conjugacy classes in the weight basis (momentum space)
3. **G21clasR** contains all the conjugacy classes in the root basis (configuration space)
4. **FormalG21clas** contains the group generators written symbolically as words in ρ,σ,τ organized in conjugacy classes with the mention of the choice of $\mathcal{Y},\mathcal{X}$ generators:

```
FormalG21clas[[1]] = {𝒴, 𝒳};                                        (0.6)

FormalG21clas[[2]] =
{{1, {ϵ}}, {3, {ρ.σ.ρ.σ.ρ.τ.σ, ρ.τ.τ.τ.σ.ρ, τ.σ.ρ.σ.ρ.τ.σ.σ}},
 {3, {σ.ρ.σ.ρ.τ.τ.σ, ρ.σ.ρ.τ.τ.σ.σ, τ.σ.ρ.τ.τ}},                      (0.7)
 {7, {σ, ρ.σ.ρ.τ.τ, σ.ρ.σ.ρ.τ.τ.σ.σ, ρ.σ.ρ.σ.ρ.τ.σ.σ, ρ.τ.τ.τ.σ.ρ.σ,
```

$$\text{fT} = \rho.\tau.\sigma.\rho.\tau.\tau.\sigma.\rho.\tau \quad ; \qquad \text{fS} = \sigma.\rho.\tau.\sigma.\rho.\tau \tag{0.10}$$

The substitution rules for the embedding are QO24A and QO24B.

2 List of routines and commands

The *Mathematica* package and its available commands

The basic commands of the present package are the following ones:

Construction of the representation of the group L168 and of its maximal subgroups and decomposition into irreps

For these tasks we have the following commands:

1. **brutcaratterL168.** This routine calculates the character of any linear representation D of the group L168 provided in the following way:
$$\text{Repra} = \{D[1],\ D[\rho],\ D[\sigma],\ D[\tau\rho\sigma],\ D[\tau],\ D[\sigma\rho]\}$$
where D[1], etc are the the matrix realization of the standard representatives of the six conjugacy classes in terms of the generators of the group, satisfying the defining relations. The routine derives the multiplicity vector encoded in the object **decompo** and it constructs the decomposition of the representation D into irreps of L168. The projectors onto the irreducible representations are named **PP168.**
2. **brutcaratter21.** This routine calculates the character of any linear representation D of the group G21 provided in the following way:
$$\text{Repra21} = \left\{ \boxed{D[e]}\ \boxed{D[\mathcal{Y}]}\ \boxed{D[\mathcal{X}^2\,\mathcal{Y}\mathcal{X}\mathcal{Y}^2]}\ \boxed{D[\mathcal{Y}\mathcal{X}^2]}\ \boxed{D[\mathcal{X}]} \right\}$$
where D[e], etc are the the matrix realization of the standard representatives of the five conjugacy classes in terms of the generators of the group $\mathcal{Y}$, and $\mathcal{X}$ satisfying the defining relations. The routine derives the multiplicity vector encoded in the object **decompo21** and it constructs the decomposition of the representation D into irreps of G21. The projectors onto the irreducible representations are named **PP21.**
3. **brutcaratter24.** This routine calculates the character of any linear representation D of the group O24 provided in the following way:
$$\text{Repra24} = \left\{ \boxed{D[e]}\ \boxed{D[T]}\ \boxed{D[\text{STST}]}\ \boxed{D[S]}\ \boxed{D[\text{ST}]} \right\}$$
where D[e], etc are the the matrix realization of the standard representatives of the five conjugacy classes in terms of the generators of the group T, and S satisfying the defining relations. The routine derives the multiplicity vector encoded in the object **decompo24** and it constructs the decomposition of the representation D into irreps of O24. The projectors onto the irreducible representations are named **PP24.**
4. **belcaratter21.** This routine requires a substitution rule **passarulla** $=\{\mathcal{Y}\to \text{D}[\mathcal{Y}],\ \mathcal{X}\to \text{D}[\mathcal{X}]\}$ and calculates the character of the corresponding representation and its explicit decomposition into irreps. The projection operators onto irreps are explicitly calculated.

Auxiliary group theoretical routines used by the package but available also to the user

Besides the basic commands described in the previous section this package contains also some general group-theoretical routines that are internally utilized but available to the user. These are

1. **generone.** Given a set of matrices named Allgroup the routine generone generates the set of all their products. Repeated use of generone arrives at a set that closes under multiplication if the original matrices were elements of a finite group.
2. **generoneName.** Given a set of matrices named AllgroupN , associated, each of them with a name, the routine generone generates the set of all their products keeping track of the non--commutative product of names. Repeated use of generoneName arrives at a set that closes under multiplication if the original matrices were elements of a finite group
3. **coniugatoL** (or **coniugatoM**, they are equivalent). If you give a set of matrices forming a finite group and you name it **gruppone**, **coniugatoL** produces the set of conjugacy classes into which the finite group is organized. The output of this calculation is named **orgclas**.

NoteBook created July 1st 2017 in Moscow on the basis of a package tested by the author for many years

Last update March 12th 2023

MATHEMATICA PACKAGE METRICGRAV

Pietro G. Fré
Professor Emeritus
University of Torino, Physics Department, Via. P. Giuria 1, 10125, Torino Italy
Home address Prof. Pietro G. Fré
Via Sant'Ilario, 4/2
15040 Pietramarazzi (AL)
Italy
email: pietro.fre@unito.it

Einstein Equations in metric formalism

In this Notebook we provide a package to calculate Einstein equations for any given metric in arbitrary dimensions using the metric formalism.

Description of the programme

What the NoteBook does

Given a n-dimensional manifold $\mathcal{M}$ whose coordinates we denote x_i and a metric defined over it and provided in the form

$$ds^2 = g_{ij}(x)\, dx^i \otimes dx^j \tag{0.1}$$

The programme extracts the metric tensor $g_{ij}\ (x)$ calculates its inverse $g^{ij}\ (x)$ calculates the Christoffel symbols $\Gamma_{ij}{}^{k}(x)$, then the Riemann tensor, the Ricci tensor and the Einstein tensor.

Initialization and inputs to be supplied

After reading the NoteBook, calculations are intialized in the following way

1. First the user types **nn** = positive integer number (which is going to be the dimension **n** of the considered manifold)
2. First the user types **mainmetric**. The computer will ask the user to supply three inputs in the following form:
   a) the set of coordinates as n-vector. That vector must be named **coordi** = $\{x_1,....,x_n\}$
   b) the set of coordinates differentials as n-vector. That vector must be named **diffe** = $\{dx_1,....,dx_n\}$
   c) the metric given as a quadratic differential that must be named ds2. The user will type **ds2**=$g_{[[i,j]]}dx^i\ dx^j$

3. After providing these inputs the user will type the command **metricresume.**

Produced outputs

1. The Christoffel symbols $\Gamma^{\lambda}_{\mu\nu}$ are encoded in an array **Gam[[λ,μ,ν]]**.

2. The Riemann tensor $R^{\lambda}_{\mu\nu\rho}$ is encoded in an array **Rie[[λ,μ,ν,ρ]]**.

3. The curvature 2-form $\mathbf{R} = d\Gamma + \Gamma \wedge \Gamma$ is encoded in an array named **RR[[λ,μ]]**.

4. The Ricci tensor $R_{\mu\rho} \equiv R^{\lambda}_{\mu\lambda\rho}$ is encoded in an array named **ricten[[μ,ρ]]**.

5. The Einstein tensor $G_{\mu\rho} \equiv R_{\mu\rho} - \frac{1}{2} g_{\mu\rho} R$ is encoded in an array **einst[[μ,ρ]]**.

The *Mathematica* code

Examples

The Schwarschild metric

In this section we exemplify the use of the package metricgrav with the case of the Schwarschild metric that we write in the following form

$$ds^2 = -\left(1 - 2\frac{\mu}{r}\right) dt^2 + \left(1 - 2\frac{\mu}{r}\right)^{-1} dr^2 + r^2 \left(d\theta^2 + \mathrm{Sin}[\theta]^2 \, d\phi^2\right) \qquad (0.2)$$

computer calculation of the Riemann, Ricci and Einstein tensors

The de Sitter metric for a manifold with positive spatial curvature

In this section we exemplify the use of the package metricgrav with the case of the de Sitter metric for a manifold of positive spatial curvature that we write as follows:

$$ds^2 = -dt^2 + \frac{\mathrm{Cosh}[H*t]^2}{H^2} \left(\frac{dr^2}{1 - r^2} + r^2 \left(d\theta^2 + \mathrm{Sin}[\theta]^2 \, d\phi^2\right)\right) \qquad (0.3)$$

computer calculation of the Riemann, Ricci and Einstein tensors

```
nn = 4;
mainmetric
```

```
coordi = {t, r, θ, ϕ};
diffe = {dt, dr, dθ, dϕ};
```

$$ds2 = -dt^2 + \frac{\text{Cosh}[H*t]^2}{H^2}\left(\frac{dr^2}{1-r^2} + r^2\left(d\theta^2 + \text{Sin}[\theta]^2\, d\phi^2\right)\right);$$

```
metricresume
```

I resume the calculation

{Null}

```
MatrixForm[einst]
```

$$\begin{pmatrix} \frac{3H^2}{2} & 0 & 0 & 0 \\ 0 & \frac{3\,\text{Cosh}[H\,t]^2}{2(-1+r^2)} & 0 & 0 \\ 0 & 0 & -\frac{3}{2}r^2\,\text{Cosh}[H\,t]^2 & 0 \\ 0 & 0 & 0 & -\frac{3}{2}r^2\,\text{Cosh}[H\,t]^2\,\text{Sin}[\theta]^2 \end{pmatrix}$$

```
scalaron
```

$6\,H^2$

Manifold Geometry in Vielbein formalism

Pietro G. Fré
Professor Emeritus
University of Torino, Physics Department, Via. P. Giuria 1, 10125, Torino Italy
Home address Prof. Pietro G. Fré
Via Sant'Ilario, 4/2
15040 Pietramarazzi (AL)
Italy
email: pietro.fre@unito.it

What this Notebook does

In this Notebook we provide a package to calculate the geometry of an arbitrary manifold in arbitrary dimension and with an arbitrary signature using the Vielbein formalism and the intrinsic components of all tensors.

Instructions for the user

The inputs

In order to initialize the calculation the user has to type five lines of inputs providing the following information

1) the dimension n=**dimse**

2) the set of coordinates an n-vector = **coordi**

3) the set of differentials, an n-vector = **diffe**

4) the set of vielbein 1-forms a n-vector = **fform**

5) the signature of the space as n-vector = **signat** of +/- 1.

Activating the calculation

After providing the above information the user will start the calculations by typing **mainstart**

The obtained outputs

The MATHEMATICA NoteBook calculates the following objects:

1. The contorsion $c^i{}_{jk}$ defined by the equation $dV^i = c^i{}_{jk}\, V^i \wedge V^k$ and encoded in an array $\mathbf{contens}_{[[i,j,k]]}$.
2. The spin connection 1-form ω^{ij} defined by the equation $dV^i - \omega^{ij} \wedge V^k \eta_{jk} = 0$ and encoded in an array $\omega_{[[i,j]]}$.

3. The intrinsic components of the spin connection defined by the equation $\omega^{ij} = \omega^{ij}_k$ and encoded in a tensor `ometen[[i,j,k]]`.
4. The curvature two-form R^{ij} defined by the equation R=dω -$\omega\wedge\omega$ and encoded in a tensor RF[[i,j]].
5. The Riemann tensor with flat indices, defined by $R^{ij} = \mathrm{Rie}^{ij}{}_{pq}\, V^p \wedge V^q$ and encoded in an array `Rie[[i,j,a,b]]`.
6. The Ricci tensor with flat indices defined by $\mathrm{Ric}^i_p = \mathrm{Rie}^{iq}{}_{pq}$ and encoded in an array ricten[[e,b]].

The MATHEMATICA Code

■ Input of the Differential Form package and symbol protection

■ The routines of this package

Main

You start this programme by typing **mainstart**

routine mainstart

Spin connection

This routine is devised to calculate the intrinsic components of the spin connection once the contorsion tensor as already been calculated

$dV^i = c^i{}_{jk}\, V^i \wedge V^k$. The package is named **spinpackgen.**

routine spinpackgen

Routine curvapackgen

This routine is devised to calculate the curvature two form and the Riemann tensor in a general situation for an arbitrary dimensional manifold and with the vielbein depending on all the coordinates. You can start this programme only after having computed the spin connection via the package **spinpackgen**

routine curvapackgen

calculation of the contorsion for general manifolds

This routine is calculates the external differential of the Vielbein $dV^i = c^i{}_{jk}\, V^i \wedge V^k$. It is named **contorgen**

routine contorgen

Example

de Sitter space in the coordinates for spatial sections of positive curvature.

In this example we calculate once again the geometry of de Sitter space using the same coordinates that are used for the same example calculated with metricgrav. Using the Vielbein formalism the homogeneous nature of the manifold is better revealed. Indeed the intrinsic components of the Riemann and Ricci tensor come out constant.

```
dimse = 4;
```

$$\text{fform} = \left\{dt, \frac{\text{Cosh}[H*t]}{H} * \frac{dr}{\sqrt{1-r^2}}, \frac{\text{Cosh}[H*t]}{H} * r * d\theta, \frac{\text{Cosh}[H*t]}{H} * r * \text{Sin}[\theta] * d\phi\right\};$$

```
coordi = {t, r, θ, ϕ};
diffe = {dt, dr, dθ, dϕ};
signat = {-1, 1, 1, 1};

mainstart

==============================

Welcome this is the Vielbeingrav package that calculates
geometry of a manifold in Vielbein formalism

Your space has dimension n = 4

You gave me the following data
```

$$\text{vector of 1-form vielbein} = \left\{dt, \frac{dr\,\text{Cosh}[H\,t]}{H\sqrt{1-r^2}}, \frac{d\theta\, r\,\text{Cosh}[H\,t]}{H}, \frac{d\phi\, r\,\text{Cosh}[H\,t]\,\text{Sin}[\theta]}{H}\right\}$$

```
vector of coordinates = {t, r, θ, ϕ}

vector of differentials = {dt, dr, dθ, dϕ}

 I resume the calculation and I evaluate the contorsion

I calculate the exterior differentials of the vielbeins

--------------------

I finished!

Next I calculate the inveverse vielbein

Done!

I resume the calculation of the contorsion

 I calculate the contorsion c[i,j,k] for

i = 1

i = 2

i = 3

i = 4

I have finished!

The result, encoded in a vector dV[[i]] is the following:
```

$$dV[1] = 0$$

$$dV[2] = H\,\text{Tanh}[H\,t]\, V[1] \wedge V[2]$$

$$dV[3] = 2\left(\frac{1}{2} H\,\text{Tanh}[H\,t]\, V[1] \wedge V[3] + \frac{H\sqrt{1-r^2}\,\text{Sech}[H\,t]\, V[2] \wedge V[3]}{2\,r}\right)$$

dV[4] =

$$2\left(\frac{1}{2}\,\mathrm{H\,Tanh[H\,t]}\,\mathrm{V[1]}\wedge\mathrm{V[4]}+\frac{\mathrm{H}\sqrt{1-r^2}\,\mathrm{Sech[H\,t]}\,\mathrm{V[2]}\wedge\mathrm{V[4]}}{2\,r}+\frac{\mathrm{H\,Cot}[\theta]\,\mathrm{Sech[H\,t]}\,\mathrm{V[3]}\wedge\mathrm{V[4]}}{2\,r}\right)$$

The contorsion is encoded in a tensor named contens

Now I can begin the calculation of the spin connection

I start

I resume the calculation of the spin connection

----

the result is

$\omega[12] = \mathrm{H\,Tanh[H\,t]\,V[2]}$

$\omega[13] = \mathrm{H\,Tanh[H\,t]\,V[3]}$

$\omega[14] = \mathrm{H\,Tanh[H\,t]\,V[4]}$

$$\omega[23] = -\frac{\mathrm{H}\sqrt{1-r^2}\,\mathrm{Sech[H\,t]\,V[3]}}{r}$$

$$\omega[24] = -\frac{\mathrm{H}\sqrt{1-r^2}\,\mathrm{Sech[H\,t]\,V[4]}}{r}$$

$$\omega[34] = -\frac{\mathrm{H\,Cot}[\theta]\,\mathrm{Sech[H\,t]\,V[4]}}{r}$$

Task finished

The result is encoded in a tensor ω[[i,j]]

Its components are encoded in a tensor ometen[i,j,m]

I calculate the Riemann tensor

I tell you my steps :

a = 1

b = 1

b = 2

b = 3

b = 4

a = 2

b = 1

b = 2

b = 3

b = 4

a = 3

b = 1

b = 2

b = 3

```
b = 4
a = 4
b = 1
b = 2
b = 3
b = 4
Finished
-------------------------------------
Now I evaluate the curvature 2-form of your space
I find the following answer
```

R[12] = H^2 V[1] ∧ V[2]

R[13] = H^2 V[1] ∧ V[3]

R[14] = H^2 V[1] ∧ V[4]

R[23] = H^2 V[2] ∧ V[3]

R[24] = H^2 V[2] ∧ V[4]

R[34] = H^2 V[3] ∧ V[4]

```
The result is encoded in a tensor RF[i,j]
Its components are encoded in a tensor Rie[i,j,a,b]
-------------------------------------
 Now I calculate the Ricci tensor
1 1   non zero
2 2   non zero
3 3   non zero
4 4   non zero
I have finished the calculation
 The tensor ricten[[a,b]] giving the Ricci tensor
 is ready for storing on hard disk
```

$$\text{Ricci}[a,b] = \begin{pmatrix} \frac{3H^2}{2} & 0 & 0 & 0 \\ 0 & \frac{3H^2}{2} & 0 & 0 \\ 0 & 0 & \frac{3H^2}{2} & 0 \\ 0 & 0 & 0 & \frac{3H^2}{2} \end{pmatrix}$$

```
----------------------------
{Null}
```

Bibliography

[1] E. Galois, "Oeuvres mathématiques d'Évariste galois," *J. Math. Pures Appl.*, vol. XI, pp. 381–444, 1846.

[2] A. Cayley, "On the theory of groups, as depending on the symbolic equation $\theta^n = 1$," *Philos. Mag.*, vol. 7, no. 1854, pp. 123–130.

[3] C. Jordan, *Traité des substitutions et des équations algébriques*. Gauthiers-Villars, Parıs, 1870. Reedición: Gabay, Parıs, 1989.

[4] E. Betti, "Sulla Risoluzione delle Equazioni Algebriche: Memoria," *Ann Sci. Mat. Fis.*. Rome, February and March 1852.

[5] A. Hurwitz, "Über algebraische Gebilde mit Eindeutigen Transformationen in sich," *Math. Ann.*, vol. 41, no. 3, pp. 403–442, 1892.

[6] F. Klein, "Ueber die Transformation siebenter Ordnung der elliptischen Functionen," *Math. Ann.*, vol. 14, no. 3, pp. 428–471, 1878.

[7] U. Stammbach.

[8] A. Cayley, "A memoir on the theory of matrices," *Philos. Trans. R. Soc. Lond.*, vol. 148, pp. 17–37, 1858.

[9] N. J. Higham, "Cayley, Sylvester, and early matrix theory." MIMS EPrint 2007.119. http://www.manchester.ac.uk/mims/eprints, 2007.

[10] J. J. Sylvester, "A demonstration of the theorem that every homogeneous quadratic polynomial is reducible by real orthogonal substitutions to the form of a sum of positive and negative squares," *Philos. Mag. Ser. 4*, vol. 4, no. 23, pp. 138–142, 1852. http://www.maths.ed.ac.uk/~aar/sylv/inertia.pdf.

[11] H. Grassmann, *Die Lineale Ausdehnungslehre, ein neuer Zweig der Mathematik*. Otto Wigand, Leipzig, 1844.

[12] G. Peano, *Calcolo Geometrico secondo l'Ausdehnungslehre di H. Grassmann, preceduto dalle operazioni della logica deduttiva*. Fratelli Bocca Editori, Torino, 1888.

[13] S. Lie, "Zur theorie des Integrabilitetsfaktors," *Christiana Forh*, pp. 242–254, 1874.

[14] S. Lie, *Theorie der Transformationsgruppen*, vol. I. B. G. Teubner, Leipzig, 1888. Written with the help of Friedrich Engel.

[15] S. Lie, *Theorie der Transformationsgruppen*, vol. II. B. G. Teubner, Leipzig, 1890. Written with the help of Friedrich Engel.

[16] S. Lie, *Theorie der Transformationsgruppen*, vol. III. B. G. Teubner, Leipzig, 1893. Written with the help of Friedrich Engel.

[17] S. Lie, *Vorlesungen über continuierliche Gruppen mit geometrischen und anderen Anwendungen*. B. G. Teubner, Leipzig, 1893. Written with the help of Georg Scheffers.

[18] W. Killing, "Die Zusammensetzung der stetigen endlichen Transformationsgruppen," *Math. Ann.*, vol. 33, no. 1, pp. 1–48, 1888.

[19] S. Helgason, "A centennial: Wilhelm Killing and the exceptional groups," *Math. Intell.*, vol. 12, no. 3, pp. 54–57, 1990.

[20] É. Cartan, *Über die einfachen Transformationsgruppen*. Leipz. Ber., 1893.

[21] É. Cartan, *Sur la structure des groupes de transformations finis et continus*, vol. 826. 1894. Thése, Paris, Nony.

[22] H. Weyl, *The Classical Groups: Their Invariants and Representations*. Princeton University Press, 1939.

[23] J. E. Humphreys, *Reflection Groups and Coxeter Groups*. Cambridge University Press, 1992.

[24] E. B. Dynkin, "The structure of semi-simple algebras," *Usp. Mat. Nauk, N. S.*, vol. 2, no. 4(20), pp. 59–127, 1947.

[25] C. Ehresmann, "Les connexions infinitésimales dans un espace fibré différentiable," *Colloque de Topologie, Bruxelles*, pp. 29–55, 1950.

[26] C. Ehresmann, "Sur la théorie des espaces fibrés," *Colloque de Topologie Algébrique du C. N. R. S., Paris*, pp. 3–15, 1947.

[27] C. Yang and R. Mills, "Conservation of isopotic spin and isotopic gauge invariance," *Phys. Rev.*, vol. 96, pp. 191–195, 1954.

https://doi.org/10.1515/9783111201535-020

[28] P. G. Fré, *A Conceptual History of Space and Symmetry: From Plato to the Superworld*. Springer, 2018.
[29] M. M. Bronstein, J. Bruna, T. Cohen, and P. Velickovic, "Geometric deep learning: grids, groups, graphs, geodesics, and gauges," 2021, arXiv:2104.13478.
[30] T. S. Cohen and M. Welling, "Group equivariant convolutional networks," 2016, arXiv:1602.07576.
[31] J. E. Gerken, J. Aronsson, O. Carlsson, H. Linander, F. Ohlsson, C. Petersson, and D. Persson, "Geometric deep learning and equivariant neural networks," 2021, arXiv:2105.13926.
[32] M. C. N. Cheng, V. Anagiannis, M. Weiler, P. de Haan, T. S. Cohen, and M. Welling, "Covariance in physics and convolutional neural networks," 2019, arXiv:1906.02481.
[33] J. Aronsson, "Homogeneous vector bundles and g-equivariant convolutional neural networks," *Sampl. Theory Signal Process. Data Anal.*, vol. 20, no. 10, 2022. arXiv:2105.05400.
[34] T. Cohen, M. Geiger, and M. Weiler, "A general theory of equivariant CNNs on homogeneous spaces," 2018, arXiv:1811.02017.
[35] E. Fyodorov, "The symmetry of regular systems of figures," *Proc. Imp. St. Petersburg Mineral. Soc., Ser. 2*, vol. 28, pp. 1–146, 1891.
[36] F. G. Frobenius, "Uber lineare Substitutionen und bilineare Formen," *J. Reine Angew. Math.*, vol. 84, pp. 1–63, 1878.
[37] R. Gilmore, *Lie Groups, Lie Algebras, and Some of Their Applications*. Courier Corporation, 2012.
[38] S. Helgason, *Differential Geometry and Symmetric Spaces*. Academic Press, 1962.
[39] N. Jacobson, *Lie Algebras*. Courier Corporation, 1979.
[40] J. F. Cornwell and J. Cornwell, *Group Theory in Physics*, vol. 1. Academic Press, London, 1984. Part A.
[41] G. Fano, *Metodi Matematici della Meccanica Quantistica*. Zanichelli, Bologna, 1967.
[42] M. Nakahara, *Geometry, Topology and Physics*. CRC Press, 2003.
[43] F. Klein, "Vergleichende Betrachtungen über neuere geometrische Forschungen," *Math. Ann.*, vol. 43, no. 1, pp. 63–100, 1893. (Also: Gesammelte Abh., Vol. 1, Springer, 1921, pp. 460–497).
[44] M. Tinkham, *Group Theory and Quantum Mechanics*. Courier Corporation, 2003.
[45] E. Wigner, *Group Theory: And Its Application to the Quantum Mechanics of Atomic Spectra*, vol. 5. Elsevier, 2012.
[46] V. Heine, *Group Theory in Quantum Mechanics: An Introduction to Its Present Usage*. Courier Corporation, 2007.
[47] H. Weyl, *The Theory of Groups and Quantum Mechanics*. Courier Corporation, 1950.
[48] A. P. Cracknell, *Group Theory in Solid-State Physics*. Halsted Press, 1975.
[49] R. S. Knox and A. Gold, *Symmetry in the Solid State*. WA Benjamin, 1964.
[50] H.-W. Streitwolf, *Group Theory in Solid-State Physics*. Macdonald and Co., 1971.
[51] M. S. Dresselhaus, G. Dresselhaus, and A. Jorio, *Group Theory: Application to the Physics of Condensed Matter*. Springer Science & Business Media, 2007.
[52] S. Coleman, *Aspects of Symmetry: Selected Erice Lectures*. Cambridge University Press, 1988.
[53] H. Georgi, *Lie Algebras in Particle Physics: From Isospin to Unified Theories*, vol. 54. Westview Press, 1999.
[54] A. Das and S. Okubo, *Lie Groups and Lie Algebras for Physicists*. World Scientific, 2014.
[55] R. Slansky, "Group theory for unified model building," *Phys. Rep.*, vol. 79, no. 1, pp. 1–128, 1981.
[56] A. Miller, "Application of group representation theory to symmetric structures," *Appl. Math. Model.*, vol. 5, no. 4, pp. 290–294, 1981.
[57] S. Sternberg, *Group Theory and Physics*. Cambridge University Press, 1995.
[58] H. F. Jones, *Groups, Representations and Physics*. CRC Press, 1998.
[59] H. J. Lipkin, *Lie Groups for Pedestrians*. Courier Corporation, 2002.
[60] P. Ramond, *Group Theory: A Physicist's Survey*. Cambridge University Press, 2010.
[61] T. Inui, Y. Tanabe, and Y. Onodera, *Group Theory and Its Applications in Physics*, vol. 78. Springer Science & Business Media, 2012.
[62] A. Zee, *Group Theory in a Nutshell for Physicists*. Princeton University Press, 2016.
[63] J. Stillwell, *Naive Lie Theory*. Springer Science & Business Media, 2008.

[64] P. Woit, "Quantum theory, groups and representations: An introduction," *Department of Mathematics, Columbia University*, 2015.
[65] A. Barut and R. Raczka, *Theory of Group Representations and Applications*. World Scientific Publishing Co Inc., 1986.
[66] J. Elliott and P. Dawber, *Symmetry in Physics, Vols. 1 and 2*. Macmillan, 1979.
[67] R. Hermann, *Lie Groups for Physicists*, vol. 5. WA Benjamin, New York, 1966.
[68] B. G. Wybourne, *Classical Groups for Physicists*. Wiley, 1974.
[69] M. Hamermesh, *Group Theory and Its Application to Physical Problems*. Courier Corporation, 1962.
[70] G. James and M. W. Liebeck, *Representations and Characters of Groups*. Cambridge University Press, 2001.
[71] P. Fré and A. S. Sorin, "Classification of Arnold–Beltrami flows and their hidden symmetries," *Phys. Part. Nucl.*, vol. 46, no. 4, pp. 497–632, 2015.
[72] P. G. Fré, *Advances in Geometry and Lie Algebras from Supergravity, Theoretical and Mathematical Physics Book Series*. Springer, 2018.
[73] U. Bruzzo, A. Fino, and P. Fré, "The Kähler quotient resolution of $\mathbb{C}^3/\Gamma$ singularities, the McKay correspondence and D=3 $\mathcal{N} = 2$ Chern–Simons gauge theories," *Commun. Math. Phys.*, vol. 365, pp. 93–214, 2019. arXiv:1710.01046.
[74] U. Bruzzo, A. Fino, P. Fré, P. A. Grassi, and D. Markushevich, "Crepant resolutions of $\mathbb{C}^3/\mathbb{Z}_4$ and the generalized Kronheimer construction (in view of the gauge/gravity correspondence)," *J. Geom. Phys.*, vol. 145, 103467, 2019, 50 pp.
[75] M. Bianchi, U. Bruzzo, P. Fré, and D. Martelli, "Resolution à la Kronheimer of $\mathbb{C}^3/\Gamma$ singularities and the Monge–Ampère equation for Ricci-flat Kähler metrics in view of D3-brane solutions of supergravity," *Lett. Math. Phys.*, vol. 111, June 2021.
[76] C. Nash and S. Sen, *Topology and Geometry for Physicists*. Elsevier, 1988.
[77] S. Mukhi and N. Mukunda, *Introduction to Topology, Differential Geometry and Group Theory for Physicists*. 1990.
[78] E. Schmutzer, *Symmetrien und Erhaltungssätze der Physik*. 1972.
[79] P. G. Fré, *Gravity, a Geometrical Course*, vols. 1 and 2. Springer Science & Business Media, 2012.
[80] M. J. Greenberg and J. R. Harper, *Algebraic Topology, A First Course*. The Benjamin/Cummings Publishing Company, 1981.
[81] P. Dennery and A. Krzywicki, *Mathematics for Physicists*. Harper & Row, New York, Evanston, and London, 1967.
[82] H. Cartan, *Théorie élémentaire des fonctions analytiques d'une ou plusieurs variables complexes*. Hermann, Paris, 1961.
[83] V. S. Varadarajan, *Lie Groups, Lie Algebras, and Their Representations*. Springer Science & Business Media, 2013.
[84] R. Gilmore, *Lie Groups, Physics, and Geometry: An Introduction for Physicists, Engineers and Chemists*. Cambridge University Press, 2008.
[85] P. G. Fré and M. Trigiante, "Chaos from symmetry: Navier Stokes equations, Beltrami fields and the universal classifying crystallographic group," 2022, arXiv:2204.01037.
[86] J. Etnyre and R. Ghrist, "Contact topology and hydrodynamics: Beltrami fields and the Seifert conjecture," *Nonlinearity*, vol. 13(2), pp. 441–458, 2000.
[87] R. Ghrist, "On the contact geometry and topology of ideal fluids," in *Handbook of Mathematical Fluid Dynamics*, vol. IV, pp. 1–38, 2007.
[88] R. Cardona, E. Miranda, and D. Peralta-Salas, "Euler flows and singular geometric structures," *Philos. Trans. R. Soc. A, Math. Phys. Eng. Sci.*, vol. 377, 20190034, September 2019.
[89] M. Eva and C. Oms, "The geometry and topology of contact structures with singularities," 2021, arXiv:1806.05638.
[90] R. Cardona and E. Miranda, "On the volume elements of a manifold with transverse zeroes," *Regul. Chaotic Dyn.*, vol. 24, pp. 187–197, March 2019.

[91] R. Cardona and E. Miranda, "Integrable systems and closed one forms," *J. Geom. Phys.*, vol. 131, pp. 204–209, September 2018.
[92] V. Guillemin, E. Miranda, and A. R. Pires, "Symplectic and Poisson geometry on b-manifolds," *Adv. Math.*, vol. 264, pp. 864–896, October 2014.
[93] J. Pollard and G. P. Alexander, "Singular contact geometry and Beltrami fields in cholesteric liquid crystals," 2019, arXiv:1911.10159.
[94] E. Miranda and C. Oms, "The singular Weinstein conjecture," *Adv. Math.*, vol. 389, pp. 1–41, 2021.
[95] E. Miranda, C. Oms, and D. Peralta-Salas, "On the singular Weinstein conjecture and the existence of escape orbits for b-Beltrami fields," *Commun. Contemp. Math.*, vol. 24, 2150076, March 2022.
[96] J. E. Humphreys, *Introduction to Lie Algebras and Representation Theory*. Springer, 1972.
[97] A. Adem and R. James Milgram, *Cohomology of Finite Groups*. Springer, 2004.
[98] P. Fré and P. Soriani, *The $N = 2$ Wonderland: From Calabi–Yau Manifolds to Topological Field Theories*. World Scientific, Singapore, 1995.
[99] P. Fré and A. S. Sorin, "The Weyl group and asymptotics: all supergravity billiards have a closed form general integral," *Nucl. Phys. B*, vol. 815, pp. 430–494, 2009.
[100] P. Fré, F. Gargiulo, and K. Rulik, "Cosmic billiards with painted walls in non-maximal supergravities: a worked out example," *Nucl. Phys. B*, vol. 737, no. 1, pp. 1–48, 2006. doi:10.1016/j.nuclphysb.2005.10.023, arXiv:hep-th/0507256.
[101] P. Fré, A. S. Sorin, and M. Trigiante, "Integrability of supergravity black holes and new tensor classifiers of regular and nilpotent orbits," *J. High Energy Phys.*, vol. 04, 015, 2012.
[102] P. Fré, A. S. Sorin, and M. Trigiante, "Black hole nilpotent orbits and Tits Satake universality classes," 2011, arXiv:1107.5986.
[103] F. Cordaro, P. Fré, L. Gualtieri, P. Termonia, and M. Trigiante, "$\mathcal{N} = 8$ gaugings revisited: an exhaustive classification," *Nucl. Phys. B*, vol. 532, no. 1-2, pp. 245–279, 1998. doi:https://doi.org/10.1016/S0550-3213(98)00449-0, arXiv:hep-th/9804056.
[104] P. Fré, V. Gili, F. Gargiulo, A. S. Sorin, K. Rulik, and M. Trigiante, "Cosmological backgrounds of superstring theory and solvable algebras: Oxidation and branes," *Nucl. Phys. B*, vol. 685, pp. 3–64, 2004.
[105] P. Fré, A. S. Sorin, and M. Trigiante, "The c-map, Tits Satake subalgebras and the search for $\mathcal{N} = 2$ inflation potentials," *Fortschr. Phys.*, vol. 63, pp. 198–258, 2015.
[106] E. Brown and N. Loehr, "Why is $\mathrm{PSL}(2,7) \approx \mathrm{GL}(3,2)$?," *Am. Math. Mon.*, vol. 116, no. 8, pp. 727–732, 2009.
[107] G. Fano, "Sui postulati fondamentali della geometria proiettiva in uno spazio lineare a un numero qualunque di dimensioni," *G. Mat.*, vol. 30, pp. 106–132, 1892. http://www.bdim.eu/item?id=GM_Fano_1892_1.
[108] B. L. Cerchiai, P. Fré, and M. Trigiante, "The role of $\mathrm{PSL}(2,7)$ in M-theory: M2-branes, Englert equation and the septuples," *Fortschr. Phys.*, vol. 67, no. 5, 1900020, 2019.
[109] R. C. King, F. Toumazet, and B. G. Wybourne, "A finite subgroup of the exceptional Lie group G2," *J. Phys. A, Math. Gen.*, vol. 32, no. 48, pp. 8527–8537, 1999.
[110] P. Fré, "Supersymmetric M2-branes with Englert fluxes, and the simple group $\mathrm{PSL}(2,7)$," *Fortschr. Phys.*, vol. 64, no. 6-7, pp. 425–462, 2016. arXiv:1601.02253 [hep-th].
[111] P. Fré and M. Trigiante, "Twisted tori and fluxes: a no go theorem for Lie groups of weak G2 holonomy," *Nucl. Phys. B*, vol. 751, no. 3, pp. 343–375, 2006. doi:10.1016/j.nuclphysb.2006.06.006, arXiv:hep-th/0603011.
[112] C. Luhn, S. Nasri, and P. Ramond, "Simple finite non-Abelian flavor groups," *J. Math. Phys.*, vol. 48, no. 12, 123519, 2007. arXiv:0709.1447 [hep-th].
[113] G. Ruppeiner, "Thermodynamic curvature measures interactions," *Am. J. Phys.*, vol. 78, pp. 1170–1180, November 2010. arXiv:1007.2160.
[114] G. Ruppeiner, "Thermodynamic curvature from the critical point to the triple point," *Phys. Rev. E*, vol. 86, 021130, August 2012. arXiv:1208.3265.

[115] G. Ruppeiner, A. Sahay, T. Sarkar, and G. Sengupta, "Thermodynamic geometry, phase transitions, and the Widom line," *Phys. Rev. E*, vol. 86, 052103, November 2012. arXiv:1106.2270.
[116] G. Ruppeiner, "Thermodynamic curvature: pure fluids to black holes," *J. Phys. Conf. Ser.*, vol. 410, 012138, February 2013. arXiv:1210.2011.
[117] G. Ruppeiner, P. Mausbach, and H.-O. May, "Thermodynamic R-diagrams reveal solid-like fluid states," 2014, arXiv:1411.2872.
[118] G. Ruppeiner and A. Seftas, "Thermodynamic curvature of the binary van der Waals fluid," *Entropy*, vol. 22, no. 11, 1208, October 2020. arXiv:2009.12668.
[119] V. Lychagin, "Contact geometry, measurement and thermodynamics," in *Nonlinear PDEs, Their Geometry and Applications. Proceedings of Wisla*, vol. 18, pp. 3–54. Springer Nature, 2019.
[120] V. Lychagin and M. Roop, "Phase transitions in filtration of real gases," 2019, arXiv:1903.00276.
[121] V. Lychagin and M. Roop, "Steady filtration of Peng–Robinson gas in a porous medium," 2019, arXiv:1904.08387.
[122] V. Lychagin and M. Roop, "On higher order structures in thermodynamics," *Entropy*, vol. 22, 1147, October 2020. arXiv:2009.02077.
[123] A. Kushner, V. Lychagin, and M. Roop, "Optimal thermodynamic processes for gases," *Entropy*, vol. 22, 448, April 2020. arXiv:2003.01984.
[124] P. A. M. Dirac, "Quantised singularities in the electromagnetic field," in *Proceedings of the Royal Society of London A: Mathematical, Physical and Engineering Sciences*, vol. 133, pp. 60–72. The Royal Society, 1931.
[125] L. Euler, *Methodus Inveniendi Lineas Curvas Maximi Minimive Proprietate Gaudentes Sive Solutio Problematis Isoperimetrici Latissimo Sensu Accepti*. Springer Science & Business Media, 1952. (Ed. Caratheodory Constantin, reprint of 1744 edition), ISBN 3-76431-424-9.
[126] L. Castellani, R. D'Auria, and P. Fré, *Supergravity and Superstrings: A Geometric Perspective. Vol. 1: Mathematical Foundations, Vol. 2: Supergravity, Vol. 3: Superstrings*. 1991.
[127] L. Andrianopoli, B. Cerchiai, R. D'Auria, A. Gallerati, R. Noris, M. Trigiante, and J. Zanelli, "$\mathcal{N}$-extended D = 4 supergravity, unconventional SUSY and graphene," *J. High Energy Phys.*, vol. 2020, 84, January 2020.
[128] A. W. Knapp, *Lie Groups Beyond an Introduction*, vol. 140 of *Progress in Mathematics*. Springer Science & Business Media, 2013.
[129] M. Henneaux, D. Persson, and P. Spindel, "Spacelike singularities and hidden symmetries of gravity," *Living Rev. Relativ.*, vol. 11, 1, 2008. doi:10.12942/lrr-2008-1. arXiv:0710.1818 [hep-th].
[130] P. Fré, F. Gargiulo, J. Rosseel, K. Rulik, M. Trigiante, and A. Van Proeyen, "Tits–Satake projections of homogeneous special geometries," *Class. Quantum Gravity*, vol. 24, no. 1, pp. 27–78, 2006. doi:10.1088/0264-9381/24/1/003. arXiv:hep-th/0606173.
[131] L. Andrianopoli, R. D'Auria, and S. Ferrara, "U-duality and central charges in various dimensions revisited," *Int. J. Mod. Phys. A*, vol. 13, no. 3, pp. 431–492, 1998.
[132] L. Andrianopoli, R. D'Auria, S. Ferrara, P. Fré, and M. Trigiante, "R-R scalars, *U*-duality and solvable Lie algebras," *Nucl. Phys. B*, vol. 496, no. 3, pp. 617–629, 1997. arXiv:hep-th/9611014.
[133] L. Andrianopoli, R. D'Auria, S. Ferrara, P. Fré, R. Minasian, and M. Trigiante, "Solvable Lie algebras in type IIA, type IIB and M-theories," *Nucl. Phys. B*, vol. 493, no. 1–2, pp. 249–277, 1997. arXiv:hep-th/9612202.
[134] P. Fré, "Gaugings and other supergravity tools of p-brane physics," in Proceedings of the Workshop on Latest Development in M-Theory, Paris, France, 1–9 February, 2001. arXiv:hep-th/0102114, 2001.

About the author

Pietro G. Fré, born in Alessandria (Italy) in 1952, is Professor Emeritus of Theoretical Physics at the University of Turin. He was Dean of the Department of Physics until his retirement, effective 1 November 2022. He was Director of the Arnold–Regge Center for Algebra, Geometry and Theoretical Physics from 2017 to 2022. From 1996 to 2000 he was President of the Italian Society of General Relativity and Physics of Gravitation (SIGRAV). He was Full Professor of Theoretical Physics and Director of the Particle Physics Sector of SISSA from 1990 to 1996. He was a fellow of the CERN Theoretical Division and a postdoctoral fellow at the California Institute of Technology and Bielefeld University in the early years of his career. He has authored about 340 scientific publications, including nine monographs. He has made outstanding contributions not only to the fields of supergravity and string theory with applications to cosmology, black-hole physics, and holographic gauge/gravity correspondence, but also to the algebraic structures underlying these theories. Of particular note are the introduction of the rheonomic formulation (with R. D'Auria), the independent invention under the name of "Cartan integrable systems" of free differential algebras and their extension with fermionic p-forms, the D'Auria–Fré algebra, which later evolved into the A-infinity algebras, the geometric formulation of the theory in 11 dimensions (M-theory), the geometric systematization of the $N = 2$ supergravity Lagrangian (with S. Ferrara et al.), the discovery of some of the homogeneous Sasaki-like varieties in seven dimensions, and the first development of their holographic equivalents by quivers. Since 2001, Prof. Fré has intensively collaborated with groups in Ukraine and Russia. In 2002 he spent a sabbatical year in Kiev and taught a long course on supergravity at the Bogolyubov Institute of the Ukrainian Academy of Sciences. He was a frequent visitor to the Joint Institute of Nuclear Research in Dubna from 2003 to 2008, and he became Scientific Attachée of the Italian Embassy in Moscow in 2009, a position which he held for 8 years, until September 2017. During his time in Russia, in addition to his diplomatic activities, he continued his scientific activities, collaborating in particular with Prof. Alexander Sorin, his frequent coauthor, and his teaching activities by teaching an advanced course in modern mathematical physics at the Moscow Engineering Physical Institute (MEPhI) for three successive years. He was also awarded two honorary degrees: one by the Ivanovo State Technical-Chemical University, the other by the Joint Institute of Nuclear Research in Dubna. Prof. Fré, who has had a considerable number of PhD students, among whom many are already Full and Associate Professors at various universities, also devotes himself to literature, and is the author of a number of novels and essays of philosophical-historical character.

https://doi.org/10.1515/9783111201535-021

Index

https://doi.org/10.1515/9783111201535-022

www.ingramcontent.com/pod-product-compliance
Lightning Source LLC
Chambersburg PA
CBHW060955210326
41598CB00031B/4831
9783111200750